Chemical Analysis for FORENSIC EVIDENCE

Chemical Analysis for
FORENSIC EVIDENCE

ARIAN VAN ASTEN

Professor in Forensic Analytical Chemistry and
On-Scene Chemical Analysis, van't Hoff
Institute for Molecular Sciences, Faculty of Science,
University of Amsterdam, Amsterdam, the Netherlands

ELSEVIER

Elsevier
Radarweg 29, PO Box 211, 1000 AE Amsterdam, Netherlands
The Boulevard, Langford Lane, Kidlington, Oxford OX5 1GB, United Kingdom
50 Hampshire Street, 5th Floor, Cambridge, MA 02139, United States

Notices

Knowledge and best practice in this field are constantly changing. As new research and experience broaden our understanding, changes in research methods, professional practices, or medical treatment may become necessary.

Practitioners and researchers must always rely on their own experience and knowledge in evaluating and using any information, methods, compounds, or experiments described herein. In using such information or methods they should be mindful of their own safety and the safety of others, including parties for whom they have a professional responsibility.

To the fullest extent of the law, neither the Publisher nor the authors, contributors, or editors, assume any liability for any injury and/or damage to persons or property as a matter of products liability, negligence or otherwise, or from any use or operation of any methods, products, instructions, or ideas contained in the material herein.

~~ISBN: 978-0-12-820715-4~~

For information on all Elsevier publications visit our website at https://www.elsevier.com/books-and-journals

Publisher: Susan Dennis
Acquisitions Editor: Charlotte Rowley
Editorial Project Manager: Hilary Carr
Production Project Manager: Paul Prasad Chandramohan
Cover Designer: Vicky Pearson Esser
Cover Design Suggestion: Arian van Asten

Typeset by TNQ Technologies

Transferred to Digital Printing 2022

To my family and my parents, ***Kuna*** *and* ***John Portier*** *whose love and care have allowed me to pursue my dreams*

To my father-in-law ***Ron Dekker*** *who I admire for his resilience and positive attitude*

In loving memory of my wonderful and caring mother-in-law ***Emmy Dekker de Groot***

To my in-laws, ***Rozan Dekker****,* ***Rob****,* ***Liese*** *and* ***Emma Coppen*** *for being such dear friends*

To my lovely wife ***Francis Dekker*** *and my kids* ***Debra*** *and* ***Timo van Asten****, it is such a great privilege and joy to have you in my life!*

Contents

About the author

Arian van Asten was appointed in 2018 as a full professor in Forensic Analytical Chemistry and On-scene Chemical Analysis at the **University of Amsterdam** (**UvA**). He is also the co-director of the **Co van Ledden Hulsebosch Center (CLHC)**, the Amsterdam Center for Forensic Science and Medicine named after the famous Dutch forensic pioneer who lived and worked in Amsterdam at the beginning of the 20th century. Additionally, van Asten is the director of the **Master's program in Forensic Science (MFS)** at the UvA. This 2-year MSc program in forensic science is unique in the Netherlands and annually admits 30-40 domestic and international students. Prior to his appointment, he worked for 12 years at the **Netherlands Forensic Institute** (**NFI**) as a member of the management team, department head and R&D, and complex/international case coordinator. He obtained his Ph.D. in analytical chemistry in 1995 at the UvA and received the **Kolthoff award** from the Royal Dutch Chemical Society for his thesis in 1996. Van Asten worked as an analytical chemist, laboratory head, perfume specialist, project manager, and department head in the Dutch chemical industry from 1995 until 2006. He is a member of the editorial board of Forensic Chemistry and sits on the scientific advisory board of the Netherlands Institute for Conservation, Art and Science (NICAS) and the National Forensic Centre (NFC) of Sweden. He chaired the scientific committee of the EAFS (**European Academy of Forensic Science**) 2012 conference that was held in The Hague, the Netherlands, and was a member of the scientific committee of EAFS 2022 that was hosted in Stockholm, Sweden. His research interests include rapid, mobile, and on-scene chemical analysis in forensic science and other fields, chemical identification of explosives, drugs of abuse and NPS, chemical profiling and forensic intelligence of explosives, chemical warfare agents, drugs of abuse and ignitable liquids, forensic toxicological analysis in alternative matrices, chemical imaging in forensic science and forensic applications of comprehensive gas and liquid chromatography. van Asten has published over 60 scientific papers in the field of (forensic) analytical chemistry, coordinates various national and EU-funded forensic science projects, and currently supervises several Ph.D. projects in forensic analytical chemistry. Since the academic year 2018—2019, he teaches the academic course **Chemical Analysis for Forensic Evidence** (**CAFE**) which is attended by MSc students in Forensic Science and Analytical

Chemistry at the University of Amsterdam. The material that he developed for the CAFE course was inspired by the idea to give the students a realistic view of how analytical chemistry is used by forensic experts on the basis of his work experience at the NFI while also presenting the underlying fundamental analytical and forensic principles (a forensic analytical chemistry framework). The course material that emerged from this idea formed the basis of the book.

Preface

Analytical chemistry has brought me a wonderful, rewarding, and exciting career after obtaining my Ph.D. in polymer analysis in 1995 at the University of Amsterdam (UvA). This career is as diverse as analytical chemistry itself, which is why I appreciate this branch of chemistry so much (in addition to its applied nature). However, it was not until 2006 that I became involved in forensic science for which my lovely wife Francis Dekker is "to blame". After working for over 10 years in the Dutch industry as an analytical chemical expert and laboratory manager, she spotted an interesting job posting at the Netherlands Forensic Institute (NFI) in the newspaper. The institute was seeking a new head for its Forensic Chemistry department and out of curiosity I decided to apply. The job interview that followed abruptly and fundamentally changed my career as I instantly fell in love with the renowned institute and forensic science, *i.e.*, the use of science to solve the crime and assist criminal justice. While working at the NFI in the 12 years that followed, I have noticed that the forensic field often seems to have this effect on scientists, many of the new talents that join forensic laboratories describe this long-lasting "infatuation", which can even outweigh other important work-related incentives such as income. Although using your scientific knowledge to contribute to society is definitely a strong motivator, I have come to the conclusion that the intellectual challenge to "crack the puzzle" and "solve the crime" is just an important factor. During my forensic years at the NFI, I became more and more involved in forensic R&D, innovation strategies, and collaboration with national and international academic partners. This also catalyzed my own involvement in forensic analytical chemistry and brought me back to the academic environment, 17 years after finalizing my Ph.D. In 2012 a dream came true when I was appointed professor on a special chair in Forensic Analytical Chemistry in the team of prof Peter Schoenmakers at the van 't Hoff Institute for Molecular Sciences of the Science Faculty of the UvA. Through a number of these special chairs, the UvA and the NFI embarked on a valuable collaboration that boosted forensic science and education in the Netherlands. My involvement in academia initiated once more an unforeseen step in my career, in 2018 I was appointed as a full professor at the UvA on a chair in Forensic Analytical Chemistry and On-Scene Chemical Analysis. From an NFI scientist and manager visiting academia, the roles were reversed as I became a UvA professor with strong ties to the forensic field and criminal justice system in the Netherlands. Part of this transfer also involved my appointment as director of the MSc program in Forensic Science at the Institute for Interdisciplinary Studies. This 2-year international MSc curriculum at the UvA is unique in the Netherlands and educates many new and talented forensic advisors, experts, scientists, and managers in the Dutch criminal justice system. As program director and full university professor, my involvement and interest in forensic science education were fueled, and in the fall of 2018, I taught and coordinated my first full academic course in forensic science. As the successor of prof. Peter Schoenmakers who initiated and ran the successful and popular CAFE (Chemical Analysis for Forensic Evidence, hence credits for the name of this book go to Peter!) course for many years, I decided to thoroughly revise the course material. The idea was to base the adapted course on the extensive knowledge and practical experience within the NFI, providing the students with a view of how forensic analytical chemistry is applied by "real" experts in "real" cases. With the full support of prof. Schoenmakers, I started working on a new set-up and new course material for the CAFE course. Somewhat surprisingly, a review of available literature did not yield a suitable book for the course. Although many excellent books on forensic analytical chemistry have been published, none was fully suited for a course that aims to provide a comprehensive overview at the MSc level and at the same time wants to present a general framework on why and how analytical chemistry is used in forensic laboratories to assist criminal investigations. There appear to be two types of books in this area, books that focus on providing main forensic principles and basic scientific concepts usually at an undergraduate level, and books (often edited with different authors for the various chapters) that aim to provide exhaustive overviews of specific methods or forensic expertise areas including extensive lists of literature references. This outcome triggered me in developing my own course material with the kind support of many NFI experts. After the first edition of the revised CAFE course, it became clear that the students really liked the new approach. Without exception, they were enthusiastic and indicated that they had obtained a lot of insight and knowledge irrespective of their forensic science or analytical chemistry background. Based on this positive feedback, the options of converting the course material into a book were explored with Elsevier. International reviewers and experts in the field confirmed the idea that a book based on the CAFE course could fill a gap in the existing literature on forensic analytical chemistry and could be of international educational value. In collaboration with Elsevier, a final decision was made early 2019 to embark on yet a new and inspiring mission to write an academic educational book, my personal contribution to the field of forensic analytical chemistry. This foreword was written when I "naively" took on the challenge and probably was mostly unaware of the huge efforts that still laid ahead. The fact that you are now reading this must mean that I somehow succeeded. However, you must ultimately be the judge whether the book delivers on its promises and makes the intended contribution to the fascinating field of forensic analytical chemistry.

Arian van Asten
Amsterdam, The Netherlands,
August 2019

Acknowledgments

This book would not have been laying in front of you and you would not be reading this page right now without the initiative, advice, and support of many excellent scientists, forensic experts, and other colleagues and friends of the author. In this section, they are properly acknowledged for their many invaluable contributions. First of all this book is the logical consequence of the initiative of **University of Amsterdam** (**UvA**) professor (and world-renowned analytical chemist) **Peter Schoenmakers** to start a course on forensic analytical chemistry. The name of the book should also be attributed to Peter as he came up with the name of the course, **Chemical Analysis of Forensic Evidence**, which (he has a unique talent for constructing project acronyms) in Amsterdam is known by the students as the "**CAFE**" course. The course is very popular with our students and when I took over in the academic year 2018–2019, I got full support from Peter to renew the content which ultimately led to the idea to convert the course material into a book.

Undoubtedly, the forensic framework as I present it in this book has been shaped and molded by the **Netherlands Forensic Institute** (**NFI**) where I worked for over 12 years. I have followed specialized forensic courses, preceded countless forensic expert exams, been instructed by experienced forensic experts, have seen case work up close, coordinated multidisciplinary investigations, and managed many R&D projects and national and international collaborations at this famous and renowned Dutch forensic institute. Additionally, the NFI has fully supported the realization of this book and has kindly allowed me to use material from their in-house photo collection and to consult experts. I deeply appreciate the contribution of my many NFI colleagues (some will be explicitly credited below) and the support of the R&D director of the institute, dr **Annemieke de Vries**.

Much of the concepts and methods discussed throughout the chapters are derived from academic research and associated scientific publications in the period 2010–2020. These scientific studies are part of the research conducted by Ph.D. students under my supervision. With respect to work presented on the analysis and chemical profiling of explosives, credits are due to **Hanneke Brust** and **Karlijn Bezemer** who successfully defended their Ph.D. theses in 2014 and 2020, respectively. In the context of this research also the contribution of **Antoine van der Heijden**, a specialist in energetic materials at TNO and professor by special appointment at the Technical University Delft, must be recognized. The various chapters on the analysis and characterization of illicit drugs and NPS (New Psychoactive Substances) rely heavily on the work and expertise of **Ruben Kranenburg** (Amsterdam Police Laboratory) and **Jennifer Bonetti** (Virginia Department of Forensic Science), both experienced experts in illicit drug analysis. Ruben Kranenburg provided much of the GC-MS and GC-IR data discussed in **Chapters 4** and **8**. Jennifer Bonetti was the first to publish on chemometric analysis of GC-MS data to distinguish drug isomers as discussed in **Chapter 8**. Both will defend their Ph.D. work in the coming years. With respect to the part on chemometrics and data analysis, discussions with **Age Smilde**, professor in Biosystems Data Analysis at the University of Amsterdam, helped me to expand my knowledge of multivariate data analysis. The final chapter on innovation has been inspired by the work within the Co van Ledden Hulsebosch Center together with **Maurice Aalders**, professor in the field of Forensic Biophysics at the Amsterdam University Medical Centers. Dr. **Rene Williams** of HIMS (van 't Hoff Institute for Molecular Sciences, Faculty of Science, UvA) kindly assisted with the creation of the molecular structures of the cobalt thiocyanide complexes discussed in **Chapter 4**.

At the NFI several scientists and forensic experts have provided valuable feedback, material, and insights that have been used throughout the chapters. They are acknowledged below according to their area of expertise. **Ivo Alberink, Peter Vergeer**, and **Marjan Sjerps**, forensic statisticians at the NFI, helped me to better understand the statistical aspects of sampling, the various aspects of measurement uncertainty, the details of score and feature-based models, and the validation of *LR* methods. With **Michiel Grutters** I discussed the details of fire debris analysis and the sample preparation methodology involved. The many valuable discussions with forensic illicit drug analysis expert **Jorrit van den Berg** aided in the realization of **Chapter 4**. I have always extensively collaborated with the Explosions and Explosives team of the NFI, the work presented on forensic explosives analysis is based on many conversations with **Mattijs Koeberg**, **Annemieke Hulsbergen**, **Jan Dalmolen**, **Eric Kok**, and **Rikus Woortmeijer**. **Dick-Paul Kloos** provided me with documentation and a deeper insight with respect to the trace analysis of drugs, medicines, and metabolites in biological samples. Forensic toxicologists **Ingrid Bosman** and **Rogier van der Hulst** helped to raise my understanding of the toxicological interpretation of analytical data and the DUI legislation in the Netherlands. I discussed the fascination field of forensic micro traces and their elemental and isotopic analysis with **Gerard van der Peijl**, **Wim Wiarda**, **Peter Zoon**, **Erwin Vermeij**, **Alwin Knijnenberg,** and **Jaap van der Weerd** and the interdisciplinary aspects of forensic investigations with **Jan de Koeijer**.

The initial concept version of this book has been carefully studied and read by **Shirly Montero** of Arizona State University. I am extremely grateful that she was willing to invest so much of her time in thoroughly reviewing this book. Her many sharp observations both in terms of forensic content and writing style have significantly improved the final version that is now in front of you.

Disclaimer:

This book has been written exclusively by the author. The acknowledged experts and scientists have made this endeavor possible and have been consulted and sometimes have provided figures and other material. This does not necessarily mean that they and their institutions support the content or the conclusions of the author. Any criticism or objections regarding this book should be directed to the author who is solely responsible for the content of this academic course material.

Reader guideline

Readers are advised to study this guideline before diving into the forensic content of the various chapters. The chapters have a given structure and are written in a specific manner to promote involvement and understanding. First of all, each chapter starts with a section called '*What will you learn?*' This introduction provides a brief outline and lists the learning objectives for the chapter. These chapter objectives contribute to the overall learning objectives. Furthermore, each chapter ends with a '*Further reading*' paragraph that provides a short list of books and articles for students who want to know more about the topics discussed.

Additionally, rather than undisputedly accepting all the views of the author by digesting large amounts of text at once, students are stimulated to form their own opinions and thoughts. This is realized through a series of questions that form a natural thread in each chapter and focus attention on crucial aspects. To maximize understanding, readers should allow themselves time to think about these questions and formulate their own answers before continuing. These questions are therefore clearly marked:

> Frames with the question mark icon contain a question.
>
> Sometimes these questions are part of a fictive case description.
>
> Form your own thoughts and formulate possible answers before continuing.

If in the subsequent discussion answers will be suggested and if this results in very important insights this can be highlighted in a separate frame:

> Frames with the light bulb icon provide an answer to a question.
>
> The associated insights are very important and require extra attention.
>
> Did your answer match these insights?
>
> If so, great! If not, do you understand why?

Occasionally, frames are also used for small excursions in which persons, instruments, techniques, or scientific aspects are discussed in more detail. These frames are indicated by a zooming glass:

> Frames with a zooming glass icon contain information on a side topic.
>
> This could be a person, instrument, method, science aspect, or expertise area.
>
> This provides interesting additional knowledge.

Throughout the regular text terminology and abbreviations are sometimes highlighted in **bold**. This is to emphasize that an important method, technique, concept, or definition is introduced. Readers are advised to pay a bit more attention to and try to remember these terms.

Learning objectives

This book provides a comprehensive overview of how analytical chemistry is used in state-of-the-art forensic laboratories to provide objective, valuable information from physical evidence to assist reconstructions in criminal investigations. Although analytical chemistry is extremely versatile in terms of methods and instrumentation, its application in forensic science is governed by general principles. The main goal of the author is to present these principles through a logical framework allowing readers to have a fundamental insight into the added value of forensic analytical chemistry. A thorough understanding of the framework will allow academic students (at the BSc and MSc level in Chemistry and Forensic Science), scholars (Ph.D. students, scientists), forensic experts, and laboratory professionals alike to make sound decisions in their future forensic endeavors. Studying this book will assist readers, both novices and experts in forensic analytical chemistry, in writing valuable proposals, developing robust methods and validating them for casework, and in deciding on innovation projects and equipment investments. With a focus on the overarching forensic analytical chemistry framework, it is a deliberate choice of the author not to discuss the basic chemistry principles in detail. Hence, readers should have a basic (BSc level) knowledge of chemistry and more specifically analytical chemistry. Knowledge of forensic science at a similar level is also recommended although associated principles will be extensively discussed. Additionally, this book does not contain exhaustive overviews of forensic analytical chemistry methods and the latest developments in chemistry-based forensic expertise areas. To ensure readability, comprehensive lists of literature references are not included. Only a limited number of references are provided per chapter, which serves as a suggestion for further reading. Essential references are occasionally given as a footnote. In addition, figures, tables, and equations are not numbered as this material is logically introduced as part of the storyline. This has been a deliberate choice of the author to increase readability.

After studying the material published in this book, readers with the required background knowledge are able to

- define **forensic analytical chemistry**, describe the underlying framework, and understand the key aspects involved (*i.e.*, sampling, qualitative analysis, quantitative analysis, chemical profiling, chemometrics, statistics, evidential value, quality, chain of custody, reporting, and innovation)
- list the main **forensic expertise areas** where analytical chemistry plays an important role, the questions from the **criminal justice system** addressed in these areas, and the associated analytical chemical methods and instruments that are used to provide the answers
- understand how **analytical chemistry** is applied in **criminal investigations** and what information is provided and how this information can assist in **forensic attribution** and **reconstruction**
- evaluate the **forensic potential** of novel analytical chemistry methods, define requirements for a successful forensic application and formulate an independent opinion on the **added value** of these new methods
- formulate meaningful **R&D projects** to develop new or improve existing forensic analytical chemistry methods based on the needs of the **criminal justice system**
- understand the **Bayesian framework** for **evidence evaluation** and its application in **forensic analytical chemistry**
- study **correlation** in multivariate chemical profiling datasets and apply basic **chemometrics** to obtain valuable forensic information from such data
- write a basic **forensic casework** report based on forensic analytical chemistry findings

CHAPTER 1

An introduction to forensic analytical chemistry

Contents

1.1 What will you learn?

In this first chapter, the field of forensic analytical chemistry will be introduced. At first glance, this field is "*nothing more*" than the application of analytical chemistry methods within a forensic laboratory requiring experts skilled in chemical analysis. However, there is more than meets the eye, by merging concepts, looking at case examples, studying questions of legal interest, and introducing forensic principles, it will be shown that forensic analytical chemistry represents a special branch with quite unique ways of working.

After studying this chapter, readers are able to

- define **analytical chemistry** and **forensic science** and combine these concepts to describe the scope of **forensic analytical chemistry**
- describe questions of interest to **criminal law** and understand how these questions govern the chemical analyses conducted by the forensic experts
- list the basic **principles of forensic science** and provide examples involving chemical analysis of physical evidence

Chemical Analysis for Forensic Evidence
ISBN 978-0-12-820715-4
https://doi.org/10.1016/B978-0-12-820715-4.00001-8

1.2 Definitions

When entering a new field or even when working in it for many years, it can be very useful to occasionally take a step back and think about the underlying reasons and motives of the associated activities. What do **analytical chemistry** and **forensic science** entail and what is the **added value of chemical analyses in the criminal justice system**? An often very useful step to increase your understanding to a higher, more abstract level is to try to write down a concise and yet comprehensive and precise definition of a certain science area or fields of expertise you are active in. This will also allow you to explain to others what you do and why, even if they are laymen. In the case of **forensic analytical chemistry**, the topic of this book, the additional challenge is that the worlds of analytical chemists and forensic scientists have to be merged. This is literally what happens when trained analytical chemists obtain a job at a forensic laboratory such as the Netherlands Forensic Institute (NFI). They enter a new world in terms of the application of their skills as over time they also become experienced forensic experts. Interestingly, the reverse situation in which a trained forensic scientist acquaints him or herself with analytical chemistry is much less frequently observed in forensic laboratories. Let us therefore start with an attempt to define the work of the analytical chemist:

What would be your definition of analytical chemistry?

As **analytical chemistry** comprises of a very broad range of activities, methods and instruments, many definitions and descriptions of this hybrid, interdisciplinary, and even somewhat elusive branch of chemistry exist. A generic description is given below but this is by no means the only valid description nor can this condensed text be attributed to the author as many very similar variations exist:

Analytical chemistry is the study of the separation, identification, and quantification of chemical compounds in natural and man-made materials.

Initially, unraveling the composition of materials and establishing the presence of chemical compounds in samples involved the use of so-called **wet-chemical methods**. These classical approaches are based on a chemical reaction in solution (hence the term "wet-chemical") to induce a color formation or a color change in the presence of a given analyte. In this way, the human eye is used as a detector based on our, for mammals rare, ability to see and distinguish colors. In **qualitative analysis**, the main aim is to demonstrate the presence of a compound of interest. As will be demonstrated in this book, these so-called **colorimetric test reactions** still play an important role in contemporary **forensic analytical chemistry**, especially in situations where the presence of certain chemicals needs to be established rapidly and on the scene. When the amount of a certain compound in a material is also an important consideration, a **quantitative analysis** can be performed through a **titration** method. A solution of a known concentration of a special reagent, a so-called titrant, is then gradually added to the sample solution. The volume at which a predefined color change (sometimes promoted through the presence of an **indicator**) is visually registered can then be used to accurately determine the concentration of an analyte.

The onset of many technological, electronical, microengineering advances in the 1950s and later the transistor, IC (integrated circuit) and computer revolution in the 1980s had an immense impact on analytical chemistry. Advanced instrumentation became available for highly specific and detailed chemical analyses based on a wide range of detection principles, exploiting the entire electromagnetic spectrum (including **spectroscopic techniques** like UV-vis, IR, FTIR, NIR, Raman, AAS, AES, XRD, and XRF) and introducing powerful new techniques such as **Nuclear Magnetic Resonance** (NMR is in principle also a spectroscopic technique) and **Mass Spectrometry** (MS, see all full technique names in the Abbreviations section). In addition to these techniques, the introduction of separation or so-called chromatographic techniques greatly enhanced the capabilities of analytical chemists. Initially, this was limited to volatile and semivolatile compounds with **gas chromatography** (GC) but from the 1980s it also became useful for polymers and other involatile and thermolabile compounds with the introduction of **liquid chromatography** (LC) and related techniques. The combination of compound separation and subsequent spectroscopic or mass spectrometric characterization completed the instrumental revolution in analytical chemistry. So-called **hyphenated systems** (*e.g.*, GC-MS, LC-UV, GC-IR, LC-MS, and GC-AES) allow detailed and sensitive compositional analysis of complex mixtures by the temporal separation of the mixture constituents, allowing chemical identification and quantification of the pure compounds as they enter the detector sequentially. The latest additions to the analytical chemical "toolbox" include powerful chemical imaging systems allowing spatially resolved chemical analysis (e.g., Raman microscopes, MA-XRF) and comprehensive, multidimensional chromatography (GCxGC-MS, LCxLC-MS) providing ultimate resolution and capacity in compound separation. These developments also triggered extensive instrument automation making the chemical analysis and **sample pretreatment** less laborious and allowed laboratories to greatly increase sample numbers and throughput. Instruments with **autosamplers** often enable 24/7 chemical analysis providing immediate results when the analyst enters the laboratory in the morning. The consequence of these developments is that the tasks of the laboratory expert has significantly shifted toward data analysis, interpretation and reporting, equipment maintenance, and quality control as the number of analyses and the amount of data per analysis continue to increase. With respect to data analysis, **chemometrics** has emerged as a new expertise area in analytical chemistry to create information and knowledge from data by using advanced (often multivariate) mathematical computer methods. Correct interpretation and representation of the results also

Analytical chemistry in a nutshell

Qualitative analysis (what?)

- Chemical identification of pure compounds
- Determination of mixture constituents
- Detection of trace analytes in samples

Quantitative analysis (how much?)

- Determination of analyte content or compound purity
- Establishing mixture composition
- Measuring trace levels of contaminants

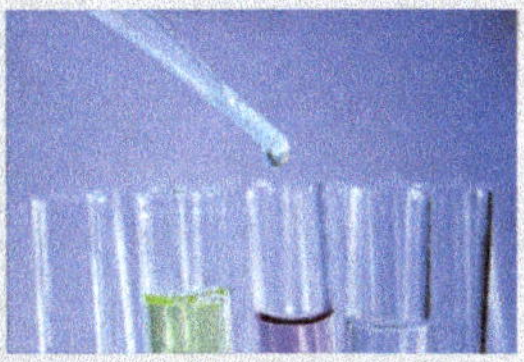

Classical, "wet-chemical" methods

- Colorimetric test reactions
- Titrations

Instrumental methods

- Spectroscopy, NMR, mass spectrometry
- Chromatography (GC, LC, CE, SEC, FFF)
- Electrochemistry (potentiometry, voltammetry)
- Gravimetry (including TGA)
- Calorimetry (including DSC)

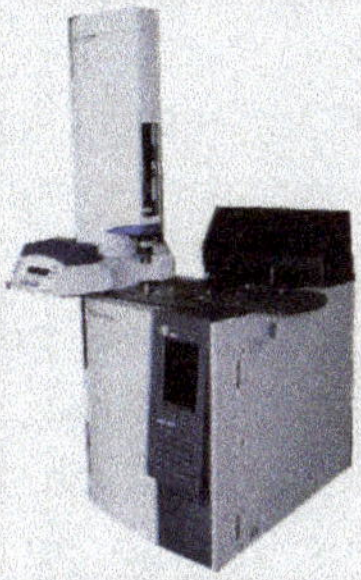

Sampling

- Head space sampling and trapping
- Liquid–liquid and solid phase extraction
- Sample dissolution and destruction
- Sample clean-up and preconcentration
- Representative, noninvasive, sterile ...

Automation

- Automated sample preparation
- Automated analysis and autosamplers
- Automated data analysis and reporting

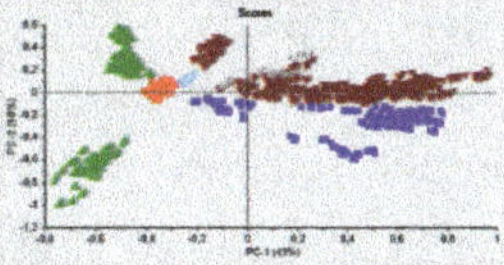

Data analysis: chemometrics and statistics

- Calibration
- Error analysis
- Data processing and (multivariate) analysis
- Databases

Quality

- Sample management (chain of custody)
- Method validation
- Quality control
- Certification and accreditation

require robust **statistical analysis** to provide insights with respect to uncertainty and the magnitude of random and systematic errors. Laboratories for which the reliability and robustness of their analytical chemical findings is of the utmost importance (e.g., hospital laboratories, forensic institutes, process analysis teams) make significant investment in quality systems, have dedicated staff working on **quality control**, and maintain quality **certifications** and **accreditations** granted and checked by authorized inspection bodies. In a forensic setting, these quality requirements also include detailed record keeping of the whereabouts and custody of physical evidence, the so-called **chain of custody**. In this introductory chapter, there is no need to discuss these techniques and aspects in detail; all these topics will naturally emerge in the various chapters as the forensic analytical chemistry framework unfolds. Instead, the basic aspects governing analytical chemistry are put in perspective below.

Now an intrinsic characteristic of analytical chemistry is its applied and "serving" nature, this field only exists by the virtue of the information-need of others. There is no use in chemically analyzing and characterizing a sample if it is not preceded by a relevant question from a colleague, client, company, institute, industry, or even society in general. The starting point of the analytical chemist is this question as he/she starts to develop methods and use instrumentation to find the answers through chemically characterization. Analytical chemistry thus aims to provide useful chemical information that allows professionals to take correct decisions and undertake the right actions. The analytical chemist thereby increases and ensures the quality of the work of these professionals. Now in the criminal justice system objective, evidence-based information is provided by the forensic scientist. So, the next step is to take a closer look to the definition of forensic science. Again, many descriptions exist as also this field is very diverse and interdisciplinary by nature. It is by no means the goal of the author to provide a comprehensive overview of existing definitions and associated scientific discussions. Instead, having students coming up with their own definitions is much more fun and interesting:

What would be your definition of forensic science? ?

If we again opt for a definition that is both concise and yet comprehensive we could arrive at the following statement:

> Forensic science is the application of a broad spectrum of sciences to answer questions of interest to a legal system.

Now interestingly, if both definitions are merged suddenly a very elegant (short, accurate, and comprehensive) description of **forensic analytical chemistry** emerges:

> Forensic analytical chemistry entails the separation, identification, and quantification of chemical compounds in natural and man-made materials to answer questions of interest to a legal system.

However, having a fancy definition for one's field of expertise is only a first step in a complex journey. The description above provides a generic frame that still does not explain why and how forensic analytical chemists do what they do. The analytical chemistry part has already been discussed allowing the reader to form a rough picture what it takes to chemically characterize compounds in complex samples. What needs to be discussed next are the questions of interest to a legal system. What are these questions in the criminal justice system, which professionals are asking them and what answers do they seek? This will be addressed in the next paragraph.

1.3 Questions of interest to a legal system

In civilized countries, the rule of law ensures a safe and just society. Laws describe the social rules of engagement and uphold and protect fundamental rights of citizens. Disputes are settled in court to break the vicious cycle of vigilantism, retaliation, and violence. In democratic systems adhering to the **trias politica** principle (separation of powers), the judiciary or judicial branch interprets and applies the laws defined by the legislative branch. In criminal law, the **public prosecution office** represents society as it indicts and prosecutes suspects of crimes. During the criminal investigation and court proceedings, suspects receive legal counsel as **defense lawyers** try to demonstrate that the indictment has not been proven beyond a reasonable doubt or is legally flawed. In the inquisitorial system as applied in the Netherlands, the verdict in a case is given by the **judges** of the **magistrate office**. In the adversarial system, a society representation in the form of a jury delivers the ultimate decision as the judge guards the court proceeding and decides on the **admissibility of the evidence** presented by the prosecution and defense council. Within the **criminal justice system**, the role of the **police** is to enforce the law. In this role, they also conduct criminal (tactical) and crime scene (technical) investigations under the supervision and leadership of the public prosecution office. The police also have the authority to arrest, detain, and interrogate suspects albeit under strict legal regulations. Special police reports of the findings serve as evidence in court and are used to incriminate or exonerate suspects.

On an individual level, a criminal court case has a severe impact on suspects, victims, and their relatives hence establishing the truth is of the utmost importance. On a generic level, correct verdicts based on convincing and legally sound evidence are essential for a just and safe society. Wrongful convictions or exonerations greatly undermine the rule of law and sense of justice. This is the rationale behind forensic science services and forensic case work as part of criminal investigations. By studying the **physical evidence**, the so-called **silent witnesses** at the scene of crime, objective information can be obtained that can be of crucial importance for an accurate **reconstruction** of the events. This assists the police and legal professionals and raises the quality of their actions and decisions.

However, forensic case work cannot simply be initiated by the forensic expert on the basis of the evidence alone. The evidence has been collected by the crime scene officers of the police for a reason in a given context of a case and on the basis of available tactical information. Hence, the physical evidence is typically accompanied by a formal request to the forensic institute with a question or a set of questions. Such requests typically originate from the team of police officers and public prosecutors investigating the case although the forensic investigation can also be requested by the defense council or magistrate office. At the initial stages after the discovery of a (potential) crime, the investigation is typically coordinated by the police. As valuable leads can quickly disappear over time, the police officers want to reconstruct the events as quickly and accurately as possible as this increases their chances of solving the case. Forensic investigations can assist in the attempt to answer the so-called **5 + 1 key questions** in a given case:

Questions of interest to a legal system (1) ?

(With What?)
Why?
What?
Where?
When?
Who?

Addressing these basic questions is essential for a precise reconstruction of the events. It should be noted that although for some incidents it is obvious that a crime has been committed (*e.g.*, when various witnesses have observed a lethal incident), the investigation of the police can also result in the conclusion that no crime was committed (*e.g.*, when somebody has committed suicide). Now consider the following two questions to better understand the reconstruction framework:

1. Why is the "Where" question of importance as the police is conducting an investigation on a given crime scene?
2. What is the reason that the "Why" question is usually not addressed by forensic experts studying the physical evidence and forensic trace material?

1. When a body is found, this does not necessarily mean that the victim was killed on this location! Quite often bodies are transported by perpetrators to hidden locations or put into clandestine graves to prevent discovery. An important aspect of the reconstruction is then to determine where the victim was killed.
2. Motive is a very important part of the police reconstruction and the indictment by the public prosecutor. However, physical evidence can typically not provide insight in the state of mind of a perpetrator. This is the realm of forensic psychiatry. Forensic toxicology and digital forensics experts can sometimes provide information that is important for the "Why" question. For instance with respect to the use of certain medication or drugs of abuse (insanity plea) or the use of certain Google search terms prior to the crime (manslaughter vs. premeditated murder).

When the police have completed the reconstruction including tactical information, witness and suspect statements, the crime scene investigation and the evidence reported by the forensic experts, the criminal investigation enters a new phase. If the dossier is unequivocally incriminating a suspect, the public prosecutor will officially indict the suspect and initiate a court case. The indictment specifies the crimes for which the suspect will be prosecuted. Now a very fundamental legal principle is that someone can only commit a crime when the action is clearly described as such in the law and that the associated law articles were in effect (ratified) prior to actions of the suspect. The indictment will therefore not only contain a summary of the facts and findings but will also relate these to relevant penal code articles. The main task of the judge in the inquisitorial legal system is to establish whether the indictment is **lawful** and thus fulfills all the legal requirements. Additionally, the indictment needs to be proved **beyond a reasonable doubt** indicating that on the basis of the facts, findings and evidence the judge is convinced that the suspect committed the crimes as described in the indictment. If these two requirements are met, the judge will convict and will arrive at a sentence based on the legal articles and the context of the case. The judge will formulate these deliberations in a final verdict. It is important to grasp that these legal aspects of a criminal investigation are of crucial importance to the work of the forensic expert and the questions they address:

Questions of interest to a legal system (2)

Is the indictment lawful?

Is the indictment proven beyond a reasonable doubt?

Therefore, forensic questions are dictated by the law!

The fact that also the work of the forensic analytical chemist is fully dictated by criminal law and the associated articles in the penal code is best demonstrated through practical examples. Throughout this book such practical examples will involve case descriptions that are fictive but are based on the daily work of forensic institutes such as the NFI and as such are highly realistic.

A case example

Two police officers apprehend two men who are fighting in front of a cafe and bring them over to the police station. In the jacket of one of the men, the officers find a plastic bag containing roughly 10 g of a white powder. The man refuses to explain the origin of the material. The officers suspect the white powder to be a synthetic drug and send the evidence for analysis to the forensic laboratory.

Do we perform a qualitative analysis only or do we include a quantitative assessment?

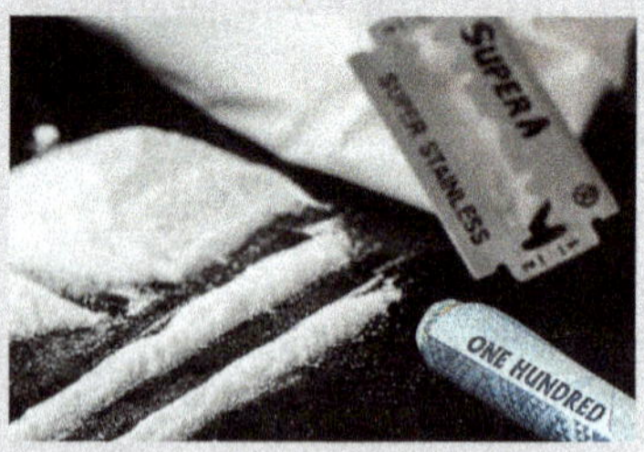

Interestingly and maybe to some readers somewhat surprisingly, the answer to this question is not found in the laboratory but rather in the law library! If the Dutch Illicit Drug act is taken as an example and when we translate the corresponding article, the course of action becomes obvious:

Dutch illicit drug act—Article 3

"It is forbidden for any of the substances listed as illicit drug to

1. transport these substances to or from the Netherlands
2. grow, manufacture, modify, process, deliver, provide, or transport these substances
3. possess these substances
4. to produce these substances"

In principle a qualitative analysis will be sufficient!

In two separate and dynamic lists, the Dutch government indicates which substances are considered to be "illicit drugs" (The two lists differentiate between so-called soft and hard drugs and this has an effect on the severity of the punishment). However, for the question at hand this is irrelevant.

The point is that the law forbids the possession of any of these listed substances without providing any quantitative limit. This means that any possession, no matter how low the amount or concentration, is punishable by law. Of course, there are many cases especially related to the production and trade of illicit drugs where amounts do matter and are of great importance to establish a reasonable sentence. However, if the indictment is related to article 3, there is no legal incentive to determine the amount of forbidden substance in the sample. Just demonstrating that an illicit substance is present is sufficient. The way this is typically done in the forensic illicit drugs laboratory is through the application of an indicative colorimetric test (e.g., the cobalt thiocyanate test or so-called Scott's test for cocaine) in combination with a qualitative screening with GC-MS (gas chromatography with mass spectrometric detection). As such laboratories typically handle thousands of samples and requests on an annual basis, performing a legally unnecessary quantitative analysis would increase the lead times of the investigation and would reduce the total number of cases that could be processed. The aspects of the chemical identification of drugs of abuse will be discussed in more detail in **Chapter 4**.

An interesting hypothetical situation arises when analysis would reveal that the sample only contains a trace amount of a forbidden substance such as cocaine. In principle, any amount constitutes a criminal offense but from a practical perspective it is clear that a suspect had no intent to use or sell drugs when he or she is in the possession of let's say 1 μg or a sample containing 1 ppm (part-per-million). Possibly, the defendant in that case was not even aware of the "possession" (although such traces could of course indicate that the suspect has handled forbidden substances). When such details are not described in the law then a practical common standard will have to emerge from jurisprudence and legal debate, for example, through rulings from the courts of appeal or the supreme court.

Now that we have seen a clear example of a "law-directed" qualitative chemical analysis, it is interesting to try to come up with an example where quantitation is required:

A case example

Can you come up with an example where a quantitative analysis of a compound is needed to determine whether a suspect has committed an offense?

Dutch traffic act—article 8 – section 2

"It is forbidden to drive a vehicle after consuming alcoholic beverages such that

1. the alcohol content in breath exceeds 220 μg ethanol per liter exhaled air
2. the alcohol content in blood exceeds 0.5 mg/mL"

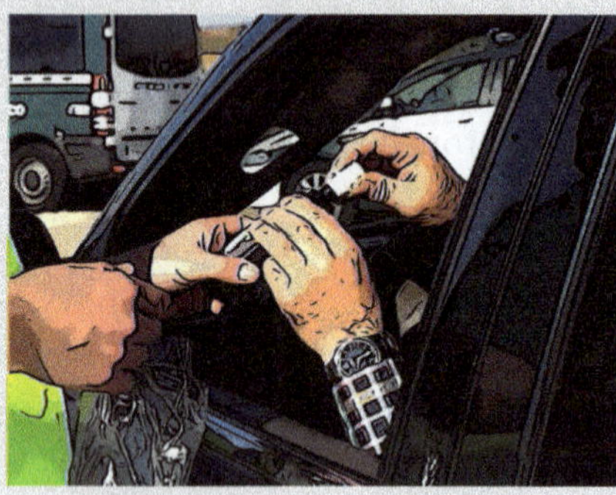

Many countries have similar DUI (driving under the influence) laws and there is international accordance between experts at what levels of alcohol consumption the driving capabilities are significantly impaired and can lead to dangerous traffic situations. These levels have consequently been entered in a legal framework and thus necessitate an accurate quantitative analysis of ethanol in human breath and blood. The process usually starts with a roadside breathalyzer test when a driver is stopped by the traffic police either during a random check or as a result of dangerous car maneuvers. These breathalyzer tests provide an indicative quantitative result based on the electrochemical oxidation of ethanol to acetic acid using oxygen in the air. If a positive test result is obtained, that is, a level exceeding the legal limit is found, the next step usually involves taking a blood sample from the suspect for an accurate assessment of the alcohol content. The blood alcohol level can for instance be established using head space sampling in combination with a gas chromatographic analysis.

An interesting situation arises when the accurate blood alcohol analysis would yield a result that is very close to the legal limit. For example, a duplicate analysis giving an outcome of 0.499 mg/mL—not guilty!—and 0.501 mg/mL—guilty!, respectively. The challenge the forensic expert is facing here is that statistics and scientific laws of probability do not "sit well" with the absolute decision that must be taken by the judge. However, as measurement uncertainty will always exist and random and systematic errors are part of any quantitative analysis, somehow these worlds must meet to arrive at a justifiable approach. This "legal limit dilemma" is discussed in more detail in **Chapter 5**.

These two examples clearly illustrate how criminal law essentially directs the questions for the forensic experts and the forensic analytical chemical methods applied. These examples have in common that the compounds of interest are directly related to the crime as described in the law. This means that the possession or use of such compounds is inherently part of the criminal activity. As a result, these compounds can be found in the law, either in the form of addendums (e.g., lists of illicit drugs) or directly in the penal law code (e.g., DUI of alcohol). However, forensic chemical analysis has much more to offer and can provide crucial information to solve a case. Almost any material can be chemically characterized for forensic attribution and reconstruction purposes. This includes physical evidence that has been used to commit a crime but is in itself not directly crime related and physical evidence that has no direct relation to the crime but has been created, transported or left as a result of the criminal activities. In the final example of this paragraph, it will be illustrated how the chemical analysis of evidence can assist in the reconstruction of a crime and ultimately the identification of the perpetrator.

A case example

In a tragic case of suspected arson, two fatalities occur. Police officers on the scene find an almost empty jerrycan containing traces of what seems to be gasoline. Unfortunately, no biological evidence or finger marks are found on the jerrycan.

What forensic chemical analysis could be employed?

Typically, in Fire Debris Analysis, Gas Chromatography with Mass Spectrometric detection is used to find traces of ignitable liquids that might have been used to deliberately start a fire. In this rare case, the forensic analytical chemists can analyze the intact gasoline residue that was probably used to commit arson. This yields a complex chromatogram showing the numerous compounds that make up the composition of this oil refined product.

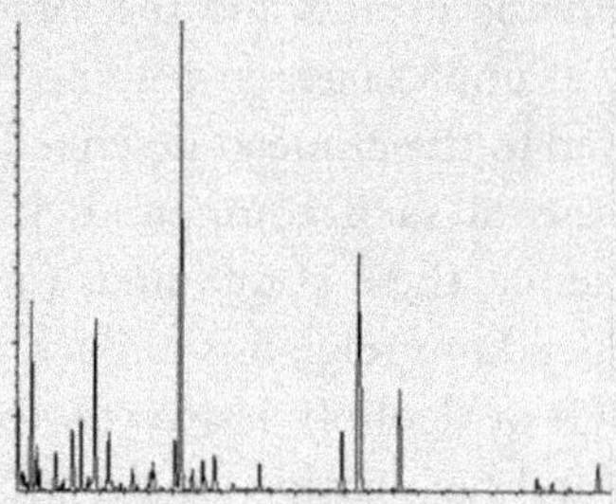

How could these findings be used by the police?

Although the main composition of gasoline is similar for various types, small variations exist as function of supplier, brand, and even batch as supplied to a gasoline station. If the assumption is made that the perpetrator obtained the gasoline at a station in the vicinity of the crime scene, a possibility exists to establish the origin of the gasoline. To that end, the police quickly obtained reference gasoline samples from nearby stations and when these samples were analyzed at the forensic laboratory, a clear "match" was found for a given gas station.

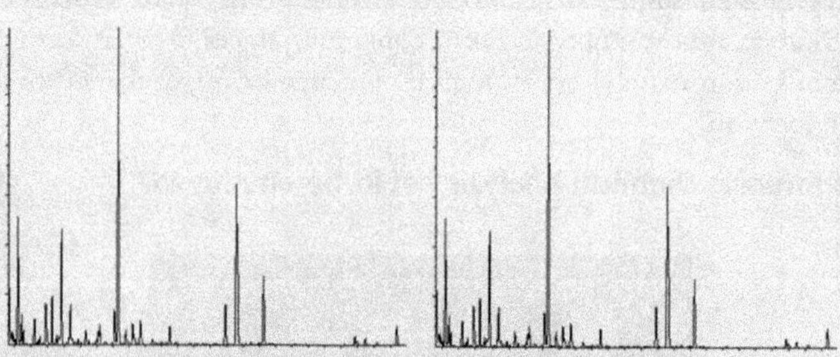

How could these findings be used by the police?

The police obtained security camera recordings from the petrol station. After chemical analysis pointed in the direction of the gasoline origin, digital video evidence now showed a family member of the victims loading gasoline in a jerrycan similar to the one found at the crime scene! This family member was apprehended and confessed to the crime after being interrogated by the police. The arson was the result of a very serious family feud.

1.4 Forensic science principles

This introduction to forensic analytical chemistry has provided an overview of analytical chemistry and has shown how the legal and law enforcement professionals strongly impact the activities in the forensic laboratory. To complete the picture, a third and final angle still needs to be considered: the forensic science principles that form the basis of forensic case work and apply to all the specific expertise or science areas employed. Forensic science is a relatively young academic discipline which emerged in many Western countries at the beginning of the 20th century. Inspired by the Sherlock Holmes novels of famous UK writer and physician Arthur Conan Doyle, scientists started to find their way to the crime scene, providing useful information from the so-called silent witnesses, the physical evidence left at the scene as a result of the criminal activities. Until then, the criminal investigation by the police was limited to witness, victim, and suspect interrogation. Initially, the first forensic investigations were conducted by physicians, pharmacists, and scientists who were consulted by the police as independent advisors. As the added value of this scientific contribution to the rule of law was quickly recognized, forensic science saw an exponential growth in the decades that followed. Nowadays, forensic case work is conducted by dedicated experts in government institutes, police laboratories, and medical facilities and are part of any serious criminal investigation. In addition, through the efforts of early pioneers and many academic scholars, forensic science also developed as a mature science area. Although the academic debate still continues whether forensic science is not merely an application of scientific principles emerging from established areas such as biology, chemistry, physics, and mathematics, the general consensus is that forensic science is an established academic discipline governed by generic principles. In this paragraph, some of these principles and their impact on the field of forensic analytical chemistry will be discussed.

In his 1963 paper entitled *The Ontogeny of Criminalistics*, USA forensic science pioneer and biochemistry professor, Paul L. Kirk stated that "***criminalistics is the science of individualization***." This famous sentence relates to the fact that the use of the term "identity" or "identification" has quite a different meaning in a forensic setting in comparison to other science areas. In a criminal investigation, there is a need to relate forensic traces such as fingermarks to a given person, a single individual that could be the perpetrator of a crime (**"biometric identification"**). However, in analytical chemistry **"chemical identification"** actually means that we establish the chemical structure of an unknown substance, it does not provide information on the origin of that substance and therefore does not indicate whether a link exists between, for example, the material found on a victim and a batch secured from a suspect. At best it can be demonstrated that no relation exists because the chemical composition is completely different. The term **individualization** was introduced by Kirk to illustrate this fundamental difference and to highlight the general aim in forensic science to develop methods with individualization capability, that is, that are able to discriminate items from the same class and ultimately connect a trace to a single individual or item. There is a natural reoccurring order in this process that can be recognized in every forensic investigation irrespective of the methodology applied. The forensic expert first needs to establish the "identity" or nature of the physical evidence. We will call this the identification step, are we indeed dealing with a blood stain, a shoe mark, a fingermark, a bullet, cocaine, or the explosive TATP? In some instances, it will be quite obvious that a shoe mark or bullet has been found on the crime scene. However, situations exist where this is much less evident. For example, when a highly deformed metal fragment is recovered from a deceased victim or a shoe mark is heavily smudged. Once the identity of the physical evidence has been established, additional steps can be taken to classify and ultimately **individualize** the evidence. Interestingly, for classification the forensic expert typically relies on **controlled processes**, either natural or man-made, that have resulted in the given **class characteristics**. A profile of a shoe mark for instance can reveal the brand and type of shoe. This requires information from the shoe manufacturer or a collection of relevant profiles allowing the expert to search the profile in a database. It also requires the forensic expert to establish that this profile is characteristic for this type of shoe, that is, that no other brands and types share this specific profile. Such findings can already be very valuable in the ongoing investigation as it can give direction to the police efforts. A crime-related shoe mark of a very expensive sneaker of size 12 most likely limits the potential perpetrator population to young males with money to spend or with a habit of stealing expensive shoes. Now let us assume that the police have arrested a suspect and that this suspect is wearing a pair of shoes of exactly this type and size. Does this mean that the shoe mark was made by the suspect (ruling out that somebody else could have worn the shoes of the suspect)? Clearly, the evidence is incriminating but it does and cannot allow the expert to discriminate between pairs of shoes of the same size and type. As many men wear this type of sneakers and the shoe size is quite common, many potential suspects exist when considering this evidence alone. Therefore, we are now looking for a method that could **individualize** the shoe mark to allow us to determine whether a link between the shoe mark and the shoes of the suspect exists. Typically, such methods focus on features of the trace that have been created by uncontrolled **random processes**. Complex random processes by their nature are highly characteristic, the probability that a random pattern reoccurs is inherently low. Readers should note that a low probability does not mean that this probability is zero as is explained below. Characteristic features such as damage to a shoe sole during use (*e.g.*, by stepping in glass or other sharp objects) can be examined by the expert and compared to a test mark of the shoe of the suspect. If several of such features exist and they match in form and position, the evidence supports the police hypothesis that the suspect was at the crime scene. What would be the a priori probability of the alternative proposition that another individual with the same shoes with exactly the same damage pattern was involved?

In their 2008 paper entitled "**The Individualization Fallacy in Forensic Science,**" Michael Saks and Jonathan Koehler stress the consequences of the erroneous assumption that a characteristic pattern is unique and hence that "a match" of such a pattern must mean that only the object or person under investigation can be the origin/donor of the trace. By assuming uniqueness, a forensic expert rules out any other potential source and this only holds when the entire relevant population has been investigated. For a fingermark and an unrestricted perpetrator population this would mean that the fingerprints of the entire world population are available and that after exhaustive comparison only the fingerprint of the suspect matched with the mark. Such data are not available, at best representative samples from scientific studies can provide statistical estimates and insight in the so-called **random match probability**, the probability that an item or person from a relevant population purely by chance would yield matching features. The lower this probability, the higher the individualization potential of the method, the more characteristic the features extracted from the forensic trace and the higher the evidential value of a match. However, no matter how high this evidential value is, a forensic expert can never claim uniqueness. Even for forensic STR (short tandem repeat) DNA profiles and the extremely low associated random match probabilities (less than one in a billion!), the forensic expert cannot state that suspect X is the donor of forensic biological trace Y. If there is a matching profile, the expert typically reports that suspect X cannot be ruled out as a donor and that the odds that a random other person not related to person X is the donor are extremely low. As will be discussed in more detail in **Chapter 6**, Bayesian statistics provide an elegant and very effective means to express evidential value, providing clear insight in the strength of the evidence while preventing the **individualization fallacy**.

How do the individualization principle and fallacy affect the work of the forensic analytical chemist? This is best discussed through an example.

Can you define the steps of detection/identification, classification, and individualization for the analysis of gasoline as described in the previous case?

"Criminalistics is the science of individualization"

Step 1—detection/identification

Paul L Kirk (USA), Forensic pioneer, May 9, 1902–June 5, 1970

The expert is faced with a transparent liquid residue in a jerrycan that smells like an organic solvent. Analysis with GC-MS will provide insight in the complex composition. Individual compounds can be identified and quantified. Fire Debris Analysis experts are well acquainted with the composition of oil distillates and refined products and they typically keep databases as to compare the findings. The overall composition matches with gasoline.

Step 2—classification

Gasoline is a mass consumer product produced and used in large volume. Many car owners use gasoline on a daily basis and acquire fuel at a gas station frequently. Gasoline is offered by several companies in different grades (e.g., in the Netherlands as E95—regular unleaded, E98—super unleaded, or E10—including bioethanol). Classification would now entail that through a detailed analysis of the GC-MS data the expert would be able to indicate the brand and type of gasoline (e.g., Shell E98). Note that the expert uses small man-made adaptations to formulations to retrieve this information from the sample. Formulations are typically carefully controlled to guarantee product consistency.

Step 3—individualization

Individualization of a product like gasoline is extremely challenging. However, the basis of gasoline is crude oil which in essence is a natural product. Crude oil is the product of bio-organic marine sediments that have accumulated over millions of years. Hence, gasoline will contain small variations in its chemical composition as a result of variations in the crude oil from which it was produced. With advanced analytical chemical techniques as GCxGC-MS (comprehensive two dimensional GC with mass spectrometric detection) or special GC-MS methods (e.g., SIE—selective ion extraction—profiles), these minor variations can be mapped and used for comparison and differentiation of gasoline samples of the same type and brand. Developing analytical chemical methods to differentiate and match samples of the same class is called chemical profiling.

How relevant is the individualization fallacy in forensic analytical chemistry and what does individualization entail from a chemistry perspective?

"Criminalistics is the science of individualization"

Individualization (fallacy) in chemical profiling

In chemical profiling, the unique object to which a forensic sample can be linked is usually a production batch. Especially man-made materials are typically produced in large volumes in continuous or batch processes. Gas stations hold large reservoirs which frequently are replenished. Whenever new gasoline is added and mixed with the remaining product a characteristic composition is created. The gasoline sample can at best be linked to the supply in the reservoir (assuming perfect mixing) at the moment of purchase. However, many customers that were acquiring the same product at the same time period as the perpetrator will have obtained gasoline with exactly the same composition. This can be seen as samples from the same gasoline supply. So the fact that chemical profiling can only provide comparison at batch level can have severe consequences especially when considering materials that are produced batchwise in large quantities. In addition, random match probabilities tend to be much higher for impurities in man-made chemical products. Hence, forensic experts need to be very careful when reporting a match in a chemical profiling investigation. Assessing the associated evidential values requires a detailed "market study" purchasing in this case many gasoline samples from different locations and different time periods to understand and document the variation of the characteristic features in a relevant population of gasoline batches.

The second forensic principle of special importance to forensic analytical chemistry is known as **Locard's Exchange Principle**. Frenchman Edmond Locard is seen as one of the founding fathers of forensic science and he realized the first ever forensic police laboratory in Lyon in 1910. His exchange principle is also known under the famous phrase "***Every contact leaves a trace.***" Interestingly, this is not how he formulated his principle, he described it in more elaborate terms referring specifically to violent crimes and the relation between the nature of the activity and the resulting traces. According to Locard, such violent crimes cannot be committed without leaving numerous signs because of the forceful nature of the act. However, this does not affect the importance of the principle and the popularity of the statement. Even today every professional working in the forensic field can cite the well-known phrase. In forensic analytical chemistry, the principle forms the basis for the investigation and characterization of **microtraces** such as glass fragments, paint chips, fibers, and gunshot residues. Chemical analysis of fingermarks can provide information on the activities of the donor such as the handling of drugs or explosives. Burglars carry tiny glass fragments on their clothing that are invisible to the human eye and have been transferred during the breaking of glass panes. In hit and run incidents with cyclists, the bike of the victim often carries paint traces from the car involved in the accident. In violent crimes and sexual assault cases, there is often a direct contact between the victim and the perpetrator. This typically leads to the mutual transfer of clothing fibers. Finding fibers originating from the victim on the suspect and vice versa can be very important for the criminal investigation especially when there was no prior relation and interaction. In a shooting incident, gunshot residue can be found on the hands and clothing of the shooter. At the same time, residue propels forward in the wake of the bullet and can be found at the point of impact. These are all examples of how the exchange principle directs the sampling and chemical analysis of microtraces.

An interesting perspective on the exchange principle emerges when no useful forensic traces are found at the crime scene or retrieved from physical evidence items. If indeed "every contact leaves a trace," this must mean that there was either no contact or that the methods applied were not effective enough. The leading motto of the Microtraces and Materials team of the Netherlands Forensic Institute is "more from less" indicating their ongoing quest for the detection, sampling, and detailed chemical characterization of the smallest of traces. Chemical profiling of such traces is extremely challenging and requires dedicated methods and high-end analytical equipment. An example of a successful microtrace profiling approach is the use of laser ablation—inductively coupled plasma—mass spectrometric (LA-ICP-MS) analysis for the elemental trace analysis of glass fragments and a whole range of other materials for that matter. Another successful development at the NFI is based on the extraction of dye from a single fiber and the detailed characterization of the extract with LC with high resolution mass spectrometry.

To conclude this introduction to forensic analytical chemistry, a final question is presented to the reader.

> Why can Locard's exchange principle also be regarded as a warning for crime scene officers and forensic experts when they conduct their investigations? ?

"Every contact leaves a trace"

Edmond Locard (France), Forensic pioneer, November 13, 1877–April 4, 1966

What holds for the criminal also holds for the forensic professional! As soon as police officers enter the crime scene, their presence will affect the scene. Every contact will create traces that are the result of the investigation and are not related the crime. When forensic experts sample evidence material, this contact will create new characteristics that alter the native state. For this reason, it is good practice to first observe the crime scene from a safe distance and agree on a course of action. After documenting the scene, the investigation is conducted with great care and in protective clothing to minimize the risk of contamination.

In the forensic institute, experts work according to a similar approach, especially important is the order of the investigations when several disciplines are involved. The order is such that non- or minimally invasive investigations are conducted first. When handling the evidence material care is taken not to cause contamination. Protective clothing is essential for anybody handling the physical evidence prior to forensic biological investigation.

Further reading

Inman, K., Rudin, N., 2000. Principles and Practice of Criminalistics: The Profession of Forensic Science. CRC Press, ISBN 9780849381270.

Kirk, P.L., 1963. The Ontogeny of criminalistics. J. Crim. law. Criminol. Police Sci. 54, 235–238.

Robertson, B., Vignaux, G.A., Berger, C.E.H., Interpreting Evidence: Evaluating Forensic Science in the Courtroom 2nd Edition, Wiley & Sons, ISBN 9781118492482.

Saks, M.J., Koehler, J.J., Rev, V.L., 2008. The Individualization Fallacy in Forensic Science, vol. 61, pp. 199–219.

CHAPTER 2

Analytical chemistry in the forensic laboratory

Contents

2.1 What will you learn?

This chapter provides a more practical view on the relevance and prevalence of analytical chemistry in the forensic laboratory. Analytical chemistry can be a critical part of the investigation but it can also have a more supportive role. Some forensic expertise areas have no association with chemical methods but sometimes chemical analysis plays a crucial yet unnoticed role. A comprehensive overview will be provided of the analytical chemical methods and instruments used in the various forensic expertise areas and the related questions that are addressed.

Chemical Analysis for Forensic Evidence
ISBN 978-0-12-820715-4
https://doi.org/10.1016/B978-0-12-820715-4.00003-1

After studying this chapter, readers are able to

- understand the relevance and prevalence of **analytical chemistry** in **forensic laboratories**
- list the main **forensic expertise areas** where **analytical chemistry** plays an important role
- formulate the questions from the **criminal justice system** addressed in these areas
- describe the associated analytical chemical methods and instruments that are used to provide the answers

2.2 Analytical chemistry in the forensic laboratory

In an average episode of one of the popular CSI TV series, a small team of forensic experts typically solves two murder cases in a 45-minute episode. They operate as true generalists, performing tasks such as following up on investigative leads, conducting the technical investigation at the crime scene, interrogating suspects, and performing a wide range of forensic case work in the laboratory. At the end, the suspect succumbs to the overwhelming evidence against him and provides a full confession confirming the reconstruction by the squad. Although entertaining and informative, this depiction is highly unrealistic and does not represent the actual ways of working of the forensic experts. Forensic laboratories have to deal with substantial case loads and need to report findings in a reasonable time frame with respect to the criminal investigation and court proceedings. As an example, the Netherlands Forensic Institute in The Hague, the key provider of forensic services in the Netherlands, annually receives roughly 50.000 requests for forensic analysis of over 100.000 pieces of physical evidence. To efficiently process such huge number of requests with a staff of roughly 500, the work is highly compartmentalized. Dedicated teams operate in specific forensic expertise areas to cover questions related to a specific crime and physical evidence type. Experts are registered by the court only for the specific area in which they conduct forensic investigations. They have a full-time job (and often also invest a considerable amount of spare time) to deal with the ongoing forensic investigations while also maintaining the strict quality norms in the laboratory, keeping up to date with new scientific developments (by reading the latest scientific literature in their field of expertise) and introducing new methods and instruments in the forensic laboratory. The required level of expertise and quality is simply impossible to guarantee when a single individual would conduct all the tasks typical of forensic investigations in a given case. With the overwhelming possibilities emerging from science and the fast rate of technological developments, providing state-of-the-art forensic information is a team effort and requires a high degree of specialization. This trend of specialization is also the result of the exponential increase in the demand for forensic science in criminal investigations that has typically been seen in many Western countries in recent decades. As forensic institutes and laboratories struggle to increase their capacity to meet the growing demand, a logical response—in addition to hiring new staff and expanding laboratory equipment—is to make the entire process more efficient. Although the associated risks of extended segmentation and detachment from the crime scene have been discussed and flagged in the forensic community (see Further Reading section), this trend has been unstoppable. The success of forensic science and the staggering increase in the case load has been most remarkable in the field of forensic biology with the introduction and subsequent evolution of forensic DNA typing.

The success and exponential growth of forensic DNA typing

The concept and technology of genetic fingerprinting was introduced by Alec Jeffreys in 1985. Subsequently, due to technological advancements based on PCR (polymerase chain reaction) and specific primers to target STRs (short tandem repeats) on different loci, STR DNA profiling quickly became the gold standard in forensic science. DNA typing can provide decisive information in a criminal investigation, especially in combination with a national DNA database containing profiles of biological traces in unsolved cases and reference profiles of convicted perpetrators. Many countries have established national DNA databases and the associated legal frameworks that dictate how DNA typing can employed to solve crime. These databases have seen a staggering growth, indicating the success of this breakthrough in forensic science. The legal and scientific toolbox for forensic genetics is continuously expanded to create new investigative opportunities (e.g., familial DNA analysis, mass screenings, mitochondrial and Y chromosomal DNA analysis, Low Copy Number DNA analysis, Next Generation Sequencing, Single Nucleotide Polymorphism).

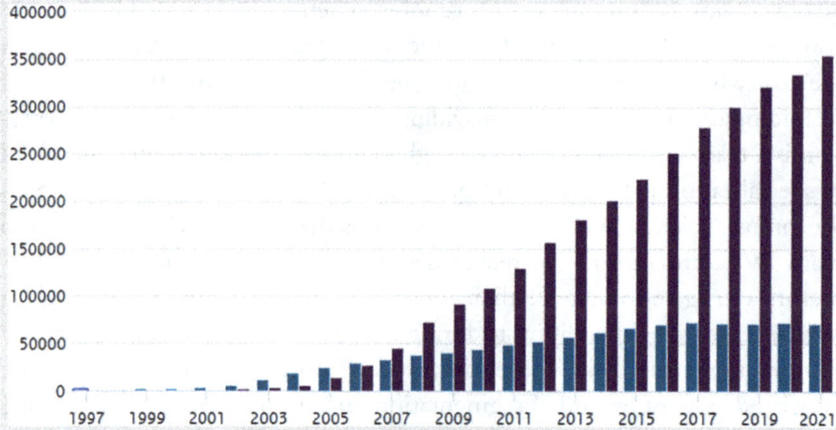

Cumulative growth of the number of STR DNA profiles in the Dutch DNA database. Trace profiles in lblue and person reference profiles in purple. In 2021, the database contained 71.653 trace profiles and 356.266 reference profiles https://dnadatabank.forensischinstituut.nl/resultaten/aantal-profielen-dna-databank-strafzaken

Which analytical chemistry method lies at the heart of the DNA profiling process?

How is it possible that these analytical procedures do not require extensive analytical expertise?

Capillary Zone Electrophoresis

Capillary Zone Electrophoresis (CZE) was introduced by analytical chemist James W. Jorgenson in 1981. This high-resolution analytical separation technique is critical to create the typical forensic DNA profile. An STR DNA profile is basically an electrophoretic separation of DNA fragments amplified with primers with variable fluorescent labels. The stacked electropherograms are recorded at different fluorescence wavelengths by the detector. Through calibration with so-called allelic ladders, the number of repeats in the amplified fragments can be established. In DNA databases, the number of repeats for each loci (two values for each pair of chromosomes) is recorded yielding a very simple list of numbers containing repeat values for all loci monitored.

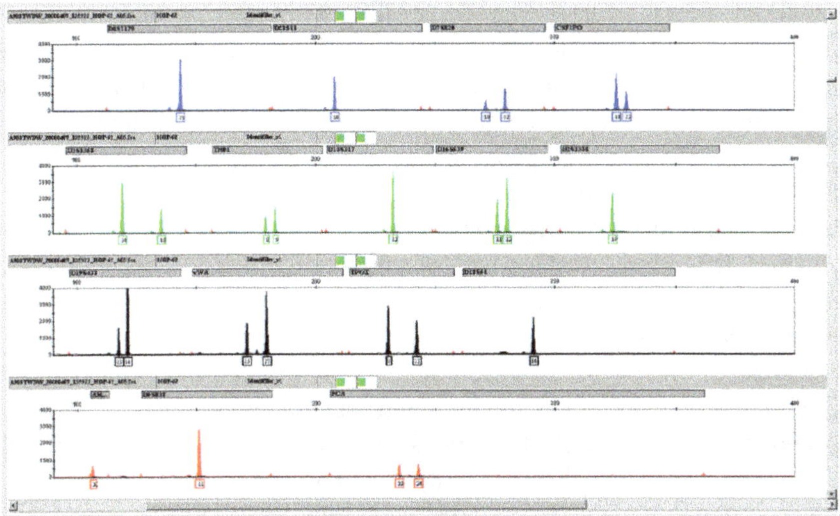

Typical STR DNA profile with allele designation for genomic DNA extracted from a cancer cell line using the *Identifiler* kit (15 STR loci + gender) on a 3130xl Genetic Analyzer (Fang et al. J Forensic Res, 2011; https://doi.org/10.4172/2157-7145.S2-005)

The fact that forensic DNA laboratories easily can generate over 100.000 of DNA profiles annually is the result of a high degree of automation. Instrument manufacturers offer complete solutions, including kits containing all the required (bio)chemicals and tailor-made equipment for sample pretreatment (DNA extraction, quantification, and amplification), CZE separation of the amplified fragments and advanced software for automated data analysis. This limits the need for trained analytical chemists in forensic DNA laboratories.

Although forensic analytical chemistry plays a critical role in creating a forensic DNA profile, DNA typing is considered to be part of the human biology domain in forensic science. Forensic DNA laboratories typically employ scientists with a genetics, biochemistry, biology, or life science background. With the automated DNA typing systems, the only analytical chemists entering the laboratory are the maintenance engineers from the equipment manufacturer. Only they are acquainted with the complex analytical chemistry that forms such a crucial part of creating the forensic DNA profile.

However, several forensic expertise areas exist where analytical chemistry has a much more prominent role and provides the main findings of the forensic investigation. The teams conducting investigations in these areas typical include several experienced analytical chemists in charge of "the labwork" and forensic experts with a strong background in analytical chemistry. Typically, this involves chemical compounds that are directly related to criminal activities or chemical traces that are the result of such activities. It will only make sense to appoint dedicated forensic expertise, capacity, and laboratory infrastructure when a specific type of physical evidence and related chemical analysis is requested frequently and in a substantial number of cases. In addition, several forensic expertise areas use forensic analytical chemical methods and instruments as part of a broader toolbox. Here, analytical chemistry supports the case investigations and teams often outsource chemical analyses to colleagues within the forensic institute. Forensic experts do not necessarily have to be experienced forensic analytical chemists but must have a solid knowledge of the chemical analysis methodology as they must be able to report all the details of the investigation and explain the findings in court.

Can you mention forensic expertise areas in which analytical chemistry plays a key role?

Can you mention forensic expertise areas in which analytical chemistry plays a supporting role?

Analytical chemistry at the heart of the forensic expertise area*

- Illicit drugs
- Forensic toxicology
- Fire debris and ignitable liquids analysis
- Explosions and explosives
- Microtraces: Gunshot residues, glass, fibers, and paint
- Forensic environmental investigations

Analytical chemistry supports the forensic expertise area*

- Fingermarks
- Questioned documents
- Crime scene investigation

*in order of decreasing case volume as typically received by national forensic institutions.

Typically, books on forensic analytical chemistry use a framework based on these forensic expertise areas, which form the chapters that are described in detail by experts in the field. The amount of expertise involved in terms of case context, questions addressed, analytical chemistry applied, and novel insights and developments can easily warrant separate books on a given area, let alone a chapter in a book. However, as described in the Learning Objectives section, it is the aim of this author to "tell the story" of forensic analytical chemistry and discuss the broader perspective. In this approach, the forensic expertise areas provide examples to illustrate the underlying principles that are addressed in the various chapters. Because it is important to understand where and how analytical chemistry is applied in forensic laboratories, the forensic expertise areas listed above will be discussed in more detail in the next paragraphs. This overview is by no means comprehensive nor detailed as this would result in an excessive chapter length. Instead the discussion of the various expertise areas focuses on the main investigations conducted and the analytical chemistry methods and instrumentation typically used to address the questions. In the lists the most frequently occurring questions and techniques are indicated in bold. Suggestions for more detailed information on a forensic expertise area basis are given in the Further Reading section.

2.3 Forensic expertise areas

Illicit drugs

Forensic experts in the field of illicit drugs combine the knowledge on how illicit drugs are produced and formulated with the skills to chemically analyze such substances. Reports with the findings of the chemical analyses play a crucial role in drug related criminal investigations. As discussed in the previous chapter, the identification of an illicit substance provides direct evidence that a crime was committed, as defined in the law of a given country such as the Dutch Opium Act. However, this field of expertise is not limited to the chemical identification of illicit drugs alone, typically the following investigations and activities are undertaken (in bold main case work and leading techniques):

- ***Chemical identification of illicit drugs (listed substances) and precursors***
- *Establishing the level of an illicit drug in a product or matrix*
- *Estimating the total weight of illicit drugs in a shipment*
- *Chemical profiling of drugs and precursor for forensic comparison*
- *Chemical analysis of dumped drug waste*
- *Chemical analysis of samples from illegal production locations*
- *Expert advice at drug waste and production crime scene investigations*

To perform these investigations, the forensic illicit drug experts rely on a number of analytical chemistry methods, techniques, and instrumentation:

- ***Colorimetric indicative tests (color tests)***
- ***GC-MS—Gas chromatography with mass spectrometric detection***
- *Raman and infrared spectrometry (also as mobile instrumentation)*
- *GC-IR—gas chromatography with infrared detection*
- *LC-MS—liquid chromatography with mass spectrometric detection*
- *NMR—nuclear magnetic resonance spectroscopy (for absolute identification)*
- *HRMS—high-resolution mass spectrometry (for absolute identification)*

Chapter 4—Qualitative Analysis and the Selectivity Dilemma, will describe the chemical identification of illicit substances in more detail. Typically, it involves an indicative test by a colorimetric chemical reaction (i.e., using a special reagent mix to induce a visually observable color change in solution in the presence of a drug or class of drug compounds) followed by a confirmatory analysis with GC-MS. **Gas Chromatography with Mass Spectrometric Detection** is the workhorse technology that is applied by virtually every drug analysis laboratory. Compounds are injected from solution into a hot injector and transported by a gaseous mobile phase through a capillary column containing a thin film of wall coated stationary phase. On the basis of volatility and stationary phase affinity, the compounds are separated in the column before entering the mass spectrometer (MS). At the entrance of the MS, molecular ionization and fragmentation is induced using electron impact (EI). This results in a mass spectrum as recorded typically in a single quadrupole and ion detector setup. EI mass spectra are known to be very specific and exceptionally reproducible. This facilitates the use of generic databases containing the spectra of thousands of compounds to identify an unknown by spectral library matching. Given the fact that most drugs of abuse are sufficiently volatile and thermo-stable, a single GC-MS method can be used to chemically identify illicit substances and related compounds rapidly and confidently.

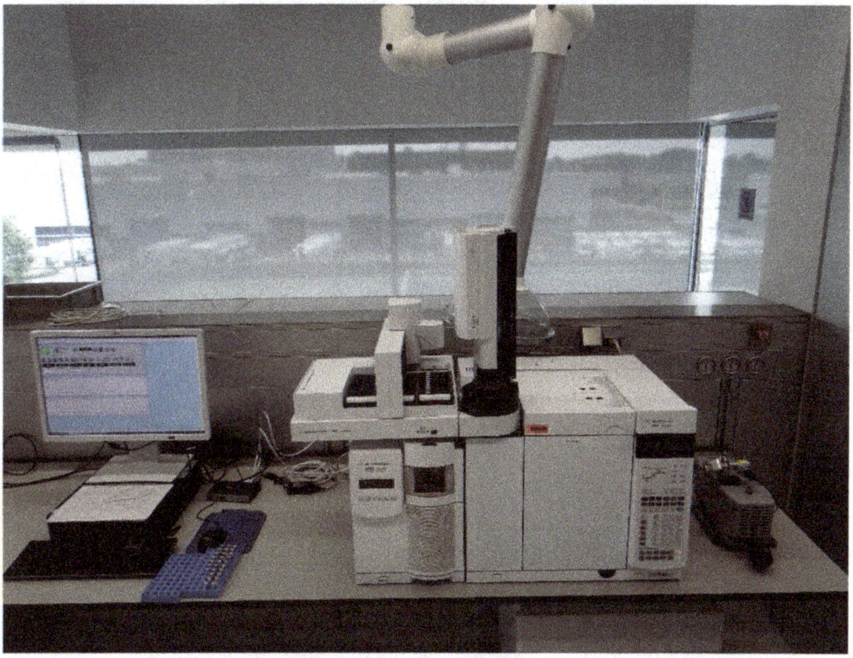

Typical GC-MS setup as used in forensic laboratories for routine chemical identification of drugs of abuse (*courtesy of Dr Jorrit van den Berg, illicit drug expert of the Netherlands Forensic Institute*). Laboratories capable of handling high volumes of case work typically have multiple dedicated instruments with autosamplers that conduct analyses 24/7 (*Note: several instrument manufacturers provide very robust instruments, the set-up shown serves as an example and does not indicate any preference*)

Forensic toxicology

Forensic toxicology combines two distinct expertise areas: pharmacology and analytical chemistry. Forensic toxicologists are often registered pharmacists and their expertise allows them to assess the effects of drugs on the human body and to understand the underlying metabolism. Typically, this knowledge is used to develop medicines and optimize dosage to improve human health and treat diseases. However, in a criminal context, the forensic toxicologist is typically interested in the abuse of drugs, medicines and toxins, and the role of these substances in criminal activities. As part of a forensic autopsy, the forensic toxicologist provides information on the presence of such compounds in the blood and tissues of the victim and indicates whether this is linked to the cause of death. However, forensic toxicological investigations are also essential for cases related to Driving under the Influence (DUI) and sexual assault involving the use of drugs (so-called date rape drugs) to sedate and incapacitate the victim. Providing useful forensic toxicological information usually requires a quantitative analysis of the compounds of interest as is explained in more detail in **Chapter 5—Quantitative Analysis and the Legal Limit Dilemma**. However, establishing trace levels of drugs and metabolites in complex biological matrices requires state-of-the-art analytical chemistry. To that end, forensic toxicologists also have a strong background in analytical chemistry and often work with dedicated analytical chemists that analyze the biological specimens in the forensic toxicology laboratory.

Typical investigations conducted in a forensic toxicological context include/entail (in bold main case work):

- ***Trace analysis of drugs, medicines, and associated metabolites in human body fluids and tissues***
- *Trace analysis of poisons and associated metabolites in human body fluids and tissues*
- *Samples analyzed include heart blood, femoral blood, and vitreous humor but depending on the context of the case also urine, saliva, hair, bone, teeth, and stomach contents can be investigated*
- *The forensic toxicological report always consists of two distinctive parts; the results of the chemical analysis and the toxicological interpretation*
- *A majority of the forensic toxicological investigations are related to the quantitative analysis of alcohol and drugs of abuse in blood in DUI cases and the quantitative screening of drugs of abuse, medicines, toxins, and associated metabolites as part of forensic autopsies*

Conducting such investigations requires advanced analytical chemistry instrumentation (in bold leading techniques and methods):

- ***LC-MS—liquid chromatography with mass spectrometric detection***
- *Main MS technique:* ***triple Quad MS*** *using* ***Multiple Reaction Monitoring (MRM)****, but also high-resolution mass spectrometric (HRMS) techniques such as TOF-MS (Time of Flight) and Orbitrap MS are used as such or in combination with Quadrupole MS.*
- *LC-MS analysis requires a suitable ambient ionization technique to eliminate the mobile phase and ionize the analytes of interest prior to MS analysis. Typically,* ***ESI (electrospray ionization****) and APCI (atmospheric pressure chemical ionization) are used in positive ionization mode*
- *GC-MS—Gas chromatography with mass spectrometric detection, typically with EI ionization and low-resolution Quadrupole MS but also HRMS detectors as TOF-MS (Time of Flight) and Orbitrap MS can be employed*
- *GC-MS analysis often requires sample derivatization to increase volatility especially for metabolites, exception is the use of head space (HS) GC-MS for the quantitative analysis of alcohol in whole blood in DUI cases*
- *Sample preparation is essential to eliminate complex biological matrix and involves a protein precipitation step*
- *Accurate quantitation requires the use of deuterated reference standards to account for analyte loss during sample preparation and variations in ionization efficiency*

LC-MS setups with ESI interfaces and triple quad MSs are most frequently employed for routine analysis in the forensic toxicological laboratory. Tailor-made methods are based on the targeted screening of a large set of compounds of forensic interest including drugs of abuse, medicines, toxins, and associated metabolites using MRM schemes. Accurate quantitative analysis requires the use of isotope labeled standards to account for analyte loss and shifts in ionization efficiency.

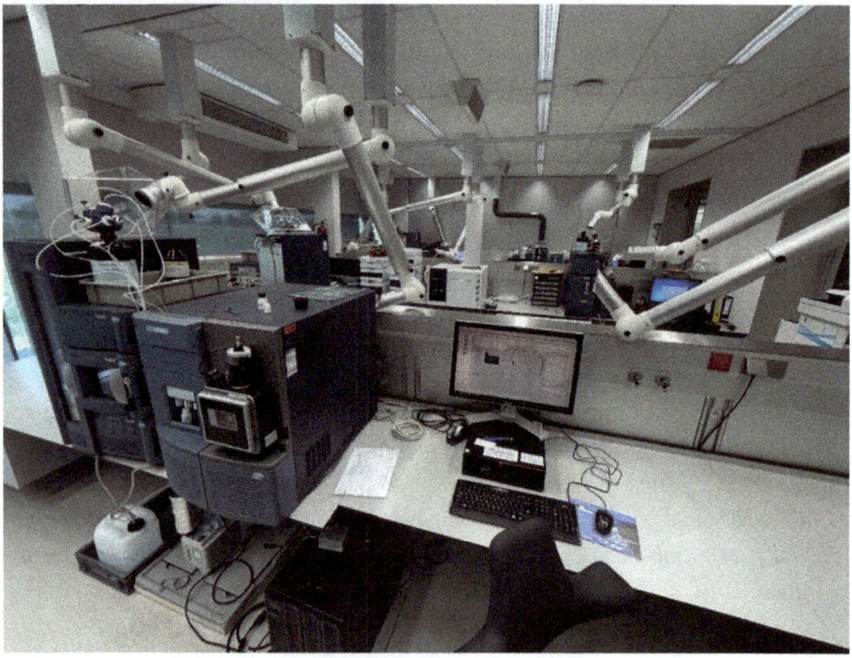

Typical LC-MS setup as used in forensic laboratories for routine forensic toxicological analysis (*courtesy of Dr. Dick-Paul Kloos of the Forensic Toxicological Laboratory of the Netherlands Forensic Institute*). Laboratories capable of handling high volumes of biological samples have several dedicated instruments with autosamplers to conduct analyses continuously (*Note: several instrument manufacturers provide very robust instruments, the set-up shown serves as an example and does not indicate any preference*)

What is the principle of triple-Q-MS and multiple reaction monitoring?

Why are deuterated internal standards needed for accurate quantitative analysis?

In triple-Q-MS, an analyte precursor ion is selected in the first quadrupole which acts as a mass filter. In the second quadrupole, which acts as a collision cell, the precursor ion is fragmented at elevated pressure. This process is termed Collision Induced Dissociation (CID). The fragments are consequently analyzed in the third quadrupole stage. For a given analyte, this is termed Selected Reaction Monitoring (SRM). This process can be completed in milliseconds (dwell time) and hence as peaks elute from the chromatographic system multiple analytes can be analyzed simultaneously by rapidly cycling SRM transitions, the so-called MRM approach.

When removing the biological macromolecular material during sample preparation, a significant amount of the analytes can be lost. Additionally, the ionization efficiency in the MS of those analytes can be strongly affected by various parameters. Sample loss and ionization suppression and enhancement can vary significantly from run to run. Proper compensation can only be provided by internal standards that are chemically identical to the analytes of interest. This requires the use of isotope labeled species.

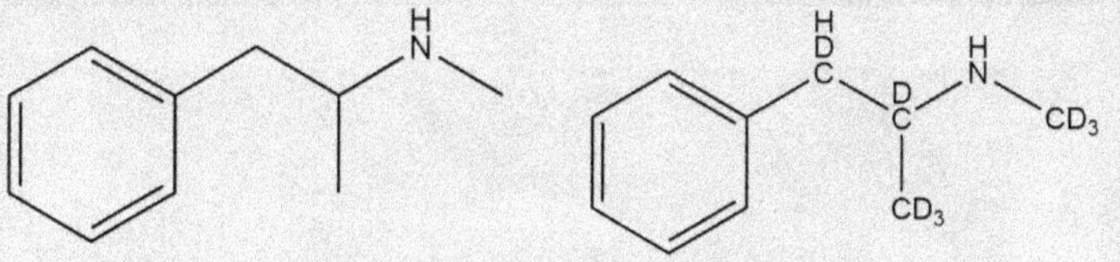

Methamphetamine (illicit drug) Methamphetamine-d8 (IS)

Fire debris and ignitable liquids analysis

Forensic investigations related to suspicious fires usually are focused on the question whether a fire was accidental or if indeed arson was committed. To address this question, objective evidence is usually provided by two forensic expertise areas. Forensic and police experts, often with an engineering background, conduct an on-scene investigation after the fire has been extinguished by the fire fighters. They try to reconstruct the fire from the remaining scene, which is a very difficult task because of the destructive nature of the fire and the subsequent invasive actions of the fire fighters. In case of arson, perpetrators typically use an ignitable liquid (e.g., petrol, gasoline or white spirit) to start a fire. Therefore, an important clue is provided if in the fire debris remnants of such accelerants are detected. To this end, the police experts at the scene take samples of the fire debris (sometimes with the assistance of specially trained dogs capable of smelling and pinpointing the presence of potential accelerant residues in fire debris) and send these samples to the forensic experts of the fire debris and ignitable liquids analysis section.

Special glass jar as used in the Netherlands to sample and analyze fire debris (*courtesy of Dr. Michiel Grutters of the Netherlands Forensic Institute*)

In the forensic laboratory, these experts conduct the following investigations (in bold main case work):

- ***Fire debris analysis,*** that is, *the detection and classification of ignitable liquid residues through chemical analysis of fire debris samples (common fire accelerants include gasoline, lamp oil, and white spirit)*
- *Chemical analysis, classification, and profiling of neat ignitable liquids (can be both bulk and trace analysis,* e.g., *traces of ignitable liquid are sometimes found on the clothing of a suspect)*

These experts have a strong background in analytical chemistry, and their work takes place in the laboratory using mostly GC-based techniques with some form of head space sampling (in bold leading techniques and methods):

- ***GC-MS—Gas chromatography with mass spectrometric detection***
- *GC-FID—Gas chromatography with flame ionization detection*
- *GCxGC-MS—Comprehensive 2D gas chromatography*
- *The fire debris matrix requires sample preparation prior to analysis, most laboratories apply some form of head space analysis of a fire debris sample,* for example, *static head space with direct injection, dynamic head space using sorbent traps or SPME (Solid Phase Micro Extraction). Samples are often heated in a closed environment to increase the sensitivity of the analysis.*

Because sample preparation is so critical in the search for ignitable liquid residues in fire debris, **Chapter 3—Sampling and Sample Preparation** will include a detailed discussion on the rationale behind the preferred sampling methods in fire debris analysis.

> Why is the identification of ignitable liquid residues in fire debris samples so challenging?

In terms of sample and analyte complexity, the detection of ignitable liquid residues in fire debris can be considered as an ultimate challenge in forensic analytical chemistry:

1. Ignitable liquids typically are refined oil products and as such have a very complex chemical composition involving thousands of organic compounds
2. These organic compounds are not exclusive to one given type of product and can occur in various ignitable liquid classes
3. The matrix, that is, the fire debris, is in itself also a highly complex and variable substrate consisting of a large number of organic pyrolysis products
4. These pyrolysis products can be identical to constituents found in ignitable liquids depending on the original substrate composition and can thus create false-positive results if the forensic expert does not interpret the analytical data correctly
5. A fire is a very destructive process, depending on the course of the fire the ignitable liquid used can be fully consumed and inherently fire debris samples can have high false negative rates. This is not the result of a flawed test but rather the conditions of the fire. One should never conclude from the absence of ignitable liquid residues that the fire was accidental
6. When dealing with a fire, ensuring safety and health has the highest priority but extinguishing a fire is a very invasive activity leading to the destruction of forensic traces and significant alteration of the scene
7. This alteration continues after the fire has been ended. The investigation has to be conducted quickly as chemical reaction and biodegradation continue to affect the chemical composition of fire debris samples

Explosions and explosives

Similar to illicit drug analysis, forensic experts that investigate explosives and explosions also deal with chemicals that are directly related to criminal activities as described in the criminal law. Energetic materials contribute to the standard of living and keep societies save when their use in engineering and in a military setting is considered. However, as with many human discoveries and inventions, such materials can also be misused for criminal activities.

What kind of crimes involves the use of explosives?

In a criminal setting, energetic materials are misused to threaten or deliberately damage infrastructure, objects, and humans. Several motives and modi-operandi can be involved:

1. The use of explosives to gain illegal access to infrastructure usually in relation to theft. Examples include the use of explosives to blow up an ATM or a bank vault to steal money and other valuables
2. A targeted attack on a person by detonating a bomb in his/her vicinity through a timer device or an activation mechanism. A typical example is the use of a car bomb that is detonated when the victim starts the car
3. Terrorists use explosives to create as much damage and victims as possible to cause fear and societal destabilization. Although terrorist acts can be aimed at a specific person or institution they can also be aimed at large groups. Locations for an attack are then selected on the basis of accessibility, risks of being detected and maximum effect
4. Either real or fake explosive constructions (both military or the so-called improvised explosive devices—IEDs) to warn or threat a certain person or organization or to create chaos and social unrest. This can sometimes also relate to extortion and the abuse of explosives to shut down businesses
5. Illegal production and use of explosives out of fascination and amusement but with very serious safety risks for those involved and innocent bystanders
6. Irresponsible and incorrect use of explosives in an otherwise legal setting. For instance, neglecting very strict safety protocols when dealing with explosives can result in criminal investigations after an incident

Forensic explosives experts typically combine detailed knowledge on the nature of explosions, the chemistry of energetic materials, and the construction of explosive devices with expertise related to the chemical analysis of explosives. Forensic casework is typically split in two types, the so-called **pre-explosion** and **post-explosion incidents**. In a pre-explosion case, energetic (or hoax) materials and associated devices have been discovered and defused before activation. This is usually a task of a military Bomb Disposal Squad, and the process is referred to as bomb ordnance. In the defusing of bombs, decisions are based on safety and the preservation of forensic traces is only a secondary consideration. However, when an explosive device is successfully dismantled, the separate parts can be studied in detail in a forensic laboratory often leading to valuable forensic evidence. The forensic explosives experts focus on the chemical analysis and the reconstruction of the device. However, such investigations are often interdisciplinary in nature as these parts can also contain fingermarks and biological material from the perpetrators. Such detailed analysis is usually not possible when the experts deal with a post-explosion case. Like a fire, an explosion is a very destructive process that not only creates but also destroys forensic traces. In general, a forensic post-explosion investigation is much more challenging also in terms of chemical analysis. Only trace levels can be found of the energetic material used and large areas need to be investigated to recover fragments of the explosive device. Sample matrices can be very complex especially when fires break out as a result of the explosion and victims have sustained seriously or deadly injuries.

Forensic pre-explosion investigations typically include:

- *Chemical identification of explosives and precursors in pre-explosion cases (bulk and trace analysis)*
- *Description of intact explosive devices and associated parts*
- *Classification and functionality assessment of intact professional (military) explosive constructions* (e.g., *hand grenades and rocket propelled grenades, so-called RPGs)*
- *Characterization and functionality assessment of home-made explosives, so-called HMEs and improvised explosive devices, so-called IEDs*
- *Classification and functionality assessment of legal and illegal fireworks and associated pyrotechnic mixtures* (e.g., *black powder and flash powder)*
- *Explosive effect analysis (what if?), what would have been the impact of the explosive and the damage to infrastructure and persons if the explosive would have been successfully activated?*
- *Providing advice at the pre-explosion crime scene with respect to forensic options*

In post-explosion cases, basically the same type of questions is addressed but the overall reconstruction is much more challenging and affects the methods and techniques applied:

- *Sampling (swabbing) of items from the scene that can contain post-explosion residues*
- *Trace analysis of explosives, precursors, and explosion markers in post-explosion swabs and samples originating from the postexplosion scene*
- *Reconstruction of the explosive device used based on the chemical trace analysis and the recovery of device fragments*
- *Reconstruction of the explosion to estimate the amount of explosive material in the exploded device (usually expressed as Tri-Nitro-Toluene, TNT equivalent mass)*
- *Providing advice at the post-explosion crime scene with respect to the forensic reconstruction*

Now in contrast to illicit drug analysis, a single analytical chemical method is not capable to effectively analyze all types of energetic materials and samples encountered in forensic casework related to explosives and explosions. The main incentive for the chemical analysis is quite straight forward: identification of the energetic materials involved. However, there is a high diversity in explosives (organic vs. inorganic, crystalline pure compounds vs. mixtures) and samples (bulk vs. trace) and this requires a tailor-made strategy and the application of several instrumental techniques as is illustrated below.

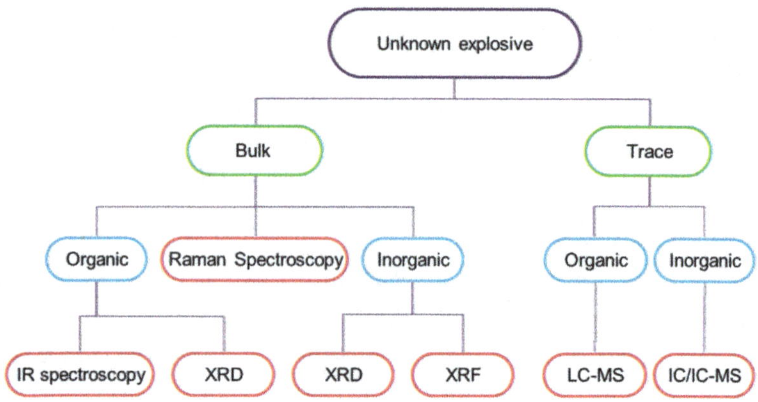

Chemical analysis framework as used by the forensic explosions and explosives experts of the Netherlands Forensic Institute. Main factors that determine the techniques applied are whether the experts deal with a bulk or trace sample, and whether the energetic material involved is expected to be of organic or inorganic nature (XRD = X-ray diffraction spectroscopy, XRF = X-ray fluorescence spectroscopy, LC-MS = liquid chromatography with mass spectrometric detection, IC = Ion Chromatography, IC-MS = Ion Chromatography with mass spectrometric detection)

In **Chapter 6—Chemical Profiling, Databases and Evidential Value** and **Chapter 7—Forensic Reconstruction through Chemical Analysis**, the chemistry, analysis, and forensic investigation of explosives are discussed in more detail.

Microtraces: gunhot residues

Gunshot residues (GSR) represent an important class of **microtraces** in the forensic investigation of shooting incidents. Where firearms examiners focus on the bullets and cartridge cases to retrieve information on the weapon(s) used, GSR experts try to retrieve particles that are invisible to the human eye (particle diameters in the nm-µm range) and that are created when the main charge in the round is ignited to propel the bullet with high speed from the barrel. These particles spread out on the hands and clothing of the shooter but are also deposited on bystanders and travel in the wake of the bullet toward the victim. If shot at close range, a more substantial GSR cloud precipitates on the victim and the amount and pattern of the GSR can thus provide information on shooting distance and angle. The elemental composition of the GSR particles can vary for different types of ammunition, and this can be used to reconstruct a shooting incident involving multiple firearms and shooters. However, GSR experts always need to consider that particles could originate from a source not related to the discharge of a firearm. Typically, particles containing the elements Pb, Ba, and Sb originate from the primer, the material that ignites the gunpowder after the round is hit by the firing pin. These so-called **PbBaSb particles** are considered to be highly characteristic for the discharge of a firearm and are not created through other environmental processes.

Typical GSR investigations include:

- *Gunshot residue detection on suspect hands and clothing of the suspects and victims*
- *Characterization of gunshot damage to clothing, items and structures*
- *Shooting distance estimation (was a victim shot at close range?)*
- *Gunshot residue attribution (linking GSR to ammunition and firearms)*
- *Shooting incident reconstruction (with forensic firearms and CSI experts)*
- *Forensic expert advice on the crime scene*

If GSR residues are detected on the hands of a suspect, does this mean that the suspect fired a gun?

If GSR cannot be found on the hands of a suspect, does this mean that the suspect was not involved in a shooting incident?

The detection and characterization of gunshot residues is challenging but the criminalistic interpretation is just as complex. This is due to the fact that GSR particles are discharged in many directions, can be transferred between objects (secondary transfer) and are easily removed from surfaces (limited persistence). When experts find particles on a person that are GSR characteristic, they typically conclude that the person has been involved in a shooting incident without specifying the specific role. Detecting GSR particles therefore does not proof that somebody fired a gun, and he could also have been a bystander. Similarly, not finding any GSR on a suspect does not mean that he did not fire a gun. This depends on the context of the case and the time between the incident and the arrest and sampling of the suspect. Over time GSR can easily be lost or removed, for example, by the washing of hands or clothes.

To be able to detect and characterize GSR, particles need to be collected from the scene, victims and suspects. When suspects of a shooting incident are taken into custody, special care needs to be taken to prevent loss of microtraces potentially present on hands and clothing items. Hands can be protected by using plastic bags and clothing items can be confiscated.

Sampling is typically done using special GSR stubs. These stubs, made of an adhesive and conductive material, are applied several times on a surface to be sampled, for example, the hand of a suspect. In this collection process, spatial information is lost but stubbing allows for effective analysis of relatively small numbers of microparticles. The work-horse technique used by GSR experts is without a doubt **Scanning Electron Microscopy** (SEM). Tailor made methodology enables the automated analysis of the stub surface, during which the software detects GSR particles, and after completion of the analysis an overview is provided indicating the particles of interest on the stub. The particles of interest are selected on the basis of their elemental composition as determined with EDX (**Energy-dispersive X-ray spectroscopy**). The principle is very similar to XRF but in SEM-EDX, the electron beam is used (as opposed to X-rays) to remove an inner electron from an atom. The resulting rearrangement of the electronic configuration leads to the emission of X-rays with characteristic energy levels. By detecting these X-rays during SEM analysis and measuring their energy level, the presence and relative abundance of elements are indicated.

SEM-EDX setup and GSR stubs used at the Netherlands Forensic Institute (*courtesy of Dr. Alwin Knijnenberg, Forensic GSR expert at the NFI, note: several instrument manufacturers provide very robust instruments, the set-up shown serves as an example and does not indicate any preference*)

However, GSR experts have a broader analytical chemical toolbox at their disposal during the forensic investigation (in bold main instrumental method used):

- *Use of stubs and foil to sample GSR particles*
- ***SEM-EDX—Scanning Electron Microscopy with Energy Dispersive X-ray analysis (enabling elemental analysis of detected particles)***
- *Colorimetric methods to visualize GSR clouds to estimate shooting distance (sodium rhodizonate to detect Pb and Ba containing particles, dithiooxamide to detect Cu traces and Zincon to visualize Cu, Ti, and Zn)*
- *micro- and macro-XRF imaging (X-ray fluorescence imaging for the non-invasive spatial analysis of GSR patterns)*
- *LC-MS—Liquid chromatography with mass spectrometric detection for organic GSR analysis*

GSR experts work in close collaboration with Firearm experts, what risk is involved in this collaboration?

Contamination!

Firearm experts inspect and investigate firearms. Part of their work is conducted at a firing range where they fire test rounds to create reference bullets and cartridge cases for forensic comparison. As a result of these activities, experts can carry large amount of GSR particles on their skin and clothing. When objects with a very high concentration of GSR particles come into contact with evidence items with trace GCR amounts, a contamination can easily occur. At the Netherlands Forensic Institute, persons entering the shooting range are not allowed to be in the GSR lab on the same day.

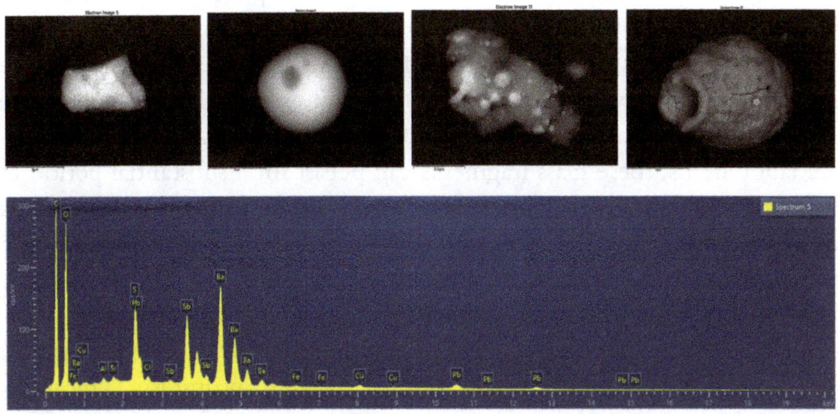

Collection of SEM photos of typical GSR particles (particle diameters range from 5 to 50 μm). The SEM-EDX spectrum of the particle on the left shows the characteristic primer elements Ba, Pb and Sb (courtesy of Dr. Alwin Knijnenberg, GSR expert at the Netherlands Forensic Institute)

Microtraces: glass, paint, and fibers

Other important microtraces in addition to GSR include glass, paint, and fibers. Such traces can provide important evidence linking a perpetrator to a crime scene or a victim. The support for such a link becomes even stronger when there is physical evidence of a mutual exchange, *that is,* when microtraces that could originate from a suspect are found on the scene and vice versa. Hence, **Locard's Exchange principle** as discussed in **Chapter 1** is a leading factor for forensic microtrace experts with respect to the materials they investigate and the analytical chemistry methods they apply. Ideally, sampling and instrumental analysis technology enables detailed chemical characterization of the smallest of particles. In principle, a wide range of microtrace materials could be of forensic interest and could assist in the criminal investigation depending on the circumstances of the crime. However, there are number of microtrace types that occur more frequently and can potentially provide very valuable forensic information. In case of GSR, this microphysical evidence is directly related to the criminal activity (although firing a gun can also be fully legal, *e.g.* by police officers or in hunting activities by licensed individuals). A similar situation exists more of less for glass fragments, although these microtraces do not originate from a tool or object that is so directly associated with a felony as a firearm. Glass fragments typically play an important role in burglary and robbery cases where perpetrators smash windows and showcase panels to gain access to houses, shops, and valuables. When rocks, hammers, or other objects are used to break glass panes, tiny glass fragments are spread out both in a forward and backward direction. As a result, small glass fragments are deposited on the clothing of a perpetrator, and depending on the texture of the fabric items, these glass fragments can persist for a substantial period of time after the criminal activities (again it should be noted that smashing a window can also be the result of a fully legal action, *e.g.* when someone tries to save an individual from a burning house or car). Because burglary and robbery cases represent a significant percentage of all criminal investigations, glass microtrace analysis is an important forensic expertise area that can be found in nearly all national forensic institutes. Typical investigations include:

- *Glass fragment collection from clothing of "breaking and entering" suspects*
- *Glass fragment classification, establishing the origin of the glass (float glass, drinking bottle, car window) from its elemental composition and physical features*
- ***Forensic glass comparison of glass fragments from "breaking and entering" suspects*** versus ***reference glass at a crime scene (e.g., a smashed window)***
- *Forensic glass comparison of glass fragments from "breaking and entering" suspects* versus *reference glass from an unsolved case database*
- *Complex case reconstruction through characterization of glass fragments from various sources (household glass, displays) and quantitative glass analysis (number of fragments)*

For a long time, forensic comparison of glass was based on accurate refractive index (RI) measurement and microscopic inspection. By compiling substantial databases of RI values for various types of glass, the evidential value of a matching RI value could be assessed. Nowadays, forensic experts apply various instrumental analysis techniques to characterize glass through its chemical composition and the presence of trace elements. Such techniques must be adapted and optimized to be able to analyze the tiniest of fragments. The overall forensic toolbox for glass analysis thus consists of a suite of methods:

- *Microscopic inspection of glass fragments*
- *RI measurement (GRIM—matching RI method)*
- *SEM-EDX—Scanning electron microscopy with energy dispersive X-ray analysis*
- *micro-XRF—X-ray fluorescence spectroscopy*
- *LA-ICP-MS—laser ablation inductively coupled plasma mass spectrometry*

Interestingly, these methods are non- or minimally destructive (LA-ICP-MS, with this technique a UV laser ablates small amounts of glass from the surface for trace element analysis) and enable further investigation of the physical evidence. Although not all forensic laboratories employ LA-ICP-MS for forensic glass analysis, this technique is used very successfully at the Netherlands Forensic Institute. Small glass fragments are characterized by establishing a quantitative profile of elemental trace impurities. These impurity profiles can be measured with high accuracy and precision by using special glass standards doped with trace elements. Furthermore, trace element composition in glass is not monitored during production and hence the profiles can be very characteristic depending on the original glass source.

Another important mictrotrace type frequently encountered as a result of criminal activities is paint deposit. As discussed before paint transfer can also be the result of a totally benign action or an unintentional accident. However, paint evidence is often valuable in cases where criminals have used a crowbar, car, truck, or other painted objects to illegally gain access to properties and valuables. If the object itself is painted, then often cross-transfer occurs during high-energy impact providing important evidence of the mutual contact. Paint transfer also plays an important role in the investigation of so-called hit-and-run traffic accidents. Deserting a wounded victim after a collision is considered a crime for obvious reasons and initially the main interest of the Police is to identify the car/motorcycle and the driver. For example, the investigation of paint residues on the victim or his/her vehicle or bike can be used to classify the car of the perpetrator (brand, model, year of production). Once a suspect has been found, the forensic chemical analysis focuses on the comparison at source level (could the trace material originate from the car of the perpetrator?). It should be noted that in this type of cases also glass fragments are often found at the incident scene and that this can also constitute important evidence. The combination of glass and paint findings can further strengthen the overall (either incriminating or exculpatory) evidence.

Typical paint microtrace investigations are listed below:

- *Finding and securing paint traces from items and objects*
- *Recognizing the type and model of a vehicle in "Hit and Run" incidents*
- *Recognizing the type and model of a vehicle in "ATM raids"*
- ***Forensic paint comparison**: establishing a link between paint traces found on a crime scene and reference paint (car, motor, bike, tool, spray paint)*

Paint evidence is usually first found, described, and characterized by visual inspection using a microscope. Paint material can consist of multiple layers depending on its source, the painting of modern cars, for example, includes several treatments and applications of separate layers. Dried car paint thus consists of several layers to protect the metal and create special visual effects (such as a metallic look). If these layers can be chemically characterized individually, the overall evidence of the paint investigation can be significantly increased. To this end, several advanced instrumental techniques can be applied. In line with the analysis of glass fragments, a prerequisite for these methods is that they are able to handle the minute size and amount of typical paint traces:

- *Microscopic visual analysis and color measurement of paint traces*
- *Comparing color and appearance against databases for car/object paints*
- *Microscopic spectroscopy and hyperspectral imaging (UV-Vis, fluorescence, Raman, FTIR)*
- *Py-GC-MS—Pyrolysis gas chromatography with mass spectrometric detection for the analysis of polymer binders in paint layers*

The combination of microscopy with spectroscopy is of high relevance for forensic paint analysis. Microscopy allows microvisual assessment but also facilitates the spectroscopic analysis at microlevel allowing the chemical characterization of individual layers. To this end, Raman microscopes are found very frequently in forensic microtrace laboratories. However, microspectroscopy does not typically allow analysis at trace level and provides only evidence on the general composition of paint layers. A more detailed analysis is feasible when using pyrolysis GC-MS but at the expensive of sample consumption.

Cars used in ATM raids and robberies are often dumped by perpetrators directly after the crime. If such a car is found for which crime involvement is quite obvious, would it then still be important to investigate paint and glass traces of the car at the crime scene? And why/why not?

Yes, this is still quite important! One needs to be very careful to label items as evidence that have not been located at the crime scene but somewhere else. Hence, it is important to establish whether the alleged car was indeed used to create the damage at the crime scene. Potential suspects could now be found through a connection to the vehicle (e.g., DNA evidence, fingermarks, fibers). However, when the car was stolen care has to be taken not to consider traces from the original owner as crime related.

Finally, fibers also represent a class of microtraces of particular interest to forensic scientists. Forensic fiber evidence plays an important role in many types of crime, especially in the investigation of violent crimes and sexual assaults. **Locard's Exchange Principle** dictates that due to the intensity of the contact between the assailant and the victim, valuable physical evidence such as fibers from clothing or human biological traces is exchanged. The interest in fibers is not so much related to the crime itself but is the result of the pervasive presence of fabrics in modern society, in clothing, furniture, and carpets. As a result, almost invisible to the human eye, single fibers are "everywhere" and are shed, spread, transferred, and get stuck via individuals, objects, and through contacts continuously. Therefore, this also occurs when crimes are committed. At the same time, the case context plays an important role in forensic fiber investigations. Transfer of fibers from and to the clothing of a victim will also occur during every day, nonviolent interactions.

Typical investigation conducted by forensic fiber experts include:

- *Sampling, finding and securing fibers from clothing and items for further analysis*
- *Textile damage analysis, characterization and reconstruction (could the damage observed been caused by a given object?)*
- *Fiber and textile classification: determining textile type/model (fiber composition and type and application in textiles)*
- *Forensic fiber comparison: establishing a link between fibers found on a victim, suspect or crime scene, and reference fibers from a textile item*
- *Fiber fusion: reconstructing traffic accidents (during high impact contact clothing fibers can get embedded in plastic matrices)*
- *Flock fibers: linking suspect to crimes through specially marked fibers (using fibers as a track and trace tool in tactical police investigations)*

Finding "odd" fibers in a "sea" of other fibers (e.g., from the clothing of a victim) is a delicate task requiring a high degree of expertise and experience. Experts spend a great deal of time behind a microscope. Fibers are removed (sampled) from a substrate, typically by using adhesive tapes. These tapes contain a lot of material of the substrate and hopefully also trace fibers from other sources of interest to the investigation. These fibers need to be removed and collected by the expert. Typically, selected fibers are collected on slides and further investigated using microscopy-based techniques. Fibers are either fully synthetic (e.g., nylon, polyester, elastane) or have an animal or plant origin (e.g., wool, cashmere, cotton). Either way, fibers can always be considered as man-made materials, even natural fibers are being processed and often chemically and physically modified and dyed in fabric production. In forensic fiber comparison, finding characteristic features is therefore challenging. Fibers are produced and processed in huge volumes (high intrinsic background levels) and fiber microtraces represent very little material for investigation. Hence, forensic fiber experts are continuously looking for methods that allow the differentiation of fibers within a given type (classification). Such methods are especially needed for fibers that extremely common such as cotton used in denim. Forensic fiber laboratories apply a diverse and dynamic set of techniques using state-of-the-art equipment:

- *Taping and microscopic inspection of tapes to find and secure fiber evidence*
- *Microscopic analysis of textile damage*
- *Microscopic visual analysis (incl polarized light) and color measurement of fibers*
- *Microscopic spectroscopy (UV-Vis, fluorescence, Raman, FTIR)*
- *Py-GC-MS—Pyrolysis gas chromatography with mass spectrometric detection*
- *LC-PDA-MS—Liquid chromatography with photo diode array and mass spectrometric detection for the profiling of fiber dyes*

Much of the analytical chemical instrumentation used for the forensic investigation of paint traces can also be used for fibers; hence forensic institutes often combine these expertise areas. An interesting extra option that is applicable to fibers but not so much to dried paints is the extraction and characterization of dyes. Although instrumentally and methodologically challenging, the extraction of dye from a single fiber creates interesting options for fiber comparison especially when degradation characteristics can be exploited. The Netherlands Forensic Institute has developed a special method based on miniaturized dye extraction (even including the extraction of reactive dyes) followed by LC-PDA-MS analysis. This methodology is successfully applied in case work as is illustrated in more detail in **Chapter 11 — Innovating Forensic Analytical Chemistry**.

Collection of microscopy photos of (from left to right) glass, paint, and fiber traces. The glass particle is surrounded by sand and dust particles and has a diameter of roughly 500 μm. The resin embedded paint chip shows a three layered car paint consisting of a ground layer (white), the main color layer (red) and a clear coat finish each with a layer thickness of approximately 50 μm. The photo on the right shows the close-up of a red acrylic fiber and several white cotton fibers, the fibers have a diameter of roughly 20 μm *(courtesy of Dr. Jaap van der Weerd, forensic microtrace expert at the Netherlands Forensic Institute)*

Forensic environmental investigations

The criminal investigation of environmental crimes, for example, the illegal dumping or processing of waste streams for economic gain, is in general challenging and legally very complex. At the same time, the damage of these crimes to society can be very substantial and can persist for decades. Environmental crimes often have an international aspect, especially when it involves the transport and dumping of waste by ships. Dumping waste out of financial motives is intentional, but environmental crimes can also occur due to unintentional accidents in combination with reckless actions and/or negligence. A "side-effect" of the illegal production of drugs of abuse is the dumping of waste in rural areas, forests, surface waters, and sewer systems.

Perpetrators can sometimes also be charged with environmental crimes in addition to violating criminal drug laws. In this setting, the dumping of waste is the result of other criminal activities and motivated by the need for these activities to remain undetected. Forensic environmental expertise covers a wide range of investigations:

- *Crime scene support at environmental incident sites*
- *Environmental sampling with personal protective equipment*
- *Forensic advice with respect to environmental sampling*
- *Forensic investigation of environmental waste incidents*
- *Forensic investigation of work-related health incidents*
- *Forensic assessment of environmental damage of dumping of waste from the illegal production of drugs of abuse*
- *Forensic investigation of illegal agricultural manure dumping*

As environmental crimes typically involve the dumping or processing of chemical waste streams, forensic analytical chemistry plays an important role in forensic environmental investigations. Given the diversity in chemical composition of samples and the complexity of the matrices involved, a substantial analytical chemical toolbox is required including state-of-the-art sampling and sample preparation methods:

- *Sampling and sample preparation techniques (SPME, liquid extraction, head space sampling, microwave digestion)*
- *GC-MS—Gas chromatography with mass spectrometric detection*
- *GCxGC-MS—Comprehensive 2D gas chromatography*
- *LC-MS—Liquid chromatography with mass spectrometric detection*
- *Raman and infrared spectrometry*
- *XRF—X-ray fluorescence spectroscopy (heavy elemental analysis)*
- *ICP-MS—inductively coupled plasma mass spectrometry (heavy elemental analysis)*

Environmental forensics is regarded as a highly specialized "niche" area and within forensic institutes often small expert teams conduct such investigations. However, accurate reconstruction requires extensive knowledge of chemical processes and environmental analytical chemistry. Therefore, it is not uncommon that certified and accredited laboratories specialized in environmental chemical analysis receive samples and requests from forensic institutes and law enforcement agencies. Such laboratories usually provide services to both industry and governmental institutions.

Fingermarks

Fingermarks (traces) and fingerprints (reference prints taken from suspects under controlled conditions) represent a very important class of so-called biometric traces in forensic science. The highly characteristic nature of the intricate ridge patterns found on the fingers, palms, toes, and soles of primates was already recognized thousands of years ago. The use of fingermarks for biometric identification in criminal investigations was suggested for the first time in 1880 by Henry Faulds.

A fingermark is a biometric trace, a trace with the potential to identify the human donor. Is Biometric identification similar to Chemical identification from a forensic perspective?

No, Biometric Identification is forensically not the same as Chemical Identification! This different perception of identification in forensic science in comparison with other science areas was the basis of the "criminalistics is the science of individualization" statement by Paul Kirk.

Chemical Identification establishes the compound (cocaine in the example) or composition but does not provide information on the origin of the sample. It answers the "what" question but does not indicate the original source.

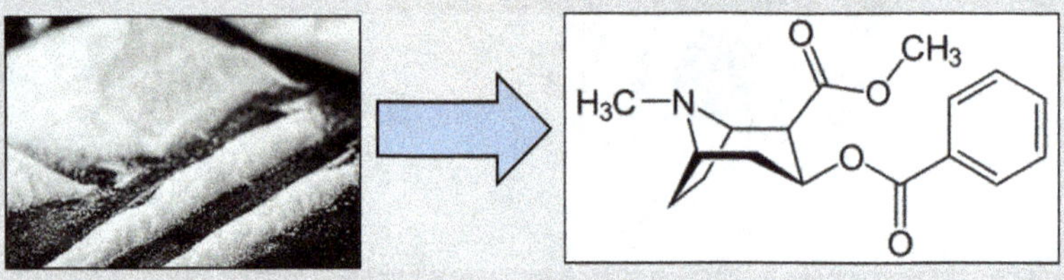

Biometric Identification links a forensic trace to a specific individual. It addresses the "who" question. If a ridge pattern is truly unique (remember the individualization fallacy!), this means that we can point out the donor of that fingermark at the exclusion of the entire human population.

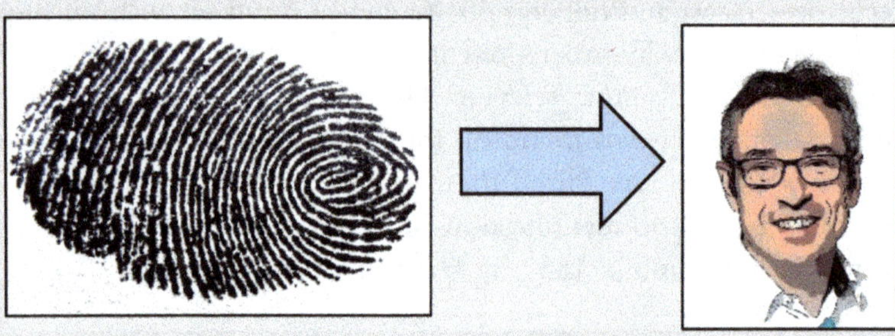

Nowadays, forensic finger mark comparison is an essential tool to solve crime and all major forensic laboratories and law enforcement agencies extensively use this expertise. Fingermarks detected and recovered from the crime scene by crime scene officers (CSOs) are entered in large databases and compared against reference prints from individuals using so-called **AFIS** systems (Automated Fingerprint Identification Systems).

The forensic dactyloscopic investigation typically includes:

- *Visualization of latent fingermarks (chemical and physical development) (Ninhydrin, Cyanoacrylate, Vacuum Metal Deposition)*
- *Capturing (gel lifting) and photographing of fingermarks*
- *Forensic fingermark comparison (dactyloscopy): linking a suspect to a crime scene*
- *Finding unknown suspects through an AFIS fingermark database search*
- *Forensic reconstruction: activity analysis and fingermark dating*

The role of chemistry and analytical chemistry in the field of forensic fingermark investigations is in principle of a secondary, supporting nature. Fingermarks on a crime scene or an object are often **latent** (i.e., hidden/concealed to the human eye). Optical techniques can be applied to create contrast allowing the dactyloscopic expert to study and photographically record the ridge pattern. This includes the use of UV light sources, optical filters, and fluorescence. At the crime scene, fingermarks are often "dusted" by CSOs, graphite powder is then applied to the mark and the particles sticking to the fatty residues visualize the latent trace after which the mark can be transferred to a suitable tape or a gel substrate. In addition, a wide range of chemicals and chemical processes are used to "develop" marks on various surfaces. The most commonly applied techniques are the ninhydrin method (colorimetric detection based on the reaction with amino acids), vacuum metal deposition (VMD, a physical development method), and cyanoacrylate fuming (cyanoacrylate polymer formation initiated at the ridge surface, also known as the "superglue" method).

In recent years, the advancement of analytical chemistry and instrumentation has provided new opportunities in the field of forensic fingermark investigation. Where the ridge pattern provides opportunities to identify the donor, chemical analysis can provide information on when the mark was placed (fingermark age determination) and what activities were undertaken by the donor (interpretation at activity level). Traces of exogenous compounds like explosives and drugs of abuse can indicate the handling and/or use of these materials by the donor. If the chemical analysis can be conducted noninvasively, both investigations can be conducted on a single mark providing a wealth of forensic information. However, although very promising scientific studies are being presented in forensic scientific literature, analytical chemical methods should ideally be compatible with the visualization processes used to detect latent marks. Surely, fingermarks first need to be detected before they can be investigated in more detail!

Although not used in regular day-to-day casework yet, the forensic analytical chemical toolbox for the chemical characterization of fingermarks is rapidly expanding:

- *Visual analysis of fingermarks using various light sources and filters*
- *Microscopic spectroscopy and hyperspectral imaging of fingermarks (UV-Vis, fluorescence, Raman, FTIR)*
- *Hyperspectral imaging of fingermarks (UV-Vis, fluorescence, NIR)*
- *Mass spectrometric imaging of fingermarks (MALDI, DESI, TOF-SIMS)*
- *Age estimation of fingermarks with FTIR and fluorescence imaging*
- *LC-MS—Liquid chromatography with mass spectrometric detection and GC-MS—Gas chromatography with mass spectrometric detection for the chemical analysis of fingermarks for donor characteristics (lifestyle markers), donor activities* (e.g., *handling of explosives, use of drugs) and trace age estimation*

Questioned documents

Whereas the amount of digital forensic evidence has been rapidly increasing over recent decades, the reverse situation exists for the forensic expertise area of Questioned Documents. This field investigates written and printed documents and includes forensic handwriting comparison, autograph authentication, money counterfeit detection and passport and license forgery investigations. The decline of this type of forensic evidence material is related to societal changes fueled by the digital technology revolution. Classical writing with pen and paper is replaced by texting and sending email, books are read using tablets and e-readers, consent is given through digital signatures, and payments are more and more done electronically (at least in the Netherlands). However, these are relatively slow processes and written letters, printed papers, paper money, passports, driver's licenses will be around for decades to come. Handwriting will remain an important skill to learn for young children and possibly written and printed documents will be used in the future in addition to the wealth of digital means to communicate. Hence, forensic document expertise will remain of significance especially in an international context. An interesting aspect of forensic document examination is the fact that evidence material itself can provide clues with respect to motive, especially when considering written text such as threat letters.

Typical Questioned Documents investigations include:

- *Visualization of indented writing*
- *Investigating document authenticity (including passports, licenses, contracts)*
- *Money counterfeit and theft (security ink staining) investigation*
- *Interdisciplinary investigations of threat letters (DNA, fingermarks, writing)*
- *Forensic investigations into document forgery*
- *Forensic signature comparison (mimicking and duplication)*
- *Forensic handwriting comparison: establishing a link between a document and a suspect*
- *Forensic document examination and comparison: establishing a link between a document and a suspect, ink dating for forensic reconstruction*

Forensic handwriting comparison is typically performed subjectively by a human expert. Attempts to fully automate this process and indicate the strength of evidence based on the characteristic features of the handwriting have never been able to completely replace the expert assessment (in line with forensic firearm investigations). In addition, several objective methods are available based on physical rather than chemical principles. Detailed investigation of a printed document for instance can indicate the printing process (inkjet vs. laser), the brand and type of printer used (classification) but can also reveal small defects that can be quite characteristic for a given printer confiscated at the home of a suspect. This requires no chemical analysis, just accurate visual registration and accurate measurements of distances and angles. However, the investigation becomes much more powerful when the questioned document expert also includes chemical analysis in the forensic toolbox, for example, by including ink analysis when investigating a potential link between a document and a printer. Because of the similarity between document ink and dyes used in fibers and paint, the framework for the chemical analysis of questioned documents bears resemblance and is based on noninvasive micro- and macrospectroscopy for ink and document characterization in combination with chromatography-mass spectrometry techniques for trace analysis. Spectroscopic analysis is usually performed with tailor-made hyperspectral imaging equipment. Additionally, LA-ICP-MS and Isotope Ratio Mass Spectrometry (IRMS) can also be interesting techniques to obtain characteristic elemental and isotopic ink and paper profiles.

Hence, a broad range of analytical methods can be applied in questioned documents investigations:

- *ESDA—Electrostatic detection apparatus*
- *Microscopic visual analysis and physical measurements of documents*
- *Microscopic spectroscopy and hyperspectral imaging (spectral scanner): UV-Vis, fluorescence, Raman, FTIR*
- *Micro- and macro-XRF—X-ray fluorescence imaging*
- *LA-ICP-MS—inductively coupled plasma mass spectrometry (paper, ink)*
- *IRMS—Isotope ratio mass spectrometry (paper, ink)*
- *Py-GC-MS—Pyrolysis gas chromatography with mass spectrometric detection (paper, ink)*
- *LC-MS—Liquid chromatography with mass spectrometric detection (ink analysis)*
- *GC-MS—Gas chromatography with mass spectrometric detection (ink dating)*

Crime scene investigation

After the discovery of a (potential) crime, the CSOs are in charge of the local technical investigation. This typically includes determining the boundaries of the crime scene (what is included and excluded in the investigation), meticulous photographic recording and documentation of the observations and the detection, recovery, and sampling of relevant traces. Special expertise areas that are related to the crime scene investigation are BPA (Blood Pattern Analysis) and forensic archeology (crime scenes involving buried human remains). Latent blood traces (after attempts by perpetrators to remove blood stains) can be detected through the use luminol, a chemiluminescent agent that is catalyzed by the iron present in hemoglobin. Detailed characterization of trace material and evidence items is normally conducted by experts at the forensic laboratory under controlled conditions. However, recent instrumental developments also create options for chemical analysis directly at the crime scene. Having such capabilities already available at the crime scene can be very beneficial for the criminal investigation. It can provide immediate leads and can be used for the selection of traces for further analysis. This leads to a more efficient use of scarce law enforcement and forensic expert capacity. However, a prerequisite is that sufficient quality can be assured when operated by staff with limited chemistry knowledge and under the uncontrolled conditions at the crime scene. This requires very robust analytical methods that are tested and validated under realistic circumstances:

- *Luminol application—chemiluminescent detection of latent blood traces*
- *Hyperspectral imaging for biological trace detection and dating*
- *Indicative testing of biological traces (rapid stain identification antibody tests)*
- *Indicative colorimetric testing for drugs of abuse and explosives*
- *Mobile Raman and IR for chemical identification of drugs and explosives*

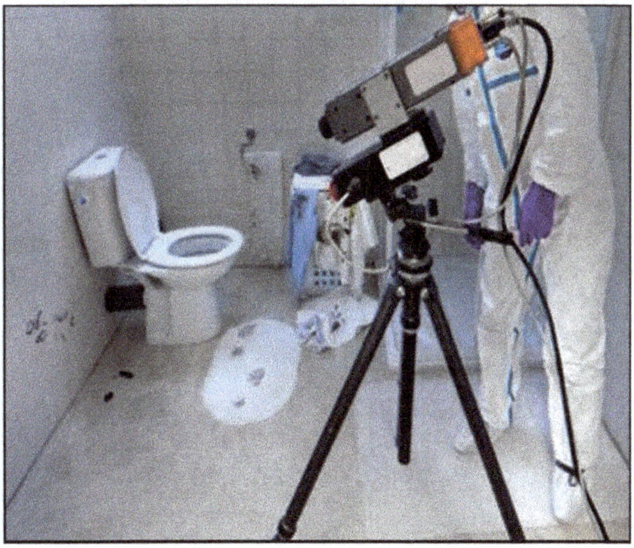

Use of a Hyperspectral Imaging System for noninvasive blood detection and age estimation at the crime scene (*courtesy of Prof. Dr Maurice Aalders, Amsterdam University Medical Centers*)

With this comprehensive overview readers are now aware of where to find analytical chemistry in a forensic laboratory and to link analytical techniques to forensic expertise areas and case work investigations. It hopefully has become clear why certain instruments and methods are needed to characterize physical evidence and forensic traces and what answers they provide with respect to the questions in a criminal investigation. However, the insight generated at this stage only provides a so-called helicopter-view, enabling students to list useful analytical techniques without any specific knowledge (yet) of the scientific basis and practical details. In the next chapters, these critical scientific and practical aspects will be discussed in more detail.

Further reading

Butler, J.M., 2010. Fundamentals of Forensic DNA Typing. Academic Press, Elsevier, ISBN 978-0-12-374999-4.

Elkins, K.M., 2019. Introduction to Forensic Chemistry. CRC Press, ISBN 978-1-4987-6310-3.

Forensic Chemistry, Pearson New International Edition, 2013. Suzanne Bell, Pearson Education Limited, ISBN 978-1-2920-3375-4.

Houck, M. (Ed.), 2015. Forensic Chemistry. Academic Press, Elsevier, ISBN 978-0-12-800606-1.

Roux, C., Ribaux, O., Frank, C., 2018. Forensic science 2020 - the end of the crossroads? Aust. J. Forensic Sci. 50, 607—618.

CHAPTER 3

Sampling and sample preparation

Contents

3.1 What will you learn?

This chapter discusses the first and often essential step in chemical analysis: sampling and sample preparation. If sampling goes wrong, this cannot be repaired by the work in the laboratory as a representative sample is not available and the composition is unknown. Two examples from forensic practice will be studied in more detail. First, we will look into statistical aspects when decisions have to be made with respect to the size and composition of a representative sample set after the confiscation of a large illicit drug shipment by law enforcement or customs. Next, the sample preparation strategy for fire debris analysis (FDA) will be discussed. The analytical chemistry and forensic principles governing the optimal sample pretreatment strategy will be explained. With these two examples, readers will also gain insight in two important forensic expertise areas: illicit drug, and fire debris and ignitable liquid analysis.

Chemical Analysis for Forensic Evidence
ISBN 978-0-12-820715-4
https://doi.org/10.1016/B978-0-12-820715-4.00004-3

After studying this chapter, readers are able to

- substantiate the relevance of **sampling** and **sample preparation** in forensic analytical chemistry
- list the main requirements for correct **sampling** and **sample preparation**
- estimate a **representative numerical sample size** for a large seizure of **illicit drug** packages on the basis of **statistical principles**
- explain the optimal **sample preparation strategy** in forensic **FDA**

3.2 Sampling and sample preparation in analytical chemistry

Sampling and sample preparation are initial and essential steps in chemical analysis. Through these steps the analytical chemist ensures that the object of the investigation (*e.g.* a food product, a reaction mixture, an aqueous waste stream, or in the case of forensic science the physical evidence) can be presented in a form that is compatible with the instrumentation used to perform the analysis.

A one-page editorial in Analytical Chemistry (1997) of Royce W. Murray on the importance of sampling and sample preparation was entitled

There is no analysis without sampling

Do you agree with this statement?

In practice, many powerful techniques (e.g., NMR, LC-MS, GC-MS) require sampling and sample preparation to obtain useful results. However, this should be seen as a drawback and a nuisance! Sampling introduces errors and uncertainty and sample preparation could be very laborious, making the analysis costly and time-consuming. Hence, analytical chemists are constantly looking for methods that provide direct information and require no sampling and (almost) no sample preparation. Reflectance spectroscopy-based techniques can often be operated in a noninvasive/minimally invasive manner exploiting the full range of the electromagnetic spectrum to assess the chemical composition. However, without sample preparation, the options for trace analysis and assessing complex mixtures remain limited.

Some examples of chemical analysis techniques that do not require sampling and sample preparation:

- Hyperspectral imaging (HIS)
- Fluorescence imaging
- Portable Raman
- Portable NIR
- Portable XRF
- XRF Imaging (micro-XRF, MA-XRF)
- IMS (Ion Mobility Spectrometry)
- Ambient ionization mass spectrometry (DESI, DART)

Interestingly, sampling is also an important term in statistics. A statistical sample represents a (random) selection of a subset to estimate the overall characteristics of a population. From a mathematical perspective, estimation means that a sample can never provide a completely precise (i.e., no random errors) and accurate (i.e., no systematic errors) description of the population. This can only be achieved by being aware of every object in the collection. However, through sampling we can efficiently characterize a population without studying it excessively. With relatively small or infrequent sampling, we can be "quick and dirty," providing information rapidly and at low cost. However, this comes at a price of increased uncertainty. In a way, physical sampling in chemical analysis has a similar purpose and is governed by the same principles. The IUPAC Gold Book contains the following description of an analytical chemistry sample: "*a portion of material selected from a larger quantity of material. The term needs to be qualified,* e.g., *bulk sample, representative sample, primary sample, bulked sample, test sample. The term 'sample' implies the existence of a sampling error,* i.e., *the results obtained on the portions taken are only estimates of the concentration of a constituent or the quantity of a property present in the parent material. If there is no or negligible sampling error, the portion removed is a test portion, aliquot or specimen. The term 'specimen' is used to denote a portion taken under conditions such that the sampling variability cannot be assessed (usually because the population is changing), and is assumed, for convenience, to be zero. The manner of selection of the sample should be prescribed in a sampling plan.*" (IUPAC. Compendium of Chemical Terminology, second ed. (the "Gold Book"). Compiled by A. D. McNaught and A. Wilkinson. Blackwell Scientific Publications, Oxford (1997). Online version (2019-) created by S. J. Chalk. ISBN 0-9678550-9-8. https://doi.org/10.1351/goldbook).

What requirements can we formulate for sampling and sample preparation in a forensic analytical chemistry context?

And when these requirements are not met what kind of errors can result from this?

Requirement	Associated error	Forensic examples
Representative (sampling)	Incorrect quantitative or qualitative analysis	Reported amount of active ingredient in a drug mixture is too high/too low (illicit drugs) Ingredient is present but not reported as such (fire debris analysis)
Minimally invasive (preparation)	Formation of compounds False positive	Metabolite from a drug is formed during sampling and preparation of a blood sample (forensic toxicology)
Hygienic/sterile (sampling)	Contamination False positive	Lab GSR residue is transferred to a case stub during the handling of the stub (gunshot residues)
Nondestructive (preparation)	Sample loss Wrong analysis order	No DNA profiling after toxicological analysis because the entire sample is consumed (forensic biology)
Stable (sampling)	Degradation False negative	No ignitable liquid residue detected in stored fire debris samples (fire debris analysis)
Traceable (sampling)	Sample swap False-positive/negative	Explosives sample is not related to a case/suspect but is wrongly labeled (forensic explosives analysis)

Sampling and sample preparation requirements and associated errors with forensic examples when these requirements are not met.

In forensic science, the dimension of the samples can vary greatly depending on the physical evidence, the expertise area, the investigation requested, and the case at hand. This also has a big impact on protocols and sample preparation. In the domain of illicit drugs and explosives bulk quantities can be encountered. With such substantial amounts, sample consumption and invasive analysis methods are less critical. However, storage, representative sampling and sample destruction (after analysis has been finalized) can be challenging. Furthermore, handling large quantities of drugs and explosives can create health and safety risks. In contrast, expertise areas such as gunshot residues, glass, fibers, and paint typically deal with microtraces. Occasionally, forensic illicit drugs and explosives experts are also asked to investigate trace material (*e.g.* potential residues on the clothing of a suspect). With such minute amounts, detecting and securing traces while preventing contamination can be complex. Options for sample preparation are often limited for microtraces and require dedicated, miniaturized protocols to prevent excessive dilution. In addition, the analytical strategy and the order of analysis in case of an interdisciplinary investigation need to be carefully considered, especially when the amount of sample is insufficient to allow multiple attempts. The third class of samples encountered in forensic case work involves trace level (in contrast to trace amounts) analysis in complex matrices. Examples include drugs of abuse and metabolites in human whole blood as investigated by forensic toxicologists, residues of ignitable liquids in post-fire samples (FDA), trace levels of energetic materials and degradation products in post-explosion swabs, and toxic compounds in environmental samples. Given the matrices and the low levels of analytes of interest, sample clean-up and analyte preconcentration is often required prior to chemical analysis. The ultimate challenge in this respect is presented to the forensic expert when trace analysis is requested for microtrace evidence material. The options can then be limited even when applying state-of-the-art analytical chemistry. Aiming to get "more (chemical information) from less (physical evidence)" ultimately has its limits, both in terms of options for chemical analysis as forensic relevance (the lower the concentration levels the more difficult it usually is to discriminate criminal activities from secondary, crime-unrelated occurrences).

3.3 Statistical sampling protocols: how many samples do we analyze?

A case example

A large quantity of what appears to be cocaine is discovered in a container in a big harbor. Two suspects have been apprehended who came to collect the material. The public prosecutor who is in charge of the criminal investigation wants to know how many of the units contain cocaine in order to estimate the total amount involved.

In total 100 packages of a weight of roughly 500 gr each have been found.

What is your sampling plan?

For criminal investigations into drug smuggling and illegal production, the total amount involved is of importance with respect to the indictment and sentence. The above described case example is frequently encountered by forensic experts in illicit drugs analysis. Given the high workload in drug analysis laboratories, not all individual samples that are confiscated in a big smuggling case can be analyzed. On the other hand, performing an analysis of a single unit is clearly insufficient to declare the presence of an illegal substance for the entire shipment.

This dilemma is discussed in detail in the UNODC report entitled "Guideline on Representative Drug Sampling" (ISBN 978-92-1-148241-6, 2009). This document is used as a standard by many forensic drug analysis laboratories and was produced in collaboration with the Drugs Working Group of the European Network of Forensic Science Institutes (ENFSI). The content discussed in this paragraph is directly derived from this UNODC report and based on the same assumptions:

- The sample properties reflect the entire population, and
- A truly random selection of samples (each sample has an equal probability of being selected), and
- The content of a single unit is homogeneous;

Given these assumptions, what should we do when of the 100 packages, 25 seem to be visually different?

We need to create two populations and sample these populations separately! Taking a representative (in terms of numbers or composition) sample of the entire set is incorrect which quickly becomes evident when it turns out that one subset does not contain drugs or a different active ingredient mix.

Given these assumptions, what should we do if the material in the units seems visually very coarse and inhomogeneous?

1. We need to make sure to process a sufficient sample quantity for analysis
2. We need to take several subsamples from a single unit to determine the associated variation in the analysis as a result of the inhomogeneity
3. If this inhomogeneity significantly affects the analysis, we could consider to homogenize the sample prior to analysis

When considering a commonsense approach, various sampling schemes can be devised and employed. Several options are listed in the following table (n = sample size, N = total number of items, x = fraction ($0 < x < 1$), y = arbitrary factor ($y < N$), a,b,c = arbitrary factors):

Approach	Advantages	Disadvantages
$n = N$	100% certainty	Excessive sample size for large N
$n = x \cdot N$ $(n = 0.1 \cdot N)$	Simple, easy to understand	Sample size too small for small N Excessive sample size for large N
$n = x \cdot \sqrt{\frac{N}{y}}$ $(n = 0.5 \cdot \sqrt{N})$	Widely accepted	Sample size too small for small N but still excessive for large N
$n = y + x \cdot (N - y)$ $(n = 20 + 0.1 \cdot (N - 20))$	Discovery of sample heterogeneity	Excessive sample size for large N
If $N<a$ $n = N$ *If* $a<N<b$ $n = c$ *If* $N>b$ $n = \sqrt{N}$	Internationally recognized method (a=10,b=100,c= 10)	Excessive sample size for large N
$n = 1$	Minimum amount of work	Limited information, risk of error

Practical schemes applied for representative sampling of large illicit drug seizures.

The benefits of such a practical approach are clear as such methods are easy to explain and straight forward to apply by law enforcement at the scene. However, this down-to-earth solution also has a downside, as arbitrary sampling schemes lack a sound statistical foundation no information with respect to uncertainty can be provided. When it is decided not to analyze all the packages of a drug shipment, the associated sampling uncertainty could be quite relevant when the overall amount of the material is of importance with respect to the indictment. Without statistical insight, it is also impossible to optimize the sample size when balancing "effort required" versus "information provided."

Therefore, applying a statistical approach in sample selection is recommended even though this increases the complexity. Objective decisions and clear statements when it comes to sampling schemes for big drug seizures can be made on the basis of the so-called **Hypergeometric Distribution:**

$$p(x) = \frac{\binom{N_1}{x} \cdot \binom{N - N_1}{n - x}}{\binom{N}{n}}$$

This formula should be read as "the probability (p) of finding x positives in a subset of n items of a total set of N items with N_1 positives (and thus $N-N_1$ negatives)." The so-called binomial coefficient corresponds to the following **factorial**-based calculation:

$$\binom{N}{n} = \frac{N!}{n! \cdot (N - n)!}$$

where

$$n! = n \cdot (n - 1) \cdot (n - 2) \cdot \ \dots \cdot \ 2 \cdot 1$$

This distribution is related to the well-known **Binomial Distribution** but without replacement of the randomly drawn items from the sample set. This reflects the actual situation on a crime scene when police officers at the instruction of the illicit drug experts select and set aside samples for further analysis in the forensic laboratory. The hypergeometric distribution can now be used to estimate the number of samples that have to be analyzed such that at least K (= $k.N$) units are positive with $(1-\alpha).100\%$ confidence (at an α of 0.05 such statement could be made with 95% confidence and with an α of 0.01 with 99% confidence). This should be explained as being correct in 95% or 99% of the cases that at least k.100% (e.g., 90% for $k = 0.9$) of the units in the confiscated shipment would contain illicit drugs. This only holds (and hence this is assumed as part of the sampling plan) when all the units in the sample set test positive. This relates to the overall hypothesis that the entire seizure consists of illicit drugs, an assumption that is generally made when dealing with drug smuggling cases. Perpetrators go to great length to conceal drugs but typically do not include deliberate "negatives" in the shipment. In the table below, corresponding sample sizes are given for various α and k values:

	$\alpha = 0.95$			$\alpha = 0.99$		
N	**$k = 0.5$**	**$k = 0.7$**	**$k = 0.9$**	**$k = 0.5$**	**$k = 0.7$**	**$k = 0.9$**
10	3	5	8	4	6	9
20	4	6	12	5	9	15
30	4	7	15	6	10	20
40	4	7	18	6	10	23
50	4	8	19	6	11	26
60	4	8	20	6	11	28
70	5	8	21	7	12	30
80	5	8	22	7	12	31
90	5	8	23	7	12	32
100	5	8	23	7	12	33
200	5	9	26	7	13	38

Sample size sheet based on the hypergeometric distribution (for the assumption that all samples contain the listed substance), with α, confidence and k, fraction.

So considering the case at hand involving a total seizure of 100 suspect packages of uniform size, weight, and appearance, a random selection of 23 or 33 samples would have to be analyzed to report with 95% or 99% confidence, respectively, that at least 90% of the samples contain illicit drugs.

With the police experts and the public prosecutor leading the investigation, it is agreed to analyze 23 of the 100 seized samples. To surprise of the forensic experts analysis shows that one of the 23 samples does not contain illicit drugs.

How does this affect the statistical analysis and what should be done?

Based on the **hypergeometrical distribution,** the anticipated conclusions can be adjusted in two ways; 1) the confidence associated with the same fraction of drug positive samples should be lowered (in this specific case from 95% to 77%) given the outcome of the analysis or 2) the fraction of positive samples needs to be reduced while maintaining the same confidence level (in this specific case the k value is 0.84 instead of 0.9). This latter approach seems to make more sense both from a practical and legal point of view. If beforehand a specific number of negative samples is anticipated, then the sample size should be increased to be able to report the same fraction of positive samples with the same level of confidence as is illustrated in the table below.

	$\alpha = 0.95$						$\alpha = 0.99$					
	$k = 0.5$		$k = 0.7$		$k = 0.9$		$k = 0.5$		$k = 0.7$		$k = 0.9$	
N	**1 neg**	**2 neg**	**1 neg**	**2 neg**	**1 neg**	**2 neg**	**1 neg**	**2 neg**	**1 neg**	**2 neg**	**1 neg**	**2 neg**
10	5	7	7	9	10	—	6	7	8	9	10	—
20	6	8	10	13	17	20	8	10	12	14	19	20
30	7	9	11	14	22	27	8	11	14	17	25	29
40	7	9	12	15	26	32	9	11	15	18	30	35
50	7	10	12	16	29	36	9	12	16	20	34	41
60	7	10	12	16	31	39	10	12	16	20	38	45
70	7	10	13	17	32	41	10	12	17	21	40	48
80	7	10	13	17	34	43	10	12	17	21	42	51
90	7	10	13	17	35	45	10	13	17	21	44	54
100	7	10	13	17	36	46	10	13	17	22	46	56
200	8	11	14	19	40	53	10	13	18	24	54	67
500	8	11	14	19	42	58	10	14	19	24	59	75
1000	8	11	14	19	45	59	11	14	19	25	62	78
5000	8	11	14	19	46	61	11	14	20	25	64	81
10000	8	11	14	19	46	61	11	14	20	25	64	81

Sample size sheet based on the hypergeometric distribution (for the assumption that all samples but one or two contain the listed substance), with α, confidence and k, fraction.

As the analysis has revealed one negative sample, one of the experts suggests to add 13 new samples to the original set of 23 to arrive at the correct sample size of 36 to be able to report the same fraction at the same confidence (assuming that all these sample contain the illicit drug).

Is this a correct procedure?

Although this seems a very logical approach, it is statistically not correct!

The approach requires a decision to be taken on the expected number of negative units before determining the sample size. The results of the analysis will then determine the proportion in relation to the confidence. Adjusting the sample size on the basis of the analytical outcomes is not allowed.

In legal indictments concerning drug smuggling cases, the total amount of drugs involved is of great importance. Also, in this case, the severity of the crime and the associated legal charges are related to the economic value of the shipment and the resulting overall societal damage it brings.

Given the fact that the analysis shows a near 100% cocaine purity, what would the forensic experts report in terms of the overall amount smuggled?

What needs to be assessed in the sample set and how needs measurement uncertainty be addressed?

In addition to the cocaine test and chemical purity analysis, the amount of powder in each of the 23 packages will have to be weighed using a balance. Inevitably this leads to variation and thus measurement uncertainty. In this case, the expert can resort to straight forward statistical procedures by establishing the mean and standard deviation of the amount of cocaine per unit on the basis of the entire samples set. Assuming a normal distribution of the given population, the well-known **student *t*-distribution** can be used to estimate the associated confidence interval when analyzing a relatively small sample set (n) of that population:

$$\overline{X} - \frac{s}{\sqrt{n}} t_\alpha \le \mu \le \overline{X} + \frac{s}{\sqrt{n}} t_\alpha$$

where $\overline{X}$ is the average weight of a single unit in the sample set, s is the standard deviation of the measurements, μ is the average weight in the population, and t_α is the student t value at confidence interval α at d_f (n-1) degrees of freedom (see table below, with a sample set of 23 and thus a d_f value of 22, the t value is 2.074 or 2.819 when working with a 95% or 99% confidence interval, respectively). This approach is valid as long as the standard deviation is not excessively large compared to the average mass per unit, as a rule of thumb the **RSD** (relative standard deviation) should not exceed 10% (i.e., $(s/\overline{X}) < 0.1$).

d_f	$\alpha = 0.95$	$\alpha = 0.99$	d_f	$\alpha = 0.95$	$\alpha = 0.99$
1	12.706	63.657	18	2.101	2.878
2	4.303	9.925	19	2.093	2.861
3	3.182	5.841	20	2.086	2.845
4	2.776	4.604	21	2.080	2.831
5	2.571	4.032	22	2.074	2.819
6	2.447	3.707	23	2.069	2.807
7	2.365	3.499	24	2.064	2.797
8	2.306	3.355	25	2.060	2.787
9	2.262	3.250	26	2.056	2.779
10	2.228	3.169	27	2.052	2.771
11	2.201	3.106	28	2.048	2.763
12	2.179	3.055	29	2.045	2.756
13	2.160	3.012	30	2.042	2.750
14	2.145	2.977	40	2.021	2.704
15	2.131	2.947	60	2.000	2.660
16	2.120	2.921	120	1.980	2.617
17	2.11	2.898	∞	1.960	2.576

Student t distribution with d_f, degrees of freedom (= sample set − 1) and α, confidence.

The overall weight of the cocaine shipment (W) can now easily be obtained by multiplying the average and standard deviation per unit with the total number of confiscated packages N:

$$N\overline{X} - \frac{Ns}{\sqrt{n}} t_\alpha \le W \le N\overline{X} + \frac{Ns}{\sqrt{n}} t_\alpha$$

After weighing the 23 units in the sample set, the average weight was found to be 485 gr with a standard deviation of 43 gr. With one negative test in a sample set of 23, it was already concluded with 95% confidence that at least 84% of the total seizure of 100 packages are expected to contain cocaine. Chemical purity analysis shows that the cocaine is almost 100% pure. Combining this information and applying the t-distribution leads to the following 95% confidence interval for the overall weight of smuggled cocaine: 39.2 − 42.3 kg. In court, uncertainty is typically interpreted in favor of the suspect (incriminating forensic evidence should never be overestimated). Hence, in this case, the courts should consider an overall magnitude of the cocaine shipment of 39 kg.

3.4 Sample preparation: ignitable liquid residue sampling in fire debris analysis

In paragraph 3.2, the forensic requirements for sampling and sample pre-treatment have been discussed. Sample preparation is inevitable for forensic samples that cannot be directly analyzed by the preferred instruments to extract the chemical information needed by the forensic experts to unravel identity and origin. Sample preparation schemes can be extensive and elaborate when considering trace levels of compounds of interest in complex chemical and biochemical matrices. This applies for instance in forensic toxicology where forensic experts need to accurately determine the level of drugs of abuse and associated metabolites in human matrices such as urine and whole blood (forensic toxicological analysis will be discussed in more detail in **Chapter 5—Quantitative Analysis and the Legal Limit Dilemma**). Sample preparation also plays a crucial role in the chemical analysis and forensic interpretation in the area of Fire Debris Analysis (FDA). In this paragraph, the rationale for the accepted sampling and sample preparation schemes in FDA will be explained and discussed step by step giving insight in the choices made in this important field of forensic chemical analysis. These insights are in part generic allowing the reader to propose and discuss suitable sampling and sample preparations methods given the evidence material (type, amount, and chemical complexity), the questions at hand (identification, classification, and/or individualization) and the preferred analytical instrumentation.

What is the preferred method to analyze ignitable liquids and why?

Gas Chromatography with Mass Spectrometric (GC-MS) and Flame Ionization Detection (GC-FID)

Gas Chromatography (GC) was conceived and developed in the fifties and sixties of the previous century. In this period, a number of critical technical developments in both fields resulted in mass spectrometers that could be coupled to GC instruments. Over a period that spanned several decades, the design of the GC-MS instrument has been further optimized into today's workhorse technique known for its excellent selectivity, sensitivity, and robustness. Also from a budget perspective GC-MS is preferred over liquid chromatography coupled to mass spectrometry (LC-MS) as long as compounds of interest are sufficiently volatile and thermostable and the sample matrix does not interfere with the analysis or contaminates the instrument.

In a typical contemporary GC-MS setup, a sample solution first needs to be evaporated by injecting a small amount (in the order of a few microliters) in a heated injection port or so-called split-splitless injector. Once in the gaseous phase, the compounds are transported by a carrier gas (nitrogen, hydrogen, or helium, typically at a flow rate of a few mL/min) to the capillary column. The capillary column (typical dimensions include a column length of 30−60m an internal diameter of 0.32 or 0.25 mm and a stationary phase film thickness in the range of 0.1−5 μm) is placed in the oven of the GC for accurate temperature control. GC analysis is normally conducted with temperature programming starting at a relatively low temperature to retain and separate the more volatile compounds in the sample. As the temperature is increased, the less volatile species enter the mobile gas phase and are transported to the mass spectrometer that is coupled to the column. The capillary column and the temperature program leads to a very efficient separation as the various sample constituents elute in order of decreasing volatility (boiling point, vapor pressure). This separation enables the MS analysis of individual compounds even for complex sample mixtures.

Continued

The compounds eluting from the capillary GC column enter the high vacuum region of the mass spectrometer. For MS characterization, compounds need to be ionized and in GC this is typically done through electron impact (EI) ionization. Through a high energy (70 eV) electron beam, electrons are ejected from a fraction of the analytes leading to the formation of positively charged molecular ions which undergo further fragmentation resulting in the formation of several lower mass fragment ions. Establishing the mass and intensity of the ions is in most GC-MS systems done with a quadrupole mass spectrometer. By applying specific RF and DC voltage settings on four parallel metallic rods, stable ion trajectories are created as function of mass to charge ratio (m/z). At specific settings, only ions of given m/z can pass the rods and reach the detector. In scan mode, the RF/DC settings are varied while the presence of ions as function of m/z is monitored and recorded as detector intensity. An MS spectrum can thus be obtained in a fraction of a second.

EI fragmentation is very reproducible (also between different instruments) and very compound specific. These two features enable the construction of comprehensive databases of EI mass spectra of reference compounds. These EI-MS databases can be used to identify unknown compounds on the basis of a library match, that is, a measured spectrum that is near identical to the spectrum of compound X in the library.

Accurate quantitative analysis in GC-MS requires compound-specific calibration because the ionization efficiency and thus the detector signal at a given concentration depends on many factors including analyte characteristics. In contrast, a flame ionization detector (FID) has an almost universal response for organic compounds. In the FID, an ion current is measured as analytes eluting from the analytical column are burned in a hydrogen flame. For compositional analysis of oil distillates, the combination of GC-MS and GC-FID is therefore very powerful. However, quantitative aspects are usually less relevant in forensic FDA.

When the total recorded detector response of all ions at any given time during the separation is summed and the total intensity is plotted as function of retention time, a so-called GC-MS TIC (Total Ion Current) chromatogram is obtained. Such a chromatogram gives an indication of the total number of compounds present in the sample. For a typical gasoline and diesel sample, these GC-MS TIC chromatograms are depicted below and a few typical compounds as identified with the MS have been numbered and indicated.

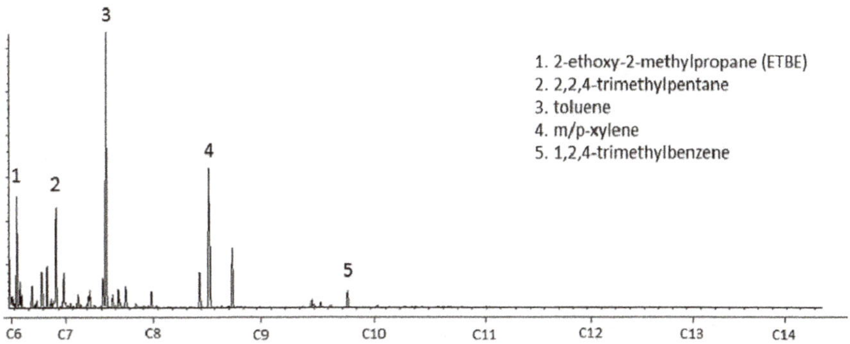

GC-MS TIC chromatogram of a typical gasoline sample, retention time on the x-axis has been converted to an n-alkane retention index *(source: Netherlands Forensic Institute.)*

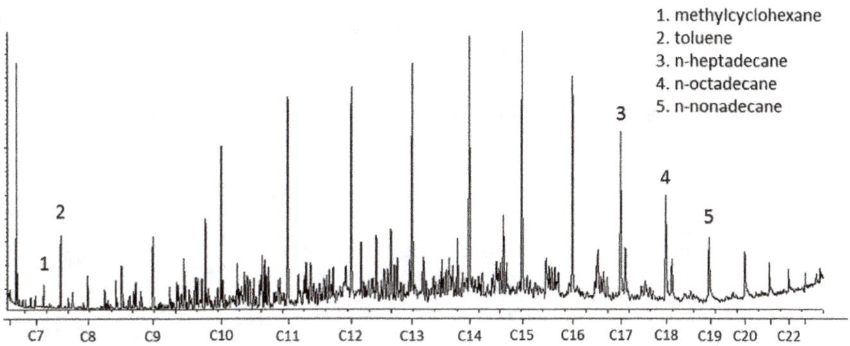

GC-MS TIC chromatogram of a typical diesel sample, retention time on the x-axis has been converted to an n-alkane retention index *(source: Netherlands Forensic Institute.)*

Retention indices *(I)* are a way to calibrate and anchor retention times making the chromatogram less affected by changes in instrumental parameters such as column dimensions, temperature program, and gas flow. This calibration is performed by analyzing a mixture of n-alkanes covering the entire volatility range. The retention index can be calculated from the alkane compounds eluting just prior and directly after the compound of interest:

$$I_i = 100\left[n + \frac{log(t_i) - log(t_n)}{log(t_{n+1}) - log(t_n)}\right]$$

where I_i is the retention index of compound i, t_i is the retention time of compound i which is eluting between two alkane standards n ($I_n = 100n$) and $n+1$ ($I_n = 100.(n+1)$).

The National Center for Forensic Science (NCFS) of the University of Central Florida hosts an international database of ignitable liquids accessible at: http://ncfs.ucf.edu/internationaldb/

This database consists of a large collection of GC-MS data of ignitable liquids from various laboratories giving insight in the typical composition of various product classes.

The GC-MS chromatograms show the chemical complexity of ignitable liquids. This is caused by the crude oil from which most ignitable liquids are produced. Although these products are produced in large quantities, the natural variation observed in crude oil allows for individualization at batch level. Although chemical profiling has proven to be a valuable tool when intact material is available in a criminal investigation, this approach is not used in FDA.

Why is the task of the expert in FDA directed toward identification and classification only?

In FDA, the expert needs to find traces of ignitable liquids in a complex sample that was taken after the fire. The fire debris consists of burnt material from the scene and constitutes a very complex and highly variable chemical matrix. Furthermore, ignitable liquid has been evaporated and consumed during the fire leaving only trace amounts, if any. Trace amounts that also differ in composition with respect to the original material. Therefore, in a fire debris sample, the impurity profile required for individualization is typically lost. The main questions addressed by the FDA expert are therefore:

1. Does the fire debris sample contain ignitable liquid residues?
2. If so, of what class of ignitable liquid?

Before discussing in more detail sensible sample preparation protocols for fire debris samples for ignitable liquid trace analysis with GC-MS, it is important to address the **classification framework** for ignitable liquids. Due to the huge diversity of oil distillate products, this is not as straight forward as for other types of physical evidence. A special **ASTM** (American Society for Testing and Materials) standard exists for FDA that includes classification schemes for reporting findings. The classification framework has two dimensions: chemical composition (type of product) and volatility (boiling point). Volatility is specified in three classes: Light (C4–C9), Medium (C8–C13) and Heavy (C9–C20+). Product classes include gasoline, petroleum distillates (e.g., kerosene, diesel, jet fuel and white spirit), isoparaffinic products (e.g., some paint thinner formulations), aromatic products (e.g., some cleaning formulations), napthenic/paraffinic products (e.g., lamp oils), normal alkanes (e.g., pentane, hexane and heptane solvents), oxygenated solvents (e.g., alcohol and ketone based formulations), and one generic class to include all remaining specialty products (e.g., turpentine and product blends).

We now want to apply GC-MS to look for the presence of ignitable liquid residues in an actual fire debris sample. This requires a suitable sample preparation strategy. Fire debris cannot be introduced in delicate analytical equipment as such.

Would you perform a solvent extraction of the fire debris to analyze the extract or would you take a sample of the headspace in the jar containing the fire debris.

What has your preference and why?

Although solvent extraction methods have been and sometimes still are used in FDA, most laboratories prefer a headspace sampling method.

As fire debris is a highly complex and unknown sample matrix, solvent extraction can lead to the injection of compounds in the GC-MS system that are not volatile enough and contaminate and degrade the column and the equipment. This leads to instrument down-time and repairs and potentially false-positive results due to contamination. Headspace samples are relatively clean and only consist of volatile compounds not capable of contaminating the instrumentation.

Head space sampling can be performed in several ways as will be illustrated and discussed in this paragraph. The first distinction that can be made is a so-called **static head space** versus a **dynamic head space** approach. In static head space, a fixed volume of a head space is sampled, processed, and analyzed. In dynamic head space, a flow is purged over a sample and analytes in the gas stream are subsequently collected and analyzed. Both principles are illustrated below.

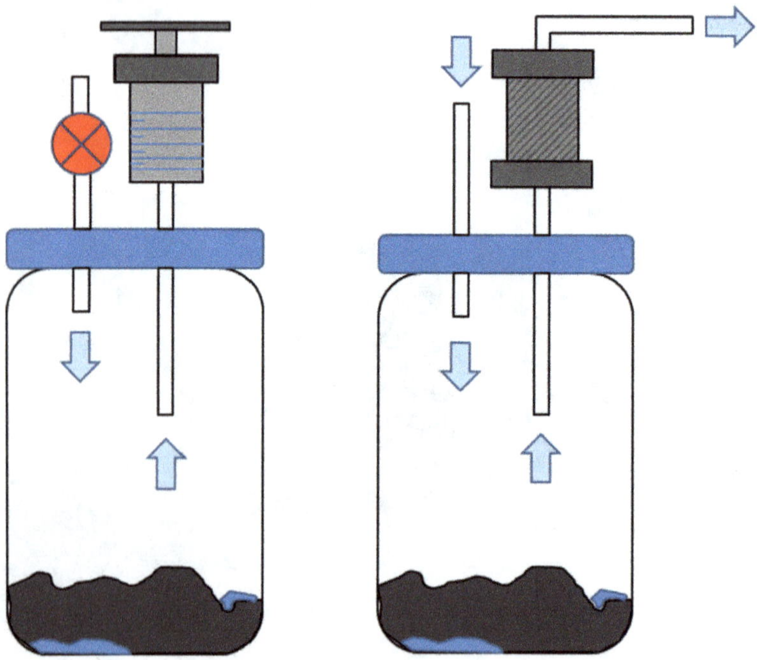

Schematic representation of static (left) and dynamic (right) head space sampling.

Static head space is intrinsically a simple and straight forward approach. Another advantage is that the sample composition directly reflects the head space composition. Direct injection of a given volume with the use of a gas-tight syringe is feasible but is limited in terms of maximum volume that can be injected on a GC-MS system. For a splitless injection (a split injection does not make sense as that would mean that a major part of the head space sample would be vented), the limited volume of the liner in the injector in combination with a 1-2 mL/min gas flow quickly results in increased injection times and associated peak broadening for the more volatile compounds. Dynamic head space provides additional sensitivity as analytes can be exhaustively purged and collected from the fire debris. This also provides options for robust quantitative analysis as collection is not affected by sampling conditions. However, the analytes need to be collected from the purge flow and this requires the use of trap with a suitable sorbent (e.g., active carbon or Tenax). The analytes of interest than need to be collected from the trap through thermal desorption or solvent extraction. This makes dynamic head space more laborious, and these additional steps each can lead to loss of analyte, contamination, and measurement uncertainty.

Would you prefer dynamic or static head space for FDA and why?

In FDA, there is a need for sensitivity as in some samples ignitable liquid residues can be present at trace level. However, the complex and unknown matrix is complicating matters. Headspace samples also contain volatile compounds from the fire debris and these compounds can be very similar to the ignitable liquid constituents. In a purge and trap strategy, huge amounts of matrix compounds can be trapped causing trap overload and strong background signal in the analysis.

Additionally, headspace content can vary significantly from sample to sample and static headspace sampling provides more flexibility. Finally, static headspace only consumes part of the sample allowing follow-up analyses if needed.

For these three reasons, Static Head Space sampling is preferred in FDA.

Several methods exist for the static head spaces sampling of fire debris samples. The starting point for all these approaches is a container with fire debris sampled at the incident scene. Often these closed containers are heated prior to sampling to stimulate the release of volatile and semivolatile ignitable liquids residues to the confined head space. After this step, forensic laboratories employ different procedures to sample and collect these head space traces for GC-MS analysis.

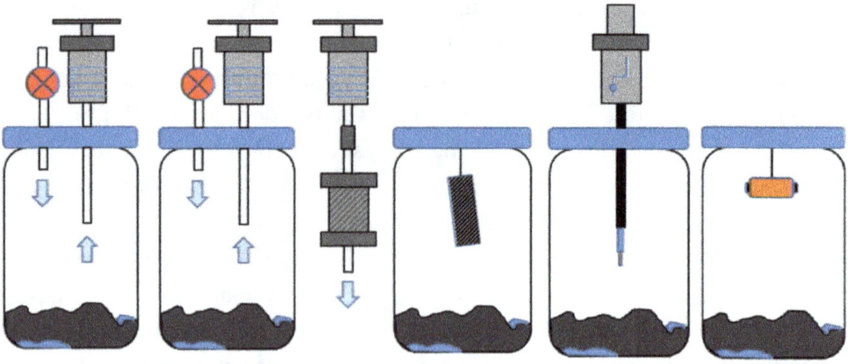

Schematic representation of various static head space sampling strategies, respectively, from left to right: direct sampling, direct sampling with trapping, activated carbon strip sorption, SPME (solid phase microextraction), and SBSE (stir bar sorptive extraction).

Each approach has benefits and drawbacks, but all methods can be and are confidently and effectively applied for FDA while meeting the required quality criteria for forensic case work (as discussed in **Chapter 9—Quality and Chain of Custody**).

When applying direct head space sampling (**DHSS**), a fixed volume is withdrawn from the head space with a gas-tight syringe and directly injected in the GC-MS system. This is a simple and straight forward approach but has reduced sensitivity because of the limited volume that can be injected in the GC-MS system. To increase sensitivity, a larger volume of the head space can be sampled but this requires an additional step to trap the analytes of interest prior to chemical analysis. This is done by direct head space sampling with trapping (**DHSST**). Effective trapping of ignitable liquid residues is obtained with activated carbon or Tenax traps. The use of activated carbon requires an organic solvent to extract the analytes of interest from the sorption material. Many forensic laboratories used CS_2 (carbon disulfide) for this purpose, an effective extraction medium but unfavorable with respect to health risks for the lab personnel. Additionally, the use of extraction solvent in part nullifies the sensitivity gain by handling a large head space volume as only a part of the extract can be injected on the GC-MS (extraction volumes are typically in the ml range, whereas GC injection volumes are limited to a few μl). For these reasons in recent years, forensic laboratories applying DHSST methods have moved to the use of Tenax trap instead of activated carbon. Tenax is the trade name of poly(2,6-diphenyl-p-phenylene oxide), PPPO, a polymer material with excellent sorption and thermal properties. Its high thermal stability enables the direct thermal desorption of collected organic trace material. When a GC-MS instrument is equipped with a thermal desorption unit and an autosampler for Tenax traps, solvent extraction is no longer required. This boosts sensitivity and makes the analysis less laborious. Tenax traps can also be reused (activated carbon is typically considered as single-use material and activated carbon traps are made from glass) as long as cross-contamination is prevented by additional thermal desorption steps at elevated temperatures. This reuse of traps also reduces the costs and environmental footprint of the analysis.

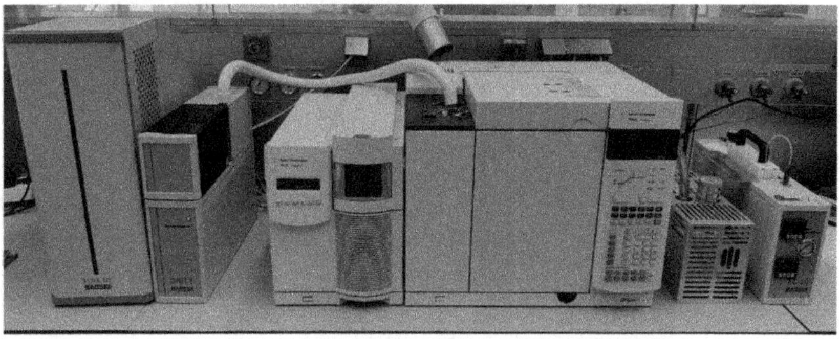

GC-MS instrument with automated thermal desorption system for Tenax traps *(courtesy of Dr Michiel Grutters, forensic FDA expert of the NFI, note: several instrument manufacturers provide very robust instruments, the set-up shown serves as an example and does not indicate any preference by the NFI.)*

Sorbents can also sample the head space directly by placing traps in the sample containing the fire debris. In this way, there is no need to transfer a fixed volume of the head space to the trap and this typically boosts sensitivity. An approach that is employed by many forensic laboratories in the United States is based on the use of metal tins similar to paint containers. The fire debris sample is placed in the tin and a strip of activated carbon (**ACS**) is then fixated on the inner side of the lid. The lid is closed, and the sample is heated to facilitate the accumulation of compounds of interest in the head space of the container such that these compounds can adsorb on the trap. The trap is then retrieved followed by solvent extraction and GC-MS analysis. This is a single-use sampling strategy that prevents cross-contamination and that is still cost effective due to the use of low-cost materials.

A very elegant method for interfacing fire debris sampling with GC-MS analysis is **SPME**. This extraction technique, introduced in 1990 by J. Pawliszyn and coworkers, is based on the application of a fiber containing a small polymer film. These films consist of materials also used as stationary phase in capillary GC column such as polydimethylsiloxane (PDMS), divinylbenzene (DVB), and polyethylene glycol (PEG) that have very good thermal properties resulting in low column bleed in GC-analysis. These polymers remain in a fluid-like state which means that analytes do not adsorb but rather absorb, that is, dissolve in the polymer film. Application of a thin film on a fiber in a syringe-like support facilitates the direct thermal desorption of sorbed analytes in the split-splitless injector of the GC instrument. SPME analysis can also be automated when the autosampler is equipped with an SPME sampler and a station to contain head space samples. However, such head space samples are usually limited in container size. Sensitivity in SPME is ultimately determined by the equilibrium distribution of the ignitable liquid residues between the fire debris, the head space, and the polymer film on the SPME fiber. When considering an aqueous sample (please note that this is not representative for a typical fire debris sample) and SPME head space sampling (so no direct contact between the fiber and the liquid), the amount n in the fiber film of a given analyte present at concentration C_0 in the aqueous sample is given by

$$n = \frac{C_0 \cdot V_1 \cdot V_2 \cdot K}{K \cdot V_1 + K_2 \cdot V_3 + V_2}$$

where V_1 is the volume of the SPME polymer film, V_2 is the volume of the liquid sample, V_3 is the head space volume, K is the partition coefficient of the analyte between the polymer film and the liquid sample, and K_2 is the partition coefficient of the analyte between the head space and the liquid sample.

Interestingly, there is a direct relation between the amount sampled and the analyte concentration which indicates that SPME can be used for quantitative analysis as long as the extraction conditions, such as temperature, are carefully controlled to maintain constant partition coefficients. The equation also shows that the amount extracted is analyte dependent because the partition coefficients depend on compound characteristics such as volatility and polarity. In general, sensitivity increases for compounds that show a very high affinity for the SPME polymer film in combination with a low affinity for the original sample medium. The consequence of the use of SPME for FDA is that the pattern of compounds is different from the head space of neat ignitable liquids and this needs to be taken into account by the forensic expert.

SPME sensitivity can typically be boosted by increasing the amount of polymer film (or the polymer film thickness) on the SPME fiber. In this way, more analyte is extracted at a given concentration in the polymer film. This is also the main idea behind using the sorptive stir bar for head space extraction. The principle and the materials used are similar to SPME but with much more polymer material. Typically, the stir bar is used in direct contact with aqueous solutions to simultaneously stir the sample and extract the analytes. However, using much thicker SPME fibers or **SBSE** is ultimately limited by mass transfer leading to excessive sampling times to reach equilibrium. For head space sampling, this is especially true for less volatile compounds with a high affinity for the extraction polymer. The transfer from the sample to the head space then becomes the rate determining step and reaching equilibrium becomes unpractical. Alternatively, exposure times can be limited but then sampling is conducted under nonequilibrium conditions. This requires even better control of the sampling parameters to ensure sufficient robustness and reproducibility.

The characteristics, benefits, and drawbacks of the various static head space extraction techniques for FDA are summarized in the Table below.

Method	Ease of use	Speed & Throughput	Cost	Sensitivity	Robustness	Interpretation Complexity	Contamination risk
SHSS	++	+	++	-	0	++	+
AC	-	--	0	++	0	+	+
Tenax	+	+	-	++	+	+	-
ACS	++	-	+	+	+	-	++
SPME	+	0	-	+	-	--	-
SBSE	+	0	-	++	0	--	-

Overview of static head space methods for fire debris analysis with ++, very favorable; +, favorable; 0, neutral; -, unfavorable; --, very unfavorable; *AC*, Static Head Space Sampling with trapping on Activated Carbon; *ACS*, Direct head space sampling with activated carbon strips; *SBSE*, Direct head space sampling with sorptive stir bars; *SHSS*, Static Head Space Sampling; *SPME*, Direct head space sampling with Solid Phase Micro Extraction fibers; *Tenax*, Static Head Space Sampling with trapping on Tenax.

The preferred method at the Netherlands Forensic Institute is based on static head space sampling with the trapping of a variable head space volume on a Tenax adsorption tube. These stainless-steel tubes with Tenax sorbent can be directly inserted in an autosampler and are thermally desorbed for 5 min at 280 °C followed by GC-MS analysis. The optimal volume to be sampled is determined by direct injection of a small head space volume on a GC-FID instrument. In this way, fire debris samples with both trace and excessive amounts of head space compounds can be processed effectively with the same method. Prior to sampling, the fire debris as collected in 2650 mL glass jars by the Dutch police at the fire scene is heated for 4 h at 70 °C to boost sensitivity. The GC analysis is performed on a PDMS capillary column of 25 m in length and with an internal diameter of 0.25 mm and a film thickness of 0.33 μm. After thermal desorption, the column is kept for 3 min at 35 °C followed by a 10 °C/min ramp to 250 °C and this temperature is maintained for 6 min. Helium is used as the carrier gas at a constant flow of 0.8 mL/min.

With this method a typical case work result is shown below.

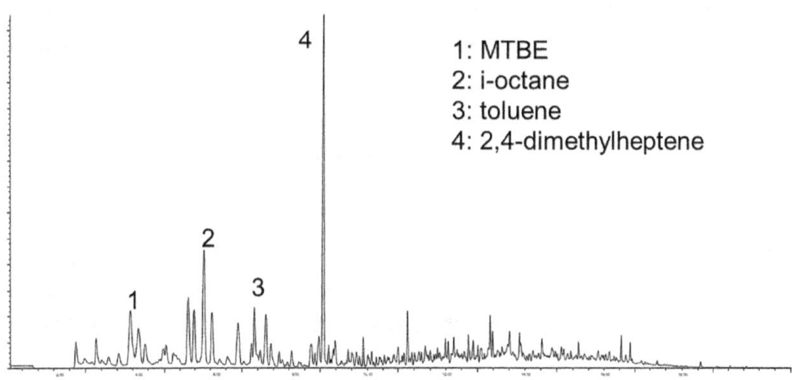

TD-GC-MS TIC chromatogram of a typical fire debris sample after static head space sampling and trapping on a Tenax tube *(Source: The Netherlands Forensic Institute)*

The experts of the NFI concluded that this fire debris sample contained traces of gasoline. Although toluene is a marker for gasoline, the most dominant compound in the head space, 2,4-dimethylheptene, is not found in this ignitable liquid.

What is the origin of 2,4-dimethylheptene?

How do the experts arrive at their "positive for gasoline" conclusion?

FDA experts often see the compound 2,4-dimethylheptene in fire debris head space because it is a well-known pyrolysis of polypropylene, a very common polymer in households. To interpret GC-MS data correctly, it is very important that experts are aware of pyrolysis processes of furniture materials, electronics, and other frequently occurring materials. It is also known that some of these products create head space profiles that are quite similar to ignitable liquid patterns. This is not that remarkable considering the fact that many man-made materials are produced from precursors obtained from crude oil. Experts of the NFI and the German BKA (Bues Kriminal Amt) have published a recent book on how to confidently identify ignitable liquid residues in fire debris sample and prevent false-positive results. Readers interested to know more about this forensic expertise area are encouraged to study this work in detail (Further Reading). The many examples illustrate the complexity associated with data interpretation also because of the variability encountered in fire debris composition and thus background head space composition. For this reason, the automation of ignitable liquid detection and classification is notoriously challenging. Although good advances are reported by forensic scientists using advanced data analysis and chemometric tools, all forensic institutes involved in FDA rely on their experts to reach a final conclusion. However, such a verdict remains subjective and human experts are prone to make interpretation errors as has been demonstrated in realistic and challenging proficiency tests (see **Chapter 9—Quality and Chain of Custody**).

As illustrated above, the overall head space composition consists of a complex mixture of ignitable liquid residues (for most positive samples) and pyrolysis products of the burnt substrates. However, a powerful tool for the forensic expert is so-called **Selected Ion Extraction** (SIE). With SIE, one or more m/z values are selected and plotted as function of retention time. In this way, the presence of certain classes of compounds can be visualized. Signals at m/z 43, 57, 71, and 85 are specific for n-alkanes, the mass difference of 14 is indicative for the loss of a CH_2 group. Indicative for aromatic compounds is the presence of a peak at m/z 91 in the mass spectrum, and this is the mass of the well-known tropylium ion ($C_7H_7^+$). For the current sample, these alkane and aromatic SIE profiles are depicted below. On the basis of these SIE profiles, the forensic experts can be more confident in their interpretation although also these profiles will consist of both ignitable liquid and substrate pyrolysis compounds.

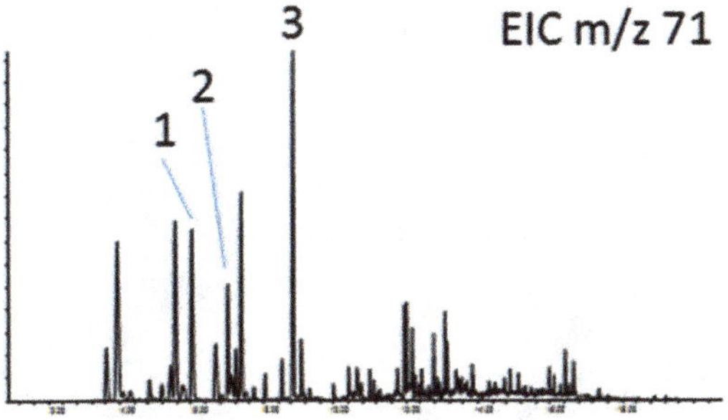

TD-GC-MS SIE chromatogram at m/z 71 showing the presence of alkane species (1 = n-heptane, 2 = 2,3,4-trimethylpentane and 3 = 2,4-dimethylheptene) *(Source: The Netherlands Forensic Institute)*

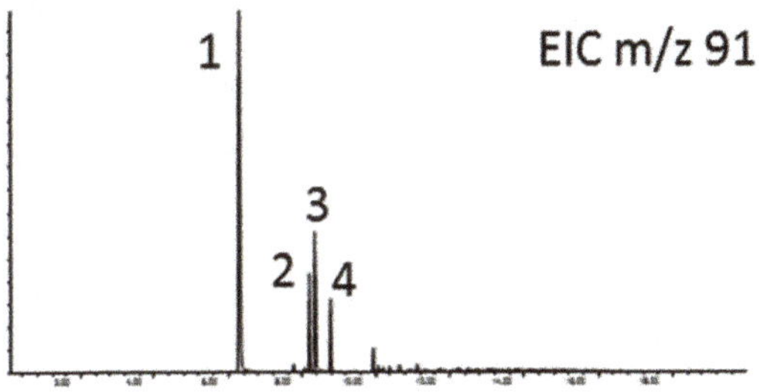

TD-GC-MS SIE chromatogram at m/z 91 showing the presence of aromatic species (1 = toluene, 2 = ethylbenzene, 3 = m/p-xylene and 4 = o-xylene) *(Source: The Netherlands Forensic Institute)*

We have seen how advanced sampling and analysis strategies are applied to detect minute traces of ignitable liquids in fire debris. Because of the complex and variable chemical matrix and the destructive effects of the fire, data interpretation is conducted by experienced forensic experts to determine whether an ignitable liquid residue is present.

If convincing analytical evidence is found that an ignitable liquid is indeed present in the fire debris sample, does this also mean that arson was committed?

Finding an ignitable liquid residue in fire debris does not mean that arson was committed (finding no ignitable liquid residue also does not mean that no arson was committed). This strongly depends on the circumstances and context of the fire.

Further reading

Guideline on Representative Drug Sampling, 2009. UNODC, ISBN 978-92-1-148241-6.

Identifying Ignitable Liquids in Fire Debris, A Guideline for Forensic Experts, Jeanet Hendrikse, Michiel Grutters and Frank Schäfer, 2015. Academic Press, ISBN 9780128043165 eBook ISBN: 9780128043875.

Standard Test Method for Ignitable Liquid Residues in Extracts from Fire Debris Samples by Gas Chromatography-Mass Spectrometry, ASTM E1618 – 19, 2019. https://www.astm.org/Standards/E1618.htm.

CHAPTER 4

Qualitative analysis and the selectivity dilemma

Contents

4.1 What will you learn?

In **Chapter 1**, the legal need for qualitative (chemical identification) versus quantitative (establishing the amount of a given compound) chemical analysis was introduced. In this chapter, we will focus on qualitative analysis in forensic chemistry and discuss the importance of selectivity when applying standardized methods for large volume screening. Selectivity in chemical identification is essential to prevent false-positive results, and this will be demonstrated with the NPS (New Psychoactive Substances) challenge for laboratories that perform illicit drug analysis. In illicit drug analysis, a false-positive, *that is,* identifying an unlisted compound as an illicit substance, can directly result in a wrongful conviction and therefore needs to be prevented at all cost. Recent scientific developments to tackle this challenge and solve the selectivity dilemma will be discussed.

Chemical Analysis for Forensic Evidence
ISBN 978-0-12-820715-4
https://doi.org/10.1016/B978-0-12-820715-4.00002-X

After studying this chapter, readers are able to
- list the main forensic expert areas and types of evidence material for which **qualitative analysis** (**compound identification**) is of importance
- describe the methodology applied by forensic **illicit drug** experts to chemically identify **listed substances** in a large **case load** setting
- understand how the **NPS challenge** creates a **selectivity dilemma** in **high volume screening** of forensic drug analysis laboratories
- explain novel approaches in forensic science to meet the **NPS challenge** and restore selectivity while maintaining **efficiency**, **speed,** and **throughput** in the laboratory

4.2 Qualitative analysis in forensic chemistry

We have seen in **Chapter 1** that the choice of the forensic expert to perform a quantitative or a qualitative analysis strongly depends on the legal framework and the context of the case. Qualitative analysis, *that is,* establishing the presence of a compound in evidence material, usually is a key activity in the following forensic expertise areas:
- Illicit Drugs/Drugs of Abuse
- Explosives and Explosions
- Gun Shot Residues (GSR)
- Fire Debris Analysis (FDA)
- Chemical threat and war agents
- Chemical identification of unknowns (e.g., powder in threat letters)

Although the aim is similar, forensic chemists in these fields are dealing with quite different evidence materials. As discussed in **Chapter 2**, this leads to a high degree of specialization and the use of dedicated sampling strategies, sample preparation methods, analytical instrumentation, data analysis, and reporting. However, some instrumental analysis equipment can be used in a broader context. For instance, LC-HR-MS (liquid chromatography with high resolution mass spectrometry) can be used to identify explosives, drugs of abuse, and a wide variety of other chemical compounds. Also, spectroscopic techniques like Raman and IR (infrared) can be used for a wide range of forensic applications provided that enough material is available for analysis, the composition is not too complex and the compounds of interest are present at a relatively high concentration. An interesting question for forensic laboratories is whether to provide analytical chemical services centrally or maintain dedicated equipment per forensic expertise area. The latter prevents contamination and is typically better from a case throughput and equipment availability perspective. However, this also leads to additional equipment (depreciation) costs especially considering the fact that in many forensic institutes equipment usage is limited, meaning that instruments are often in standby mode waiting to be used. Another negative consequence of separate expert laboratories is that the use of specialized, high-end, and very expensive equipment is limited. The significant equipment investment required is only possible when the instrumentation would be used in an interdisciplinary manner. For that reason, a mixed model is often observed where a substantial investment in high-end equipment is done centrally with dedicated operators that support forensic experts from several expertise areas. Another option is that the instrument use is coordinated by the forensic experts from one field but that they also process requests from various institute colleagues. In the Netherlands Forensic Institute both variants occur, with dedicated scientists specialized in the interdisciplinary use of high-resolution mass spectrometry and GSR experts maintaining SEM (scanning electron microscopy) instruments but also supporting other colleagues requesting elemental microtrace analysis.

The amount of material available for the qualitative analysis also greatly affects the choices made and instruments used by the forensic experts. When considering drug analysis and pre-explosion cases where the intact explosive charge is available for analysis, **bulk amounts** can be available to the experts. This could even introduce challenges with respect to representative sampling as discussed in **Chapter 3**. However, having a lot of evidence material provides benefits with respect to amount required for analysis and the use of (several) destructive methods that consume the sample. There are also no concerns with failed analyses and having material available for quality control, or confirmation and checks by other laboratories. This situation is very different for GSR experts that typically work with **trace amounts**, a given number of μm range particles on a stub for which the elemental composition needs to be determined. The benefit of SEM is that the analysis is not destructive, once the stub has been analyzed the particles are still there to be characterized with other techniques and by other institutes. If the main technique of GSR experts was destructive, critical physical evidence is lost in the analysis process and there is only one attempt "to get it right." Any error in the process would then mean that valuable evidence is irreversibly lost. This requires very strict quality control and documentation also because there is no option by the defense council to check the outcomes by requesting an analysis by an independent laboratory. When dealing with trace material also the use of multiple techniques needs to be carefully thought through, as analysis needs to be performed sequentially (with bulk amounts, subsamples can be taken and analysis can be run in parallel as long as each subsample is representative). Totally nondestructive, low-risk (*e.g.* in terms of contamination) analyses are performed first. For destructive techniques, a choice must be made of the most informative method given the other data available and the context of the case. Destructive analysis can only be initiated when the outcomes of all the prior investigations are confirmed to be satisfactory. In the Netherlands like in many other countries, performing a destructive analysis eliminating all the available evidence material requires consent from the public prosecutor leading the criminal investigation. Prior to this important decision, the legal professional involved needs to be carefully informed by the forensic experts on the added value of the analysis and its ramifications. It is important to note that trace levels of evidence material can also occur in illicit drug and explosives cases. For instance, when experts find traces of drugs on clothing items or fingermarks of a suspect or when residues are found in a dismantled illegal site allegedly used for the production of improvised explosive devices (IEDs). Under such circumstances, the trace aspects discussed above also apply to the drug or explosive analysis. Standard approaches used when bulk amounts are available no longer apply and the forensic drug and explosive experts must create dedicated plans that could include analytical instrumentation that they do not use frequently and require support from colleagues from other fields.

Trace levels of compounds of interest in evidence material also present specific challenges and require specific strategies for chemical analysis. The difference between trace amounts and trace levels need to be appreciated in this respect. For trace amounts the relative concentration of the compounds of interest can be very high, for example, when considering a powder residue of cocaine. The challenge lies in the fact that the overall mass of cocaine available for analysis is limited. When dealing with trace levels of analytes in bulk samples the main issue is with the sensitivity of techniques. If a technique is known to have a relatively high detection limit, a suitable analyte enrichment/purification step needs to be included as part of the sample pretreatment. An additional complication arises when the evidence material is chemically complex containing a lot of other (dominant) compounds thus representing a challenging sample matrix. The latter situation—trace levels in bulk samples with a complex matrix—typically applies for fire debris samples. Consequently, as discussed in **Chapter 3**, extensive and dedicated sample pretreatment strategies have been developed for Fire Debris Analysis (FDA). In FDA, an additional complication in the detection and classification of ignitable liquids arises from the fact that the expert needs to look for several representative compounds, as these products are complex (petrochemical) mixtures themselves.

The most challenging conditions exist when the expert is facing **trace levels of compounds of interest in trace amounts of evidence material of complex composition**. Such conditions can apply for forensic explosive experts when dealing with a post-explosion investigation. At the crime scene, swab samples are taken which are sent to the laboratory for further analysis. These swabs are subsequently extracted with the aim to detect and identify trace amounts of intact explosive or associated degradation products in order to establish the composition of the explosive charge used to cause the explosion. Such an extract can contain matrix compounds originating from the explosion and resulting fire debris. This ultimate analytical chemical challenge can nowadays be confidently addressed with LC-HR-MS. The high-resolution mass spectrometer combines the required selectivity and sensitivity, whereas liquid chromatography limits the chemical background by compound separation resulting in the sequential introduction of compounds in the high vacuum of the mass spectrometer. The LC system also has an important function to prevent contamination of the mass spectrometer.

Finally, in this introduction paragraph, it should be noted that although qualitative and quantitative analysis are presented as well-defined tasks and are discussed in this book in separate chapters, in forensic practice this categorization is much less obvious and a gray area exists. Forensic case work often involves elements of qualitative and quantitative analysis and analysts in forensic laboratories are usually well equipped and trained to do both. Identification of drugs of abuse and explosives is frequently also combined with a quantitative analysis to address the "how much" question. In a criminal investigation, it is often of importance to understand how much drugs were smuggled or produced or how big the damage could have been if the explosive was successfully detonated. In fire debris analysis, the relative amounts of compounds in the head space can be indicative of the associated ignitable liquid. Even in GSR analysis the number of particles can be of importance when addressing questions on activity level (see **Chapter 9—Reporting in the Criminal Justice System**), for example, to differentiate a bystander from the shooter in a shooting incident.

Can we do a quantitative analysis without a qualitative analysis?

No, in principle we first need to establish and assure the chemical identity before we can measure the level of the compound in the sample. Quantitative analysis methods usually are based on qualitative analysis principles for which appropriate calibration schemes are developed with the use of reference standards. It does not necessarily mean that a quantitative method is also suitable for compound identification when considering samples with unknown composition. However, the selectivity in a reliable quantitative method must be assured, meaning that we need to be sure that the measured feature can be attributed to the analyte of interest and not to other compounds in the sample. Because the composition of evidence material is always unknown (in contrast to for example factory samples produced under well controlled and recorded conditions), selectivity can only be assured through compound identification.

In the remainder of this chapter, the principles of qualitative analysis and the selectivity challenge when applying screening methods in laboratories dealing with high case loads will be discussed for the field of **illicit drug analysis**. Typically, ample material is available for analysis and the concentration of the psychoactive substances in the powder and tablet samples is high. In light of the prior discussion, the experts are dealing with bulk amounts with "bulk levels" of the compounds of interest and this typically does not pose restrictions with respect to the chemical analysis. The main challenge in the drug analysis laboratory is in the number of cases, at the NFI the forensic drug experts typically report over 5000 cases annually. To successfully handle such a substantial case load, a laboratory requires a very efficient process that provides acceptable lead times and prevents growing backlogs. Part of this process is an automated screening for the chemical identification of drugs of abuse using highly standardized methods. This chapter deals with the challenge to assure selectivity in chemical "mass screening," considering a changing illicit drug market through the accelerated introduction of new designer drugs or the so-called **NPS**.

4.3 Chemical identification of illicit drugs

In **Chapter 1**, we have seen that the rationale for a "*qualitative analysis only*" to demonstrate the presence of drugs of abuse is based on the illicit drug law in the Netherlands. This law states that it is forbidden to possess "substances listed as illicit drugs" without stating any associated quantities or levels. What the law does not substantiate is which compounds are "listed as illicit drugs" and how this is legally established. Like many other countries, the Netherlands maintains lists of compounds that are classified as illicit drugs under the illicit drug law. These lists are not fixed and change over time to reflect changes in availability and the presence of drugs of abuse in society. From a judicial perspective, it is efficient to detach the dynamic list of illegal substances from the actual legal framework and hence changes to the list can be processed as administrative measures of the government ("Maatregelen van Bestuur" in Dutch). In the Netherlands, two lists are defined in the illicit drug law, the so-called **hard drugs** are compiled in "**List 1**" and a second category consists of **soft drugs** compiled in "**List 2**". The difference between soft and hard drugs in the Dutch legal framework is based on the severity of the associated health risks for users (*e.g.* the risk to get addicted or die from an overdose) and the detrimental effects to society. However, it should be noted that defining a psychoactive substance as a hard of soft drug is a somewhat ambiguous process as user abuse habits may vary. As the effects to society are considered to be more detrimental for hard drugs, the associated sentence after conviction is more severe. The Dutch illicit drug law states that for crimes involving List 1 compounds, perpetrators can be convicted to a maximum of 6 years in prison whereas for List 2 compounds this maximum sentence is "only" 2 years. Of course, the actual sentence given by a judge depends on the conditions of the case and the criminal track record of the suspects involved. The existence of lists of banned substances in Western countries originates from several 20th century United Nations (UN) conventions initiated by the United States of America (USA). In 1961 a "*Single Convention on Narcotic Drugs*" was organized, in which several psychoactive substances were classified and listed in four schedules for which international control was agreed to limit production, distribution and use. A decade later, in 1971 the "*Convention on Psychotropic Substances*" was held to include a range of new, mostly synthetic psychoactive substances (stimulants, depressants and hallucinogens) in the framework of international control. These substances became increasingly popular in the seventies, an era of expanding non-medical, recreational use of drugs. Finally, in 1988, the "*Convention Against Illicit Traffic in Narcotic Drugs and Psychotropic Substances*" led to an even stricter international framework requiring countries to impose national measures to prevent the production, trafficking, and sales of psychoactive substances. This convention was held in a time frame in which the USA declared their "*War on Drugs*" in an attempt to control the drug related rise of organized crime and public health issues (*e.g.* the crack epidemic). Many countries like the Netherlands drafted their national illicit drug laws such that their international obligations from ratifying the UN conventions were met. Article one of the Dutch illicit drug law even makes a direct reference to the corresponding UN treaties.

Can you mention any compounds that are listed as "hard drugs" (List 1) and listed as "soft drugs" (List 2) in the Netherlands?

Can you guess how many compounds are currently listed as illicit "hard drugs" (List 1) and listed as "soft drugs" (List 2) in the Netherlands?

Frequently occurring "List 1 compounds" in illicit drug case work in the Netherlands include plant derived substances like **cocaine** (either as Chloride salt or free base, crack cocaine) and **heroine** (di-acetylated morphine) and fully synthetic phenylethylamine-based compounds such as **amphetamine** (speed), **methamphetamine** (crystal meth), and MDMA (XTC, ecstasy). The hallucinogen **LSD** (Lysergic acid diethylamide) is also listed as a hard drug in the Netherlands. Molecular structures of these compounds are given below. In 2020, the list of hard drugs in the Netherlands contained over 200 entries.

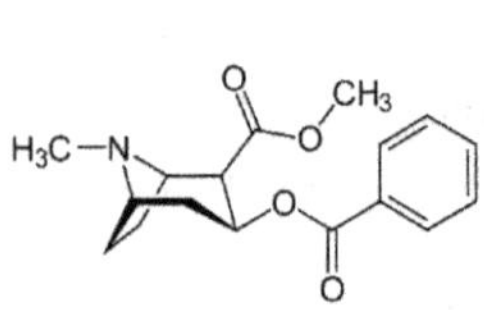

Cocaine

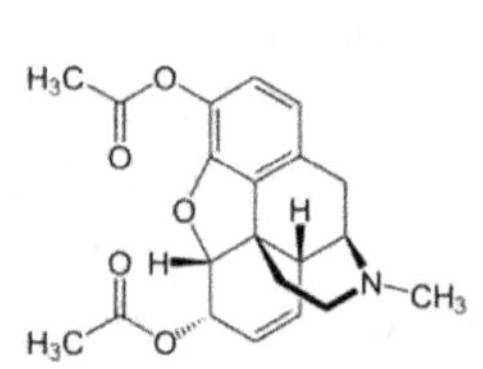

Heroine

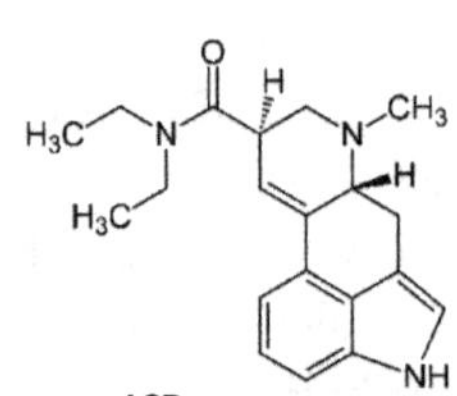

LSD

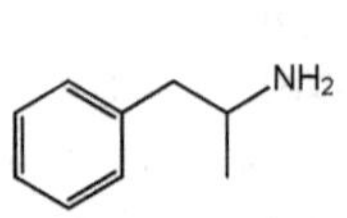

Amphetamine

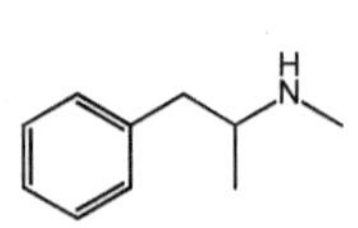

Methamphetamine

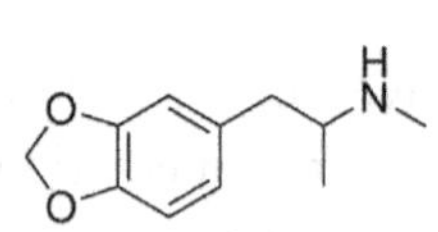

MDMA

The most familiar example of a "soft drug" and representative for the Dutch approach of tolerating minor drug offences involving List 2 compounds is cannabis (marijuana). The active compound in this natural product from the cannabis plant is **THC** (tetrahydrocannabinol). The leaves and branches of the Catha Edulis (Qat, Khat) can also be found on the list of soft drugs. The active compound **Cathinone** in this plant material in its pure form is, however, considered a List 1 hard drug. The list also contains over 90 species of mushrooms (so-called paddos) that contain the psychedelic compounds **psilocine** or its phosphorylated counterpart **psilocybine**. Many of the synthetic compounds that are listed as soft drugs belong to the class of **benzodiazepines**. These compounds have strong tranquilizing effects and are used as regular medicines to treat anxiety and insomnia. The use of these medicines on prescription by a doctor or psychiatrist and obtaining this medication by patients from a registered pharmacy is therefore fully legal. However, because of their potent psychoactive effects, these medicines are also sold and used as illicit drugs. Well-known benzodiazepines in the List 2 of soft drugs include Oxazepam and Temazepam. In 2020, in addition to the mushroom species, more than 75 compounds and natural products were registered as soft drugs in the Netherlands. Structural formulas of some of the List 2 compounds described are given below.

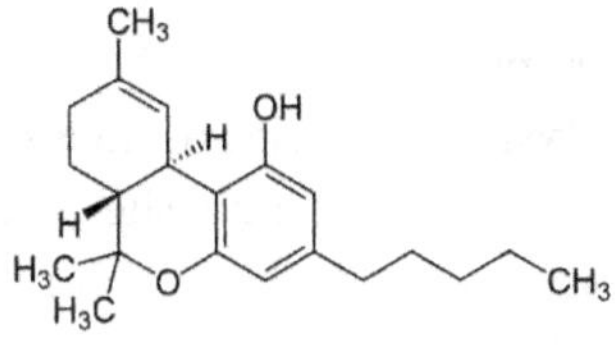

THC

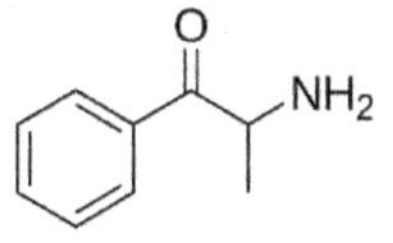

Cathinone

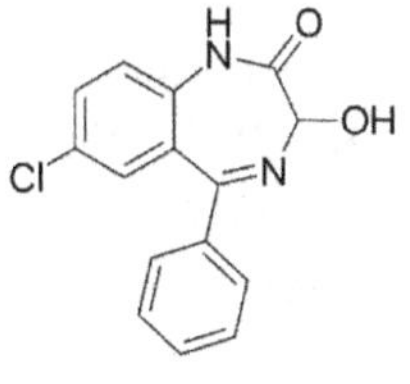

Oxazepam

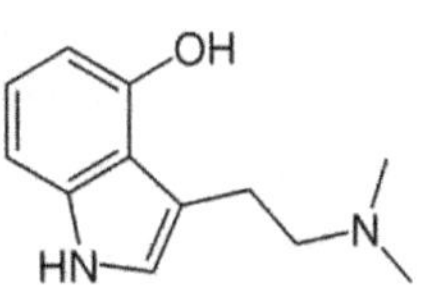

Psilocine

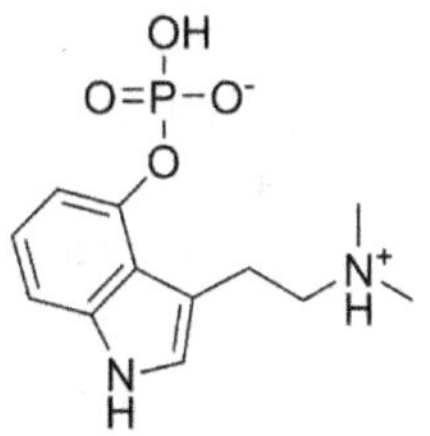

Psilocybine

With legal clarity on the compounds involved in illicit drug analysis, the associated chemical identification task for the forensic institutes and police organizations is now clear. A large number of suspect samples need to be analyzed on a daily basis for the presence of listed compounds. This list of compounds is dynamic and can contain over 300 individual chemical entities. The chemical analysis must be efficient and rapid given the high case load but at the same time must be sufficiently selective to unambiguously demonstrate the presence of an illicit drug. Especially in relatively simple cases involving the individual possession of a small amount of suspect material, a false-positive result (erroneously reporting the presence of a List 1 or List 2 compound on the basis of the results of the chemical analysis) will almost certainly lead to a wrongful conviction. The whole reason for applying chemical analysis in criminal investigations is to accurately reconstruct events to assist law enforcement, so having somebody found guilty on the basis of an error in the forensic laboratory is a worst-case scenario that should be prevented at all cost. Designing a robust and efficient framework for the chemical identification of drugs of abuse is furthermore challenging because of the diversity and unknown nature of the confiscated evidence material. Samples can consist of multiple compounds (cutting agents, additives, tableting agents), can be contaminated, or can contain novel compounds not encountered before in case work. An internationally recognized guideline for the chemical analysis of evidence materials for the presence of illicit drugs is presented by the USA-based **Scientific Working Group for the Analysis of Seized Drugs (SWGDRUG)**. This team of forensic scientists and experts publishes and regularly updates recommendations for laboratory standards for illicit drug analysis. Recognizing that different analytical techniques can be employed by forensic laboratories for compound identification, the SWGDRUG guideline classifies these techniques according to three categories with different degrees of selectivity. Category C techniques provide only class or group-type information, category B provides "selectivity through chemical or physical characteristics," and category A is reserved for analytical methods that provide molecular structural information of compounds. In the table below, an overview of analytical techniques is given including their SWGDRUG assigned class in the June 2019 recommendations of the working group (studying the SWGDRUG document is highly recommended to scholars with an interest in forensic illicit drug analysis as it covers a broad range of topics that need to be considered in this field of expertise, including a code of conduct, education and training and quality measures, more details can be found in the Further Reading section).

Category	Method	Comments
A	Infrared spectroscopy (IR)	Also includes Fourier transform IR (FTIR)
	Raman spectroscopy (Raman)	
	Mass spectrometry (MS)	
	Nuclear magnetic resonance spectroscopy (NMR)	
	X-ray diffractometry (XRD)	Provides information on crystal structure
B	Capillary electrophoresis (CE)	
	Gas chromatography (GC)	
	Ion mobility spectrometry (IMS)	
	Liquid chromatography (LC)	
	Supercritical fluid chromatography (SFC)	
	Thin-layer chromatography (TLC)	
	Ultraviolet/visible spectroscopy (UV/Vis)	For a relevant wavelength range
	Microcrystalline tests	
	Macroscopic and microscopic Examination of cannabis	For cannabis only
C	Colorimetric tests	Chemical reaction with a color reagent
	Fluorescence spectroscopy	
	Immunoassays	
	Melting point	
	Pharmaceutical identifiers	Product identifiers can be a valuable source of information but can be copied in counterfeit products

SWGDRUG classification of analytical methods for the identification of drugs of abuse (June 2019 version)

SWGDRUG provides several rules for designing a suitable chemical analysis scheme for qualitative illicit drug analysis:

- When a laboratory employs a Category A technique, robust chemical identification requires the use of at least one other technique from Category A, B, or C provided this additional technique focuses on a different chemical or physical feature
- When a laboratory does not employ a Category A technique at least three different techniques are required for robust chemical identification of which two need to be from Category B
- Hyphenated techniques combining a chromatographic and a spectrometric method can account for two different techniques in the analysis scheme
- The presence of an illicit drug can be reported when in a given analysis scheme all methods are applicable for the sample considered and provide positive test results that satisfy all quality norms

Which analysis scheme is most frequently used by forensic laboratories for the chemical identification of illicit drugs?

Colorimetric testing (Category C) in combination with Gas Chromatography (Category B)—Mass Spectrometry (Category A).

A compound is identified on the basis of a positive colorimetric test result (color formation after addition of a reagent) and a matching retention time and mass spectrum when compared to data for a standard or reference in a database for that compound.

The combination of colorimetric tests and GC-MS allows experts to efficiently process a large case volume while ensuring sufficient selectivity to confidently report the presence of a controlled substance. Often the colorimetric test also serves a selection process, samples are only analyzed with GC-MS when the expected color is formed, and the test is positive. This further improves the speed of the process and reduces the cost as expert capacity and instrument time is not wasted on samples that do not contain illicit drugs. However, the downside of this approach is that false-negative results on the colorimetric test will not be discovered as no further chemical characterization is conducted. This potential weakness could be exploited by informed criminals by adding materials to their drug mixtures that deliberately incapacitate the colorimetric test. However, such degree of sophistication in the drug manufacturing process is seldomly observed in forensic case work.

Despite concerns regarding selectivity, colorimetric tests are extremely effective especially in the first phase of a criminal investigation. Procedures are straightforward typically only requiring the addition of liquid reagents to a small amount of suspect sample. The analysis is rapid as the chemical reactions occur almost instantly. Furthermore, no expensive detectors are necessary as the color change can be visually observed by the individual conducting the analysis. These features allow for analysis outside the forensic laboratory by non-experts. To prevent direct handling of suspect materials and the (often aggressive) reagents in solutions, several manufacturers offer single-use colorimetric drug tests in tailor-made pouches in which reagents can be released by breaking seals after applying external pressure. Additionally, such devices include special strips and swipes that enable sampling of minimal amounts without the need for manual contact. Single-use assays prevent potential false-positive results due to contamination and are very cost-effective with prices in the range of 1−10 euro/USD per test. Instrumental alternatives such as portable Raman or IR spectroscopy typically have negligible consumable costs but require an investment cost of up to 30.000 euro/USD per instrument. So only after 3000−30000 analyses, the portable instrument is going to provide "return on investment" not considering maintenance and repair costs and operator training. Another benefit of the single use colorimetric tests is that they can be used by a much wider population of law enforcement and security personnel (30.000 tests can be widely distributed in contrast to a single portable instrument). A more recent development to further increase the quality of colorimetric testing in the field is the use of smartphones to objectively record the color formation and to provide a quality framework to document and link the case at hand to the evidence obtained with the colorimetric test (see also **Chapter 9—Quality and Chain of Custody**). By using the camera on the smartphone, the suspect sample can be photographed and the test result can be objectively recorded. By using a colored reference object, fluctuations in light conditions can be corrected for in a semi-calibration procedure. Finally, the communication capabilities of smart phones can be exploited to send the data directly to the organization's secure cloud/server environment. In this way, findings obtained in the field and under challenging conditions are directly archived and do not require extensive administration by the officers on the scene. At a later stage, the results can be processed in the official legal documentation such as forensic and police reports. An overview of the most common color reactions used by law enforcement, security personnel, customs offices, and forensic laboratories is given in the table below.

Order	Name	Chemistry	Drug of abuse	Positive test result
1	Marquis	Formaldehyde + sulfuric acid *Broad spectrum test*	MDMA/MDEA Amphetamine/ Methamphetamine Heroine	Black Orange Purple
1	Scott/ ruybal	Cobalt thiocyanate + hexachloro platina acid in phosphoric acid	Cocaine (Levamisole, Lidocaine) Heroine	Blue Green
1 > 2	Simon's	Acetaldehyde + natrium nitroprusside	Metamphetamine/ MDMA (secondary amine)	Blue
1 > 3	Duquenois- Levine	(1) vanillin + acetaldehyde (2) sulfuric acid	THC (marijuana/ hashish)	Purple
1 > 4	Ferric chloride	Ferric chloride hexahydrate in water	GHB (phenols)	Dark brown

Colorimetric test reactions typically used to screen unknown suspect illicit drug samples by law enforcement and forensic laboratories.

Typically, in forensic laboratories the Marquis and Scott/Ruybal test are applied as a first step of testing. Depending on the outcomes of these two tests follow-up test can be conducted to finetune the colorimetric indication. Although some of these color reactions date back to the 19th century, the chemistry involved is not always fully understood and the reaction products formed are not always known. Creating color requires the formation of conjugated systems or transition metal complexes. In conjugated systems, delocalized electrons in overlapping p-orbitals reduce the energy involved in electronic transitions. If the wavelength of the associated photon absorption is within the visible range of the electromagnetic spectrum (400–800 nm), the selective reflection or transmission will yield a typical color as perceived in human vision. The remarkable colors that can be associated with transition metal coordination complexes are related to electron d-orbital transitions from the HOMO (highest occupied molecular orbital) to the LUMO (lowest unoccupied molecular orbital). These transitions and thus the perceived colors can be strongly affected by the oxidation state of the metal and the ligands. Specific coordination chemistry and chemistry induced color changes provide a degree of selectivity to these types of colorimetric tests. The specific blue color in the Scott test, a positive test result indicating the presence of cocaine (this strong visual feature is also often exploited in films and TV series) is related to the cobalt (Co) complex depicted below.

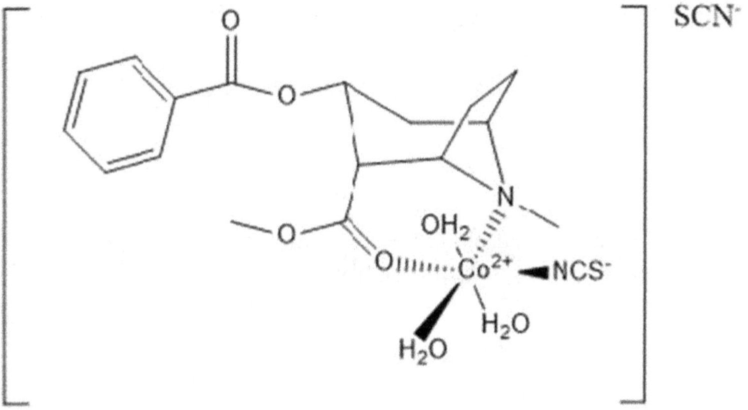

Cobalt cocaine complex giving the distinct light blue color of a positive Scott's test. *(Structure: courtesy of Dr Rene Williams, van 't Hoff Institute for Molecular Sciences, UvA)*

Despite the selective coordination chemistry involved, also the Scott test is known for potential false-positive outcomes. Frequently occurring additives such as lidocaine and levamisole are known to yield a similar blue color formation. Although related to cocaine adulteration, these compounds are not listed and hence possession of these substances is in itself not a criminal offense according to the illicit drug law. Lidocaine is also used as a local anesthetic to treat skin irritations or as a sedative by dentists, so this compound is widely available in society for reasons that are totally unrelated to illicit use.

The antihistamine diphenhydramine is also a compound known to cause a false-positive result for the Scott test for cocaine. This potentially could lead to the arrest of an innocent person that happens to be allergic. Given the structure below can you chemically understand the blue color formation by drawing the Cobalt coordination complex?

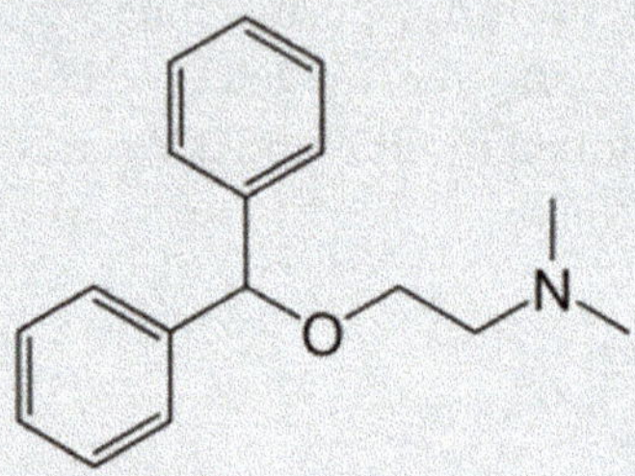

Diphenhydramine has an oxygen atom and a tertiary amine group in a geometric position to form a similar Co complex.

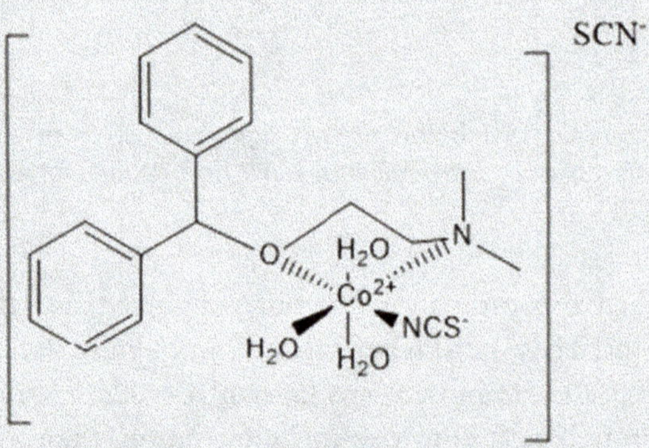

Structure: courtesy of Dr Rene Williams, van 't Hoff Institute for Molecular Sciences, UvA

These false-positives outcomes for frequently used chemicals are inevitable in colorimetric testing. Some of these tests target compound classes such as primary and secondary amines and exhibit an even lower selectivity. This explains the indicative status of these tests and hence the need for confirmation in the laboratory with more robust techniques such as Gas Chromatography-Mass Spectrometry (GC-MS) which will be discussed in more detail next.

GC-MS was introduced in the 1950s and through numerous technological developments quickly became a work-horse technology from the 1970s onwards in the analytical chemistry laboratory. Contemporary laboratories typically house numerous GC-MS systems and several instrument suppliers offer extremely robust and reliable instruments. For forensic laboratories processing large volumes of alleged drug samples, it is not uncommon to have five or more dedicated GC-MS instruments that are used "24/7" with the use of autosamplers and automated analysis. The principles of GC-MS were already discussed in paragraph 3.4 of **Chapter 3** as this technique is also the main method of choice for fire debris analysis. In illicit drug analysis, extensive sample preparation is usually not required as sample composition is less complex and the amount of the psychoactive substances (when present) is relatively high. These significant levels are directly related to the desired effects or efficacy of production and trafficking. It simply does not make sense for drug users and manufacturers to use, produce, and smuggle trace amounts of drugs. Occasionally, complex matrices can be encountered in illicit drug analysis, for example, when drugs are consumed as constituent in food products or are mixed in complex matrices to prevent detection during transport. This typically requires additional sample preparation steps prior to GC-MS analysis. However, regular case samples can simply be dissolved in a suitable organic solvent. A small volume of solvent is subsequently injected on the **split/splitless injection port** of the instrument. As the injector is maintained at a high temperature (normally in the range of 200–250°C), the sample constituents evaporate in the so-called liner and are transported by the carrier gas to the capillary GC column. Typically, the injector is operated in **split mode** because of the relatively high concentration of the compounds of interest. This means that most of the evaporated sample is "vented" via the split flow. The relative amount of sample transported to the capillary GC column can be adjusted through the split flow in relation to the column flow. This is controlled in the GC instrument by adjusting the flow resistance in the split valve. With a split gas flow of 25 mL/min and a column flow of 1 mL/min, a split ratio of 1:25 is realized and roughly 4% (1/26th) of the sample injected is analyzed. However, at complete and rapid evaporation, this relatively small amount is fully representative for the sample. The benefits of split mode include rapid injection, minimal pressure fluctuations, limited contribution to band broadening, minimal risk of sample overloading and control of sensitivity through split flow settings. In **splitless injection mode,** maximum sensitivity can be achieved as the entire sample is transferred in gaseous state to the GC column. However, this requires careful optimization and refocusing of the analytes at the start of the analysis to prevent excessive band broadening. In splitless mode, the time needed to transfer the content of the liner to the analytical column can take up to several minutes because the total flow in the liner now equals the column flow which is typically in the mL/min range. For semi- and nonvolatile compounds, refocusing can be achieved through a relatively low start temperature in the temperature program in combination with a sufficient stationary phase film thickness. However, solvent condensation must be carefully controlled. In splitless injection, also the liner volume needs to be large enough to fit the entire evaporated sample volume at the given pressure regime in the injector. If the liner volume is insufficient, sample expansion can lead to pressure increase and contamination of the injector. As sensitivity is usually less critical in the analysis of street drug samples, split injection is therefore much more straight forward and the preferred mode for injection for illicit drug analysis in forensic laboratories. As GC-MS is such an important technique in analytical chemistry, it is also very cost-efficient and can be exploited at relatively low cost per analysis. The robustness of the instrumentation minimizes repairs and down-time, provides a high sample throughput while assuring the quality of the results. Consequently, analysis and reporting can be automated to a large degree. A schematic representation a contemporary GC-MS instrument and a split/splitless injector and is given below.

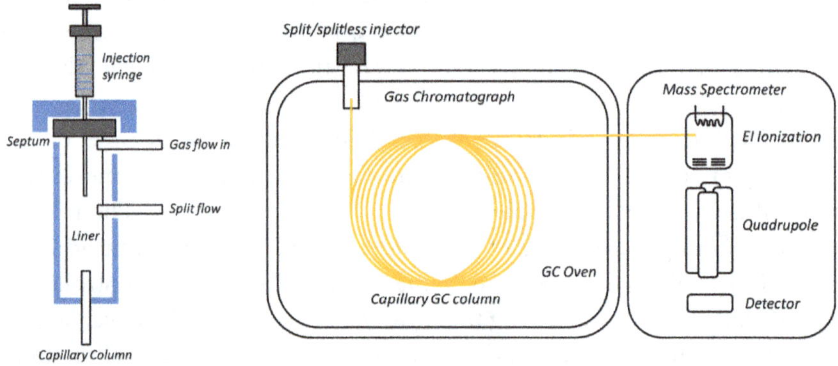

Schematic representation of a GC-MS instrument (right) with details of a split/splitless injector (left)

The qualitative screening of drugs of abuse does generally not require specific conditions for the GC-MS analysis. Regular columns with respect to length, internal diameter and stationary phase type and thickness can be used in combination with standard settings with respect to the mobile phase type (He or N_2), pressure and flow and the oven temperature program. One complication that arises is that many drugs contain a primary or secondary amine functionality in the molecular structure. This feature is especially apparent in low molecular weight synthetic phenethyl amine compounds like amphetamine, methamphetamine, and MDMA. These amine compounds are known to show adsorption on active sites in the GC injector, capillary column and transfer liners causing "bad" peak shapes (*e.g.* peak tailing) especially when these compounds show limited retention and elute early in the temperature program. Furthermore, these compounds are often produced and sold as HCl salts (with a positive charge on the amine moiety) and this hampers evaporation in the split/splitless injector of the GC instrument. To improve the chromatographic analysis, some forensic laboratories derivatize the samples prior to analysis to reduce the reactivity and improve the GC-MS analysis. Derivatization also increases the selectivity of the overall procedure. However, this approach requires some laboratory benchwork for the lab staff and given the high case volume this results in a significant on-cost with respect to capacity, time, and consumables. A simpler approach to prevent HCl salt related issues is to add a base to the sample solution (*e.g.*, in methanol) such as sodium bicarbonate. The salt will not dissolve in the organic solvent and hence will not contaminate the GC system as it settles at the bottom of the sample vial. Due to the presence of a base, the ammonium form of the drug will be converted to its uncharged amine form when the suspension of the basic salt and organic solvent is shaken. A slightly more sophisticated approach is a solvent-solvent extraction procedure where the drug sample is dissolved in water and the active amine compounds are subsequently extracted to an immiscible organic solvent (*e.g.* DCM) after adding a strong base to the aqueous solution.

With an effective and efficient GC-MS method in place, let us now take a closer look at an actual confirmation of the presence of cocaine in a street sample after chemical analysis by the forensic laboratory.

The sample was sent to the forensic laboratory after a positive Scott test performed by police officers in the field. Would you now as a forensic drug analysis expert repeat the colorimetric test or would you directly proceed with the GC-MS analysis?

According to the SWGDRUG guidelines, the confirmation of the presence of cocaine needs to be based on colorimetric testing (Category C) technique in combination with GC-MS (Category B and Category A). The field test was conducted under uncontrolled conditions and not within the quality framework of the laboratory. Therefore, illicit drug analysis laboratories always conduct their own colorimetric test to confirm the positive test result that was obtained by the police officers at the time of the arrest. Only in this way the findings can be presented as forensic evidence in court.

In this case, the positive test result in the form of the typical blue color formation was also found in the laboratory as is clearly indicated in the photograph that was taken of the test result to document the finding in the forensic file.

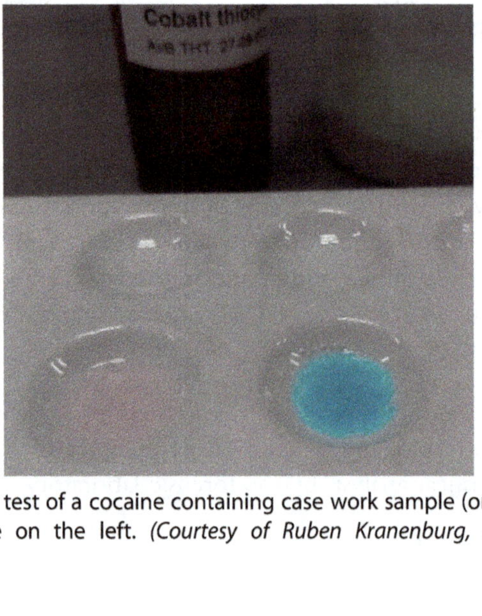

Result of the Scott test of a cocaine containing case work sample (on the right) and of a blank reference on the left. *(Courtesy of Ruben Kranenburg, Amsterdam Police Laboratory)*

The sample was now analyzed with GC-MS according to the standard methods applied by the Amsterdam Police. Confirmation of the presence of cocaine or any other listed substance for that matter is based on the retention time in the GC separation and the Electron Impact mass spectrum as compared to reference spectra of listed substances stored in an electronic library. As discussed in **Chapter 3**, GC-MS is known for the very reproducible EI (electron impact) mass spectra obtained for organic compounds. Hence, large commercial EI-MS libraries can be employed to identify unknown compounds irrespective of the "make and model" of the GC-MS instrument. Furthermore, dedicated libraries can be constructed containing compounds of interest that are regularly analyzed. In this case, such libraries would consist of EI-MS spectra of all listed substances, known precursors for drug synthesis and associated compounds. In terms of stability and robustness of the mass spectra, the analysis of standards in the same sequence as the case work samples is not required. However, without the use of retention time indices or calibration procedures, such standard mixtures are analyzed to confirm a matching retention time. Including reference samples and standards is also considered good laboratory practice because it provides a check whether the instrument is working according to specifications and the analyses have been conducted properly. Degrading sensitivity, contamination issues (deteriorating peak shapes), and gas leaks occasionally occur and can be adequately addressed by the analysts, and such problems are typically revealed by studying the results of the control samples and standards. Below the GC-MS total ion chromatogram (TIC) is depicted of our unknown sample and the same chromatogram obtained for a standard solution containing seven drugs of abuse including cocaine. Clearly there is a compound peak observed in the casework sample that has the same retention time as the cocaine reference in the standard mix.

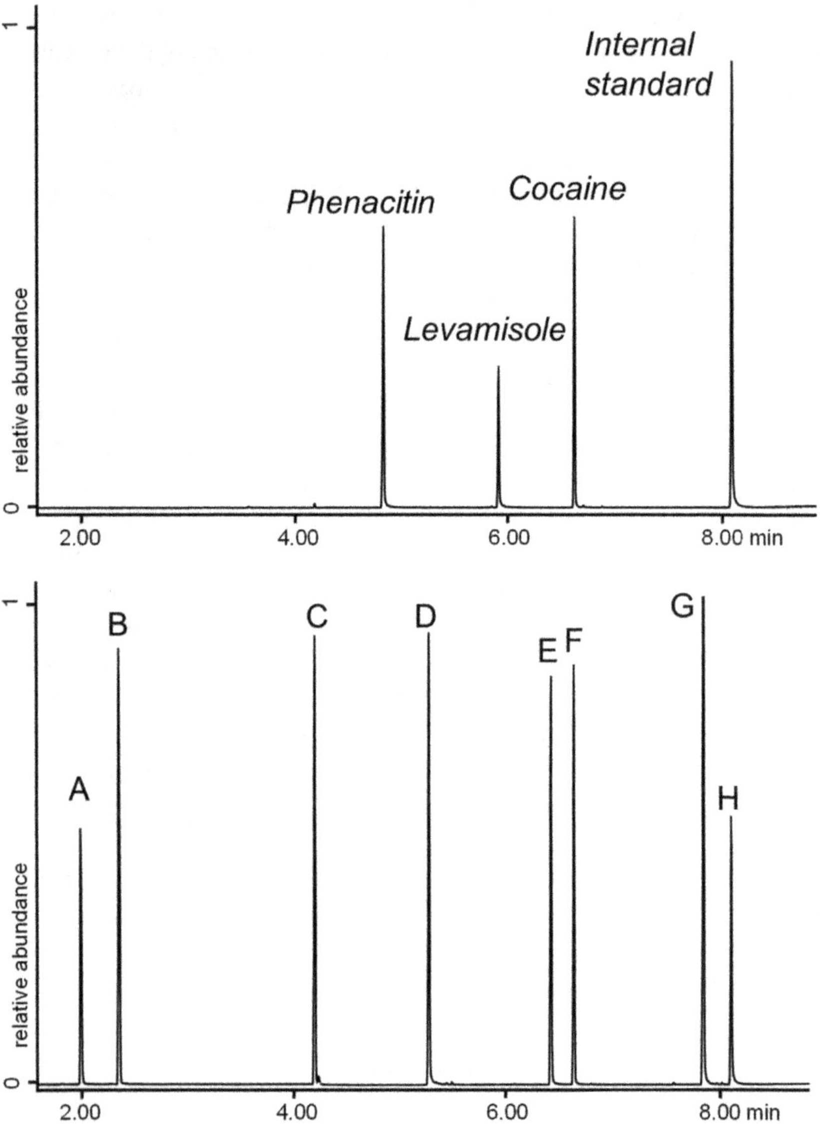

GC-MS TIC chromatograms of the case work sample (top) and reference standard mixture (bottom). Elution order: Amphetamine (A), Methamphetamine (B), MDMA (C), 2C–B (D), methadone (E), cocaine (F), heroin (G), internal standard C-28 (H); GC method: 1 μL split injection at 300°C, column: 15 m; 0.25 mm ID; 0.25 μm film (5% phenyl/95% polydimethylsiloxane, oven program: 100°C (1.5 min), 30°C/min until 300°C. *(Courtesy of Ruben Kranenburg, Amsterdam Police Laboratory)*

We have a positive Scott test under laboratory conditions and a peak at a retention time that is expected for cocaine. The final confirmation now needs to come from further inspection of the mass spectrum. The measured mass spectra from the compounds in the case sample and the corresponding library match scores are depicted below.

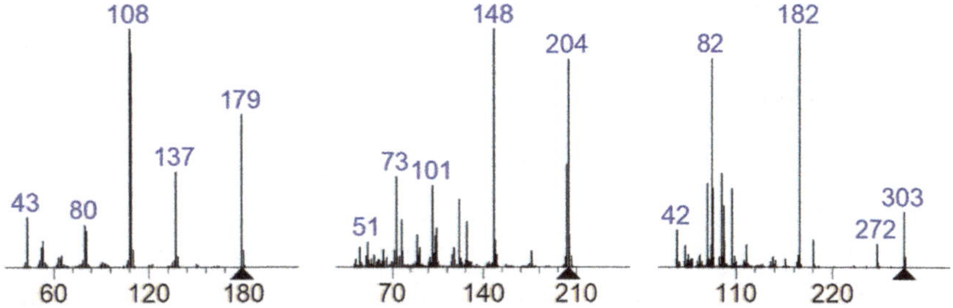

EI-MS spectra of the unknown constituents in the case sample with the assigned compounds and the corresponding library match score. A perfect match with the library mass spectrum results in a score of 100. *(Courtesy of Ruben Kranenburg, Amsterdam Police Laboratory)*

The molecular mass of cocaine ($C_{17}H_{21}NO_4$) is 303 and indeed a peak at relatively low intensity can be seen at this mass (the low intensity signals at higher m/z values are related to ^{13}C isotopes, this is explained in more detail in **Chapter 7—Forensic Reconstruction through Chemical Analysis**). Electron impact mass spectra are normalized to the intensity of the so-called **base peak**, the m/z value for which the highest intensity is measured. In the depicted spectrum for cocaine, the base peak is found at m/z 182, the measured signal is set at a maximum intensity of 100%, and all other signals are scaled accordingly. Electron impact is considered a relatively "hard" ionization technique, the amount of energy transferred by bombarding the compound of interest in the vacuum of the mass spectrometer with accelerated (typically at 70 eV) electrons is substantial. As a result, the molecular ion that is formed by the ejection of a single electron (denoted as $[M^{\cdot}]^+$) from the original compound is highly unstable and undergoes extensive fragmentation. This results in a characteristic spectrum made up of numerous charged fragments of lower m/z. The fragmentation also results in the formation of neutral species, but these fragments cannot be detected by the mass spectrometer. Because of the extensive fragmentation, molecular ion peaks exhibit a low abundance in the mass spectrum and are often even completely absent for aliphatic compounds. Aromatic compounds typically exhibit a more dominant molecular ion peak in the spectrum because delocalization of the unpaired electron makes the ion more stable. The EI spectrum now acts as a "fingerprint" with a high compound specificity and this spectrum can be compared to a reference spectrum or searched against a large collection of library spectra. In the latter process, the analyst instructs the GC-MS software to compare the measured spectrum against all the reference spectra in a given library (or several libraries). After completing the comparison algorithm, the software provides a ranked list of potential library candidates with the most similar reference mass spectra. The degree of similarity is usually expressed with a so-called match score. A value of 100 corresponds to the perfect match, indicating that there is no measurable difference between the two spectra. Typically, match scores that provide a strong indication that the unknown is the reference compound considered are in the range of 90–100. Despite the robustness of the EI mass spectrum, small differences always occur between measured normalized ion intensities especially when considering low intensity signals near the detection limit. For match scores below 90, spectral differences become more significant and are most likely of a systematic and not of a random nature because the unknown is not the reference compound suggested by the software. Despite the sophisticated and automated process of EI-MS library searching, ultimately the analytical chemist needs to visually inspect the spectra and make a final decision with respect to the identity of the unknown compound.

Why can GC-MS-based compound identification not be fully automated and is human expertise required for final confirmation? Can you think of three reasons?

1. A match score does not say anything about the selectivity of the mass spectrum. Some compounds have highly detailed EI mass spectra. As a result match scores can be a bit lower but a matching pattern can be obvious to the human eye. However, some compounds due to their structure yield mass spectra with only a few, low m/z fragments. Such spectra are not very informative and much less characteristic. A human expert can take this into account when studying the spectra.
2. The reverse situation can also exist, that is, a relatively high match score but visually clearly different spectra because for instance low abundance m/z patterns do not match. In this case, such patterns have a lower impact on the comparison algorithm than more abundant m/z signals.
3. The software does not consider the "chemical context", for example, are there more compounds in the ranked list with a very high match score? If so, which of these compounds is the most likely candidate? Is a compound known to have isomers that occur frequently in casework? If so, do these isomers yield very similar spectra?

In the case of our sample, the expert compares the measured mass spectrum with that of the cocaine reference sample and concludes that the two spectra appear identical. Furthermore, the library search yields a ranked list with cocaine at the number one position with a match score of 99.4. There are no other compounds with a match score over 90 making cocaine the only plausible candidate. On the basis of three independent results (positive Scott test, a peak with a GC retention time in the expected time window and a matching EI-MS spectrum), the forensic illicit drug experts now arrives at the final conclusion that the sample contains the listed hard drug cocaine. The results are summarized in a forensic report and sent together with the expert interpretation to the legal and police professionals handling the case. On the basis of this evidence, the suspect will almost certainly be convicted for illegal activities as described in the illicit drug law (assuming that the physical evidence was obtained lawfully, and no other administrative and laboratory errors have been made). It should be noted that there are very few examples where the forensic findings are of such direct importance to the outcome of a criminal investigation. The reason for this is that the evidence in this case is directly linked to the crime as described in the law. In other words, demonstrating the presence of a substance in a sample through chemical analysis immediately proves a criminal activity as any action (even possession) involving a listed drug is forbidden by law. The only nuance here could be a situation in which the defendant was not aware of the presence of the material, *for example,* in a situation where somebody is used to smuggle drugs, either unknowingly or in a situation of coercion. The criminal investigation then focuses on the plausibility of such a statement by the defendant. The corresponding sentence will depend on the context of the case (*e.g.*, amount of drugs involved) and the background of the suspect (*e.g.*, criminal track record). However, in the Netherlands the sentence can never exceed 6 years in prison for a felony involving hard drugs.

In the next paragraph, we will see how the rise of NPS, has created issues in the forensic laboratory in the last decade with respect to GC-MS selectivity in illicit drug analysis. You will learn how the forensic drug experts successfully addressed these challenges by expanding their "toolbox" for chemical analysis.

4.4 The NPS challenge: addressing the selectivity dilemma

The combination of colorimetric testing with GC-MS analysis has been used by forensic laboratories for decades for the robust chemical identification of listed illicit drugs. This framework not only provides the required selectivity to prevent false positive outcomes but is also highly efficient allowing forensic laboratories to process high case volumes, limit backlogs and perform the analyses in a cost-effective manner. However, in recent times, the global rise of **NPS** has forced forensic drug analysis experts to adapt and expand their methods for chemical analysis. Because of the global scale and impact of the NPS phenomenon, international organizations as the **UNODC** (United Nations Office on Drugs and Crime) and the **EMCDDA** (European Monitoring Center for Drugs and Drug Addiction) play a vital role in collecting information from individual countries to provide insight and oversight. The UNODC in a recent Smart Update communication entitled "Understanding the synthetic drug market: the NPS factor" (see Further Reading section) effectively summarized the NPS challenge:

"Since 2009, new psychoactive substances (NPS) have captured the attention of the international community and transformed the global synthetic drug market. The rapid emergence of new substances is certainly unparalleled. What is more striking is the evolution of a market where psychoactive effects were derived from a limited number of closely related chemical structures and innovation mainly featured the adaptation of synthesis routes, including the use of alternative precursor chemicals, to one where the desired psychoactive effects are obtained from hundreds of different substances with diverse chemistry. This dramatic transformation has implications for effective monitoring, understanding and control of synthetic drugs and their precursor chemicals."

How would you define a New Psychoactive substance?

What have NPS in common with listed substances and what sets it apart?

The UNODC uses the following definition of NPS:
substances of abuse, either in a pure form or a preparation, that are not controlled by the 1961 Single Convention on Narcotic Drugs or the 1971 Convention on Psychotropic Substances, but which may pose a public health threat.

In this context, the term "new" refers to the introduction of the compound as substance of abuse, it does not necessarily indicate that the compound itself is new or was recently discovered.

As the lists of controlled substances at a national level as part of illicit drug laws are directly related to the UN conventions, NPS represent compounds with psychoactive effects similar to illicit drugs, which are abused in a similar fashion as illicit drugs but that are not under legal control.

Can you mention the reasons for the exponential increase in NPS both in terms of amount and diversity of compounds encountered?

The most important reason for the NPS surge is the fact that new compounds are initially not controlled by countries working with banned substance lists. A very important principle in criminal law is that you cannot be convicted for something that is not defined as such in the law. Hence, producing, transporting, selling, and using NPS are not a criminal act under the illicit drug law as long as the compounds are not appearing on the lists. For this reason, NPS are sometimes also referred to as "Legal Highs". However, the production and use of NPS clearly has the same purpose and therefore after discovery countries typically start a process of adding NPS to their lists. The result is a perpetual cycle, once an NPS has become a controlled substance, the drug manufacturers focus on a new, similar compound which again is uncontrolled. This leads to a "cat and mouse" game between criminals and the authorities and fuels the continuous emergence of new compounds entering the drug market.

However, producing NPS requires a more extensive knowledge of chemical synthesis and access to a wide range of raw materials. Consequently, the production and sale of NPS has a strong international character with compounds being produced in country A, then shipped to country B and distributed to countries C, D, and E. The Internet plays a very important role in these processes and producers and users arrange shipments and orders through the Dark Web or Dark Net.

In 2020 the UNODC reported a total of 950 NPS in 120 individual countries in their "Current NPS Threats" update. These NPS cover a wide range of psychoactive effects and are categorized by the UNODC in six classes based on their "mode of action":

- Classic hallucinogens (CH)
- Dissociatives (D)
- Sedatives/hypnotics (S/H)
- Stimulants (S)
- Synthetic cannabinoid receptor agonists (SCRA)
- Synthetic opioids (SO)

The molecular structures from some frequently occurring NPS in each category are depicted below, including their trade name.

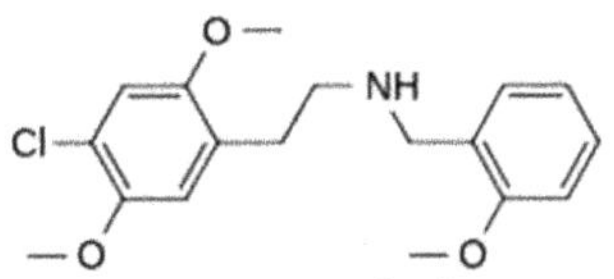

25C-NBOMe (CH)

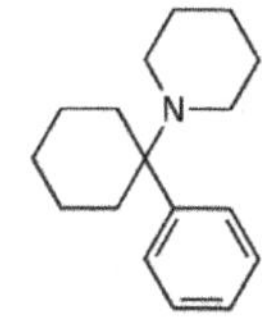

PCP, 'angel dust' (D)

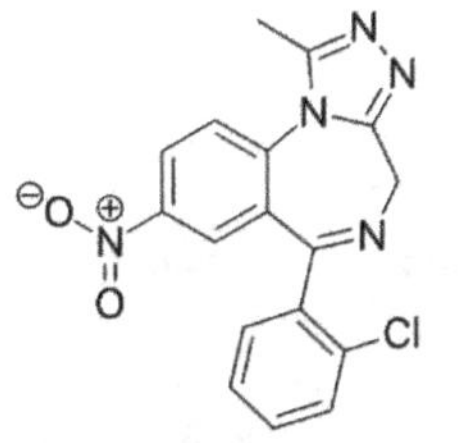

Clonazolam (S/H)

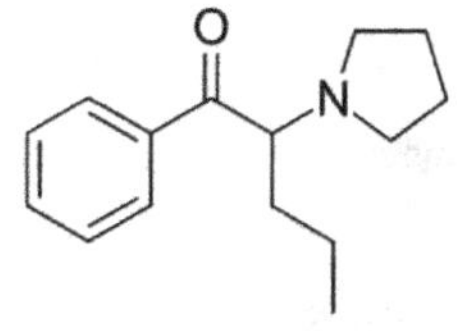

α-PVP, **'Flakka'** (S)

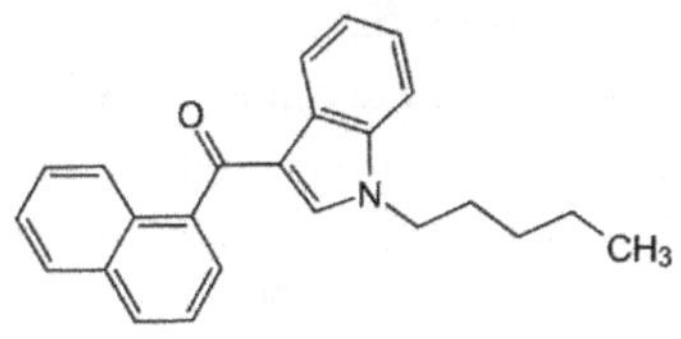

JWH-018 (SCRA)

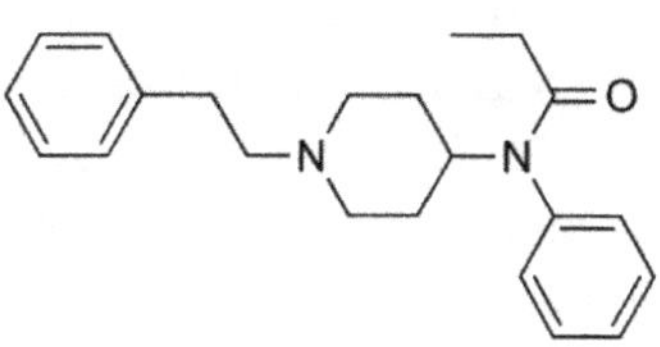

Fentanyl (SO)

The stimulant Flakka is just one example of many NPS belonging to the synthetic cathinone family. Other frequently encountered synthetic **cathinones** are mephedrone, MDPV, methylone, and methedrone. Mephedrone is a variant of methyl cathinone, whereas MDPV and methylone have structural features similar to MDMA. However, these are just a few examples of over 20 cathinone variants reported to date.

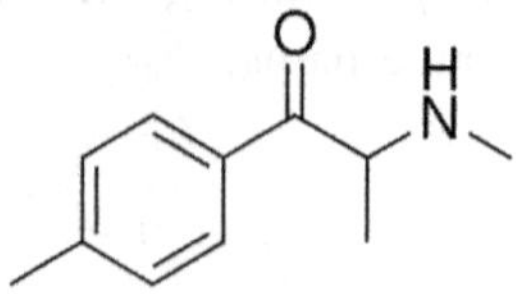

Mephedrone, 4-MMC (S – Cathinone class)

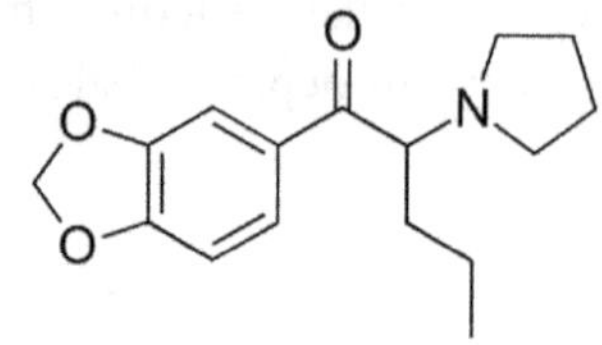

MDPV (S – Cathinone class)

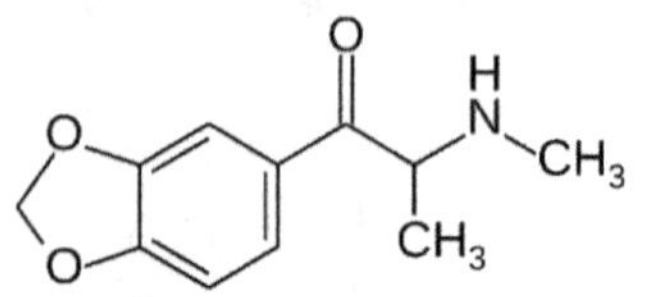

Methylone (S – Cathinone class)

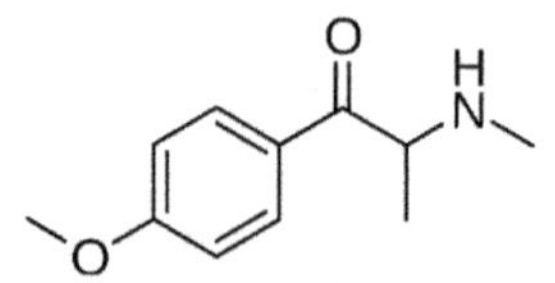

Methedrone (S – Cathinone class)

The synthetic opioid **fentanyl** is another example of a compound for which many structural variants exist. Fentanyl-based NPS are one of the causes of the current opioid overdose crisis in the USA. According to the NIH (National Institute on Drug Abuse) in 2018 on average 128 US citizens died daily as the result of an opioid overdose. Fentanyl-related compounds are extremely potent pain killers acting in a similar way as morphine and heroine. The most potent of the fentanyl family is **carfentanyl**, used legally to sedate large wildlife animals such as elephants. A dose of 2 mg of fentanyl is sufficient to cause an overdose death of a human adult and carfentanyl is reported to be 100 times more potent then fentanyl (source: Wikipedia). For this reason, the handling of fentanyl by law enforcement and forensic experts can be very dangerous without extensive safety and personal protection procedures in place. Fentanyl-related compounds are for this reason not only considered as drugs of abuse but also as potential chemical warfare agents. Because of their potency, fentanyl-related compounds are relatively cost-efficient to produce and transport. Illicit drug experts now frequently report fentanyl "spiked" heroine samples. Drug addicts are often not aware of the presence of such a powerful psychoactive compound resulting in the overdose epidemic. According to the UNODC, in 2019 more than 50 fentanyl variants have been discovered by forensic laboratories of which nearly 40 were not controlled at that time. Some molecular structures of fentanyl analogues are depicted below.

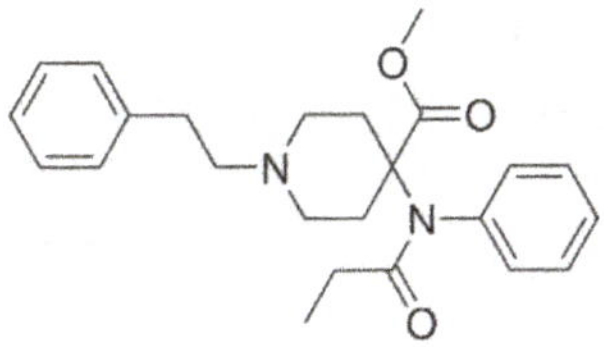

Carfentanyl (SO – Fentanyl class)

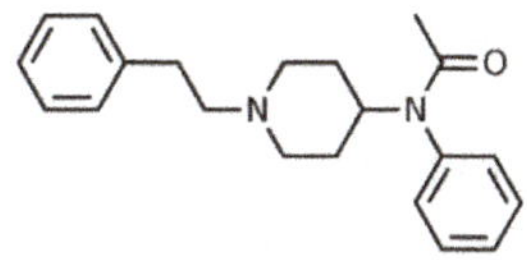

Acetylfentanyl (SO – Fentanyl class)

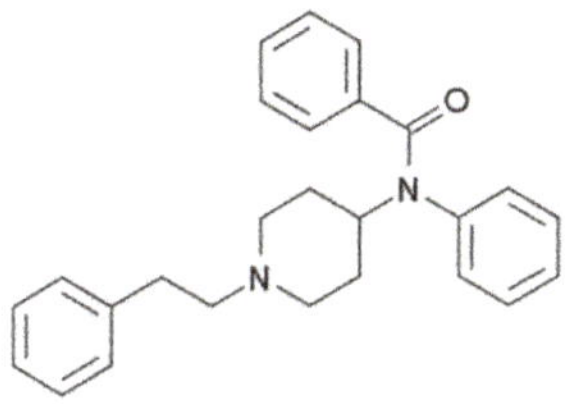

Benzoylfentanyl (SO – Fentanyl class)

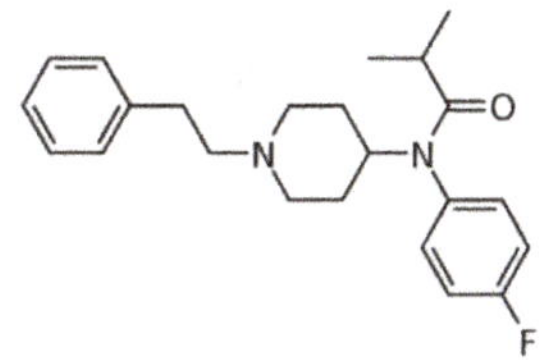

4-FIBF (SO – Fentanyl class)

With a clear picture of the NPS phenomenon and its strong impact on the global illicit drug market, we now shift our attention to the consequences for the forensic laboratories. In a relatively short time period, the illicit drug analysis expert is now facing a case load that consists of samples that could contain compounds that have not been encountered before, for which no standards exist and for which no reference mass spectra are available in the GC-MS libraries. Compounds that could additionally also yield quite similar color formation in colorimetric tests as expected for controlled substances. With the range of psychoactive substances rapidly expanding, compounds with a different legal status (i.e., controlled vs. uncontrolled) might structurally be almost identical. This includes structural isomers of equal molecular formula and mass which only differ in the position of a single functional group. Although the use of colorimetric tests in combination with GC-MS has a high intrinsic selectivity and meets SWGDRUG approval, the rapid increase in designer drugs could result in cases where the chemical identification protocol fails to properly distinguish drug analogues. The associated chemical analysis challenge and ways to address the issue will be discussed in the remainder of this paragraph on the basis of the three isomers of **Fluor-Amphetamine** (FA), an amphetamine-based stimulant.

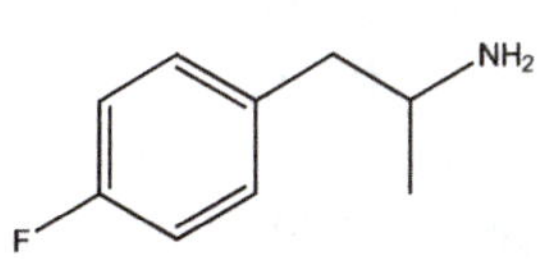

4-FA (S – amphetamine class)

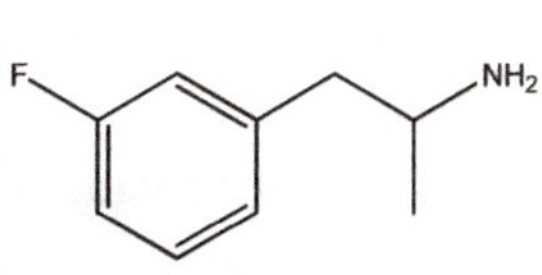

3-FA (S – amphetamine class)

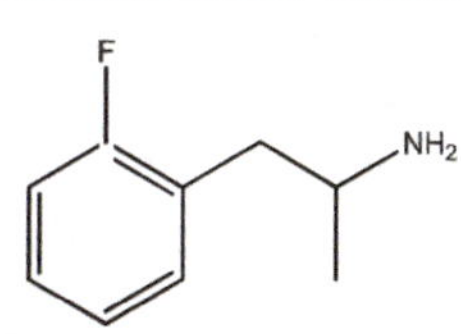

2-FA (S – amphetamine class)

In the Netherlands, the compound 4-FA (4-fluoro-amphetamine) has been registered as a hard drug since 2017. However, the structural isomers 3-FA and 2-FA, which are occasionally encountered in Dutch case work, were still uncontrolled in 2020. Therefore, it is of critical importance for illicit drug experts to correctly identify these isomers in drug samples to prevent wrongful convictions or exonerations. The picture below shows an overlay of total ion GC-MS chromatograms of the three FA isomers.

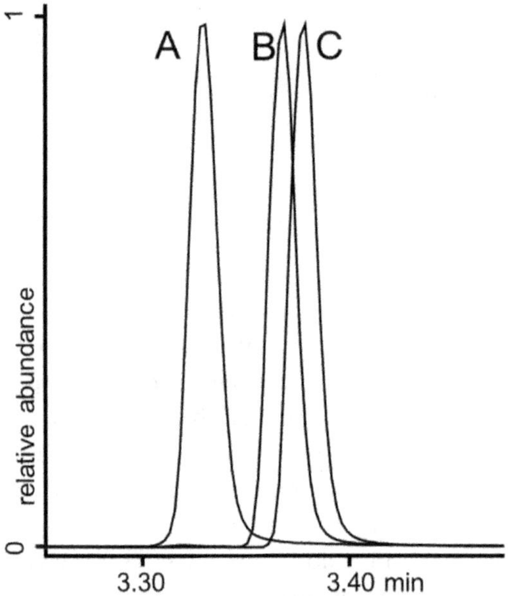

GC total ion chromatogram of 2-fluoroamphetamine (A); 3-fluoroamphetamine (B); 4-fluoroamphetamine (C) *(Courtesy of Ruben Kranenburg, Amsterdam Police Laboratory)*

The three isomers elute in a time window of 0.1 min or in other words 6 s. For a robust assignment of the correct isomer on the basis of retention time, either the separation must be improved and/or the retention times must be calibrated through the use of standards or retention indices. Although this is feasible, it should be noted that screening methods must be able to identify an extensive number of listed substances in a rapid and efficient manner to prevent backlogs and excessive costs. When applying a rapid, broad screening method, the correct identification of the FA ring isomer will be quite challenging. However, in GC-MS the main part of the selectivity in compound identification is usually provided by the EI mass spectrum. The following spectrum for a 4-FA reference standard is obtained:

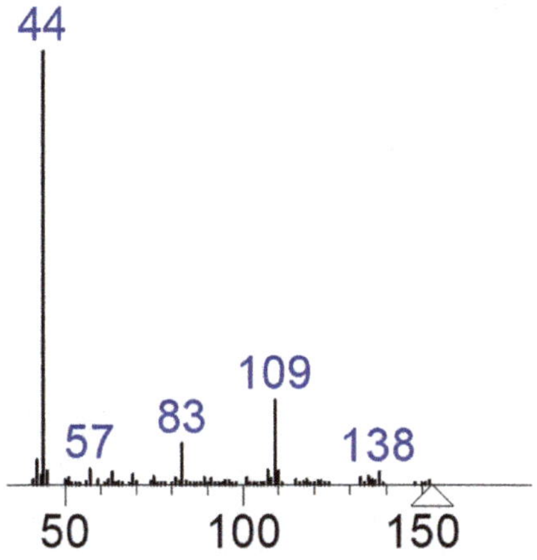

The 70 eV EI mass spectra of 4-fluoro-amphetamine *(Courtesy of Ruben Kranenburg, Amsterdam Police Laboratory)*

In comparison with the MS spectrum of cocaine, the spectra of amphetamines are much less characteristic and are dominated by abundant low molecular mass base peaks that frequently co-occur for various isomers and analogues. To assess whether the fluoro amphetamine isomers can be distinguished with mass spectrometry, we need to unravel the structure of the main ions in the spectrum.

The molecular mass of fluoro amphetamine is 153 Da. The molecular ion at m/z 153 is hardly visible in the EI-MS spectrum, which is quite common for aliphatic compounds or aromatic compounds with alkyl chains.

Can you assign the fragments at:

m/z = 44, m/z = 83, m/z = 109 and m/z = 138?

Tip: unsubstituted aromatic compounds often have a peak at m/z = 91 in the mass spectrum, this is the well-known tropylium ion formed from the Ph-CH2+ ion fragment. Elimination of acetylene from the tropylium ion then often also results in a peak at m/z = 65.

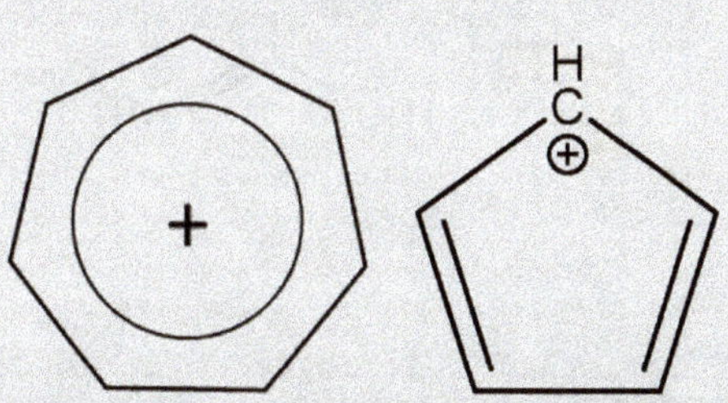

m/z = 138

The mass difference of 15 Da versus the molecular weight suggests the loss of a CH_3 group, and this is very often observed in EI mass spectra.

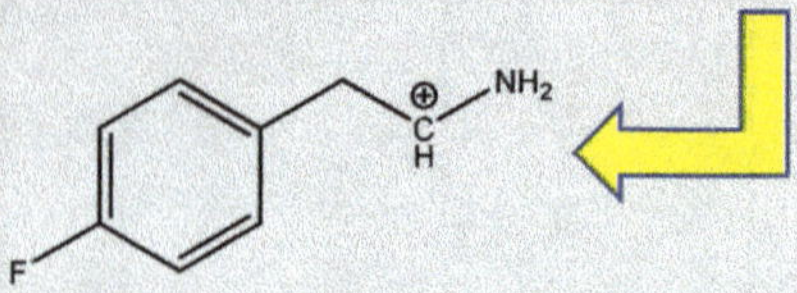

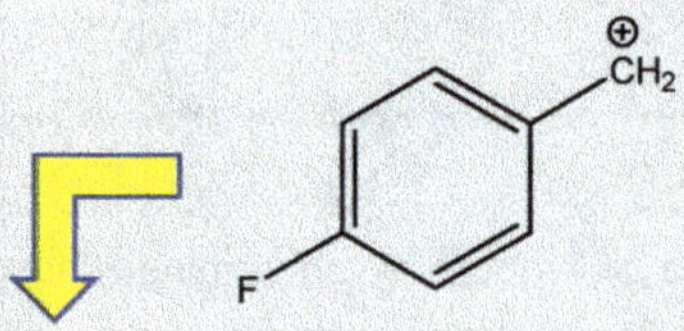

m/z = 109

The mass difference versus the tropylium ion is 18 Da which equals the mass difference between H and F

m/z = 83

The mass difference with the m/z = 65 fragment is again 18 Da indicative of the presence of fluorine.

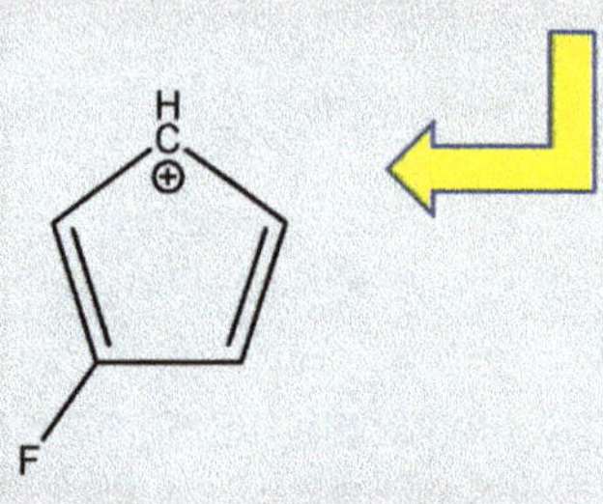

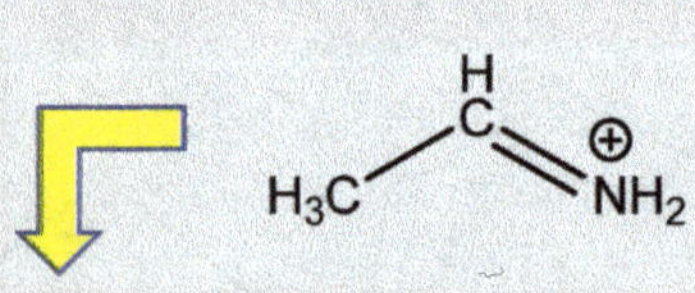

m/z = 44

This fragment corresponds to the primary amine moiety in the molecule that splits off in an α cleavage (the C atom adjacent to the amine).

Given this fragmentation pattern and the identified ions, do you expect the EI-MS spectra of 2-FA and 3-FA to be very different from 4-FA?

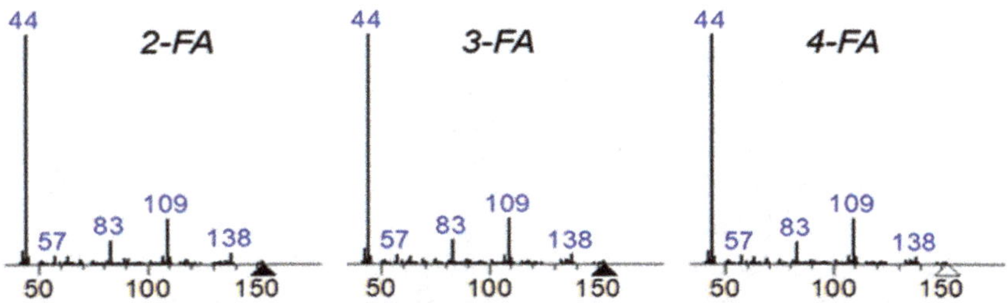

The 70 eV EI mass spectra of 2-, 3-, and 4-fluoro-amphetamine *(Courtesy of Ruben Kranenburg, Amsterdam Police Laboratory)*

The results shown above clearly illustrate the serious problem that exists with differentiating the fluoro amphetamine ring isomers when relying on colorimetric tests in combination with GC-MS. However, given the legal framework, it is essential that the forensic illicit drug expert reports the correct isomer. The introduction of designer drugs thus forces forensic laboratories to expand their analytical capabilities.

What techniques are generally applied for establishing the chemical structure of unknown compounds?

Chemical identification of unknowns is usually based on spectroscopic techniques like NMR (Nuclear Magnetic Resonance), (FT)IR (Fourier-Transform Infrared), Raman, UV, XRF (X-ray fluorescence), XRD (X-ray diffraction) in combination with exact mass measurements with high-resolution mass spectrometry.

Final confirmation is obtained when the material is synthesized and made available as a reference standard .

Why is the application of spectroscopy techniques not always suitable for drug analysis in a forensic laboratory dealing with a high case load?

Can you mention three reasons?

1. Cost of analysis (especially NMR and FT-ICR-MS)
2. Speed of analysis (difficult to automate)

   But the most important reason:
3. Street samples are often mixtures, containing adulterants, cutting agents, impurities, and contaminations. This yields mixed spectra which are much more complex to interpret!

Forensic institutes typically do not have direct access to very expensive and complex equipment like **NMR** and **FT-ICR-MS**. Sometimes forensic experts can access these techniques through academic or industrial contacts, but such analyses remain rare for regular casework. In contrast, benchtop **FT-IR** instruments are affordable, easy to operate (when using an **ATR—Attenuated Total Reflection**—cell, powders can be analyzed as such without the need for any sample pretreatment) and provide detailed spectral data. Therefore, most illicit drug analysis laboratories routinely use infrared spectroscopy. However, mixtures result in complex spectra that are composed of the weighed sum of the spectra of the individual ingredients. Such mixed spectra can only be successfully searched against a reference drugs library if the spectral features of the listed substance dominate the spectrum. This is the case when the active ingredient level is very high (very pure samples only containing adulterants at low level) or when the compound of interest exhibits strong and characteristic IR absorptions. However, case samples are known to be diverse in terms of composition and active ingredient level. Frequently this limits the use of IR spectroscopy to identify the correct isomeric structure as is illustrated below for the cocaine sample that was discussed earlier.

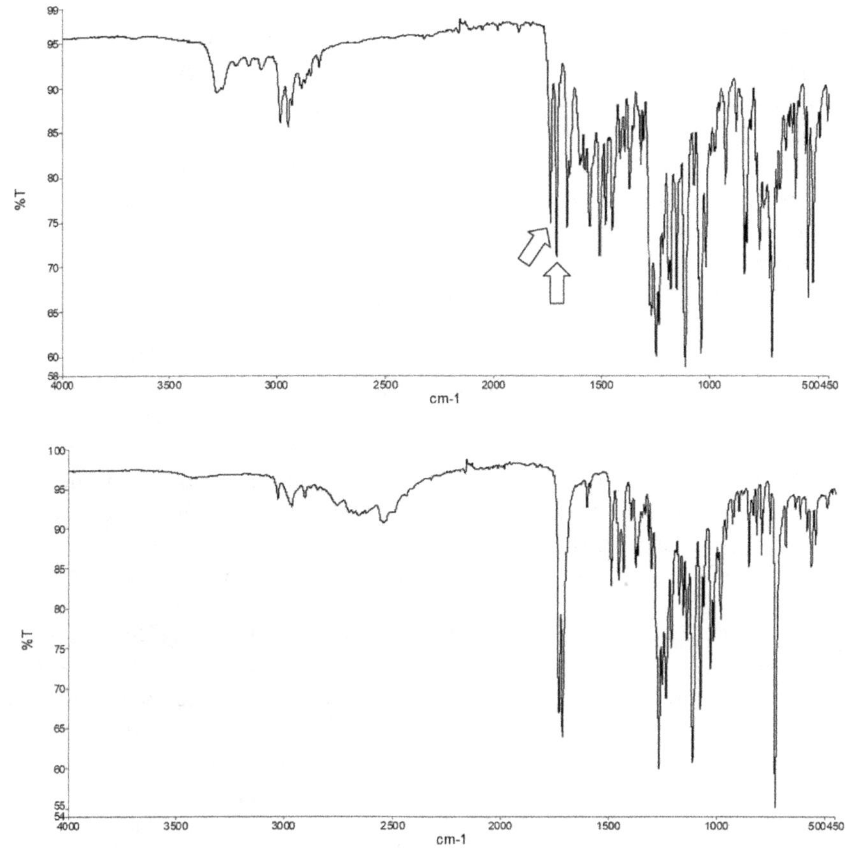

ATR-FTIR spectra of the cocaine case work sample (top) and Cocaine HCl standard (bottom). Only the two absorption bands indicated by the arrows can be attributed to cocaine in the mixture *(Courtesy of Ruben Kranenburg, Amsterdam Police Laboratory)*

If spectroscopy does not provide the ideal solution to tackle the selectivity issue due to the NPS phenomenon, what options remain for the forensic drug analysis laboratories?

Like GC-MS, we need to couple a separation technique to spectroscopy to obtain the spectra of the pure compounds even if present in a mixture with other compounds.

The most powerful combination that is available is GC-IR (Gas Chromatography with InfraRed detection) and although the interfacing of GC with IR is not easy, manufacturers offer robust and affordable GC-IR solutions and this type of equipment has been installed in many drug analysis laboratories. There are two types of technology used to couple GC to IR.

The first approach is based on a vapor phase IR cell where the GC eluent enters a heated elongated glass tube, the so-called light pipe, through which IR radiation is sent. The non-absorbed IR radiation is consequently detected at the other end of the light pipe to reconstruct the gas phase IR spectrum. Typically, the IR is operated in FT (Fourier Transform) mode to enhance speed and sensitivity. In FTIR, the incident beam contains all wavelengths in the IR domain at variable intensities. The so-called interferograms can be converted back to IR spectra using the Fourier Transform algorithm.

Although vapor phase IR is a versatile and robust instrumental solution for GC-IR, the systems based on cryogenic trapping followed by FT IR transmission spectroscopy are more frequently encountered in the forensic drug analysis laboratory. In these GC-IR systems, the compounds of interest are fixed on a cooled ZnSe disc that is IR transparent. As the disc is rotating and the column exit position is changed radially, the time-resolved separation is "recorded" in a spiral track on the disc. The IR spectra of areas of interest are subsequently recorded. Advantages of this approach include compatibility with traditional IR databases, higher spectral resolution and tunable sensitivity through FTIR spectra collection and sample stacking (collecting the material from multiple GC runs of the same sample to increase the absolute amount of compounds on the disc).

Using **GC-IR** enables experts to measure "pure" IR spectra of NPS compounds of interest. It is known that IR spectra, specifically in the **IR fingerprint region**, show significant differences for aromatic ring isomers as vibrations are determined by the exact geometry of the molecule. As is illustrated below and contrary to the GC-MS results discussed earlier, the GC-IR spectra of 2-FA, 3-FA, and 4-FA show clear differences. Using a combination of GC-MS and GC-IR now fully restores the confidence in the drug identification process. The EI-MS spectra indicate the presence of fluoro amphetamine. As the expert knows that GC-MS alone is insufficient to determine the FA isomer type, he/she now conducts a follow-up

GC-IR analysis (it is not necessary to conduct a GC-IR analysis for all cases and types of drugs, *e.g.*, for cocaine the selectivity offered by GC-MS is sufficient) to establish the position of the Fluor atom on the aromatic ring. A spectral library of reference standards can be constructed to confidently assign the correct isomer. Interestingly, when dealing with an unknown NPS or a compound for which no certified standard is available, modeling software is available to adequately predict IR spectra on the basis of proposed chemical structures. In computational chemistry, **Density Functional Theory (DFT)** is applied to estimate the quantum states of multielectron systems in molecules by considering external potentials and the energy involved in electron—electron interactions. Being able to postulate a chemical structure of an unknown compound by comparing the measured IR spectrum to quantum mechanical calculations is of great value given the rapid growth of NPS compounds in the illicit drugs market.

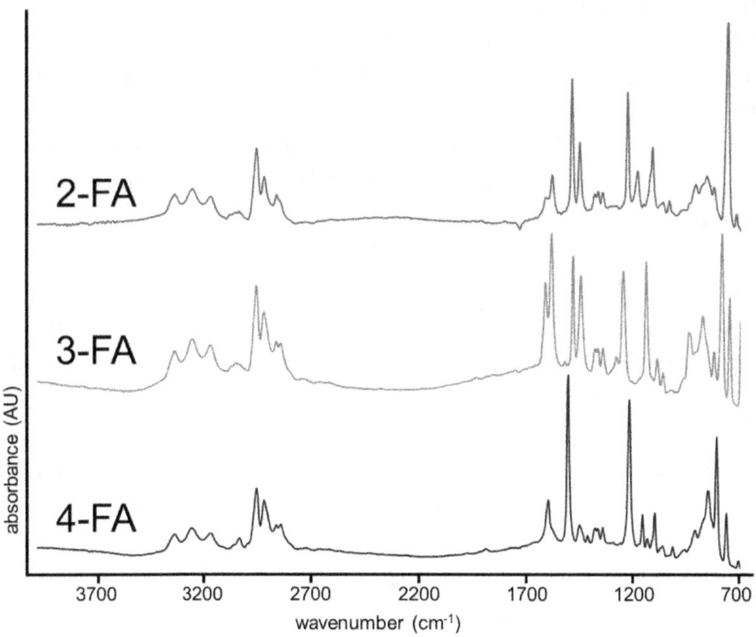

GC-IR spectra of 2-FA, 3-FA, and 4-FA. The transmission IR spectra were acquired on a cryogenic disc trapping system *(Courtesy of Ruben Kranenburg, Amsterdam Police Laboratory)*

Although most forensic drug analysis laboratories have chosen GC-IR to expand their chemical identification capabilities, some interesting other options for NPS isomer differentiation have recently been proposed in the forensic scientific literature. One very appealing approach involves the chemometric data analysis of EI-MS spectra. Rather than working with spectral libraries, hit lists, match scores, and visual inspection by experts, the robustness of spectra can be exploited to link small differences to isomeric forms. Of course, using information that is already available through the GC-MS analysis conducted for every case work sample is very attractive for the smaller forensic laboratories that have limited budget and expertise to expand the analytical instrumentation. However, revealing this "hidden information" in mass spectra requires extensive knowledge of data analysis and chemometrics. Developing and explaining the complex computational methodologies involved presents additional challenges as will discussed in **Chapter 8—From Data to Forensic Information and Insight using Chemometrics**.

Another instrumental option for drug isomer differentiation is provided by the novel **VUV (Vacuum Ultra-Violet**) detector for Gas Chromatography. The term vacuum in VUV relates to the extended wavelength range for which the UV spectrum of compounds of interest can be measured. In contrast to common UV detectors for liquid chromatography, which are limited by the presence of solvents, the VUV detector for GC can measure UV spectra from 125 to 420 nm. Extending the lower wavelength limit ensures that all organic compounds, also aliphatic species, can be detected as at these high energies also electrons involved in σ bonds can be excited to higher energy states. Although VUV spectra typically contain less information with respect to the corresponding molecular structure, the VUV detector for GC provides several advantages compared to GC-IR. These advantages include increased sensitivity comparable to regular MS and FID detectors, good linearity of the detector response allowing accurate quantitative analysis, improved instrumental robustness as no moving parts or complex interfaces are required and therefore also very accurate and reproducible spectral data. As VUV spectra can be measured with minimal random and systematic variation, small differences in the spectra of drug isomers can be used to identify the correct isomeric form. Although for some drug classes, such as for instance cathinones, the spectral isomer variation is obvious and can visually easily be discerned, the VUV spectra for the fluoro amphetamines discussed in this chapter are actually very similar. However, a library search always results in the correct assignment because the small differences that exist in the spectra of 2-, 3-, and 4-FA are still very significant in relation to the measurement variation observed. The combination of GC-VUV and GC-MS therefore also offers sufficient selectivity for NPS identification although VUV offers limited options for identifying unknown isomers on the basis of computed spectra.

Although the forensic experts have found satisfactory laboratory solutions to chemically identify listed substances including the growing numbers of NPS with the required confidence as needed for the courts, this does not stop the vicious circle and the growing health risks associated with the use of NPS for which the psychoactive and toxicological effects on the human body are relatively unknown. The reason for this is that the growth in NPS in recent times is related to the legal framework and is not affected so much by the chemical identification process in the forensic laboratory.

How could the NPS challenge be addressed legally, that is, what legal solutions could exist to make the production and international trade of designer drugs less attractive?

The legal root cause is linked to the use of lists of banned substances. So, the legal solution to the problem is to introduce another system that is no longer based on specific compounds and exact chemical structures. Several countries have therefore adapted their illicit drug legislation framework in recent times. Typically, two approaches can be distinguished:

1. Define the effect and the intended use irrespective of the chemical structure

This approach has been adopted by the UK in the New Psychoactive Substances Act from 2016. From the Home Office Document "Review of the Psychoactive Substances Act 2016" published in 2018 (see Further Reading section):

> *"The Psychoactive Substances Act 20161 (PSA) came into force on 26 May 2016 and created a blanket ban on the production, distribution, sale and supply of psychoactive substances in the United Kingdom for human consumption.*
>
> *..., the term 'new psychoactive substances' or 'NPS' is used to describe substances which fall within scope of the Psychoactive Substances Act (any substance which is capable of producing a psychoactive effect in a person who consumes it, and is not an exempted substance) ..."*

However, this course of legal action is not without its own challenges. The complexity now shifts to the exact definition of psychoactive effects and the fact that regular and legal use of such compounds (e.g., as medicine or as food stuff) must be exempted. For that reason, some other countries like Belgium and France have introduced another approach:

2. Define a generic molecular framework for different classes of NPS to immediately ban a whole range of potential isomers when a new compound is discovered

New compounds that now fall within the generic framework but have not emerged yet in the illicit drug market are banned before they are even produced. This could stop the vicious NPS cycle although the definition of such a framework is still tricky. If the framework is too specific, illicit drug manufacturers could possibly still find/design new compounds that fall outside the generic definition. However, if the framework is too general, a lot of commonly used compounds suddenly require exemption.

In the Netherlands at the time this chapter was written (June 2020), a new legislation is in preparation which merges the list and class definition approach by keeping the traditional List 1 and List 2 substances but extending this with a list (List 1a) based on generic molecular descriptors. An example of such a generic structure definition, as defined in the draft law provided for consultation is given below.

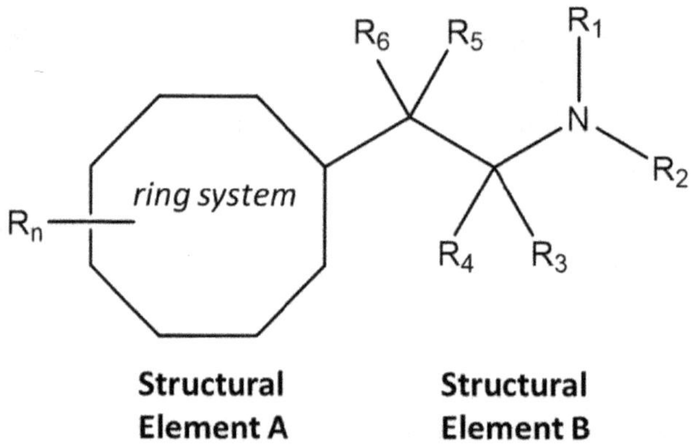

Generic framework for phenylethyl amine-based NPS in the Dutch draft Illicit drug law as proposed for consultation (extensive definition of the ring systems and R groups not shown)

Irrespective of the legal framework chosen and its impact on the NPS phenomenon, forensic laboratories involved in illicit drug analysis most likely will need to perform compound-specific chemical identification to allow correct application of the law. In other words, the courts will need to be provided with detailed chemical information by the experts in addition to his/her interpretation whether the identified compounds fall under the legal framework to arrive at the correct legal decision. Therefore, using additional techniques besides colorimetric testing and GC-MS analysis has become the new norm and not the exception. Even the smaller illicit drug case work laboratories will have to adjust to this new reality and will have to invest in new equipment and expert knowledge.

Further reading

Review of the Psychoactive Substances Act 2016, 2018. UK Home Office, 978-1-5286-0863-3. https://www.gov.uk/government/publications/review-of-the-psychoactive-substances-act-2016.

Scientific Working Group for the Analysis of Seized Drugs (SWGDRUG) Recommendations, SWGDRUG, Version 8.0, 2019-June-13. http://swgdrug.org/approved.htm.

UNODC Global SMART Update Vol. 19, Understanding the synthetic drug market: the NPS factor, March 2018. https://www.unodc.org/documents/scientific/Global_Smart_Update_2018_Vol.19.pdf.

World Drug Report 2019 (United Nations Publication, Sales No. E.19.XI.8), 2019. UNODC, 978-92-1-148314-7, eISBN: 978-92-1-004174-4. https://wdr.unodc.org/wdr2019/.

CHAPTER 5

Quantitative analysis and the legal limit dilemma

Contents

5.1 What will you learn?

After the detailed discussion of qualitative analysis in the previous chapter, we will now focus on the aspects concerning quantitative analysis in forensic chemistry. Quantitation of traces of drugs, medicines, and metabolites in human biological samples is essential in the field of forensic toxicology, and therefore, this chapter focuses on forensic toxicological chemical analysis to illustrate this concept. Because of the complex biomatrices of the samples studied in forensic toxicology, the preferred analytical method is liquid chromatography coupled to mass spectrometry (LC-MS). The principles of bioanalysis with LC-MS will be discussed including the use of robust sample preparation and calibration procedures. Principles and features of various mass spectrometer instruments will be discussed. No matter how robust the method is, variation in the measured levels will always be observed. How the forensic toxicologist deals with this

Chemical Analysis for Forensic Evidence
ISBN 978-0-12-820715-4
https://doi.org/10.1016/B978-0-12-820715-4.00009-2

inherent measurement uncertainty in relation to absolute limits as stated in the law (i.e., the legal limit dilemma) will be explained in the final section.

After studying this chapter, readers are able to

- list the main forensic expert areas and types of evidence material for which **quantitative analysis** (**compound amount and mixture composition**) is of importance
- describe the methodology applied by **forensic toxicology** experts to chemically quantify **listed substances** and associated **metabolites** in a large **case load** setting
- understand why quantitative analysis is essential for a correct interpretation of the findings by **forensic toxicologists**
- understand how **measurement uncertainty** in quantitative analysis in forensic chemistry can be correctly characterized and reported using **statistics**
- explain the nature of the **legal limit dilemma** and how measurement uncertainty can effectively be addressed in **court rulings**

5.2 Quantitative analysis in forensic chemistry

As discussed in **Chapter 4**, the analytical methodology used for chemical identification forms the basis for a quantitative analysis if required from a forensic and legal point of view. Quantitative analysis corresponds to the determination of the level (or subsequently the overall amount) of identified compounds in the evidence material. In liquids, this is expressed as the concentration of substance X in solution in mol/L (molar concentration) or g/L (mass-based concentration). If the original physical evidence is a solid, the level is typically given as the mole or mass-based percentage (wt%). Often direct determination of the level of a substance in a solid is not feasible for the analytical methodology applied. In that case, a fixed amount of the sample must be carefully weighed and dissolved in a known amount of a suitable solvent. After the concentration is determined in this sample solution, the actual level in the original material can be calculated on the basis of the sample weight. This straightforward approach is required for all methods that cannot directly analyze solids including separation methods like GC-MS and LC-MS (gas chromatography and liquid chromatography with mass spectrometric detection, respectively) that can be found in any forensic laboratory.

Can you mention analytical techniques that enable direct quantitative analysis of compounds in solid samples without the need to dissolve the sample prior to analysis?

In paragraph 3.2 in the chapter on sampling, it was shown that chemical analysis is sometimes feasible without sample preparation (and dissolving can also be regarded as a sample preparation step). When considering solids, this means that the technique considered must be capable of interrogating the sample directly either invasively or non-invasively. Typically spectroscopic techniques or mass spectrometric techniques involving lasers or particle beams can offer this functionality. Examples include:

- Raman, NIR, (ATR-)IR and UV-vis reflectance spectroscopy
- XRF (X-ray fluorescence spectroscopy)
- Neutron Activation Analysis (NAA)
- LA-ICP-MS (laser ablation-inductively coupled plasma-mass spectrometry)
- Ambient ionization mass spectrometry (DESI, DART, LAESI)
- Secondary Ion Mass Spectrometry (SIMS)

A challenge when considering quantitative analysis is the need for a suitable calibration scheme that effectively accounts for any matrix effects that depends on the other constituents present and the overall appearance of the sample. This requires either a measurement principle that is not affected by the sample composition or the use of special calibration standards that resemble the general composition of the evidence material. An example of the latter is the use of certified glass standards doped with known amounts of trace elements for elemental profiling of glass fragments with LA-ICP-MS. However, this requires knowledge on the nature of the physical evidence.

Another challenge is inhomogeneity of the material, especially for techniques probing only a relatively small sample surface or volume. This can lead to systematic errors (determined levels that significantly deviate from the actual amount) which can be mitigated by sampling and analyzing various sample locations and calculate average levels. However, this increases the overall analysis time and often results in a higher random error in the reported levels. It should be noted that these challenges do not exist or are less relevant for liquid samples.

In **Chapter 1**, the legal basis for the need for a quantitative chemical analysis was explained through the example of DUI (Driving under the Influence) legislation and accurately establishing alcohol levels in exhaled air and human whole blood. Other examples of forensic expertise areas or evidence materials for which there is a "need for quantitation" are:

- Forensic Toxicology—Drugs, medicines, and metabolites in human whole blood, urine, vitreous humor, saliva, sweat, hair, bone
- Illicit Drug Analysis—Quantity of drugs produced/smuggled
- Explosives—Amount of energetic material in an improvised explosive device (IED)
- Environmental Forensics—Nature and severity of spills and industry incidents
- Chemical Profiling (as discussed in more detail in **Chapter 6—Chemical Profiling, Databases, and Evidential Value**)

In Forensic Toxicology, quantitative analysis is performed not only for alcohol but is typically incorporated for all relevant compounds. Why and how this is accomplished in a high throughput fashion for a large set of compounds will be discussed in detail in the next paragraphs. The principles of quantitative chemical analysis will thus be illustrated for this very relevant forensic expertise area at the interface of chemistry and medicine.

However, quantitation can also play an important role in illicit drug investigations. At first hand, this seems to be in disagreement with prior messages confirming the legal basis for a "qualitative-only" approach in identifying listed substances during a high volume illicit drug screening. Indeed, the criminal law in the Netherlands only defines criminal actions and associated punishments in relation to the nature of the material (*i.e.*, hard and soft drugs as List 1 and List 2 substances, respectively).

Why are quantitative aspects sometimes still important in illicit drug cases?

Criminal law typically states the maximum sentence in relation to a crime. In the Netherlands, this is six years for a hard-drug related offense (List 1 compound) and two years for a soft-drug related crime (List 2 compound). If the allegations are proven beyond a reasonable doubt, the court will apply a sentence within this maximum range depending on the circumstances including the prior criminal records of the suspects and the severity of the crime. In a given case, the severity is determined by the magnitude and impact of the illegal actions, the amount of money involved and the overall health and environmental risks to society. For illicit drugs, this is often related to the amount of drugs produced/smuggled and the potency of the product (type and level of psychoactive substances involved). To provide the courts with relevant information to support their ruling, forensic illicit drug experts also often perform quantitative analyses. It is interesting to note that in this case, the forensic information is not used as evidence that a crime has been committed, but rather to assess its impact and therefore the appropriate punishment for the criminal acts that have already been proven by the qualitative analysis.

Other forensic questions where experts are requested to provide "numbers" in addition to chemical identity often occur in criminal investigations concerning explosives and environmental crimes. When an explosive material has been identified, forensic experts are often asked in pre-explosion investigations whether a bomb was functional and if so what the impact of the explosion would have been. This is especially relevant when home-made explosives and IEDs have been used. The functionality of a device depends on the construction, especially the ignition part, but also on the chemical composition of the main charge. The impact typically depends on the nature and weight of the main charge, that is, the amount of energetic material present in the IED. This requires weighing but also quantitative analysis into the composition of the materials used. Chemical analysis in the field of **Explosives (pre-explosion)** and **Explosions (post-explosion)** will be discussed in more detail in **Chapter 6—Chemical Profiling, Databases, and Evidential Value** and **Chapter 7—Chemical Reconstruction through Chemical Analysis**.

Investigations into potential environmental crimes are usually very complex, both in terms of the legal aspects as the forensic investigation. Such investigations usually require the knowledge and expertise of institutes and laboratories specialized in environmental analysis. In addition to the legal proceedings (could and should somebody be held accountable for the damage to the environment) also the safety, health, and environmental aspects need to be considered. Damage to the environment can result from industrial accidents which invokes liability questions. The liability issue is especially relevant when employees get injured or die during work-related incidents involving chemicals. In addition, damage to the environment can also be related to actions of criminals such as the illegal dumping and mixing of waste for economic gain, or waste generated as a result of the production of illicit drugs. Environmental samples (*e.g.* soil, surface water, chemical waste) are typically complex and of unknown composition and require tailor-made analytical methods for detailed analysis. To assess the environmental impact, the forensic experts need to unravel the compounds and amounts involved. Highly toxic compounds cause damage to the environment at low concentrations but high levels and amounts of less harmful substances can also severely disrupt ecosystems. Therefore, chemical analysis in **forensic environmental investigations** frequently also includes quantitation of the contaminants involved.

Finally, quantitative analysis is also of importance in chemical profiling studies as discussed in more detail in **Chapter 6—Chemical Profiling, Databases, and Evidential Value**. When considering chemical profiling, we go "beyond chemical identification and quantitative analysis" to study the potential relationship between materials of relevance in a criminal investigation. To investigate whether two batches of an illicit drug could have the same origin (e.g., material found on a deceased overdose victim and material retrieved at the home of a suspect), the experts focus on low level impurities in a material. Chemical profiling methods are used to generate impurity profiles that allow differentiation of materials belonging to the same class. Ideally such methods yield characteristic (fingerprint-like) but also very robust profiles. Robust in this respect indicates a good reproducibility and repeatability obtained with a straight-forward and efficient analytical method. Although a chemical impurity profile can be generated from arbitrary numbers, this robustness is usually increased when the impurities are quantified, that is, when the profile relates to the actual levels of the impurities in the evidence material. A suitable calibration scheme for instance, can effectively correct for detector response fluctuations.

For the remainder of this chapter, we will exclusively focus on chemical analysis in **Forensic Toxicology**. This involves the accurate targeted quantitative trace analysis of a substantial number of low molecular mass bioactive compounds of forensic interest in complex biological samples such as human whole blood. Such samples present extremely challenging matrices containing high molecular mass constituents such as proteins and enzymes. Additionally, for highly toxic or psychoactive compounds, trace levels in the ppb (part-per-billion) range are forensically relevant and need to be accurately determined. The technique capable of meeting these challenges and thus the work-horse methodology in this field is **LC-MS**. Therefore, the principles of LC-MS will be discussed in this chapter. Finally, we will shift our attention to what is defined as the "**Legal Limit Challenge**" and learn how forensic toxicologists address and report measurement uncertainty in relation to absolute limits in **DUI** legislation.

5.3 Forensic toxicology: trace level quantitation of small molecules in complex biomatrices

The type of compounds identified in forensic illicit drug analysis is also of interest in forensic toxicology. But while the forensic drug expert is mainly interested in raw material analysis and drug production processes, the forensic toxicologist focuses on the biological activity of these compounds in the human body after consumption. In addition, the scope of compounds is broader in forensic toxicology also including medicines and toxins. Criminal investigations involving regular medication can be related to substance abuse, such as DUI (*e.g.*, lethal accidents through antidepressant impaired driving), illegal prescriptions, and the sedation (*e.g.*, in date rape cases) and murder of victims (*e.g.*, through injection of insulin). Toxins are typically related to poisoning cases. Although much less frequent than in the past, murder attempts by adding poisonous compounds to food and drinks is an occurrence of all times and many countries have infamous cases often in a family setting. Such toxins can be man-made (*e.g.*, the assassination of former Russian FSB agent Alexander Litvinenko in 2006 with radio-active Polonium-210 or the use of VX nerve agent in 2017 to murder Kim Jong-nam, the brother of North Korean leader Kim Jong-un, at the airport of Kuala Lumpur in Malaysia) but can also be retrieved from natural sources (*e.g.*, trees of the Taxus family, very popular garden plants, are actually quite toxic to mammals including humans).

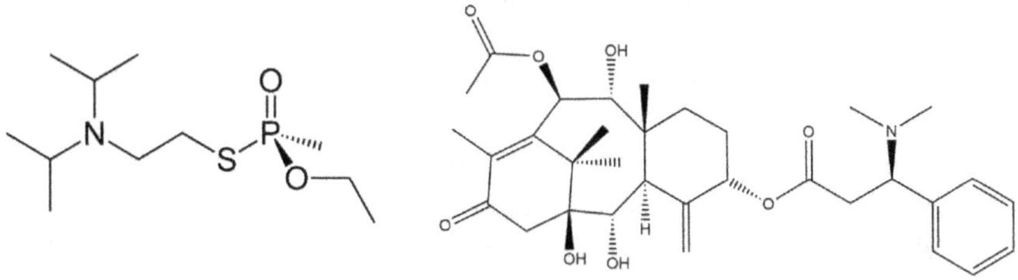

Chemical structures of the nerve agent VX on the left and of Taxin B on the right. Man-made VX is an extremely poisonous compound of the thiophosphonate family causing massive nerve failure leading to paralysis and death by asphyxiation. Taxin B has cardiotoxic properties and can cause cardiac arrest and respiratory failure if consumed in sufficient amounts *(Wikipedia)*

The field of forensic toxicology operates at the interface of forensic analytical chemistry and forensic medicine. State-of-the-art analytical chemistry and expertise in bioanalysis is needed to accurately determine the presence of drug traces in biological samples. In addition, interpretation of the analytical data requires knowledge on the bioactivity of these compounds in the human body and the associated metabolic pathways, that is, the way the human body (mostly the liver) chemically modifies exogenous compounds to mitigate harmful effects and facilitate removal. These metabolic processes further expand the scope of compounds to be considered. In addition to drugs, medicines and toxins, their metabolites also provide very relevant information to accurately reconstruct substance abuse. Some compounds are metabolized very quickly and can hardly be detected in the human body and in that case a more persistent metabolite can serve as an indicator that this compound was consumed. For compounds with a longer half-life (*i.e.* the time during which 50% of drug present in the body has been converted/removed), the ratio between the main metabolites can provide information on the time since consumption.

> In Forensic Toxicology, almost all chemical analyses are of a quantitative nature, just establishing the presence of a trace of a certain drug, medicine, chemical, or metabolite is not very useful. Why is the quantitative aspect so important?

The need for quantitative analysis in forensic toxicology becomes apparent when considering the two most important classes of casework responsible for a majority of the workflow:

- Forensic toxicological investigations as part of forensic medical cases (concerning surviving victims or suspects under influence) and forensic autopsies (concerning deceased victims)
- Forensic toxicological investigations in DUI cases

The effect of an exogenous compound is determined by its intrinsic potency and the amount consumed and subsequent levels in the human body. This essential role of concentration on substance impact was already indicated in the early Renaissance by famous Swiss physician and philosopher **Paracelsus** (Theophrastus von Hohenheim, 1493–1541) who stated: "*All things are poison, and nothing is without poison; the dosage alone makes it so a thing is not a poison*".

Copy by an anonymous 17th century painter held in the Louvre museum of a painting of Paracelsus by the Flamish painter Quentin Matsys. The name Paracelsus in Latin means "equal to Celcus" and refers to Aulus Cornelius Celsus (~25 BC-~50 AD), a Roman encyclopedist who documented views on human health, diet, medicine, and surgery in the Roman empire *(Wikipedia)*

This phrase represents a basic principle in forensic toxicology and reflects that a potentially toxic compound can be harmless when consumed at sufficiently low levels whereas substances that are generally considered to be beneficial (*e.g.*, water) can become harmful, toxic, or even lethal when consumed in excessive amounts. When studying the toxicity of a compound using test animals, lethal levels are usually expressed in so-called **LD50** values (lethal dose 50%): the amount required to kill half of the population in a given exposure test. Toxic compounds have low LD50 values, exposure to relatively small levels can already have deadly consequences. An example of such a very toxic compound is **carfentanyl** introduced in **Chapter 4**. This compound of the fentanyl class is used by park rangers and veterinarians to tranquilize large wildlife animals like elephants to enable medical checks and treatments or animal relocation. However, this compound is now also regularly found by forensic drug experts in street heroine in the USA. The addition of fentanyl painkillers to heroine is causing a staggering increase in the number of drug overdose cases as reported by the National Institute on Drug Abuse. NIDA reported an average daily death toll of 128 in 2018 due to the opioid overdose crisis in America. Heroine drug addicts, often unaware of the addition of fentanyl related compounds, die predominantly due to acute respiratory failure. Although the average lethal dose for humans is unknown, the WHO (World Health Organization) states that carfentanyl is approximately 10.000 times more potent than morphine. A mg dose of fentanyl, approximately 100 times more potent than morphine, can already have lethal consequences (source: Wikipedia). This extreme action on the human body makes fentanyl compounds also a potential chemical agent and there are indications that carfentanyl was used by the Russian security forces to sedate Chechen terrorists that were keeping hostages in a Moscow theater in 2002. The release of this strong opioid as a fine mist/aerosol in the theater prior to the rescue attempt had dramatic consequences as 125 people died of asphyxiation (source: Wikipedia). As fentanyl compounds can also be absorbed through the skin or inhaled, their potential presence in illicit drug evidence materials also requires stringent safety precautions for police officers, forensic experts, and other officials that confiscate, transport or investigate illicit drug samples. In case of an accidental exposure to a fentanyl analogue, Naloxone is a known antidote that can be administered to treat the associated breathing problems.

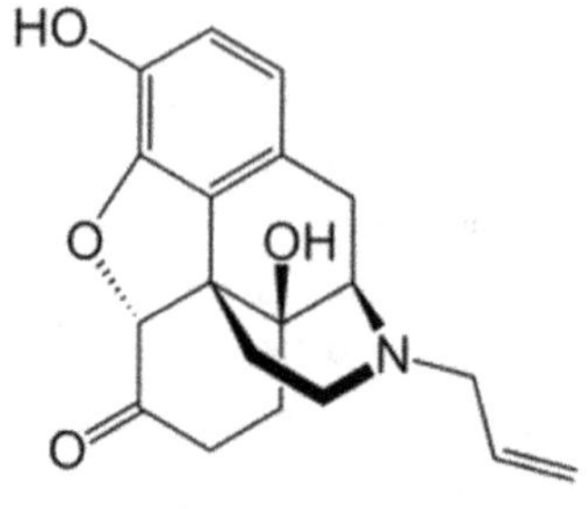

Morphine-like molecular structure of Naloxone, an opioid receptor antagonist to treat fentanyl related overdose symptoms *(Wikipedia)*

In case of a forensic autopsy, the aim of the pathologist is to determine the cause of death and to retrieve medical information that can aid in reconstructing the fatal circumstances (*i.e.*, the manner of death). If the cause of death could be related to a drug overdose or the exposure to a toxin or chemical agent, the pathologist needs the expertise of the forensic toxicologist. Based on his/her pharmacological knowledge and past case experience, the toxicologist will interpret the analytical findings reported by the forensic toxicology laboratory after analysis of the heart blood, femoral blood and vitreous humor (clear, gel-like liquid in the eye ball) samples secured during the autopsy. The prior discussion with respect to toxicity hopefully makes it perfectly clear that for a meaningful interpretation, quantitative data is essential. Only on the basis of measured concentrations of drugs, medicines or toxins and associated metabolites will the toxicologist be able to assess the role of these exogenous compounds in the death of a victim. Even with accurate data for various autopsy samples, this task remains extremely challenging as is illustrated in the graph below.

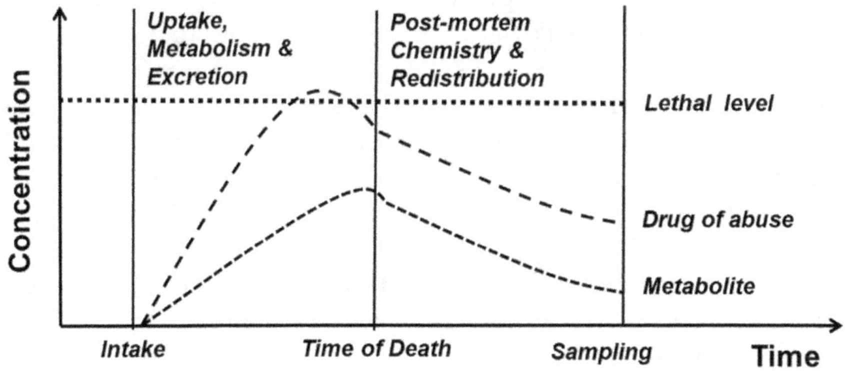

Schematic representation of the change in drug and metabolite concentration over time in a biological matrix of a drug overdose victim

The data available only reflect the levels at the time of the autopsy, but actual levels at the time of death could have been much higher especially when a substantial time has passed between the time of death and the sampling. Although autopsies in the Netherlands are typically conducted within a week after death, in some cases the actual discovery can take considerable more time depending on the circumstances (e.g., when finding human remains of a victim in a clandestine grave). For many drugs and medicines, the human metabolism is reasonably well understood on the basis of clinical pharmacological data generated in medical studies.

However, the forensic toxicologist also needs scientific insights with respect to postmortem chemistry and redistribution phenomena. Typically, forensic institutes are not allowed to use human remains for general scientific purposes. The remains must be returned as quickly and as intact as possible to the family members after the autopsy. Postmortem pharmacological studies are typically also not conducted in medical hospitals leaving the toxicologists with findings and circumstances from prior cases. Because postmortem conditions can vary greatly and the biochemistry involved is governed by microorganisms, the pharmacological events after death are much more prone to variation compared to the relatively controlled conditions of the metabolic processes in living humans (although toxicologists must also carefully consider human variation such as age, sex, health, ancestry, body mass, and drug tolerance). As a result, forensic toxicological conclusions are typically much less conclusive/definite than the reported levels of the associated compounds.

Already in the very first chapter of this book, the need for quantitative ethanol analysis for drunk driving cases was explained due to the legal limits as defined in the law. However, DUI cases nowadays often also involve the use of medications or drugs of abuse, either as such or in combination with (excessive) alcohol consumption. Criminal charges are also brought forward against perpetrators who have caused traffic accidents because their driving skills and response times were negatively affected through the use of drugs or medicines. It is the responsibility of every driver to make sure that he/she is fully capable (legally, technically, and mentally) to drive a motorized vehicle and to participate in traffic in a controlled and safe manner. This is essential to ensure the safety of others and for instance prescription medicines such as antidepressants clearly indicate that driving is not allowed while under medication. However, without predefined legal limits, the legal complexity of such cases increases. The courts now have to look into the extent to which somebodies driving abilities were impaired on the basis of toxicology findings, police reports, witness statements and possibly video material illustrating the driving behavior at the time of the accident. Additionally, without legal limits there is no means for preventive, corrective actions as part of routine traffic controls. To provide a clearer legal framework for drugs related DUI cases, an extension of the Dutch Traffic Act was introduced in the Netherlands in 2017. In this extension, legal limits were set for seven common drugs of abuse in whole blood in addition to alcohol:

Illicit drug	Compound analyzed	Legal limit Single use [μg/L]	Legal limit Combined use [μg/]
Cannabis	THC	3	1
Heroine	Morphine	20	10
Cocaine	Cocaine	50	10
Amphetamine	Amphetamine	50	25
Methamphetamine	Methamphetamine	50	25
MDMA	MDMA	50	25
GHB	GHB	10.000	5.000
Alcohol	Ethanol	500.000	200.000

Legal limits in whole blood for frequently occurring drugs of abuse in DUI legislation in the Netherlands (when multiple amphetamines are found the summed level should not exceed 50 μg/L)

Why is heroine measured as morphine in human whole blood?

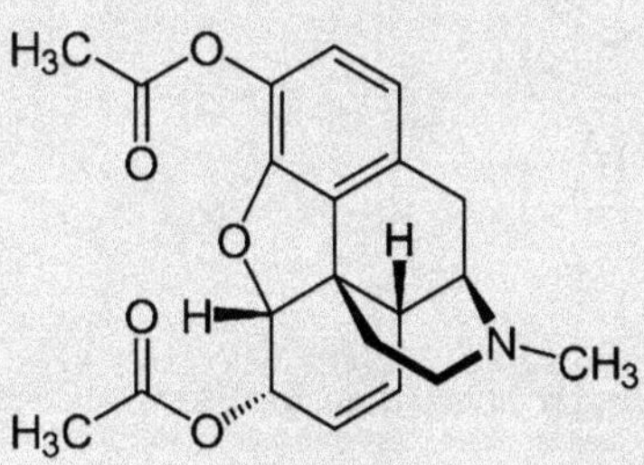

Heroine

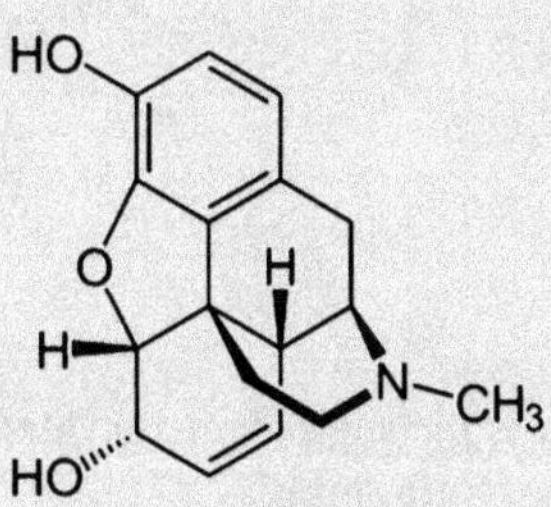

Morphine

Heroine, also known as diacetyl morphine or diamorphine, is produced from the natural opioid morphine by acetylation usually with acetic anhydride. Heroine is less polar than morphine and can therefore be injected directly into the blood stream and rapidly passes the blood–brain barrier. This makes heroine more potent than morphine but its effects wear out more rapidly. However, in the body heroin is very rapidly hydrolyzed into morphine. With a very short half-life, it is difficult to detect heroin as such in the body over time. This makes it also challenging to distinguish between the use of morphine or heroine as heroine is quickly converted into morphine. An intermediate metabolite with a longer half-life that can reveal the use of heroine is 6-monoacetylmorphine (6-MAM).

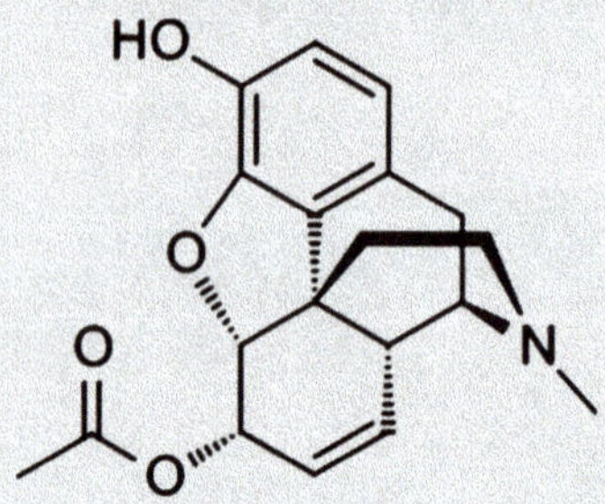

6-MAM

Why is the legal limit for THC so much lower than for the other compounds?

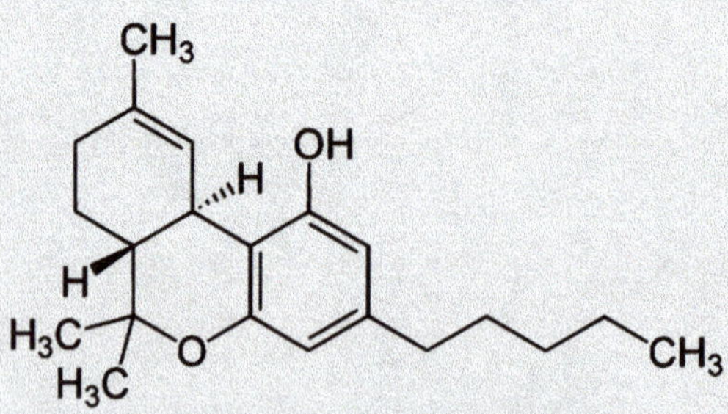

Tetrahydrocannabinol

The limits are linked to driver impairment. Experts have established that low levels of THC already significantly increase reaction times and negatively affect driving ability. Cannabis is a very frequently used recreational List 2 drug in the Netherlands and the active compound THC is quite persistent in the human body, especially when ingested orally. This can quickly result in a situation where people think they are capable of driving but actually risk a DUI offense in light of the new legal limits for drugs of abuse.

The introduction of the new legal limits in the extended Dutch Traffic Act provided a challenge for forensic toxicological laboratories. Typically, dedicated methods are used for the measurement of ethanol in human whole blood. In addition to the alcohol breathalyzer, special drug saliva tests based on antibody technology were now introduced for road-side testing by law enforcement. In line with the procedure for alcohol testing, after a positive saliva test the police will take the suspect to the police station to obtain a blood sample for an accurate laboratory analysis to establish whether levels exceed the DUI limits for drugs of abuse. This results in an increased volume of blood samples that in addition to the blood alcohol methodology now also needs to be quantitatively analyzed for trace amounts of several drugs of abuse.

Where GC-MS is the "workhorse" technique for the analysis of illicit drug and fire debris samples, LC-MS has obtained that same status in the clinical and forensic toxicology laboratory. Although many of the compounds of forensic toxicological interest can be analyzed with GC-MS in their native state (*e.g.*, drugs of abuse in an overdose or DUI case), trace analysis at ppb level in a complex human biological matrix represents numerous challenges. Extensive sample preparation is required often in combination with chemical derivatization of the more polar metabolites. Sample preparation is required to remove the high molecular weight biological compounds like proteins and enzymes that would otherwise contaminate the GC injector and column due to lack of volatility. Such compounds do not evaporate and are not transported by the gaseous mobile phase but rather precipitate and thermally degrade. As part of the sample work-up, preconcentration of the analytes of interest is also required because GC-MS typically offers (sub) ppm and not ppb sensitivity. This makes the application of GC-MS for the analysis of forensic toxicological samples laborious, costly and time-consuming. Therefore, LC-MS, first introduced in the 1970s, quickly became the preferred methodology once robust equipment became available in the 1990s. It took over 2 decades to overcome the technical challenge of interfacing a separation method based on a liquid mobile phase with a mass spectrometer requiring high vacuum condition.

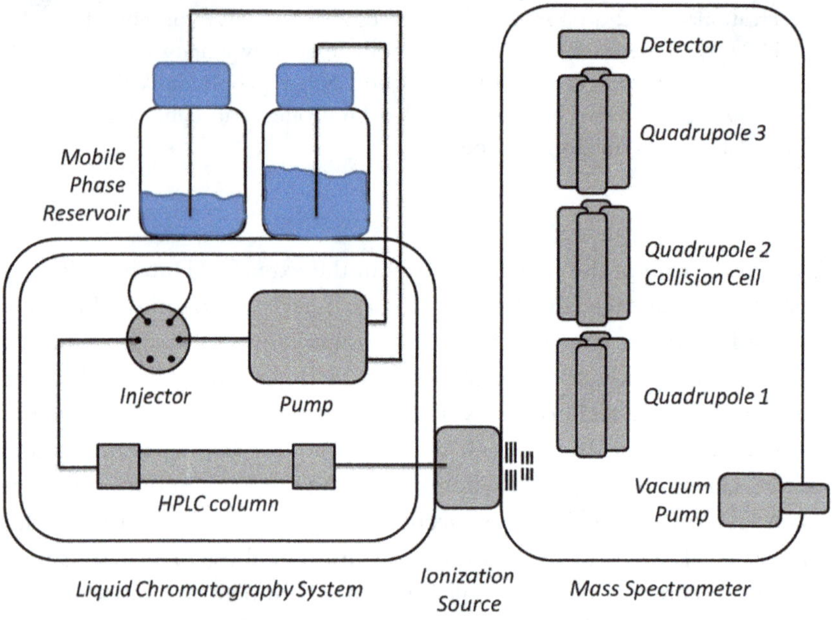

Schematic representation of an LC-MS instrument (with an illustration of a triple quad mass spectrometer, the most frequently used MS instrument)

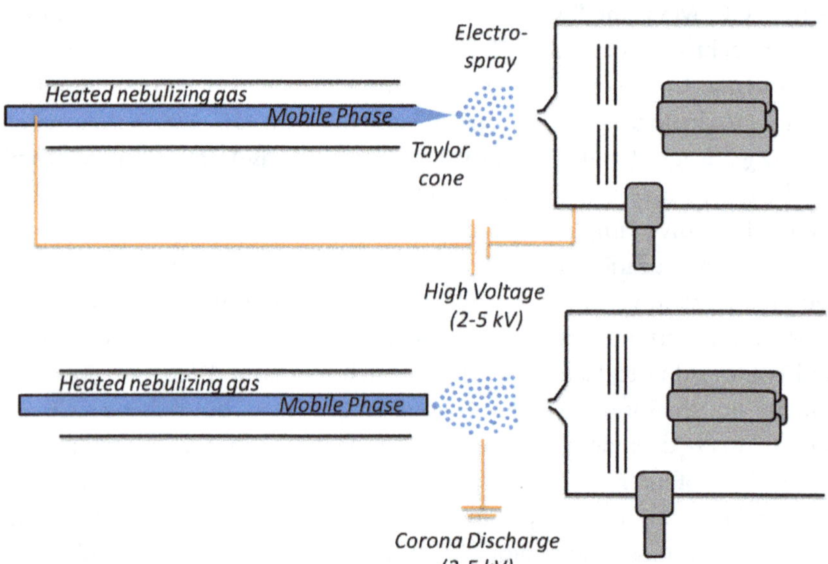

Illustration of Electrospray Ionization (top, ESI) and Atmospheric Pressure Chemical Ionization (bottom, APCI), both in positive ionization mode

The most frequently used interface involves the generation of a so-called electrospray by applying a substantial voltage (usually in the range of several kV) between the mobile phase exiting the LC-column and the inlet of the mass spectrometer. Due to the electric field, the liquid forms a so-called Taylor cone that breaks up in a fine mist of small droplets. Evaporation of these droplets is facilitated by a heated nebulizing gas (typically nitrogen) and as the charged droplets reduce in size, analytes are ionized. ESI is considered to be an **ambient ionization** technique because ionization takes place at ambient pressure and not in the high vacuum of the mass spectrometer. Applying a positive voltage creates positive ions (**positive ion mode**), whereas applying a negative voltage creates negative ions (**negative ion mode**). In a broad screening of an unknown sample, usually both modes are applied sequentially to scan for compounds that either easily form positive or negative ions. In forensic toxicology, most psychoactive compounds typically contain a nitrogen moiety (THC being an exception). Such compounds ionize well in positive mode, especially when an acidic modifier is added to the mobile phase to promote protonation and the formation of positively charged ammonium species. For compounds that do not ionize efficiently with **ESI**, such as nitrate containing organic explosives, another approach is offered by **APCI**. In this mode, ionization is facilitated by a Corona discharge in a mist of LC eluent droplets created by a heated nebulizing gas. Also APCI can be applied in positive and negative mode but is considered to be a "harsher" ambient ionization technique than ESI (often typed as a "soft" ionization technique, indicating limited fragmentation during ionization) and is better suited for analytes that do not ionize easily at atmospheric conditions. To ensure that the system can effectively process the mobile phase solvent, relatively small flow rates must be applied and the mobile phase should only consist of relatively volatile compounds. This limits the use of salts and other non-volatile modifiers in liquid chromatography when coupled to mass spectrometry.

The rapid development of LC-MS has also been accompanied with the availability of a wide range of mass spectrometric techniques with increasing capabilities in terms of sensitivity, speed, mass resolution, and data analysis. To date, the following MS instruments/modes are available for LC-MS analysis (discussed in order of increasing capability but also instrument cost):

Single quad mass spectrometer

The single quad mass spectrometer is comparable to the MS systems used in GC-MS with electron ionization. It typically offers **unit mass resolution** (indicating that it can discriminate ions with a mass difference of at least

1 Da) but without the chemical fingerprinting of compounds because of the limited fragmentation with ambient ionization. This type of instrument is relatively cheap, fast, and offers accurate quantitative analysis but is not used often in forensic laboratories because of the limited mass resolution and selectivity.

Ion trap mass spectrometer

As the name suggests, ions are trapped in this type of mass spectrometer by applying a combination of static and oscillating electric and magnetic fields. In this way, ions can be manipulated (maintained in or ejected from the trap), fragmented, and detected. Like single quad instruments, ion traps also operate at unit mass resolution but offer more selectivity because analyte ions can be fragmented in the trap and identified by analyzing the fragment ions. In MS^n mode, multiple fragmentation stages ($n = 1, 2, 3, \ldots$) can be applied for enhanced selectivity and to facilitate structure elucidation (albeit at reduced sensitivity). In MS^3 mode, the fragments ions are fragmented to characterize the "fragments of the fragments." Ion trap mass spectrometers are therefore considered as highly versatile instruments especially given the relative low instrument cost. Ion trap mass spectrometers are also known to maintain good performance at relatively high pressures in the mass spectrometer. However, they are also known for their limited **mass accuracy** (*i.e.*, the difference between the measured and accurate mass of an analyte of interest) and limited precision in quantitative analysis. This latter is a point of concern when considering forensic toxicology.

Triple quad mass spectrometer

The principle of the triple quad mass spectrometer was already briefly explained in **Chapter 2**. Basically, it consists of an arrangement of three subsequent single quadrupole units. The first unit selects the mass of interest as the ions with the selected m/z value are fragmented at elevated pressure in the second unit which acts as a collision cell. Finally, the resulting fragments are analyzed in the third quadrupole unit in combination with the ion detector. This entire process is performed very rapidly (in relation to the time scale of the liquid chromatography separation) allowing the almost simultaneous analysis of multiple compounds of interest. When a method is applied to monitor numerous target compounds in an LC-triple-Q-MS analysis, this is called **Multiple Reaction Monitoring** or **MRM**. A triple quad mass spectrometer provides unit mass resolution, good selectivity and excellent performance in quantitative analysis at a reasonable price. For this reason, LC-triple-Q-MS is currently the "work-horse" instrument for small molecule analysis in biological matrices in clinical and forensic toxicological laboratories.

Time of flight mass spectrometer

In a time of flight (TOF) mass spectrometer, ions are accelerated in an electric field and their m/z value is established by accurately measuring the time required for ions to cross a certain distance from the point of entry to the dtector. This measurement principle allows for higher mass resolution than the ion trap and quadrupole systems. Accurate measurement requires a relatively long flight path of the ions and hence TOF instruments can often be recognized by substantial tubes protruding out of the main body. However, some setups ingeniously use deflectors to increase the ion distance while minimizing the dimensions. Stand-alone TOF systems do not have MS/MS capability and thus have limited selectivity. The increased resolution offers options to distinguish **isobaric** compounds (compounds with the same nominal mass but different chemical formulas) but without fragmentation **isomeric** compounds (compounds with the same chemical formula but different chemical structure) can never be resolved as they have the same exact mass. In quantitative analysis, TOF mass spectrometers tend to perform somewhat worse than the triple quad instruments. The TOF principle allows for very rapid mass scanning and hence this type of mass spectrometer is often used for chemical imaging, *for example,* in combination with **MALDI** (**Matrix Assisted Laser Desorption Ionization**).

Orbitrap mass spectrometer

The orbitrap mass spectrometer developed by Prof. Makarov has revolutionized and introduced high-resolution mass spectrometry in regular laboratories. Its measurement principle is associated with the **Fourier-transform ion cyclotron resonance mass spectrometer** (**FT-ICR-MS**) using Fourier transformation of an oscillating current as the ions move in an orbital-like pattern around a spindle-shaped inner electrode. However, the orbitrap does not use a magnetic field and consequently it has the advantage of being a bench-top instrument. The mass resolution is not as high as that of an FT-ICR-MS but still sufficient to suggest molecular formulas from the measured exact mass. Because of this feature, the orbitrap instrument is the only bench top mass spectrometer that can be used for structure elucidation of unknown compounds (untargeted analysis). The orbitrap mass spectrometer offers unparallel mass resolution but is a very expensive instrument and is less suited for quantitative analysis. As such it also does not offer MS/MS and therefore it cannot distinguish isomeric compounds despite the very high mass resolution. No matter how high the mass resolution, without fragmentation no MS instrument will be able to distinguish isomers that share the same m/z value.

Hyphenated systems

Some of the most powerful but also costly mass spectrometers for LC-MS analysis combine some of the instrumental features discussed above. Examples of such hyphenates systems are Q-TOF (combination of quadrupole and TOF mass spectrometry) and LTQ-Orbitrap (combination of ion trap and orbitrap mass spectrometry) instruments. These approaches offer flexibility but more importantly create superior selectivity by combining MS/MS fragmentation with high resolution MS. For the qualitative analysis of post-explosion residues, the Netherlands Forensic Institute employs an LC-LTQ-Orbitrap MS method that combines sensitivity and selectivity to identify traces of intact explosives in complex environmental samples obtained from crime scene swabs. It should be noted that in this case accurate quantitation is not necessary.

The table below summarizes some of the key features of the various mass spectrometric instrument types:

MS type	Mass resolution*	Mass accuracy [ppm]	Instrument cost	Selectivity	Quantitative analysis
Ion trap	1000	100	+	+	−
Single quadrupole	1500	5	++	−	+
Triple quadrupole	1500	5	+	+	++
Time of flight	10,000	5	+/−	−	+/−
Orbitrap	100,000	0.5	−	+/−	−
FT-ICR-MS	1,000,000	0.1	− −	+/−	−

++: excellent, +: good, +/−: average, −: mediocre, − −: poor*
*mass resolution is a range indication as this parameter is dependent on the analyte mass and the specific instrument specifications and different definitions exist

For a good understanding of resolution in mass spectrometry, it is important to know the definition of this parameter. Actually, two approaches exist to express the ability of an MS instrument to distinguish two ions with closely resembling m/z values. The **IUPAC (International Union of Pure and Applied Chemistry**) organization defines mass resolution (R) for a peak with mass m and with a **FWHM (full width at half maximum**) of Δm (also indicated as the resolving power) as:

$$R = \frac{m}{\Delta m}$$

This definition only requires one ion signal to calculate the resolution. However, typically resolution indicates how well two closely situated peaks can be distinguished. This depends on the mass difference in relation to the measurement variation or the width of the mass signals (for normally distributed, Gaussian, signals this is indicated by the standard deviation). An alternative definition of resolution in mass spectrometry is therefore considering two peaks. It is based on the same formula but now R is defined as the resolving power and Δm is considered to be the mass resolution and is defined as the mass difference at which the valley of two partially coeluting mass signals (of equal intensity) corresponds to a certain fraction of the peak height. The two definitions are illustrated below.

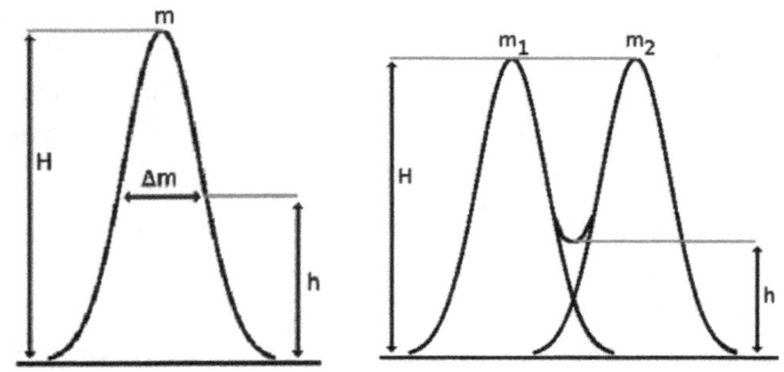

Two definitions of resolution in mass spectrometry, H is the maximum signal or so-called peak height and h is either 0.5H (left picture: IUPAC definition) or 0.1H (right picture: alternative definition) where Δm equals m_2-m_1, the mass difference for which the valley of the two peaks equals 10% of their peak heights. Please note that this definition assumes equal signals for the two ions considered, a condition which is not often met *(Wikimedia Commons, public domain picture)*

The concept of resolution is also very important in chromatography. It is very important to grasp the difference between MS and chromatographic resolution. In LC-MS, the chromatographic resolution indicates the degree to which compound zones are fully separated before they are characterized by the mass spectrometer. Ideally there is complete separation of the compound zones of the various analytes of interest in a sample. Otherwise in case of coelution the mass spectral data is made up of a contribution of both compounds. In contrast, mass spectral resolution indicates whether two ions with similar m/z values can be differentiated. Such ions could well originate from a single compound as different adducts or fragments. A frequently applied definition for resolution in separation methods is given below were t_r is the retention time of two closely eluting peaks 1 and 2 from the LC column (with 2 being the later eluting compound) and w is the peak width at the peak base (in time units) which for Gaussian peaks roughly equates to 4x the standard deviation (σ):

$$R = \frac{t_r(2) - t_r(1)}{0.5^*(w(2) + w(1))}$$

In this definition, a retention time difference Δt_r that equals the peak base width results in an R value of 1 (assuming equal band broadening for both compounds which is a reasonable assumption given the fact that these peaks have similar retention times), corresponding to almost perfectly separated peaks (2% peak overlap). If the standard deviation of the peaks would be taken as a measure of bandwidth, an R value of four would reflect full separation with minimal compound coelution. Of course, for perfectly Gaussian peaks, there is always some degree of overlap irrespective of the retention time difference. Furthermore, co-elution can be handled by a triple quad mass spectrometer as long as the precursor ions have different nominal masses. However, in method development it is advisable to optimize the LC separation and minimize co-elution as this typically makes a method more robust. A trade-off exists between chromatographic resolution and analysis time, an increase in resolution for a given column requires longer run times per sample. Consequently, when dealing high caseloads, experts are looking for sufficient resolution at minimal analysis time.

Currently, hyphenated MS systems are becoming more affordable and are used more frequently in forensic laboratories. However, to date LC-triple quadrupole MS remains the most popular combination for the target analysis of drugs, toxins, medicines, and associated metabolites ("small molecules") in biological matrices. It offers flexibility, multi-compound analysis, selectivity, sensitivity, robustness, and excellent quantitative analysis for a very decent price. In the remainder of this chapter, we will discuss the development of an LC-Triple Q MS method for the accurate targeted trace analysis of the additional compounds as defined in the DUI law in the Netherlands (THC, Morphine, Cocaine, Amphetamine, Methamphetamine, MDMA, GHB). LC-MS method development in bioanalysis consists of three stages, sample preparation, LC separation, and MS analysis:

Sample preparation

Even for liquid chromatography, the direct analysis of complex biofluids such as human whole blood is not feasible and some sample work-up is required. This already starts when the blood sample is taken from the suspect after a positive saliva test, because the natural process of blood clotting needs to be prevented. To this end, anti-coagulants like EDTA (ethylene-di-amine-tetra-acetic acid) or heparin are added. In addition, prior to LC analysis of the listed substances, the solid cellular substances (red blood cells, platelets) and high molecular weight biological matrix (proteins, enzymes, carbohydrates) need to be removed as much as possible. Extensive sample clean-up schemes can be devised to isolate and preconcentrate the low molecular weight compounds of interest. This can be based on solvent-solvent extraction or the use of **SPE (solid phase extraction)**. However, given the high caseload in DUI cases it is important to minimize the amount of time and lab capacity that is needed for sample preparation. Therefore, a very straight forward and popular approach to prepare blood samples is to add an organic solvent like acetonitrile directly to a volume of blood to "crash out" all the biopolymeric material in the sample. The remaining liquid can then directly be processed for LC-MS analysis. The downside of this approach is that the massive precipitation after the addition of the organic solvent also leads to a significant but unknown loss of the analytes of interest. To accurately account for this analyte co-precipitation **isotope-labeled internal standards** have to be added to the sample prior to the sample preparation step. These internal standards are identical to the drugs of abuse to be analyzed but differ in mass because some of the atoms have been exchanged for their stable isotopes (e.g., Deuterium (^{2}H) instead of Hydrogen (^{1}H), ^{13}C instead of ^{12}C or ^{15}N instead of ^{14}N). The fundamentals and use of isotopes in forensic investigations are discussed in more detail in **Chapter 7—Forensic Reconstruction through Chemical Analysis**. In the part on MS analysis below, the specific application of isotope labeled standards (in bioanalysis this often involves **deuterated standards** indicating the exclusive use of Deuterium) in forensic toxicological analysis is explained in more detail.

LC separation

The most common operation of liquid chromatography is the **reversed phase** (**RP**) mode with **solvent programming** (also termed **gradient elution**). In **RP-HPLC** (high performance liquid chromatography) columns are used that are tightly packed with spherical silica particles of 2–5 μm in diameter and with apolar surfaces by chemical modification. The most frequently applied RP columns are the C18 types in which octadecyl carbon chains are grafted on the silanol groups on the surface of the bare silica particles. These hydrocarbon chains facilitate hydrophobic interactions resulting in retention of apolar compounds. The extent of interaction is "tuned" by the composition of the mobile phase, more specifically the ratio of aqueous buffer and organic solvent. In solvent programming the v% of organic solvent is gradually increased until even the most hydrophobic compounds elute from the column and are transported to the detector. This gradient elution can be regarded as the LC equivalent of temperature programming in GC, it provides for an efficient separation of compounds over a vast polarity range in a short run time. The mobile phase is mixed and pumped through the HPLC column by a piston pump capable of generating the high pressures (400–500 bar) needed to create typical column flows of 0.5–1.5 mL/min. In combination with MS, mobile phase flows should be limited to minimize the amount of solvent that needs to be removed by the vacuum pumps of the detector. This can be achieved in a UHPLC system (ultra-high-performance liquid chromatography) with columns that contain particles with reduced diameter and have a reduced inner diameter and column length. In UHPLC reduced mobile phase flow rates in the range of 0.2–0.5 mL/min are applied.

When analyzing drugs of abuse with RP-HPLC-MS, the pH of the aqueous mobile phase is typically set at high values. Why are these basic conditions maintained during the separation?

As many drugs of abuse contain amine groups, a high pH ensures that the compounds are uncharged and show sufficient retention on a RP column with an apolar stationary phase. Charged species show poor retention and therefore poor separation in RP-HPLC. In combination with MS, the choice of buffer systems is quite limited to prevent salt precipitation in the high vacuum of the mass spectrometer. For a high pH aqueous phase, an ammonium bicarbonate buffer can be used.

MS analysis

When the LC separation has been optimized (in terms of resolution and analysis time), the retention (or elution) times of the compounds of interest are known. Through the use of standards, the **MRM** (**Multiple Reaction Monitoring**) method can now be constructed. An MRM method consists of a scheme of **precursor** (also known as parent) and **product** (also known as daughter or fragment) **ions** for the analytes (so-called **transitions**). As explained in **Chapter 2—Analytical Chemistry in the Forensic Laboratory**, the mass spectrometer can rapidly cycle through various combinations of precursor and product ions and as such can deal with closely eluting compounds of interest. However, for optimal sensitivity, it is best to only monitor those transitions that are relevant for a given time window in the LC chromatogram. Therefore, a table of transitions is constructed as function of elution time. For the DUI compounds, typically the mass spectrometer is operated in positive mode which means that a positive voltage is applied to generate the electrospray from the mobile phase exiting the analytical column. In the electrospray and subsequent solvent evaporation, protonated target ions are formed with a mass of 1 Da higher than the nominal mass of the native drug compounds. This protonation is indicated by the symbol $[M+H]^+$. By performing a full scan after fragmentation in the collision cell (the second quadrupole) the product ions with the most abundant signals can be selected for the MRM method. Although abundance is important with respect to sensitivity (low detection limits), product ion selection should also consider selectivity. Some fragmentation processes (*e.g.*, neutron loss of 18 Da indicative of water) are very common and could lead to relatively high background levels or potential false-positive results. In addition, low molecular weight fragments can have a high abundance but can also be nonselective because they correlate to a certain type of functional group that frequently occurs. A product ion with an m/z of 91 Da corresponding to the well-known tropylium ion indicative of a phenyl group in the molecule is for instance considered to be nonselective because of the frequent occurrence of benzene rings in organic compounds (the structure and occurrence of the tropylium ion in mass spectra of aromatic compounds is discussed in more detail in **Chapter 4—Qualitative Analysis and the Selectivity Dilemma**). So, selecting the most effective product ions for each compound of interest is an important aspect of method development. To ensure sufficient selectivity a minimum of two product ions need to be monitored for each analyte. Typically, one of these fragments is used for quantitative analysis. The abundance ratio of the two fragment ions is compared against the reference standard data as an additional check for sample matrix effects, co-elution or false-positive findings by the presence of chemically similar species, such as isomers and isobars. For the compounds for which legal limits have been set in the new DUI law in the Netherlands, the following mass transitions could be used in an LC-triple quad MS MRM method:

Illicit drug	**Compound analyzed**	**Precursor Ion ($[M+H]^+$)**	**Product Ions (quant)**	
Cannabis	THC	315	193	123
Heroine	Morphine	286	201	152
Cocaine	Cocaine	304	105	150
Amphetamine	Amphetamine	136	119	65
Methampheta-mine	Methampheta-mine	150	119	65
MDMA	MDMA	194	105	133
GHB	GHB	105	45	87

Ion selection for the LC-triple Q MS analysis of drugs of abuse in DUI cases ('quant' indicates that this ion is selected for the quantitative analysis)

As indicated in the sample preparation part, **isotope-labeled internal standards** are used to compensate for sample loss, matrix induced **ion suppression** and any other factors that affect mass spectrometer sensitivity. Such standards are chemically (almost) identical to the analytes of interest (this is not entirely true, the variable number of neutrons can to some extent for instance affect the interaction with the stationary phase of the LC column and thus lead to small differences in retention time) but at the same time can easily be differentiated in the mass spectrometer on the basis of their mass difference. A wide range of mostly deuterated standards for medicines, drugs, and frequently occurring metabolites are commercially available for clinical and forensic toxicology laboratories but including such reference materials in the analysis is typically a costly affair. Methods for high-volume casework application should therefore not only be optimized in terms of quality, speed, and throughput, but also in terms of costs. A typical set of labeled standard and related precursor and product ions is given in the table below:

Compound analyzed	Internal standard	Precursor Ion ($[M+H]^+$)	Product Ion (quant)	
THC	THC-D3	318	196	123
Morphine	Morphine-D3	289	153	201
Cocaine	Cocaine-D3	307	185	105
Amphetamine	Amphetamine-D8	144	97	127
Methamphetamine	MA-D8	158	93	124
MDMA	MDMA-D5	199	165	107
GHB	GHB-^{13}C4	109	44	91

Examples of suitable deuterated standards for the LC-MS analysis of drugs of abuse in DUI cases (Dx and 13Cx are indicating the number of deuterium and 13C atoms present in the standard, respectively)

Why do these internal standards all have multiple (D3-D9) deuterium or (C4) ^{13}C atoms?

We need to take natural isotopes in the analytes of interest into account! The general presence of deuterium is very low (<0.02%) but ^{13}C is a stable carbon isotope with a relatively high natural abundance (1.11%). If we would use D1 deuterated standards, the mass of the [M+H]+ of the standard would overlap with the [M+H]+ of the analyte molecules containing one ^{13}C atom. The more C atoms are present in the molecule the higher this single ^{13}C fraction will be.

The molecular formula of cocaine is $C_{17}H_{21}NO_4$. What fraction of cocaine molecules will exclusively exist of ^{12}C and have a nominal mass of 303 Da? With 17 C atoms in the molecule, can you calculate the fraction of molecules having one ^{13}C atom in the structure? And what would that fraction be when considering n ^{13}C atoms in a molecule containing m C atoms?

When considering a cocaine molecule consisting entirely of ^{12}C atoms, there exists only one configuration and the corresponding fraction can directly by calculated from the presence of this stable isotope in our natural environment:

$$\% \left(^{12}C\right) = 100 \cdot \left(1 - abundance\ ^{13}C\right)^{17} = 82,7\%$$

This percentage is lower than one typically expects beforehand because with so many carbon atoms in a molecule, the odds that at least one atom is a stable ^{13}C isotope are quite considerable. With increasing molecular mass and complexity of organic compounds, this ^{12}C fraction is further reduced. When considering a compound with 100 C atoms in its structure, the fraction would have dropped below 33%.

The next question of interest is how the remaining 17.3% of cocaine is composed in terms of molecules containing 1, 2, 3, …, 17 ^{13}C atoms. When considering the fraction with only one ^{13}C atom, we now need to appreciate that this isotope can be present at 17 positions in the molecule:

$$\% \left(^{13}C1\right) = 100 \cdot \left(abundance\ ^{12}C\right)^{1} \cdot \left(abundance\ ^{13}C\right)^{16} \cdot 17 = 15,8\%$$

This is directly related to what is known in mathematics as permutation calculations, the number of random arrangements of elements in a given collection. When considering n ^{13}C atoms in a molecule consisting of m C atoms, the fraction is given in generalized from by the following formula:

$$\% \left(^{13}Cn\right) = 100 \cdot \left(abundance\ ^{12}C\right)^{n} \cdot \left(abundance\ ^{13}C\right)^{(m-n)} \cdot \binom{m}{n}$$

where (see also **Chapter 3—Sampling and Sample Preparation**)

$$\binom{m}{n} = \frac{m!}{n! \cdot (m-n)!}$$

Using this formula for cocaine ($m = 17$) indicates that still 1.42% of the cocaine molecules will contain two ^{13}C isotopes, whereas this fraction drops to 0.08% when considering three ^{13}C isotopes. This is the rationale for the use of labeled standards with at least 3 Da mass difference with the analyte of interest. Only with at least three deuterium or ^{13}C atoms in the molecular structure, the internal standard will not be significantly affected by the naturally occurring isotope configurations.

Another interesting observation is that for some of the isotope labeled internal standards, the fragment ions have the same m/z value as for the associated analytes of interest (e.g., cocaine and morphine). Why does the D3 standard of cocaine yield a fragment with the same mass as the fragment of cocaine? And does this pose any issue for the quantitative analysis of cocaine as both the analyte and standard will co-elute from the LC column and enter the mass spectrometer simultaneously?

It is quite common that fragments of deuterated standards have the same mass to charge ratio as the fragments of the actual compound. This indicates that the charged fragment only contains H and no D atoms. This has no consequences for the quantitative analysis as the precursor ions still differ in mass! As the mass spectrometer rapidly cycles through the MRM transitions, the analyte and the associated internal standard are monitored separately.

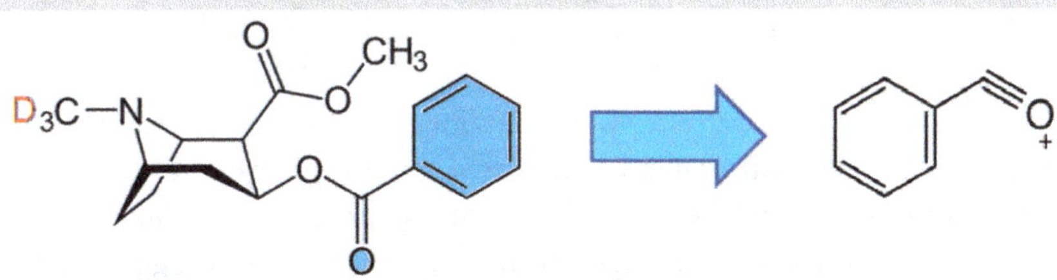

Chemical structure of D3 cocaine and the m/z 105 ion that is created from the fragmentation of both the protonated cocaine (m/z 304) and its isotope labeled standard (m/z 307)

5.4 Measurement uncertainty: addressing the legal limit dilemma

In the previous paragraph, step by step a robust, sensitive, selective, and yet relatively fast and straight forward method for the accurate quantitative analysis of drugs of abuse in human whole blood was constructed. Although requiring high-end equipment (LC-triple-Q-MS), dedicated forensic toxicology laboratory infrastructure, and substantial forensic bioanalysis expertise, the approach enables the trace analysis of all the analytes as mentioned in the new DUI law with a single method. With sufficient instrumental capacity including back-up equipment to deal with sudden instrument failures, high and variable caseloads can be managed while acceptable lead times can be guaranteed. The only aspect that now must be considered is how the findings of the quantitative analysis have to be reported to the legal professionals and judges involved in the case. We will quickly discover that this task is surprisingly challenging as forensic toxicologists have to deal with a fundamental incompatibility of science and the law. This challenge is defined as the **Legal Limit Dilemma** in the title of this chapter as it is essential for forensic experts to grasp and appreciate the issue at hand.

> The law sets an absolute limit for acceptable levels of drugs of abuse in human whole blood when driving a vehicle. The forensic toxicology laboratory has an accurate LC-MS method for quantitative analysis and reports the drugs found and the measured level in the blood. If a level is found that exceeds the legal limit, can we now state with certainty that the suspect committed a crime? What scientific dilemma is posed by the absolute limits as stated in the law?

As scientists we cannot state with certainty to the courts that the true concentration (often indicated in forensic science as the **ground truth**) of a given drugs of abuse in the whole blood of a suspect exceeds the legal limit as set in the DUI law. Although the findings of the toxicological analysis can provide strong indications that the level exceeds the limit, any chemical analysis irrespective of the robustness of the applied method will be accompanied by a certain degree of measurement uncertainty. If we would take multiple blood samples from the suspect and perform repetitive LC-MS analyses, one would not expect to find exactly the same concentration every time. Rather the levels found will show some degree of variation, this variation is small for robust and optimized methods but never fully absent. When a very large number of analyses would be performed on the same sample, typically a normal (**Gaussian**) distribution is obtained yielding an average concentration $<c>$ and a standard deviation s_c:

$$p(c) = \frac{1}{s_c\sqrt{2\pi}} e^{\left(-\frac{(c-<c>)^2}{2s_c^2}\right)}$$

The Gaussian distribution can be considered as a probability density function indicating the fraction of an infinite number of measurements that would yield a given concentration c. The highest value of $p(c)$ is obtained for the average concentration and equals the pre-exponential factor. The probability density rapidly decreases as concentrations are considered that are considerably higher or lower than the average value obtained from the measurements. However, a very important feature of the Gaussian function is that the probability will never be zero irrespective of the concentration considered. This means that the probability that the actual level of the drug of abuse in the blood of the suspect is below the legal limit can be very low but will never be zero when an average concentration is measured that is well above the legal limit. This is in a nutshell the Legal Limit Dilemma, the judge ruling in the case wants a simple yes or no answer on the question whether the drug level exceeded the legal limit, whereas fundamental scientific principles prevent the forensic expert to provide such an absolute statement.

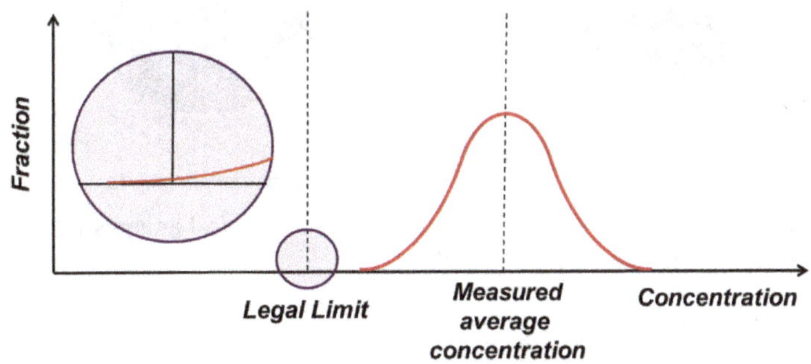

Visualization of the Legal Limit Dilemma

The normal distribution represents the random, uncontrolled variations that occur during each analysis and the standard deviation is an indication of the associated random errors. However, in addition to **random errors** each method typically also introduces a **systematic error**, that is, a consistent deviation of the measured average concentration from the actual, ground truth level. To study the systematic errors associated with a chemical analysis method control samples can be prepared by adding known levels of drugs of abuse to drug-free whole blood reference samples. As this enables the experts to understand the reasons for and minimize systematic errors, systematic errors just as random errors can typically not be fully eliminated. In analytical chemistry, the terms **precision** and **accuracy** are associated with random and systematic errors. A method with a high precision yields a low random error (low standard deviation) and a method with a high accuracy yields a low systematic error (small difference between measured and actual level):

Illustration of precision and accuracy in quantitative chemical analysis with the bull's eye representing the actual, ground truth level

> Can you mention sources of random errors that reduce the precision and sources of systematic errors that reduce the accuracy of a forensic toxicological LC-MS method?
>
> And how could we minimize these errors during method development and daily operation?

Random errors when analyzing small molecules in biological matrices can occur during pipetting and weighing of samples and standards, but can also be caused due to small variations in the HPLC separation conditions or fluctuations in MS sensitivity. Even data processing can lead to random fluctuations, for example, during manual or automated peak integration. Systematic errors can be the result of uncalibrated pipettes and balances, reduced impurity of standards, degradation of standards, samples and calibration solutions, analyte adsorption on the HPLC column and drift in MS sensitivity during sample sequences. During method optimization, such sources for random and systematic errors need to be carefully studied and minimized. Robust methods are methods for which the results remain relatively unaffected by random and systematic fluctuations during the analysis and that include sufficient checks and balances to spot emerging concerns. Such methods meet the quality criteria in terms of precision and accuracy to assure that the findings can serve as reliable and admissible evidence. The required quality framework is discussed in more detail in **Chapter 9—Quality and Chain of Custody**.

However, no matter how carefully the method is optimized and how many quality controls are included, a certain degree of uncertainty both due to random and systematic errors will always be applicable to the measured levels. It is essential that this information is included in the forensic report because only the combination of the levels found and the associated uncertainties will allow a judge to decide whether the evidence presented is convincing enough to exclude reasonable doubt. The associated uncertainties can only be assessed and reported by the forensic experts that have conducted the analyses and have interpreted the findings. It is, however, up to the legal professionals to translate these results and weigh the uncertainties within the legal and case contexts. Because scientists and legal professionals speak "a different language" and have a different background knowledge, the implications of the random and systematic measurement variations need to be explained in a scientific correct yet understandable manner. This will be discussed in more detail in **Chapter 10—Reporting in the Criminal Justice System**.

But to be able to include the required quality and reporting frameworks, the systematic and random errors associated with the reported levels of drugs of abuse in whole blood need to be established. These errors and the overall uncertainty can be assessed by using control samples. From a control sample dataset containing n analysis results, we can define the relative average systematic deviation $<d>$ from the expected level as:

$$<d> = \frac{\sum_{i=1}^{n} d_i}{n}$$

where the relative systematic deviation d from the actual level c^* for an individual measurement i is defined as

$$d_i = \frac{c_i - c^*}{c^*} = \frac{c_i}{c^*} - 1$$

The standard deviation $\langle s_d \rangle$ for $\langle d \rangle$ is given by (this follows from general mathematical properties of the variance)

$$\langle s_d \rangle^2 = \frac{s_d^2}{n}$$

where the standard deviation s_d for n individual measurements can be obtained from

$$s_d^2 = \frac{1}{n-1} \cdot \sum_{i=1}^{n} (d_i - \langle d \rangle)^2$$

The use of the term n-1 rather than n is known as **Bessel's correction** and corrects for the reduced **degrees of freedom** when using the average value of a set of n data points to estimate the standard deviation. This correction is only significant for small sample sets, if we consider a control sample set of $n = 20$, the variance s_d^2 increases 5% which further reduces to 2.6% for the standard deviation s_d.

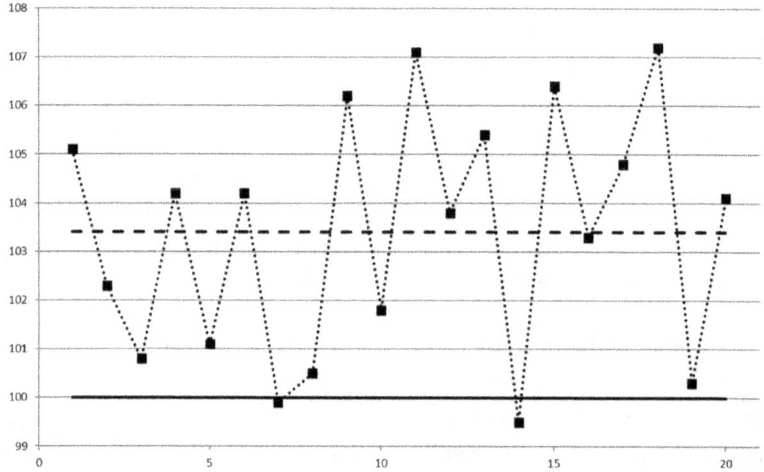

Illustration of a fictive analysis of 20 control blank whole blood samples spiked with cocaine at a concentration of 100 ppb (ug/L) showing the random and systematic errors of the LC-MS method. The dashed horizontal line represents the average level, which is 103.4 ppb for this artificial dataset

To characterize the random error associated with the LC-MS analysis, the variation of the individual measurements versus the average value for the control sample dataset needs to be considered by calculating the standard deviation s_c:

$$s_c^2 = \frac{1}{n-1} \cdot \sum_{i=1}^{n} (c_i - \langle c \rangle)^2$$

A dimensionless deviation r_c can be defined relative to the actual concentration c^* of the drug of interest in the control sample:

$$r_c = \frac{s_c}{c^*}$$

As we correct the measured level for the average systematic deviation associated with the method, the overall relative standard deviation r_t follows from the fact that variances are additive (under the assumption of independence):

$$r_t = \sqrt{r_c^2 + \langle s_d \rangle^2} = \sqrt{\frac{s_c^2}{(c^*)^2} + \frac{s_d^2}{n}}$$

From the definition of the relative systematic deviation d and the application of the mathematical variance properties (i.e., the variance of a constant is zero and multiplying a variable with a constant leads to a variance that is multiplied with the squared value of that constant), a clear relation between s_d and r_c and thus s_c can be established:

$$r_t = \sqrt{\left(1 + \frac{1}{n}\right) \cdot s_d^2} = \frac{1}{c^*} \sqrt{\left(1 + \frac{1}{n}\right) \cdot s_c^2}$$

By determining the average deviation (systematic error) from the actual concentration for a control sample set and establishing the standard deviation (random error), we are now able to provide all essential information for the LC-MS analysis of a case sample involving a blood sample of an individual suspected of driving under the influence. In general, it is good practice to use a pooled data set from control and/or calibration samples to characterize the variance associated with a quantitative chemical analysis. However, it is essential that such samples are fully representative for the case samples because otherwise certain sources of variation might erroneously be excluded. If the case sample is analyzed m times (typically samples are analyzed in triplicate), the average concentration $<c^{\#}>$ is given by

$$\langle c^{\#} \rangle = \frac{\sum_{i=1}^{m} c_i^{\#}}{m}$$

With the data from the control sample set, we are now able to correct the level for the systematic error and calculate a confidence interval from the associated the measurement uncertainty. The student *t*-distribution introduced in **Chapter 3—Sampling and Sample Preparation** can be used to establish the relative confidence interval *RCI* for a significance level α:

$$RCI = \pm \frac{t(d_f, \alpha) \cdot r_t}{\sqrt{m}}$$

As the legal question is whether the suspect drove a car with a drug concentration in his/her blood that exceeded the legal limit as set in the DUI law, we focus on the lower level of the confidence interval ($c_l^{\#}$):

$$c_l^{\#} = \langle c^{\#} \rangle \left(1 - \langle d \rangle - \frac{t(d_f, \alpha) \cdot r_t}{\sqrt{m}} \right)$$

where d_f represents the degrees of freedom equaling $n-1$ with n being the size of the control sample data set. To illustrate this approach, we now return to our fictive case sample involving the analysis of cocaine for which a DUI threshold was set at 50 μg/L whole blood. From a triplicate analysis of the case sample, an average cocaine concentration of 65.4 μg/L is obtained.

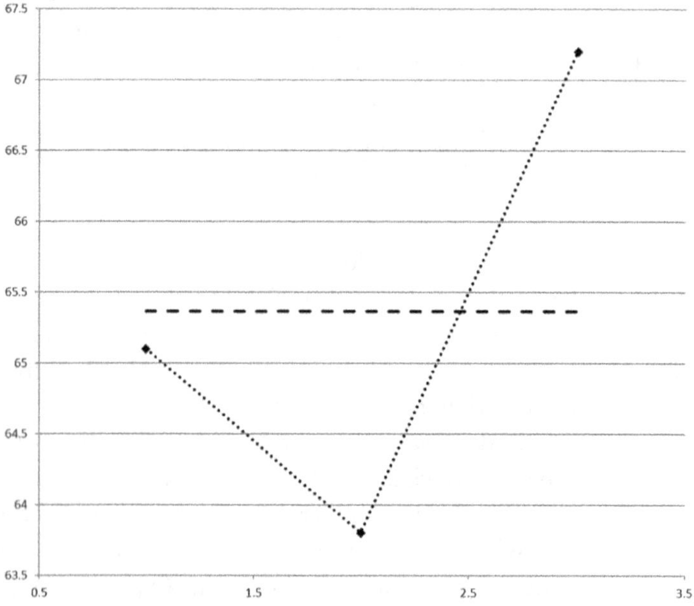

Illustration of a triplicate analysis of a whole blood case sample for which an average cocaine level of 65.4 ppb (ug/L) is found which exceeds the 50 ppb DUI threshold

For a confidence threshold α of 99%, the corresponding t value for 19 degrees of freedom ($n = 20$) is 2.861 (see student t table in **Chapter 3—Sampling and Sample Preparation**). From the control sample data set the relative average systematic deviation $<d>$ was found to be 0.03 (reported concentrations are on average 3% too high) while a relative total standard deviation of was obtained of 0.025 (2.5%). This leads to a lower level of the (two-sided) 99% confidence level of 60.4 ppb which still is well above the legal limit for cocaine. Although this constitutes very incriminating evidence, the expert cannot provide absolute certainty. At this confidence level, the percentage of a large number of repeated measurements that will yield a concentration below 60.4 ppb is expected to be very low (0.5%).

Because in DUI cases, the outcome of the forensic toxicological analysis is so directly linked to the legal verdict, we can resort to **one-sided statistical hypothesis testing** to illustrate the legal limit dilemma further. Because in criminal law someone is considered innocent until proven guilty, the **null hypothesis** (H_0) should reflect this:

H_0: *The cocaine level in the blood of the suspect while he/she was driving the car was equal to or below the legal limit of 50 ppb*

This will also be the defense position whereas the public prosecutor in the given case will formulate an *alternative hypothesis* (H_1) based on the belief that the defendant committed a DUI related crime:

H_1: *The cocaine level in the blood of the suspect while he/she was driving a car exceeded the legal limit of 50 ppb*

In the decision taken by the court based on the forensic toxicological findings, two types of errors can be made:

- **Type I error**: the null hypothesis is true but is rejected which leads to a wrongful conviction
- **Type II error**: the null hypothesis is false but is accepted which leads to a wrongful acquittal

The full judicial decision scheme is illustrated below.

ground truth / *verdict*	**H_0 is true suspect is innocent**	**H_1 is true suspect is guilty**
H_0 is accepted suspect is acquitted	No error Correct acquittal	Type II error Wrongful acquittal
H_0 is rejected suspect is convicted	Type I error Wrongful conviction	No error Correct conviction

Typically, a very conservative approach will be chosen when it comes to rejecting the null hypothesis because a wrongful conviction is generally seen as more detrimental than a false acquittal (although the victims of a crime and their family members might have a different view, acquitting suspects on the basis of legal technicalities can also lead to considerable social unrest). However, the laws of probability and statistics dictate that when considering a large number of DUI verdicts occasionally a wrong decision will be taken by the courts on the basis of the forensic toxicological findings. It is important to note that the Type I and II errors are related, selecting a very tight significance level to prevent wrong convictions automatically will result in relatively high numbers of false acquittals and vice versa. The only way to prevent erroneous legal rulings is not to provide a verdict at all, which is of course in direct conflict with the function of the courts.

An interesting exercise is to determine the t value that would yield a lower limit for the confidence interval that equals the threshold concentration as listed in the DUI law. For the example discussed in this paragraph, a lower limit of 50 ppb is obtained for a t value of 13.7. The associated α value is roughly 2e-9, which basically means that 99.9999998% of the measurements will yield a result that exceeds the legal limit. One could argue that the corresponding probability of making a type I error is so slim that it does not warrant reasonable doubt and that therefore the judge could reject the null hypothesis and find the suspect guilty of driving a vehicle with a too high level of cocaine in his/her blood. However, as the measured average level gets closer to the legal limit, the increased probability for a wrongful conviction might instigate sufficient doubt for a judge to acquit. This deliberation is exclusively to the trier of fact after being briefed by the forensic toxicologist with respect to the observed levels and associated confidence intervals and error rates. Considering the large volume of DUI cases such deliberations are often formalized at a certain stage to ensure uniform interpretation and application across the various courts.

CHAPTER 6

Chemical profiling, databases, and evidential value

Contents

6.1 What will you learn?

Until now we have discussed the application of principles and approaches of analytical chemistry that are neither specific nor exclusive to the forensic field. Structure elucidation, establishing the composition of mixtures or substance purity, and quantifying trace levels of compounds in complex samples belong to the daily tasks of the many analytical chemists that work in industrial, food, medical or environmental laboratories. However, from this point forward the use of analytical chemistry will become more dedicated to the "forensic cause", *that is,* the use of analytical chemistry to provide valuable evidence that allows law enforcement and the criminal justice system to investigate and prevent crime. More specifically, the use of impurity analysis for chemical profiling purposes will be discussed in this chapter. For instance, creating characteristic chemical impurity "fingerprints" of materials of forensic relevance allows experts to investigate whether batches of the same class of compounds and substances could be linked. This could provide answers to questions such as whether the drugs taken by the overdosed victim originates from the supply in possession of the suspected drug dealer. To address and answer such questions in a

Chemical Analysis for Forensic Evidence
ISBN 978-0-12-820715-4
https://doi.org/10.1016/B978-0-12-820715-4.00010-9

scientifically sound and robust manner, we need to introduce **Bayes rule**, a statistical framework that is ideally suited to express the evidential value associated with the forensic comparison of characteristic trace patterns and that is used by many forensic institutes across the world. In forensic science, chemical impurity profiling has mainly focused on illicit drugs given the extensive case volume, the international nature of the illegal trade, and the suitability of GC-MS to provide information on trace impurities in addition to the chemical identification of the listed substances. However, chemical profiling can in general be applied to any material, for any type of impurity, and with any type of analytical instrumentation capable of quantitatively analyzing these impurities. Because illicit drugs have been extensively discussed in the previous two chapters on qualitative and quantitative analysis, we will link the topic of chemical impurity profiling to the forensic investigation of explosions and explosives. It will be demonstrated how chemical profiling methods can be developed for intact energetic materials as secured by bomb ordnance specialists from undetonated **improvised explosive devices** (**IEDs**) at the scene of the crime and how such methods can aid the criminal investigation.

After studying this chapter, readers are able to

- list types of forensic traces and materials for which **chemical impurity profiling** can be used to investigate whether two evidence items share a common source
- describe the basic construction of an **IED** (**Improvised Explosive Device**), the function of the various parts, and list the type of forensic traces and evidence that can be secured from an intact IED
- understand how a **chemical impurity profiling** method can be developed for an organic explosive such as **tri-nitro-toluene** (**TNT**) with GC-MS
- explain the Bayesian framework for establishing evidential strength of forensic comparisons, reproduce the **Bayes formula** and define the **Likelihood Ratio** (**LR**).
- calculate **Euclidean distances** of multivariate impurity data for sample pairs, construct a **score-based model** from a representative **database** and estimate the **LR value** for a case comparison using the models constructed from the reference data
- explain the differences between a **score-based** versus a **feature-based model** for the forensic comparison of chemical impurity profiles

6.2 Criminalistics is the science of individualization

As stated in the introduction, chemical profiling based on impurity analysis goes beyond detection, chemical identification, and quantitative analysis and brings forensic analytical chemistry into the realm of criminalistics and intelligence to assist law enforcement, legal professionals, and even intelligence agencies and policy makers. Chemical profiling in forensic chemistry bears similarity to firearm and fingermark comparison and the generation of DNA profiles from biological traces. As such it can bring real added value to a criminal investigation revealing potential links between physical evidence on the basis of characteristic features in the form of impurities present. The development of analytical chemistry methods with the aim to create chemical fingerprints of materials is an activity that is rather specific to the field of forensic science and criminal justice. Similar approaches can also be found in the authentication of natural products such as essential oils used in fragrances or expensive wines from a certain region and age. However, such investigations are often also performed within the context of a criminal investigation, for instance when there is a suspicion of fraud or adulteration. It is important to realize that the development of chemical profiling strategies does not only involve method development using advanced analytical instrumentation but also requires background knowledge on the production process of materials, the construction of relevant sample collections and databases, the application of chemometrics and data modeling, and the use of statistics. All these aspects will be discussed in this chapter.

At the heart of the chemical profiling approach lies a very important principle that was already discussed in **Chapter 1—An introduction to Forensic Analytical Chemistry** and that was formulated by US forensic science pioneer Paul Kirk, as "*Criminalistics is the science of individualization*". In addition to detection/chemical identification and classification, individualization aims to link the physical evidence to its original source, for man-made materials this is typically a given batch as produced at a given moment in time and at a given production location. This is the chemical equivalent of linking a fingermark or a biological trace to a donor by comparing ridge patterns and STR DNA profiles, respectively. We should of course not forget the **Individualization Fallacy** also introduced in **Chapter 1**, reminding us that making absolute statements regarding the origin of physical evidence is typically not possible. However, through chemical profiling, our findings can support or refute the proposition that a link between materials exists. We will see that the more characteristic the impurities are, the stronger this support will become. The table below provides an overview of the analytical techniques used to address chemical profiling related questions.

Expertise area	Evidence material	Technique	Question
Illicit drug analysis	Drugs of abuse	GC-MS	Do these two batches of XTC pills come from the same illegal production lab? Is the heroine found at the overdosed victim originating from the heroine batch of the arrested heroine dealer?
Microtraces	Glass Paint Fibers	LA-ICP-MS Py-GC-MS LC-MS	Do the glass fragments found on the suspect's coat originate from the house of the burglary victim? Is the paint residue found on the bike of the traffic accident victim originating from the car of the suspect?
Explosives	Organics Inorganics	LC-MS/GC-MS ICP-MS/IC-MS	Is the intact explosive material found at the two crime scenes related, that is, originating from a single production batch? Was the explosive prepared by the suspect used in the bomb at the crime scene?
Questioned documents	Paper Ink Counterfeits	LA-ICP-MS LC-MS/GC-MS	Was the pen of the suspect used to write the threat letter received by the victims? Were the various threat letters printed on paper from a single batch?
Environmental forensics	Spills	LC-MS/GC-MS	Was the oil residue found in the North Sea spilled by the suspect's ship? Is the soil residue from the shoe of the suspect originating from the crime scene where the human remains of the victim were found?

Expertise area	Evidence material	Technique	Question
Fire debris analysis	Ignitable liquids	GC-MS	Is the paint thinner residue on the coat of the suspect coming from the bottle left on the crime scene? Was the gasoline found in the container at the crime scene obtained from this petrol station?
Materials	Tape, Plastic bags, Miscellaneous	LC-MS/ GC-MS LA-ICP-MS LA-ICP-MS	Does the tape used to bind the victim originate from the roll found in the car of the suspect? Did the garbage bag with the human remains found on the crime scene come from the role in the house of the suspect?

GC-MS, gas chromatography-mass spectrometry; *IC-MS*, ion chromatography-MS; *LA-ICP-MS*, laser ablation-inductively coupled plasma-MS; *LC-MS*, liquid chromatography-MS; *Py-GC-MS*, pyrolysis-GC-MS

Chemical profiling methods can be applied to a vast range of evidence materials to answer a broad range of questions in criminal investigations as is illustrated in the table. For the remainder of the chapter, the chemical profiling concept will be discussed and demonstrated for a fictive explosives case involving the organic explosive **TNT (2,4,6 Tri-Nitro-Toluene)**. TNT, an organic, military-grade explosive is a solid at ambient conditions with a yellow appearance. It was invented in the 19th century and was mainly used as a dye before its very favorable properties as an explosive were discovered. With a relatively low melting point of just over 80°C, TNT is easily processed and can be melt-casted into specific containers and shapes (illustration of a melting chunk of TNT at 81°C is shown below, source: Wikipedia). Despite the fact that this organic energetic material contains both the oxidizer (the NO_2 groups) and the "fuel" (the C and H atoms) in its chemical structure, it is relatively stable making it safe to handle and store. As a **secondary explosive** (this will be discussed in more detail in **Chapter 7—Forensic Reconstruction through Chemical Analysis**) it requires a **primary explosive** booster to initiate **detonation**. TNT is produced by the nitration of toluene with nitric acid in the presence of sulfuric acid. This reaction proceeds in three steps with a nitrate group being added with each step. The production of TNT leads to a considerable number and amount of side products and impurities, requiring clean-up and stabilization steps that result in a substantial chemical waste stream. Incorrect waste disposal procedures have led to environmental issues and soil pollution surrounding TNT production sites. An extensive analytical methodology has been developed to monitor TNT levels and associated breakdown products in soil. Because of the complex synthesis, TNT is not easily produced as a home-made explosive and TNT encountered in case work typically has a military or civil background. The explosive can be encountered as the main charge of, *for example,* a hand grenade that was obtained illegally. TNT is also used in combination with other energetic materials such as PETN (known as Pentolite) and ammonium nitrate (known as Amatol). The molecular structure is shown below and with a chemical formula of $C_7H_5N_3O_6$ the nominal molecular mass is 227 Da.

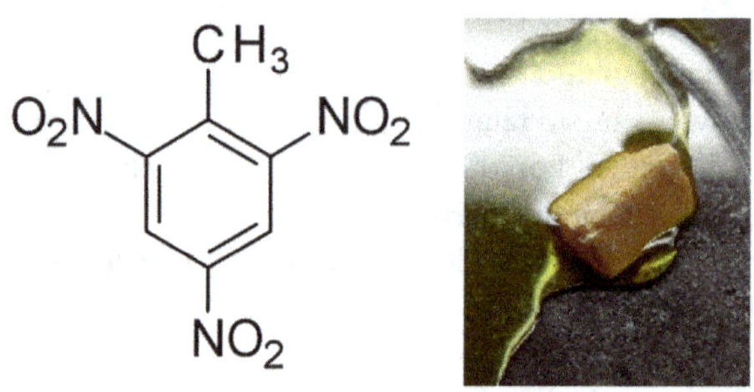

Chemical structure of TNT and melting of TNT at 81°C *(Wikipedia)*

An explosive case example (1)

The board of a well-known franchise receives an extortion letter in which payment of five million euros is demanded. The criminal(s) threaten to conduct a series of bomb attacks on their shops if their demands are not met. The company contacts the police. A few days later what appears to be an **IED (Improvised Explosive Device)** is found at one of the stores. The store is evacuated and the device is successfully defused by the **Bomb Ordnance Squad**. All the evidence is sent to the forensic institute for further investigation. The IED turns out to be a functional explosive construction capable of seriously injuring and even killing people in the store.

An IED consists of several components, can you figure out what is the function of parts 1 to 4?

If an IED can be defused and the components remain more or less intact, this offers many options for forensic investigation, what kind of traces and evidence are we looking for?

Component no 1 is the main charge of the device, the wrapping contains the energetic material. Item no 2 is the so-called blast cap, when activated it causes a small explosion, which initiates the detonation of the main charge. Part no 3 is a battery, the energy source to activate the blast cap. Item no 4 is an (old-fashioned) mobile phone, which forms the activator of the device.

By using a mobile phone, the device can be triggered from a distance (another option is a timer for a delayed activation). When the phone receives a call or message the battery is connected to the blast cap causing the detonation of the main charge.

In cases like this, it is very urgent to stop the threat and prevent a detonation of an IED at another store of the franchise. Hence the forensic investigation initially focuses on identifying the perpetrators or finding clues that will quickly lead to the perpetrators. Therefore, the following forensic investigations take priority:

1. Fingermarks and biological (DNA) traces left during the construction of the IED
2. Digital information (numbers, addresses, calls) from the phone
3. Security camera footage of the store and witness statements

An explosive case example (2)

Unfortunately, no useful DNA traces or fingermarks are found on the IED preventing the police from directly finding a potential lead via a database match. However, the phone provides several useful leads. After good tactical police work, this leads to a suspect, a former employee who recently was fired by the company. The police get a warrant to arrest him and search his house. In the house, they find (amongst other things) a package containing a yellowish material. Analysis in the forensic laboratory shows that this material is TNT, that happens to be the explosive material that was identified by the forensic explosive experts as the main charge in the IED at the store.

Does this finding mean that the suspect committed a crime?

Does this finding incriminate the suspect with respect to the extortion case?

Typically many countries do not allow individuals to produce, possess, and use energetic materials unless they have a special exemption or license related to their professional activities. This is also true for the Netherlands where the special law on firearms and ammunition states that it is forbidden to possess materials that are intended to cause a fire or explosion. Exemptions for civil use are for instance related to the use of explosives in engineering and construction activities. Without an exemption, the presence of TNT in his house means that the suspect committed a criminal act.

Finding an explosive at somebody's house is a rare event. However, there are many types of explosives and explosive mixtures. Finding TNT in both instances could be a mere coincidence but the fact alone should be considered incriminating when considering a potential connection of the suspect to the extortion case and the IED found at the store. The question that now remains is how incriminating this exactly is? Dutch forensic experts report a wide variety of organic explosives in case work. PETN is the mostly encountered explosive material followed by TNT, which is found in roughly 20%–25% of intact explosive investigations. Hence finding TNT in a criminal investigation involving explosives is not a rare event. It is clear that the presence of TNT in the IED and at the suspect's home only constitutes weak incriminating evidence.

An explosive case example (3)

As the police questions the suspect, he declares to have a fascination with explosives and he pleads guilty to illegal possession of an explosive. However, he denies having made the IED and being the person behind the extortion. Despite a comprehensive tactical and technical police investigation, there is limited direct evidence of the involvement of the suspect in the extortion case. The police team meets with the forensic explosives experts to ask, which options remain to compare the TNT in the IED with the TNT in the suspect's home.

What could be done to investigate whether the two TNT samples are connected?

We could resort to a detailed comparison of the TNT materials found at the suspect's home and in the IED. Although both materials contain high-grade TNT, differences could exist. The most straight-forward comparison can be conducted on the basis of visual and physical features. Do the samples look, feel and behave the same? On a deeper level, we can resort to chemical profiling, trying to map and quantify impurities and degradation products in the TNT samples. If the two samples originate from a single source it is expected that they would share the same impurity profile. Additionally, the isotope ratios of the light elements (C, H, N, O) could be compared. This latter approach requires the use of **Isotope Ratio Mass Spectrometry (IRMS)**, a technique, which is discussed in more detail in **Chapter 7—Forensic Reconstruction through Chemical Analysis**. IRMS is the only option to discriminate or match samples of very high purity. However, the TNT production process is suitable to create a chemical-impurity fingerprint. In the next paragraph, a tailor-made method based on GC-MS will be introduced.

6.3 A chemical impurity profiling method for the organic explosive TNT

The chemical identification of explosives is less straight-forward in comparison to illicit drugs for which GC-MS-based screening methods can be deployed. This was also addressed in **Chapter 2—Analytical Chemistry in the Forensic Laboratory**. The scheme presented in this chapter on the techniques that can be used to establish the presence of organic and inorganic energetic materials before and after an explosion does not include GC-MS. The reason for this is the limited volatility (low vapor pressure) for some compounds in relation to the limited thermostability of explosives. Given the energy stored in the molecule enabling the rapid, exponentially progressing exothermic reaction required to initiate an explosion, some of the more sensitive and involatile compounds thermally degrade rather than evaporate at elevated temperatures. As GC-MS requires compounds to be transported in a gaseous mobile phase and temperature programs are operated with temperatures up to 300°C, this technique is not suited as a broad screening technique for organic explosives. Naturally, inorganic, pyrotechnic mixtures including salts such as ammonium nitrate and sodium perchlorate are incompatible with GC-MS unless a chemical derivatization step is included to facilitate the conversion into more volatile compounds. However, GC-MS can be a suitable technique for the analysis of some of the energetic materials that are encountered in forensic investigations. The organic peroxide explosive **TATP** (**tri-acetone-tri-peroxide**), for instance, is known for its high vapor pressure and risks associated with sublimation (this primary explosive will be discussed in more detail in **Chapter 7—Forensic Reconstruction through Chemical Analysis**). In addition, also TNT is sufficiently stable and volatile to be analyzed with GC-MS and this also holds for the impurities that we want to analyze to chemically distinguish or match TNT materials.

To minimize degradation, the elution temperature of TNT and associated impurities should be as low as possible. This can generally be realized by using a short capillary GC column with a relatively large internal diameter and low film thickness of the stationary phase. However, the use of short, wide-bore GC columns poses a challenge in GC-MS due to the high vacuum imposed by the mass spectrometer. Without sufficient restriction, this will lead to very high flow rates while the column head pressure cannot be controlled by the pressure and gas flow regulator in the GC instrument. To solve this problem a so-called **restrictor** needs to be used, a short capillary column with a small internal diameter (typically up to 1 m in length with an internal diameter of 100 or 50 μm with a deactivated internal glass surface but without a stationary phase). This restrictor ensures sufficient flow resistance due to the small internal diameter. To operate a short, wide-bore GC column under normal chromatographic conditions the restrictor needs to be placed in between the analytical column and the mass spectrometer. In this way, the main pressure drop from the column head pressure to the high MS vacuum is mainly situated over the length of the restriction column. The analytes are separated in the analytical column at a mobile phase gas pressure that is almost equal to the head pressure because of the low flow restriction in the wide-bore column. However, a very interesting situation arises when the restrictor is placed between the split/splitless injector and the analytical column (see a schematic representation of the vacuum-outlet GC-MS setup below showing the positioning of the restrictor between the injector/pressure regulator and the analytical column). The analytes in the analytical column are now experiencing very high linear velocities in the vacuum regime of the mass spectrometer.

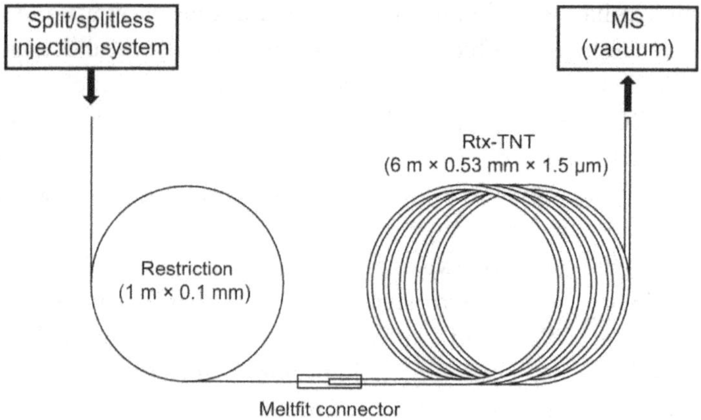

Vacuum-outlet GC-MS setup for the chemical impurity profiling of TNT *(Reprinted with permission from Elsevier from H, Brust., S, Willemse., T, Zeng., A, van Asten., M, van., Koeberg, A, van der Heijden., A, Bolck., P, Schoenmakers, 2014. Impurity profiling of trinitrotoluene using vacuum-outlet gas chromatography-mass spectrometry. J. Chromatogr. A 1374, 224–230, https://doi.org/10.1016/j.chroma.2014.11.055)*

An interesting aspect of this special mode of operation, which is named **vacuum outlet GC-MS**, is the impact of the low-pressure regime on peak broadening phenomena. The linear gas velocities at which minimal peak broadening occurs are typically 3–10 times higher compared to optimal conditions under normal GC and GC-MS conditions. This can lead to very rapid separations, *that is,* short analysis times. An additional advantage of vacuum outlet GC-MS is the relatively low elution temperatures under these conditions. This is an especially interesting feature for thermolabile compounds such as explosives. Furthermore, the rapid analysis leads to limited peak broadening and thus excellent sensitivity and relatively low detection limits. In addition, the column dimensions lead to favorable conditions with respect to sample capacity. These features are attractive from a chemical profiling perspective where we need to inject a relatively large amount of the main compound to analyze its trace impurities. The main limitation of vacuum outlet-GC MS is caused by the combination of a short column length and a relatively substantial internal diameter of the capillary column. This leads to a reduced number of plates and hence limits the resolution of the separation and the overall peak capacity (i.e., the total number of compounds that theoretically can be separated). Therefore, vacuum-outlet GC-MS is less suited for the analysis of complex samples. However, for the chemical profiling of TNT samples, the number of impurities is expected to be limited, hence this technique was shown to be a perfect solution for a very rapid and robust quantitative analysis of impurities. The results that will be shown next have been reprinted with permission from Elsevier from a 2014 paper in the Journal of Chromatography A that is listed in the Further Reading paragraph at the end of this chapter.

To develop and optimize the vacuum-outlet GC-MS method, various reference materials were obtained for potential impurities that could be present in forensic TNT samples. Because of the extensive study of environmental issues related to the production of TNT and associated waste streams, the chemical structures of these production and degradation impurities are well known: (1) 2-nitrotoluene (2-NT), (2) 3-nitrotoluene (3-NT), (3) 4-nitrotoluene (4-NT), (4) (1,2-DNB), (5) (2,6-DNT), (6) 1,3 dinitrobenzene (1,3-DNB), (7) 2,5 dinitrotoluene (2,5-DNT), (8) 2,4 dinitrotoluene (2,4-DNT), (9) 3,5 dinitrotoluene (3,5-DNT), (10) 3,4 dinitrotoluene (3,4-DNT), (11) 1,3,5 trinitrobenzene (1,3,5-TNB), (12) 2,3,4 trinitrotoluene (2,3,4-TNT), (13) 4-amino-2,6 dinitrotoluene (4-A-2,6-DNT), and (14) (2-A-4,6-DNT) (14).

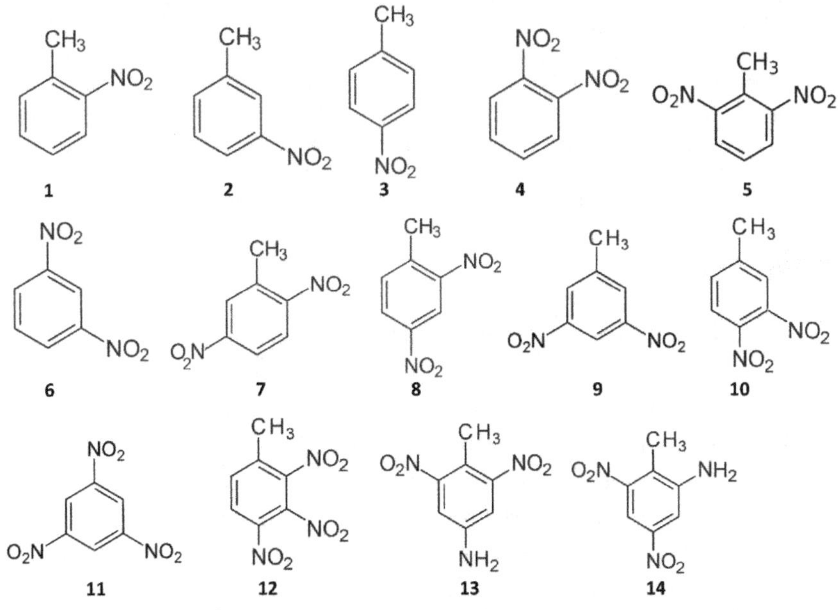

Chemical structures of known TNT impurities *(Wikipedia public domain and own work)*

With the vacuum-outlet GC-MS setup as depicted above a fast separation with sufficient resolution could be obtained as is illustrated in the chromatogram depicted below. Analysis time did not exceed 4 min and all compounds showed excellent peak shape and sensitivity. Only one significant co-elution was observed but this could easily be resolved by applying the additional selectivity offered by the mass spectrometric detector. Both compounds could be quantified without mutual interference by selecting different extracted ion traces from the full scan mass spectrum as is illustrated in the insert of the figure.

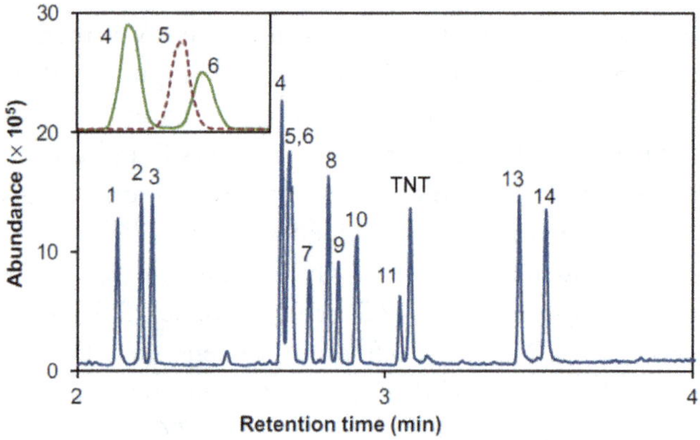

Separation of reference mixture of TNT and associated impurities (10 ppm each in ACN) with vacuum-outlet GC-MS with an injection of 1 μL at 250°C and a split ratio of 1:10, a helium inlet pressure of 276 kPa and the following temperature program: 35°C (1 min) – 80°/min – 250°C (1 min). The MS was operated in full scan mode in the range of 30–500 m/z. The insert indicates how ion extraction resolves co-eluting compounds 1,2-DNB, 1,3-DNB (m/z 168), and 2,6-DNT (m/z 165, dashed line). The main chromatogram is based on the TIC signal of the MS *(Reprinted with permission from Elsevier from H, Brust., S, Willemse., T, Zeng., A, van Asten., M, van., Koeberg, A, van der Heijden., A, Bolck., P, Schoenmakers, 2014. Impurity profiling of trinitrotoluene using vacuum-outlet gas chromatography-mass spectrometry. J. Chromatogr. A 1374, 224–230, https://doi.org/10.1016/j.chroma.2014.11.055)*

Although a fast separation of the TNT-related impurities with excellent sensitivity and selectivity was achieved, this does not automatically mean that vacuum-outlet GC-MS can be successfully used for the chemical profiling of TNT. Actual impurities as found in TNT (as opposed to environmental samples) may show distinct differences in prevalence and levels. But even more importantly we need to consider and mitigate the effect that TNT itself will be present in great excess to the traces of interest. To maximize sensitivity, we need to create very concentrated solutions for the TNT samples. However, excessive amounts of TNT will negatively affect the separation and peak shapes of the compounds of interest. Additionally, high levels of TNT in the mass spectrometer can affect ionization efficiencies, lead to detector saturation effects and lead to instrument fouling, which in turn can have a negative effect on the accuracy and precision of the quantitative analysis of the impurities and robustness of the method. Hence an optimum will exist for the TNT sample concentration with respect to impurity sensitivity and overall robustness. Unfortunately, at this optimum, (corresponding to 2 mg/mL or 2000 ppm), the vacuum-outlet GC-MS method was not detecting any impurities in the TNT samples. This situation did not improve when using SIE (selected ion extraction) from the full scan datasets indicating that the limit of detection was exceeding the actual levels (assuming that such impurities are indeed present in the samples).

Military and civil grade TNT has a very high purity with residual impurities at ppm level. Initial measurement with GC-MS at full scan revealed insufficient sensitivity. Even the application of SIE (Selected Ion Extraction) did not reveal any impurities. What quadrupole MS method could be employed to boost sensitivity? And at what "price"?

With Selected Ion Monitoring (SIM) the sensitivity can be boosted with a factor of 10–15. The quadrupole only monitors predefined masses at a given time window in which compounds of interest elute. Compared to full scan mode more analyte ion signal is collected during each cycle and this leads to a higher signal-to-noise ratio. However, in this mode, we can no longer measure the full mass spectrum of the compounds. The MS is exclusively used as a selective detector and not as a tool for compound identification. Of course, both approaches can easily be combined with a duplicate analysis of the same sample with a total scan and SIM method on a single GC-MS setup.

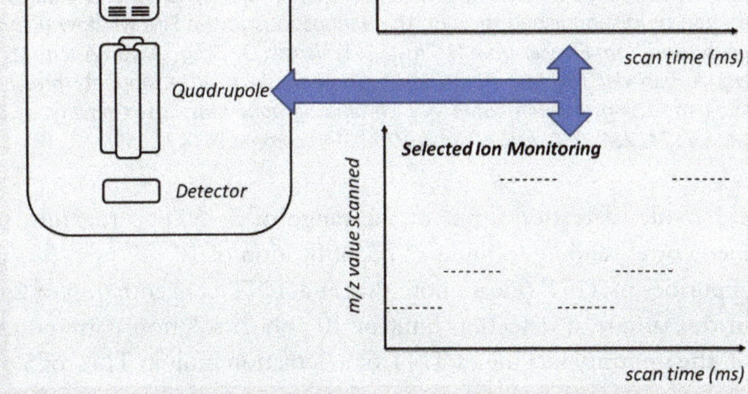

With the MS operated in SIM mode as explained above and with a tailor-made SIM method with different masses selected for different elution windows, sufficient sensitivity was obtained to detect and quantify the impurities present in TNT samples. An example of typical SIM GC-MS traces for TNT case samples is shown below.

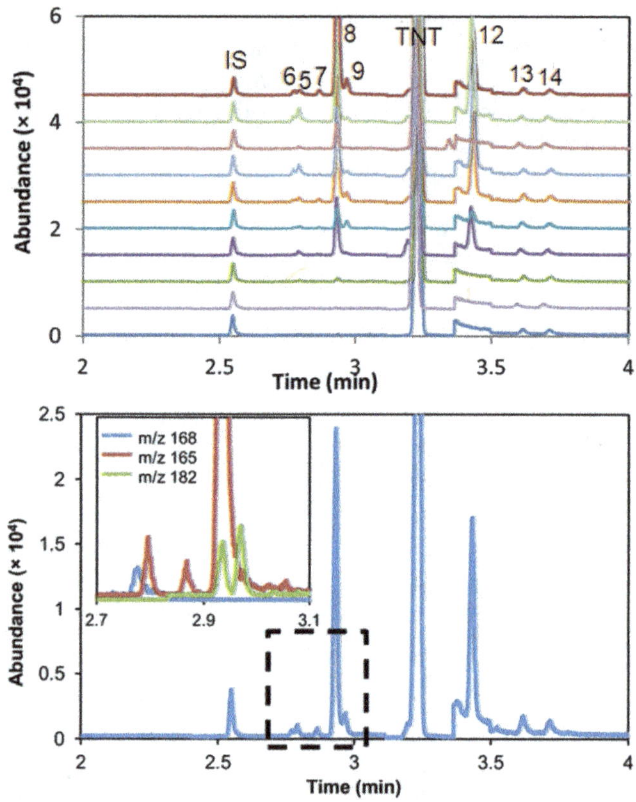

Top: GC-MS SIM trace overlay showing the impurities in 10 typical TNT samples dissolved at 2 mg/mL in ACN. Bottom: detailed view of the SIM trace of one of the TNT samples with the insert showing in different colors how 1,3 DNB and various DNT isomers can be distinguished through the various ions in the SIM window *(Reprinted with permission from Elsevier from H, Brust., S, Willemse., T, Zeng., A, van Asten., M, van., Koeberg, A, van der Heijden., A, Bolck., P, Schoenmakers, 2014. Impurity profiling of trinitrotoluene using vacuum-outlet gas chromatography-mass spectrometry, J. Chromatogr. A 1374, 224–230, https://doi.org/10.1016/j.chroma.2014.11.055)*

In SIM mode, detection limits in the range of 5–50 ppb (ng/mL) were obtained corresponding to limits of quantification of 10–17 ppm (µg/g) of the impurities in TNT (Please note that at a TNT concentration of 2 mg/mL in the sample, a detection limit of 10 ppb in solution corresponds to 5 ng of the impurity in a mg of TNT or a detection limit in TNT of 5 ppm.

As the limit of quantitation and limit of detection is typically defined as the concentrations yielding a signal of 10 times and 3 times the standard deviation of the noise, respectively, the corresponding limit of quantitation exceeds 15 ppm). For a set of 40 TNT samples from actual forensic casework, typical impurity levels were encountered in the range of 50—300 ppm. Such levels could be measured with satisfactory accuracy and precision by applying an internal standard (methyl decanoate) in combination with external calibration using reference standards. The relative standard deviation of replicate analyses was 2%—13% for the various impurities, with the highest random variations reported for those compounds eluting after TNT. It can be expected that the analysis of these compounds is hampered by the excess of TNT eluting from the GC column due to overloading phenomena. The excess of TNT and potential other matrix effects were investigated by spiking relatively pure TNT samples in the sample set with known amounts of reference standards. Recoveries in the range of 97%—115% were found indicating limiting matrix/TNT effects on the quantitative analysis of the impurities and demonstrating the validity of external calibration with reference standard solutions. However, a relatively high recovery of 135%—140% was found for the amino impurities (compounds 13 and 14). The reason for this was not well understood but for reasons that will be addressed later, there is no need to further investigate or correct this through, for example, calibration via standard addition.

The fact that we have developed a robust vacuum-outlet GC-MS method for accurate quantitative analysis of impurities in TNT does not automatically mean that we can use such an approach successfully for chemical profiling purposes in forensic case work. An essential requirement that is not controlled by the analytical chemist is that TNT samples from different origins exhibit significantly different impurity profiles. An intrinsically complex impurity profile can be very deceiving in this respect when it turns out that the production process leads to a detailed but very uniform profile or when there is a single supplier that is producing certain energetic materials on a global scale under highly controlled conditions and in huge batch sizes. This would lead to highly similar profiles for different samples only illustrating that the features analyzed cannot be used to distinguish samples. With such a generic profile, a "**match**" (*i.e.*, two very similar profiles with differences that fall within the range of the measurement uncertainty) would basically have no forensic value because a forensic case work sample would match with any reference sample, not only the material recovered from a suspect. This directly relates to the principle of Dr. Paul Kirk that was introduced in **Chapter 1—An Introduction to Forensic**

Analytical Chemistry. When considering classification, we could potentially distinguish TNT samples on the basis of certain features that relate to production processes or material grades (e.g., differentiating low, medium and high purity on the basis of the overall impurity level). Maybe we would also be able to recognize a certain producer on the basis of a typical impurity signature or link the profile to a certain brand or product mixture for a given civil or military application (often additional chemicals, such as stabilizers and plasticizers, are added to explosives to modify and optimize the material properties). This would allow us to exclude a potential link between two sources of TNT if we demonstrate that they belong to different classes. Similarly, an actual match in a case would only demonstrate that both sources belong to the same product class. Although this to some extent will be incriminating such evidence should be considered with great care because such matching class characteristics will apply to many TNT batches that have been produced according to the associated specifications. For this reason, powerful chemical profiling methods target impurities that are created during uncontrolled, random processes. Such processes could be related to degradation chemistry, which could be linked to the age and/or storage conditions of a certain material. Additionally, random impurities could originate from raw materials and other chemicals used in the process. Due to the random processes involved, the created impurity profiles could be highly characteristic for a specific batch of TNT irrespective of its class. It is this search for characteristic and uncontrolled features that triggered the insight "*criminalistics is the science of individualization*". To better understand the discrimination and thus association potential of a chemical profiling method, the forensic analytical chemist should investigate the production process, logistic aspects, and the market and trade conditions of the considered evidence type in more detail. Ideally, such information is provided by producers and suppliers even allowing the forensic expert to visit factories to understand every detail of the production process and the logistic chain. A well-known approach to show the individualization potential of a novel forensic profiling method is to demonstrate that subsequent sample batches that have been produced in a given factory under highly similar conditions can be differentiated. However, commercial companies are typically not very eager to provide detailed information to forensic institutes on their proprietary production processes. Another reason for the reluctance to collaborate is that these companies do not like to see their legal products being associated with illegal activities and criminal abuse. And finally, there is the simple "time is money" principle, assisting forensic experts requires the involvement of factory technicians and company experts for activities that have no business value. In rare events, the knowledge of impurities as measured with high-end equipment in forensic laboratories could also be useful to better understand product properties and optimize production. A collaboration with a forensic institute could then yield a win-win situation and producers might be interested to facilitate a study and provide reference materials and information for the validation of a profiling method. However, even with an interested and cooperating industry partner, the forensic experts carefully need to consider whether the information obtained is forensically relevant and demonstrate that the findings are also applicable to the materials encountered in the investigation of actual criminal activities. These materials constitute the physical evidence that is available to the expert for the forensic explosives investigation. If for instance only a certain type of TNT is preferably used by criminals or when criminals often handle energetic materials in a certain manner that could affect the impurity profile. For this reason, a very effective manner for forensic experts to understand the potential of a profiling method is to take a relevant population of samples from prior case work. Such materials are readily available when forensic experts keep a physical collection and are highly relevant because these materials are directly linked to criminal investigations. A drawback is that the expert typically is unaware of the origin of the material and does not know the ground truth when considering a potential link. Separate cases are considered to represent independent, unrelated materials in the absence of clear tactical links. However, such links might exist due to criminal network ties unnoticed by law enforcement (conversely, revealing such a link through chemical profiling might also shed new light on a case and could even provide new leads to solve a crime).

With the previous discussion in mind, let us now revisit the figure displaying the overlay of SIM traces of the 10 TNT samples from forensic case work. These cases had no apparent relation and originated from different time frames and different regions of the Netherlands. Assuming that these TNT samples are unrelated we hope to see significant differences in their impurity profiles. This is indeed the case, based on the current vacuum-outlet GC-MS method most samples can be differentiated. The traces displayed in the bottom part represent TNT samples of very high purity and matching samples that are virtually free of impurities has of course limited value. Another interesting observation is that the variation in the impurity level for the various TNT samples differs per impurity type. The two amino impurities (compounds 13 and 14) seem to be present in every sample at roughly the same level. This indicates that these compounds are less interesting from a chemical profiling perspective, if the observed differences reflect the measurement uncertainty, we are unable to distinguish TNT samples on the basis of the amino-dinitrotoluene isomers. To assess this in a more robust manner we need to transfer the measured SIM trace to a quantitative impurity profile. To this end, we need to establish the peak areas and convert these peak areas to concentrations using the external calibration curves obtained with the reference standards. Based on the amount of TNT sample that was dissolved, the measured concentrations can be converted to the amount of impurity expressed as ppm (μg/g). Each sample should be analyzed a number of times (at least in triplicate) to assess the relative standard deviation or confidence interval in the established levels. Having knowledge of the method precision is critical because it will allow the forensic expert to objectively assess whether two profiles "match" on the basis of sound statistics. The conversion of the raw GC-MS signals to quantitative impurity profiles is illustrated in the figure below. The quantitative impurity profiles shown confirm the qualitative conclusions on the basis of the SIM traces, significant differences in the impurity levels (in relation to the measurement uncertainty) are observed for the various case work TNT samples.

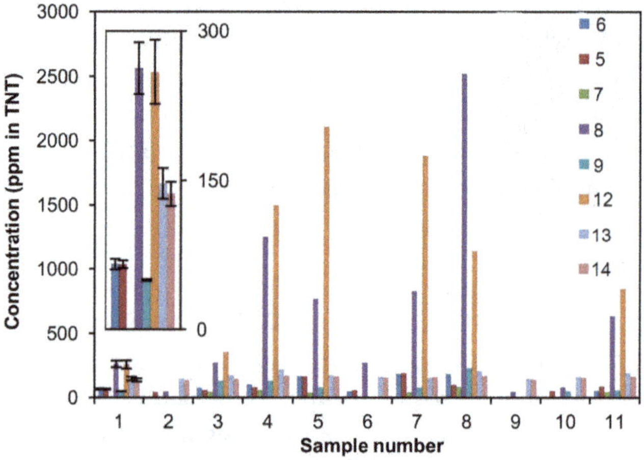

Quantitative impurity profiles for several TNT samples in the reference collection as determined with the vacuum-outlet GC-MS method. The insert shows details of the profile of the first TNT sample with error bars representing the standard deviation (n = 10) *(Reprinted with permission from Elsevier from H, Brust., S, Willemse., T, Zeng., A, van Asten., M, van., Koeberg, A, van der Heijden., A, Bolck., P, Schoenmakers, 2014. Impurity profiling of trinitrotoluene using vacuum-outlet gas chromatography-mass spectrometry. J. Chromatogr. A 1374, 224–230, https://doi.org/10.1016/j.chroma.2014.11.055)*

We have convincingly demonstrated that the vacuum-outlet GC-MS method meets all the required criteria to compare the two TNT samples in the extortion case and hopefully, this will lead to valuable forensic information to answer the question of whether the TNT retrieved from the IED could originate from the TNT found at the home of the suspect. After analysis of both samples with the new method and construction of the impurity profiles, two potential outcomes can be envisioned:

Potential outcome 1: *"the two impurity profiles show significant differences"*

Potential outcome 2: *"the two impurity profiles match"*

The forensic expert now needs to carefully consider how he/she is going to answer the question regarding the potential link between the TNT in the IED and the material found at the home of the suspect. The reader is advised to pause for a moment and try to formulate an appropriate answer for both possible outcomes. Appropriate in this context means scientifically valid and useful/valuable for the criminal investigation and court proceedings. To those contemplating potential expressions, it will quickly become clear that such a formulation (some suggestions are listed below) is far from straight-forward!

Potential outcome 1: *"the two impurity profiles show significant differences"*

- *The IED contains TNT from a source not originating from the suspect*
- *The IED most likely has not been made with the TNT of the suspect*
- *The IED could contain TNT of the suspect but from another source*
- *Somebody else made the IED (with TNT from another source)*
- *The suspect made the IED (with TNT from another source)*

Potential outcome 2: *"the two impurity profiles match"*

- *The IED contains TNT of the suspect*
- *The IED most likely contains TNT of the suspect*
- *The IED could contain TNT of the suspect*
- *The suspect made the IED*
- *Somebody else made the IED (with TNT from the suspect)*

When considering the suggested answers most readers will probably appreciate that the formulations are too absolute to be scientifically robust, too general to be of value, or reflect aspects not covered by the investigation such as activities undertaken by the suspect. If significant differences are observed between the two TNT impurity profiles (Potential outcome 1), typically very strong evidence is obtained that the explosive charge in the IED is not originating from the batch of TNT found at the home of the suspect. Excluding potential laboratory errors (contamination, degradation, erroneous registration, accidental swapping, this will be discussed in more detail in **Chapter 9—Quality and Chain of Custody**), there seems to be no plausible explanation for obtaining these results when the two TNT samples were actually linked. When considering biometric evidence like fingermarks or DNA profiles, it is evident that when the biometric features do not match the traces simply cannot come from the suspect and therefore must originate from another individual. However, in the field of forensic analytical chemistry, matters are typically a bit more complex as illustrated by the two following questions.

Question 1

If significant differences are observed in the impurity profiles of material recovered from a crime scene and reference material obtained from a suspect, does this automatically mean that these materials cannot share a common origin?

Question 2

If significant differences are observed in the impurity profiles of material recovered from a crime scene and a reference material obtained from a suspect, does this automatically mean that the suspect is not involved in the crime?

Question 1

No! It depends on the type of material, the features/impurities that are analyzed, the history of the evidence items, and the context of the case. If two fractions from the same batch are split and then experience variable conditions or undergo separate treatments, differences in composition can occur while hopefully also shared features will remain. For this reason, chemical profiling methods ideally target impurities that are constant and insensitive to for instance degradation and aging. In the presence of so-called "explainable differences", the forensic expert can still arrive at a conclusion that the evidence actually supports the linkage scenario.

Question 2

No! A possible explanation could be that the suspect had obtained different batches of TNT and used one batch to produce the IED while the other batch was intended for the construction of a second IED.

Although it is clear that in forensic material analysis the interpretation of non-matching impurity profiles is not "clear cut," the forensic expert must be extremely cautious in providing explanations for the observed differences to law enforcement and the courts. Nullifying exculpatory evidence by formulating possible reasons for the "non-match" typically ignores the most obvious explanation, that is, that the materials are not related and hence that the suspect is not involved in the crime. In a setting where undesirable outcomes from a prosecution's perspective are "reasoned away" with the help of the forensic expert, severe miscarriages of justice can occur.

In the case of potential outcome 2, we have obtained two near-identical impurity profiles, or in other words, two profiles that are indistinguishable because the differences observed are within the range of the measurement uncertainty. This finding can be considered incriminating evidence, not only is the explosive material found at the home of the suspect of the same class, but it also contains the same impurities at the same levels as the TNT in the IED retrieved from the crime scene. The suspect must be very unlucky indeed when he is actually not involved in the crime and the match has occurred purely by chance (a so-called **random match**). However, such a statement is neither scientific nor useful in court. The latter also holds for an expert playing it safe by stating that "*it cannot be ruled out that the TNT in the IED is originating from the TNT batch found at the home of the suspect*" or "*the TNT in the IED could originate from the TNT found at the home of the suspect*". The verbal impact of such vague statements probably will lead to an underestimation of the evidence by the triers of fact. However, making an absolute statement with respect to the origin of the TNT in the IED ("*The TNT retrieved from the IED is originating from the TNT batch found in the home of the suspect*") is definitely also non-scientific as we do not know how characteristic the given impurity profile is in a relevant population. In addition, we should not make the **individualization fallacy** as discussed in **Chapter 1—An introduction to Forensic Analytical Chemistry** and mistakenly label characteristic profiles as unique thereby excluding any other potential source. We simply lack the data and knowledge to make such bold statements. What is needed in case of a match is a framework for establishing and expressing the so-called **evidential strength**, the extent to which the forensic evidence supports the scenario of the prosecution (the two TNT samples are linked) over the defense proposition (the two TNT samples are unrelated). Such a framework based on so-called Bayesian statistics has been developed by forensic experts and scientists over the last decades. The reason why Bayesian inference is so well suited for reporting forensic findings in a criminal court case will be discussed in detail in the next paragraph.

6.4 Bayes theory and the likelihood ratio

It is beyond the scope of this book to discuss in detail the derivation of the well-known formula of Bayes as it is applied in forensic science (for more information readers are referred to the book *Statistical Analysis in Forensic Science: Evidential Value of Multivariate Physicochemical Data* listed in the Further Reading section). Bayesian statistics introduced by Thomas Bayes in his 1763 paper, was developed in its current form by the famous French scientist and philosopher Pierre-Simon Laplace in the early 19th century. The Bayesian framework is based on the same mathematical laws as frequentist statistics but considers an alternative view on the concept of probability. The frequentist probability is defined as the relative number of times that a certain event occurs when a given experiment is repeated indefinitely (*e.g.*, the probability of throwing a 6 with a dice). In contrast probability in the Bayesian context is considered as a degree of belief that can be "updated" on the basis of new information and (forensic) findings. The formula Bayes is based on the law of conditional probability:

$$P(A|B) = \frac{P(B|A) \cdot P(A)}{P(B)}$$

where P denotes probability, A and B indicate two events and the notation $P(A|B)$ should read as "*the probability of the occurrence of event A given the occurrence of the occurrence of event B*". If the two events are independent, then $P(A|B)$ simply equals $P(A)$. The formula of Bayes as applied in forensic science follows from this definition of conditional probability:

$$\frac{P(H_p|E)}{P(H_d|E)} = \frac{P(E|H_p)}{P(E|H_d)} \cdot \frac{P(H_p)}{P(H_d)}$$

where H_p denotes the prosecution and H_d the defense hypothesis and E indicates the evidence obtained from the forensic investigation. So, $P(H_p|E)$ should read as "*the probability of the prosecution hypothesis being true given the evidence*", alternatively $P(E|H_d)$ indicates "*the probability of obtaining the evidence given that the defense hypothesis is true*". The formula of Bayes is often abbreviated to:

$$\alpha_{posterior} = LR \cdot \alpha_{prior}$$

where α represents the odds or ratio of probabilities and LR is the abbreviation for the **Likelihood Ratio**. In essence, the formula of Bayes states that the posterior odds (the odds after the forensic investigation) equal the prior odds (the odds before the forensic investigation) multiplied by the ***LR* value** obtained for a given forensic investigation. So in terms of a legal verdict in a criminal investigation, the triers of fact (a jury or a judge) should update their initial beliefs with respect to the propositions of the public prosecutor and the defense council on the basis of the forensic expert report and the presented evidence. These initial beliefs are based on the context of the case and other information and evidence presented in court that led to the arrest of the suspect.

This concept of updating beliefs (odds) is very well suited for a legal context and for an objective contribution of the forensic expert to the criminal investigation. This becomes clear when we consider some of the consequences that follow from the Bayesian framework (this will be discussed in more detail in **Chapter 10—Reporting in the Criminal Justice System**):

- The hypotheses (also called propositions) are formulated by the public prosecutor and the defense council. Typically the public prosecutor assumes that the suspect committed a crime whereas the defense council formulates an alternative proposition indicating that the suspect is innocent
- The triers of fact (jury or judge) establish the prior odds based on the context of the case, testimonies, documentation, and other information and evidence presented by the public prosecutor and defense lawyer. These prior odds can be updated to yield the posterior odds based on the evidence presented by the forensic expert
- The realm of the forensic expert is restricted to the Likelihood Ratio, this means that the expert only reports on the probability of the evidence given the hypotheses and not on the probability of the hypotheses given the evidence. This makes sense as the expert is unaware of the context of the case and is only responsible for the forensic investigation and evaluation of the results
- The evidential strength of a forensic investigation is dependent on the propositions, the same result can correspond to different probabilities when different hypotheses are considered

The hypothesis pair considered needs to be exclusive meaning that both hypotheses cannot be true at the same time. It is not required for the hypotheses to be exhaustive (*i.e.*, $P(H_p)+ P(H_d) = 1$) although this is often the case. The *LR* value reported by the expert is a quantitative expression of the evidential strength associated with the findings of the forensic investigation. To be able to report the *LR* value the forensic expert needs to have access to that data and scientific knowledge that will allow an assessment of the probability of the findings when considering the prosecution and defense hypothesis. As we will see in the remainder of this chapter, this is not always straightforward and requires much more effort than the investigation of the physical evidence in the case at hand. But first, we need a better understanding of *LR* values and their meaning. An *LR* value can be infinitely large but can never be smaller than zero because probabilities can never have a negative value. *LR* values that are smaller than one indicate that the probability of the evidence is higher when the defense hypothesis is true which results in a denominator ($P(E|H_d)$) that outweighs the numerator ($P(E|H_p)$). This indicates that the evidence is supporting the Defense hypothesis and should therefore be considered exculpatory evidence. When the probability of the evidence becomes higher under H_p than under H_d the evidence is incriminating, which corresponds to *LR* values exceeding 1. Very high likelihood ratios are obtained when the probability of the evidence becomes very small when considering the Defense hypothesis. The working range of *LR* values is illustrated in the figure below.

Range of *LR* values for exculpatory and incriminating evidence

Can the *LR* value also be one? If so, what does this mean?

Yes, an *LR* value of one is obtained when the evidence is equally probable under both hypotheses, for example, when $P(E|H_p) = P(E|H_d)$ then the $LR = 1$. When the *LR* equals one the posterior odds equal the prior odds. This indicates that the forensic investigation does not support one hypothesis over the other and does not assist the triers of fact in their decision.

To better grasp the meaning of a given *LR* value we will do a simple thought experiment. Consider a case with a crime-related blood stain of the perpetrator in the pre-DNA era. We only have the option to determine the blood group and rhesus factor, combined this corresponds to a person's blood type. The following hypotheses are considered by the prosecutor and defense lawyer, respectively:

H_p: "*The suspect is the donor of the blood stain*"
H_d: "*Another person is the donor of the blood stain*"

The evidence (E) generated by the forensic expert entails the determination of the blood group and rhesus factor of the crime scene blood stain and comparing that against the blood group and rhesus factor of a reference blood sample obtained from the suspect (the law states that a suspect of a violent crime is required to participate in such a test).

The blood of the perpetrator (crime scene stain) is found to be "O-positive". We can consider two possible outcomes, in the first outcome the blood type and rhesus factor of the suspect deviates from this, for instance, "A-negative" (The blood group can be A, B, AB, or O and the rhesus factor can be positive or negative-yielding in total eight different blood type options).

Potential outcome 1—nonmatch situation

H_p (prosecution):	*"The suspect is the donor of the crime scene blood stain"*
H_d (defense):	*"Another person is the donor of the crime scene blood stain"*
E (evidence):	*The blood type of the blood stain is O+* *The suspect has blood type A-* *The blood types do not match*

What *LR* value do we report to the court?

What effect has this on the posterior odds?

$$P(E|H_p) = 0$$

$$P(E|H_d) = x$$

$$LR = P(E|H_p) \,/ P(E|H_d) = 0$$

The probability of finding a non-matching blood type when the suspect is indeed the donor has to be 0! In the absence of laboratory errors, this outcome can simply not be explained under H_p. This makes the probability of the evidence under H_d in a way irrelevant as long as we are sure that it cannot be 0 as well. Because its numerator is 0 the *LR* value itself will also be zero. This leads to an absolute statement as the posterior odds will be 0 as well for any given prior odds. Because the blood groups do not match, we can state with certainty that the suspect is not the donor of the blood stain at the crime scene. As the stain is crime-related and the context indicates a single perpetrator, the suspect needs to be acquitted immediately! In the Bayesian framework, an *LR* value of 0 indicates that the outcome is not possible when the hypothesis of the prosecutor is true leading to an absolute statement that the defense hypothesis therefore must be true.

Potential outcome 2—match situation

What information do we need to establish the *LR* value?

What *LR* value do we report to the court?

How would we verbally describe the evidence?

$$P(E|H_p) = 1$$
$$P(E|H_d) = \text{rmp (O+)} = 0.395$$
$$LR = P(E|H_p)\ /P(E|H_d) = 1/0.395 = 2.53$$

The probability of finding a matching blood type when the suspect is indeed the donor has to be one! In the absence of laboratory errors, this outcome is expected every time we run the blood type test. Now the interesting question is what the probability of a match is when actually the suspect is not the donor of the stain. In case any other individual can potentially be the perpetrator, we need to know the relative occurrence of blood type O+ in a relevant population. What is the probability that a random individual has a matching blood type? This is called the random match probability (rmp). Such population data is generally available, Wikipedia for instance states that O+ is the most frequently occurring blood type in the Netherlands, 39.5% of Dutch residents have this blood type. This means that the probability of the evidence when the Defense proposition is true equals 0.395 and this leads to an *LR* value of 2.53. Although this is a relatively low number it exceeds one and thus indicates that the evidence is incriminating as it supports the proposition of the prosecutor. The expert will state in court that the evidence (a matching O+ blood type) is 2.5 times more likely when the prosecution hypothesis is true than when the defense hypothesis is true. What this means for the outcome of the trial is not up to the forensic expert, the jury members or judges can now update their beliefs by multiplying the *LR* value with their prior odds. If these prior odds were very low then probably this new evidence is not strong enough to take away any reasonable doubt that might exist with the triers of fact on the involvement of the suspect.

The rarity of the feature affects the evidential value

H_p (prosecution): *"The suspect is the donor of the crime scene blood stain"*

Hd (defense): *"Another person is the donor of the crime scene blood stain"*

E (evidence): *The blood type of the blood stain is AB-*
The suspect has blood type AB-
The blood types match

What *LR* value do we report to the court?

$$P(E|H_p) = 1$$

$$P(E|H_d) = \text{rmp (O+)} = 0.005$$

$$LR = P(E|H_p)\ /P(E|H_d) = 1/0.005 = 200$$

The *LR* value increases with a factor of 80 when the matching blood type is AB- rather than O+. The reason for this is that the AB- blood type is a much more rare blood type found on average only in five out of 1000 Dutch citizens. As a result, the random match probability is much lower and for this blood type, the evidence is 200 times more likely under H_p than under H_d.

It is important to understand that for *LR* values supporting the prosecution hypothesis ($LR > 1$) the statement of the expert can never be absolute. Very high *LR* values lead to very high posterior odds (assuming that the prior odds are not 0) and very high posterior odds lead to a probability for H_p given *E* that approaches but never actually becomes 1. This is illustrated below for a situation where H_p and H_d are exhaustive and the prior odds are set at 1 (assuming an equal probability for H_p and H_p prior to the forensic investigation, this is by no means generically applicable as it assumes a 50/50 situation, "a coin flip," for the prior belief of suspect involvement).

$$P\left(H_p|E\right) = \frac{LR}{LR + 1}$$

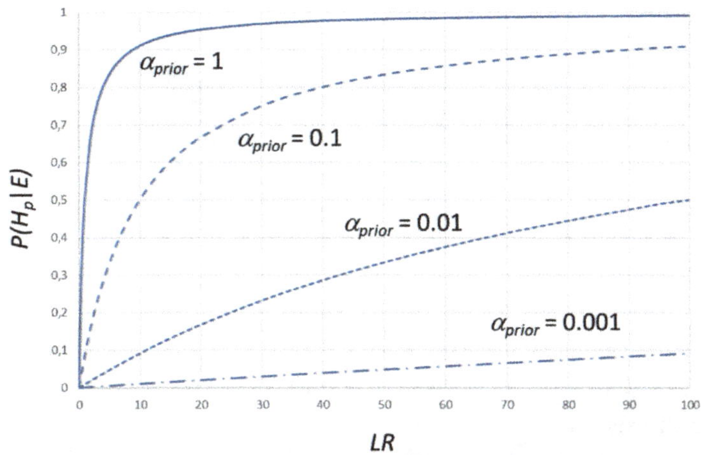

Probability of the prosecution hypothesis as a function of *LR* for various prior odds under the assumption that the hypotheses are exhaustive ($P(H_p)+P(H_d) = 1$). For prior odds of 1 the prior probability of H_p equals the prior probability of H_d

The propositions affect the evidential value

H_p (prosecution): *"The suspect is the donor of the crime scene blood stain"*

H_d (defense): *"The suspect is the donor of the crime scene blood stain but it is not crime related, the suspect cut himself shaving the day before the crime was committed"*

E (evidence): *The blood type of the blood stain is O+*
The suspect has blood type O+
The blood types match

What *LR* value do we report to the court?

$$P(E|H_p) = 1$$

$$P(E|H_d) = 1$$

$$LR = P(E|H_p) \, / P(E|H_d) = 1$$

The *LR* value is now one indicating that the findings are equally probable under both hypotheses. Basically, the defense council is confirming that the blood stain is from the suspect but they provide an alternative explanation of how it got there without the involvement of the suspect in the crime. In such a situation, the source-level information does not add any value to the evaluation of the hypotheses. The blood type analysis does not provide direction to the triers of fact.

A final important aspect of the Bayesian framework that is addressed in this paragraph concerns the combination of evidence. An important requirement for combining evidence is that the same hypothesis pair is considered. If this requirement is fulfilled, completely different evidence types can be combined even if they are conditionally dependent. However, matters simplify greatly if the evidence is independent because in that case the individual *LR* values can be multiplied:

$$\alpha_{posterior} = (LR_1 \cdot LR_2) \cdot \alpha_{prior}$$

That this "multiplication rule" can lead to very high overall *LR* values is demonstrated in forensic DNA profiling. Scientific research has shown that for the selected loci on the various human chromosomes, the STRs are fully independent. By multiplying the random match probabilities for the observed repeat combinations on the individual loci, the random match probability (rmp) of a full 15 STR profile (excluding the amelogenin gene that determines the sex of the donor) quickly becomes astronomically small as is illustrated below.

Locus	Genotype	Random match probability
D10S1248	14,16	0.087
VWA	17,18	0.122
D16S539	10,13	0.035
D2S1338	19,25	0.037
D8S1179	10,11	0.023
D21S11	28,31.2	0.038
D18S51	14,15	0.059
D22S1045	15,16	0.247
D19S433	12.2,14	0.034
THO	6,8	0.059
FGA	21,23	0.061
D2S441	10,14	0.123
D3S1358	15,15	0.091
D1S1656	12,17.3	0.046
D12S391	20,21	0.043

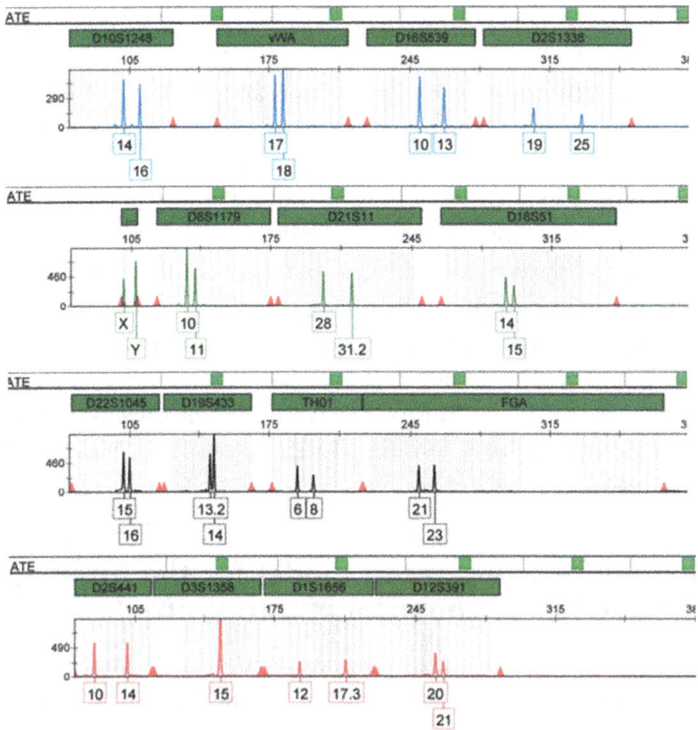

NGM Multiplex STR DNA profile of a famous NFI scientist (anonymous tribute) and the random match probabilities of his observed genotype for the individual loci

Although the *rmp* values of the individual genotypes are not very small (3%–25%), multiplication leads to a random match probability of the full STR profile of 4.5e-19! Compared to the blood type example the associated *LR* value (which equals 1/*rmp*) for a match is extremely large (2e18). With such high *LR* values, even very small prior odds will lead to a probability for the prosecution hypothesis that is almost equal to 1. Please note that versus an open population of potential perpetrators, the DNA analysis can never provide absolute proof that the suspect is the donor of the biological stain, the prior odds will be 0.999999 … This situation changes when considering a closed population such as criminal investigations for which the number of potential perpetrators is limited. A famous example often used to illustrate this, is the murder on a cruise ship. Assuming that nobody could have entered the vessel and left after committing the crime (e.g., the murder took place on the open sea and was quickly discovered), the perpetrator population is limited to all the persons present on this ship. By comparing the DNA profile at the crime scene with reference profiles of all the occupants, an absolute answer is provided by the evidence on the origin of the stain (as long as there are no identical twins onboard).

We will now return to the explosives case and the chemical profiling of TNT. In the next and final section of this chapter, ways to apply the Bayesian framework to estimate the evidential strength (*LR* value) when comparing impurity profiles will be introduced. The use and impact of Bayesian statistics by forensic experts in court testimonies and verbal statements regarding the evidence will be discussed in more detail in **Chapter 10—Reporting in the Criminal Justice System**. Additionally, frequently occurring fallacies in evidence interpretation will also be analyzed. For a more in-depth discussion of the Bayesian framework and its application in forensic chemistry, readers are referred to the book **Statistical Analysis in Forensic Science—Evidential Value of Multivariate Physicochemical Data** (see Further Reading section).

6.5 Building a score-based model for the forensic comparison of chemical impurity profiles

Showing that a robust chemical profiling method leads to significant differences in profiles between samples from different origins is not sufficient to establish the evidential value of a match in a case comparison. To be able to indicate associated random match probabilities in a similar fashion as in forensic DNA profiling, a substantial database with impurity profiles of TNT samples needs to be created. These TNT samples must represent a relevant population and typically this is the case when including case work samples. Interestingly, where in DNA profiling a relatively small subset of individuals suffices to accurately estimate rmp values in the entire population, this typically does not apply to forensic material analysis. In the study on the vacuum-outlet GC-MS method, a total of 50 TNT samples were analyzed including NFI casework samples and military reference materials from TNO. In general, it is difficult to create large forensic databases for explosives because the annual number of cases is usually limited especially compared to the illicit drug case load of a forensic laboratory. This has consequences for the modeling approach as will become clear in this section. The reference dataset that we can use for estimating *LR* values can be visualized as a "datacube" (indicating a 3D nature) with at its axes (1) TNT reference sample, (2) type of TNT impurity, and (3) the impurity level expressed as ppm in TNT. This is illustrated in the figure below. This figure shows a profile of one specific TNT sample in the database but the overall cube consists of 50 such profiles.

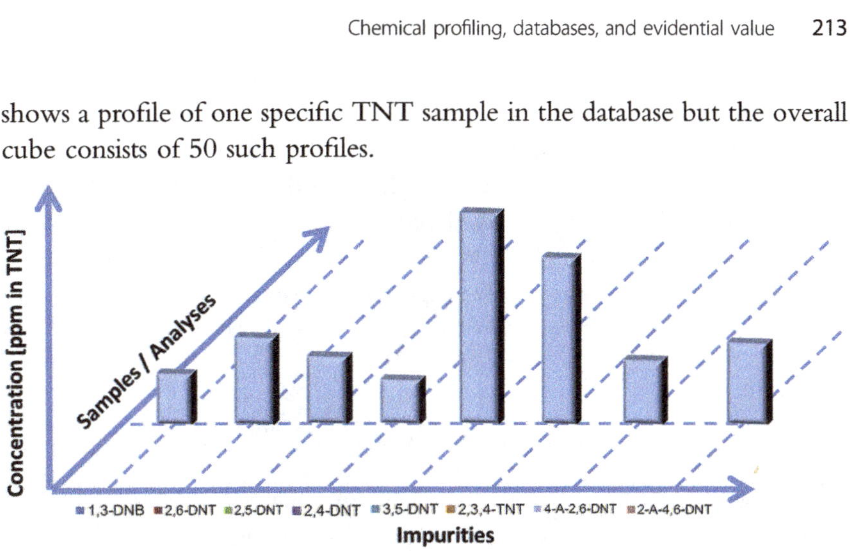

Visualization of the TNT reference database containing quantitative profiles based on 8 impurities as analyzed in 50 TNT samples

As a first suggestion to establish the random match probability of a typical profile we could resort to a simple method that is similar to the approach used in forensic DNA profiling. We could establish the distribution of the level of each impurity within the population of 50 TNT samples and fit the data to an appropriate probability distribution function. If the values show a normal distribution we could fit the data with a Gaussian function to be able to estimate the relative frequency in the TNT population for any given impurity level. It should be noted that it is by no means expected nor logical that such a distribution should be normal. However, for the sake of simplicity, this, for now, will be assumed for all impurities.

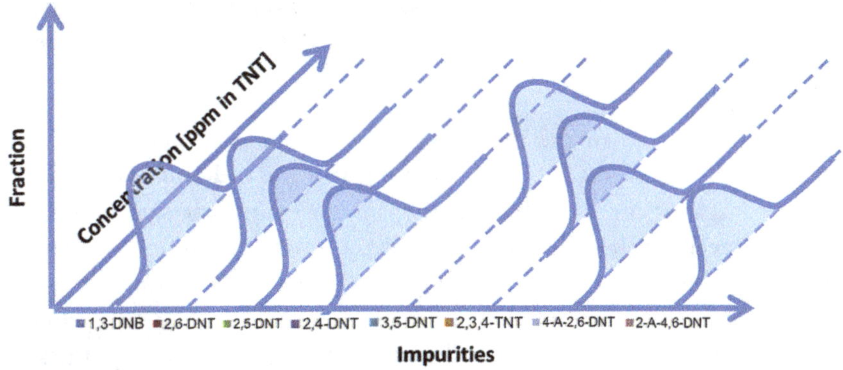

Visualization of the distribution of the levels of the 8 impurities as observed in 50 TNT samples. The data have been fitted to a normal distribution to estimate the probability of finding any given level of an impurity in the TNT population (please note that such population data do not have to be and are often not normally distributed)

We could try an approach that is similar to the rmp calculations of forensic DNA STR profiles by "simply" multiplying the rmp estimation for each individual impurity. With a total of eight impurities, very high *LR* values can be expected even if individual probabilities are relatively high. If for instance for a giving matching profile the rmp of each impurity is 0.25 (indicating that 25% of the TNT samples in the database share this impurity level), the resulting *LR* value for the full profile match would still exceed 65000!

> Unfortunately, for chemical profiling datasets, a DNA-like rmp approach cannot be applied. It will lead to *LR* values that are too high, a mistake a forensic expert should prevent at all times because it can lead to wrongful convictions. When we compare the TNT impurity profile with the STR DNA profile, there are two distinct differences. Can you list and explain these differences?

The first difference is that the number of Short Tandem Repeats (two per locus, one value for each chromosome of which one originates from your father and one from your mother) can be considered discrete whereas impurity levels are continuous. As the number of repeats can only take on a limited range of discrete values, measurement uncertainty does not affect the outcome. As long as a full profile is obtained such a profile is not expected to vary in repetitive runs. For continuous data, such as impurity levels, measurement uncertainty will lead to variation when the same sample is analyzed repeatedly as was already discussed in **Chapter 5—Quantitative Analysis and the Legal Limit Dilemma**. This has consequences for the probability of the evidence under the hypotheses considered.

As it was indicated in the previous paragraph the multiplication of *LR* values is only allowed for forensic findings that are **independent** and are evaluated against the same hypothesis pair. So before assessing individual impurity level distributions we need to demonstrate feature independence. For chemical impurities such independence is not a given, if compounds are formed during the same process, or when experiencing the same conditions, or when they are introduced into the final product as impurities of the same raw material, a significant correlation can be expected. Correlation in this context means that high amounts of impurity *X* in a sample are accompanied by increased levels of impurity *Y*. Compounds can also be inversely correlated, that is, high amounts of X co-occur with trace levels of Y and vice versa. This can indicate that compound Y is formed from a chemical reaction of X in the product or that compounds X and Y are impurities in different components of the product formulation. The degree of dependence can be assessed by calculating the **correlation coefficient** r for each pair of features in the dataset:

$$r_{xy} = \frac{1}{n-1} \cdot \sum_{i=1}^{n} \left(\frac{x_i - \langle x \rangle}{s_x} \right) \cdot \left(\frac{y_i - \langle y \rangle}{s_y} \right)$$

where n is the number of samples in the database (for the vacuum-outlet GC-MS study this corresponds to 50 TNT samples), x_i and y_i are the measured levels for the impurities X and Y in TNT sample i, $\langle x \rangle$ and $\langle y \rangle$ represent the average impurity level for all 50 samples in the database and s_x and s_y are the corresponding standard deviations. When the correlation of a feature with itself is considered, the coefficient equals one indicating perfect correlation. Perfect inverse correlation results in an r-value of -1 whereas a correlation coefficient of 0 indicates that the impurity levels show no correlation and are randomly distributed in the x-y plane. For $r = 0$ the two associated features are considered to be independent but in practice, some degree of association is usually apparent leading to an r value in the range of -1 to 1. Although the correlation coefficient is very often used to study multivariate datasets, it is very important to note that it is only sensitive to linear dependencies. Complex nonlinear correlations between features can result in low r values and therefore it is advisable to visually inspect correlation through scatter plots of the data in each feature plane.

A correlation table for the TNT impurities as measured with vacuum-outlet GC-MS is given below. Because $r_{xy} = r_{yx}$ the table can take a triangular shape to inspect all possible pairwise correlations.

13DNB	1,00							
26DNT	-0,15	1,00						
25DNT	0,44	-0,08	1,00					
24DNT	0,19	0,27	0,65	1,00				
35DNT	0,56	-0,12	0,79	0,65	1,00			
234TNT	0,29	-0,09	0,23	0,11	0,23	1,00		
4A26DNT	0,04	-0,30	0,26	0,18	0,23	-0,11	1,00	
2A46DNT	0,00	-0,20	0,15	0,06	0,06	-0,06	0,87	1,00
	13DNB	26DNT	25DNT	24DNT	35DNT	234TNT	4A26DNT	2A46DNT

Correlation table showing the pairwise correlation coefficients of the TNT impurities as measured with vacuum-outlet GC-MS, the degree of correlation is indicated by various shades of gray

Inspection of the table reveals that several impurities exhibit a significant correlation ($r > 0.5$). Interestingly, this all concerns a positive correlation with the highest r values observed for 4-amino-2,6 dinitrotoluene and 2-amino-4,6 dinitrotoluene. This should not come as a big surprise because these two impurities are chemically very similar and are most likely related to the chemical reduction of one of the nitrate moieties in TNT to an amine functional group. The correlation can be visually inspected in the scatter plot given below. At an r-value of 0.87 including both impurities in a multivariate dataset does not add a lot of additional chemical information and thus differentiation potential. If in a new TNT sample one of the impurity levels is measured, the level of the remaining isomer can be predicted with reasonable precision and accuracy using linear regression on the data below.

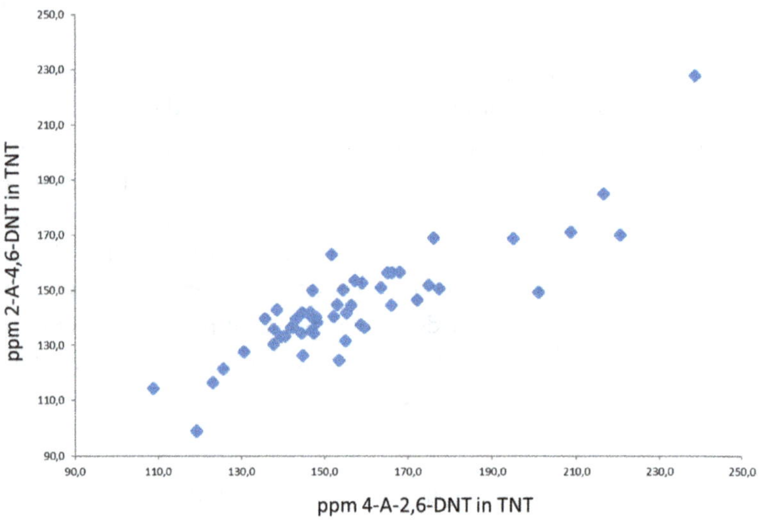

Scatter plot of the level of 4-amino-2,6 dinitrotoluene and 2-amino-4,6 dinitrotoluene in the 50 TNT samples in the database showing a high degree of correlation

As a simple rmp approach is not feasible in chemical impurity profiling, we will have to resort to more complex models to be able to assess the evidential value when comparing evidence items in a case. Two approaches have been developed by forensic statisticians, termed feature-based and score-based modeling. **Feature-based models** consider the individual features:

$$LR = \frac{f(a, b|H_p)}{f(a, b|H_d)} = \frac{f(b|a, H_p)}{f(b|H_d)}$$

where *f* represents multidimensional density functions of the feature vectors ***a*** and ***b*** under both hypotheses:

H_p: *The two TNT samples A and B originate from the same source*
H_d: *The two TNT samples A and B originate from different sources*

A feature vector for a TNT sample after vacuum-outlet GC-MS analysis is constructed from concentration values for the eight impurities analyzed. Building robust feature-based models requires complex mathematics and computations, which is beyond the scope of this book. Readers that want to study feature-based modeling in more detail are referred to the article of Bolck et al. listed in the Further Reading section as a good starting point. In the remainder of this chapter we will focus on the alternative option of building a **score-based model** for multivariate continuous chemical profiling datasets:

$$LR = \frac{f\left(s(a,b)\middle|H_p\right)}{f\left(s((a,b)\middle|H_d\right)}$$

where *s* denotes a univariate score that indicates the degree of similarity of the feature vectors. The complexity of the modeling is greatly reduced by estimating *LR* values from the degree of similarity of the multidimensional chemical profiles. However, this also comes at a cost, by not considering the characteristic nature of the individual features, information is lost. Consequently, evidential values from score-based models are usually conservative and in order of magnitudes smaller than *LR* values from feature-based models. Score-based models, on the other hand, are robust, can be applied to relatively small population datasets, and do not overstate the evidential value of matching profiles. A comparison of the score-based versus the feature-based approach is given in the table below.

Aspect	Score-based model	Feature-based model
Distribution of ...	Similarity scores	Features
Rarity of features incl?	Only indirect	Yes
Dimensionality	Univariate	Multivariate
Robustness	High	Low
Sensitivity	Low	High
Database size required	Medium	Substantial
LR values	Low to medium	High to very high

Building a substantial and representative population database for intact explosives from forensic casework is challenging. Therefore, developing a score-based model seems to be the safest option despite the fact that this will generate relatively low evidential strength, as not all the feature information is exploited. At the Netherlands Forensic Institute, a feature-based model is successfully applied in glass comparison casework. The features, in this case, consist of elemental (trace) levels measured with LA-ICP-MS (Laser Ablation-Inductively Coupled Plasma-Mass Spectrometry). With specially developed glass reference standards doped with elements at specified concentrations, the elemental levels can be measured with excellent accuracy and precision even for very small glass fragments. Such glass fragments are frequently found and retrieved from the clothing of suspects in burglary and robbery cases (or from the clothing of victims in hit-and-run cases). With the laser, samples can be taken from the fragments repetitively to measure the elemental profile very efficiently (as no sample preparation is required) and rapidly (the elemental profile as measured by the MS is created almost instantly). Elemental profiles obtained from glass fragments originating from suspects are then compared to profiles obtained from reference glass secured from the crime scene by taking glass samples from broken windows, glass vitrines, windshields, or bottles. Burglary cases typically occur very frequently and glass reference samples are very easy to obtain. In combination with the very efficient and rapid way glass samples can be analyzed with LA-ICP-MS, it is, therefore possible to build and maintain an extensive up-to-date population database with elemental glass profiles. Even with these positive conditions, a feature-based model requires meticulous validation and calibration to report trustworthy *LR* values. In the Further Reading section, three more references are provided that discuss validation and calibration of *LR* methods in more detail. Calibration in this respect entails the anchoring of *LR* values to the actual size of the population database. Feature-based methods have a tendency to generate very high evidential strengths for matching profiles that include many (>10) features. If an *LR* value (which can be considered as the inverse of the associated random match probability) is much higher than the number of samples in the population database it is clear that the odds of finding such a random match by comparing the case profile to all profiles in the database will be very small. Therefore, the estimated evidential strength can be considered as an extrapolation based on a limited number of comparisons. The calibration process anchors the evidential strength on the amount of data available for its estimation and this leads to *LR* values in the range of 1.000–10.000 for matching elemental profiles in glass comparison cases. This considerable evidential strength is used with great success in criminal investigations and has even led to the creation of a database of unsolved cases containing profiles of crime scene glass samples. As burglars often commit crimes repeatedly, glass residues found on their clothing can contain clues to past burglaries. In this way, multiple crimes can be solved when a perpetrator is arrested after a single incident.

To build a score-based model we need to find a suitable parameter that quantitatively expresses the similarity of two multidimensional chemical profiles. If the chemical profiles are considered as two points in an *N*-dimensional feature space (where *N* is the number of impurities) a distance metric such as the Euclidean, Manhattan, or Canberra distance can be used for this purpose. The smaller the distance between two points the more similar the two profiles will be. Alternatively, when using the Cosine similarity (as applied in Spectral Angle Mapping) the chemical profiles are considered as vectors and the orientation rather than the magnitude of the vectors is considered to express the degree of similarity. **Euclidean distance** is the most frequently applied distance metric in score-based modeling and it is defined as the shortest possible distance between two points in an *N*-dimensional Cartesian space. In a plane (corresponding to two features/impurities), the shortest distance *d* is given by the well-known Pythagoras's theorem. When we expand this to three dimensions an interesting progression pattern emerges as is illustrated below.

$$d = \sqrt{(x_2 - x_1)^2 + (y_2 - y_1)^2 + (z_2 - z_1)^2}$$

where x, y, and z denote the levels of three impurities in a chemical impurity profile, and subscripts 1 and 2 denote the actual values measured for two samples that are being compared in a forensic investigation. So when the distance is 0 the impurity levels for the two samples are identical, a situation that in practice will not occur as measurement uncertainty will create small differences even for repeated analysis of the same sample.

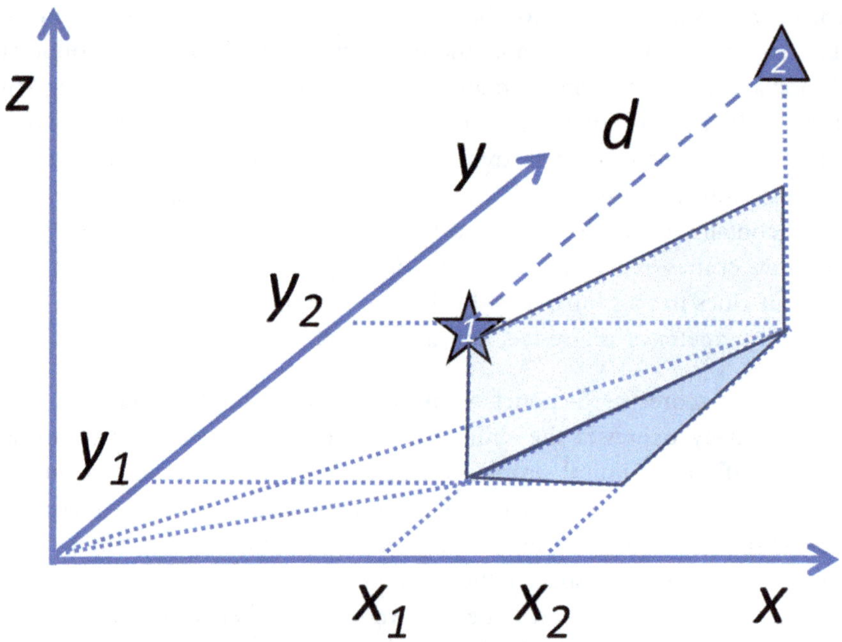

Shortest distance d between two points in a three-dimensional space where x, y, and z represent the levels of three impurities in a chemical profile and the subscripts 1 and 2 denote the values for two samples being compared in a forensic investigation

When considering the current vacuum-outlet GC-MS method, the feature space consists of eight dimensions as we are measuring the levels of eight impurities in TNT. Although such high dimensionality can no longer be vizualized, the Euclidean distance *d* can still be calculated:

$$d_{1_2} = \sqrt{\sum_{i=1}^{n} ((x_i)_2 - (x_i)_1)^2}$$

This indicates that the shortest distance between two samples 1 and 2 in the *n*-dimensional impurity space can be calculated by taking the square root of the sum of squared differences of the individual impurity levels.

So, no matter how many impurities we are measuring in a chemical profile and irrespective of the amount of correlation between these impurities, we are now able to quantitatively express the degree of similarity in a single number, the Euclidean distance. But how should this similarity metric now be used to estimate *LR* values if two samples are compared in a forensic investigation? For this, we have to resort to the population database of reference samples. If we include several samples of the same batch of TNT, we have a valuable collection of data for which we in principle know the ground truth (as stated before if we include casework samples it could be that two samples are considered to represent different origins whereas in practice the cases are linked without the investigators being aware of it). Similarity scores (the Euclidean distance *d*) can be collected for **known matches** (the **within distribution**) and for **known non-matches** (the **between distribution**) in the database. Through permutation, a substantial number of similarity scores can be created for a limited-sized database. Typically, the number of **different source** comparisons is much larger than the number of **same source** comparisons if the number of source repeats is limited. For a database of n original sources for which 3 samples are being analyzed, we can perform $3n$ same source comparisons and $n(n-1)/2$ different source comparisons. For a database of 50 sources measured in triplicate, we can extract 150 similarity scores for the within-source similarity distribution but almost 10-fold more data points (1225 to be exact) for the between-source distribution. The point to be made here is that a reference population database for a score-based model should contain enough repetitive samples from the same source to allow for an accurate prediction of the score distribution of matching pairs.

To be able to estimate the probability of obtaining a certain distance score (*d* value) for the Prosecutor hypothesis (common source) and the Defense hypothesis (different source), the within and between source distance data obtained from the numerous comparisons in the reference database need to be transformed in probability density functions. This requires two steps: (i) binning of the data to "count" the number of comparisons with a score in a given bin size (discrete range of *d*-values) allowing us to express the corresponding frequency in a histogram, (ii) converting a discrete histogram into a continuous probability density function allowing us to estimate a probability for every given *d*-value obtained from a pairwise comparison in a forensic investigation. To bin the data, a suitable bin size needs to be set. There is no single mathematical procedure to establish the optimal bin size and different bin sizes can reveal multiple characteristics of the data. As a rule of thumb, the number of bins is often set as the square root of the total number of data points. The bin size can then be obtained by dividing the difference between the maximum and minimum *d*-value by the number of bins. However, care has to be taken not to include outliers, and finding the optimum histogram with respect to a number of bins and bin size is often a process of trial and error. A disbalance in the number of data points for the within- and between-distribution could lead to different histogram settings. An illustration of a typical result is given below. For the known matches the chemical profiles are expected to be similar leading to relatively low *d* values. The distribution is clearly not normal as most scores are close to zero. Higher distances occasionally also occur due to measurement variation and/or sample inhomogeneity. For the known non-matches the *d* values are much higher because typically the profiles for two samples from different origins will contain various significant differences. However, random matches can also occur leading occasionally to relatively low distance scores for two samples from a different origin.

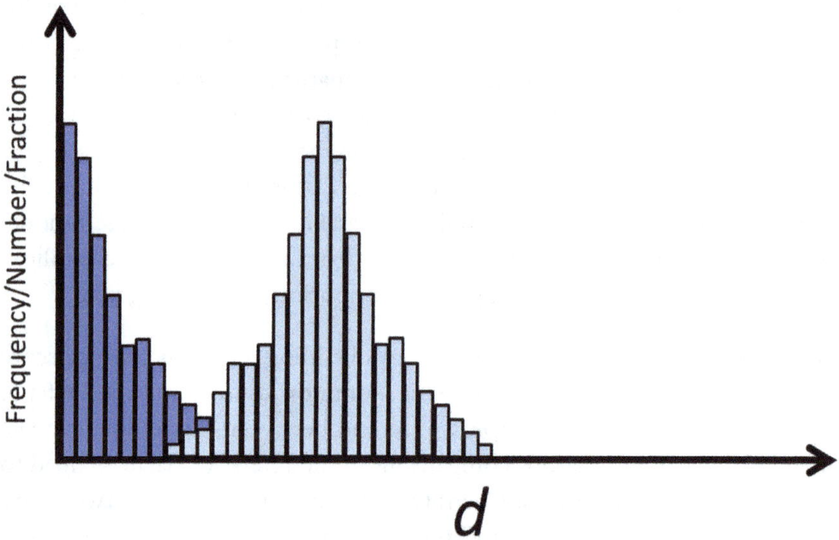

Histogram of *d*-values (similarity distance scores) for pair-wise comparisons of chemical profiles of known matches (same source, dark-colored) and known nonmatches (different source, light-colored) in a reference population database

With the histogram constructed, we now need to fit the data to create a probability density function. For data that is not normally distributed (which is often the case for this type of score metric) the **Kernel Density Estimation** (**KDE**) is often applied. This approach is explained in more detail in **Chapter 3** of the book *Statistical Analysis in Forensic Science: Evidential Value of Multivariate Physicochemical Data* (see Further Reading section). The KDE curve is constructed from a sum of Gaussian curves positioned on the histogram datapoints. The amount of smoothing is set through the standard deviation and the magnitude is corrected on the basis of the number of observations such that the overall density distribution has a peak area of 1. This process is illustrated in the figure below.

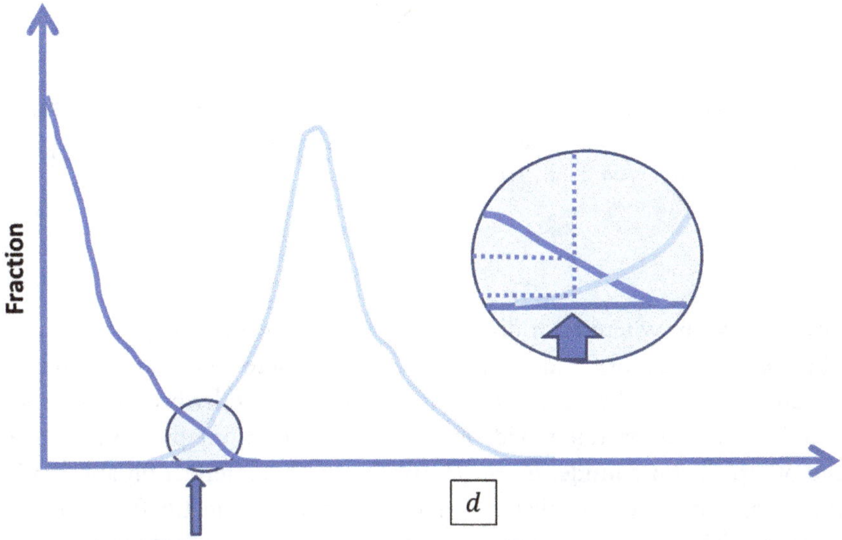

KDE (Kernel Density Estimation) probability distributions as a function of the Euclidean distance *d* obtained from the same source (dark-colored) and different source (light-colored) histograms. De zoomed-in detail shows the corresponding probabilities under the same source and different source hypotheses for a given value of *d* as obtained from a comparison in a forensic case

We are now able to estimate the *LR* value from a case comparison. First, we calculate the Euclidean distance for the two chemical profiles obtained from the analysis of the TNT in the IED and the TNT retrieved from the suspect. A low distance metric will correspond to a high degree of similarity and vice versa. For the calculated *d*-value, we retrieve the corresponding probabilities for obtaining this distance metric for the same source (public prosecutor) hypothesis and different source (defense) hypothesis using the corresponding KDE distributions. The *LR* now simply corresponds to the ratio of these two probabilities indicating how much more likely the evidence (d-value) is for the prosecutor hypothesis versus the defense hypothesis:

$$LR = \frac{f(d|H_p)}{f(d|H_d)}$$

For a given d-value the KDE same source and different source distributions intersect. What is the associated LR value when such a distance score would be obtained in a forensic comparison and what does this mean?

For the d-score where both distributions exactly overlap the probability of obtaining such a score is identical for both hypotheses. This corresponds to an LR value of 1, the evidence is equally probable under both hypotheses and the chemical profiling evidence does not support one proposition over the other. When considering the distributions, this intersection also marks those parts of the curves that correspond to what in the LR framework is termed **misleading evidence**. For the same source comparisons, the small fraction of the distribution with a value of d exceeding this intersection point corresponds to an LR value smaller than one thus erroneously supporting the Defense hypothesis. Vice versa the part of the different source comparisons distribution that yields d values prior to the intersection point leads to LR values that exceed one and thus incorrectly support the Prosecutor hypothesis. These rates of misleading evidence, obtained from the areas of these fractions marked by the point of intersection are inevitable when the curves show significant overlap, which typically occurs in score-based models for which LR values for known matches are in the range of 10—100. Only when the distributions are clearly separated, these rates of misleading evidence will become extremely small (they will never be equal to zero because no matter the distance, these summed Gaussian distributions will always show some degree of overlap but with very low associated probabilities).

Finally, all pieces of the puzzle are in place and we can now return to the question and case at hand. Can you correctly complete the overview below when the comparison of the chemical impurity profiles of TNT from the IED and the TNT found at the suspect's home leads to an *LR* value of 40?

H_p (prosecution): "*?*"
H_d (defense): "*?*"
E (Evidence): *?*
Conclusion?

H_p (prosecution): "*The TNT from the IED is originating from the batch of TNT found at the home of the suspect*"

H_d (defense): "*The TNT from the IED is originating from another source of TNT not related to the materials found at the home of the suspect*"

E (evidence): *Two quantitative impurity profiles for the two TNT samples have been measured with a new vacuum-outlet GC-MS method.*

The Euclidean distance d expresses the degree of similarity and an LR is estimated from the same source and different source KDE probability distributions constructed from a database of 50 TNT samples: $LR = P(d|Hp)/P(d|Hd) = 40$.

Conclusion:

Finding such a degree of similarity in the impurity profiles of the two TNT samples is 40 times more probable if the Prosecutor hypothesis is true than when the Defense hypothesis is true. This finding is incriminating for the suspect, not only are we dealing with the same type of explosive material but it has now also been demonstrated that both samples have a very similar organic impurity profile.

Further reading

Bolck, A., Ni, H., Lopatka, M., 2015. Evaluating score- and feature-based likelihood ratio models for multivariate continuous data: applied to forensic MDMA comparison. Law Probab. Risk 14, 243–266. https://doi.org/10.1093/lpr/mgv009.

Meuwly, D., Ramos, D., Haraksim, R., 2016. A guideline for the validation of likelihood ratio methods used for forensic evidence evaluation. Forensic Sci. Int. 276, 142–153. https://doi.org/10.1016/j.forsciint.2016.03.048.

van Es, A., Wiarda, W., Hordijk, M., Alberink, I., Vergeer, P., 2017. Implementation and assessment of a likelihood ratio approach for the evaluation of LA-ICP-MS evidence in forensic glass analysis. Sci. Justice 57, 181–192. https://doi.org/10.1016/j.scijus.2017.03.002.

Vergeer, P., van Schaik, Y., Sjerps, M., 2021. Measuring calibration of likelihood-ratio systems: a comparison of four metrics, including a new metric devPAV. Forensic Sci. Int. 321. https://doi.org/10.1016/j.forsciint.2021.110722, 110722-110722.

Zadora, G., Martyna, A., Ramos, D., Aitken, C., 2014. Statistical Analysis in Forensic Science: Evidential Value of Multivariate Physicochemical Data. Wiley, Hoboken, New Jersey, USA, 978-0-470-97210-6.

CHAPTER 7

Forensic reconstruction through chemical analysis

Contents

7.1 What will you learn?

In the previous chapter, you have been introduced to the concept of **chemical profiling** and how the quantitative analysis of impurities can allow for a comparison of materials of forensic interest. Typically, such methodology is used in criminal investigations to find case connections (*e.g.*, the same batch of material was used to commit several crimes and appears in several crime scenes) or to establish a potential link between a suspect and a crime. This approach is only feasible when "*there is something to compare*." If, for instance, no tri-nitro-toluene (TNT) was found during the search of the house of the suspect in the extortion case discussed in the previous chapter, the impurity profile for the intact TNT in the **improvised explosive devices** (IED) could not have been used to directly link the suspect to the explosive device. However, this statement is not entirely correct because the impurity profile itself can provide useful information on the state of the material and could yield valuable clues how it was made, by whom, and with what raw materials. This process of reversed chemical engineering to decipher the origin of materials falls under the umbrella of **forensic intelligence** techniques. Forensic intelligence approaches are used during the criminal investigative phase and assists law enforcement to

Chemical Analysis for Forensic Evidence
ISBN 978-0-12-820715-4
https://doi.org/10.1016/B978-0-12-820715-4.00011-0

find, evaluate, and rank potential leads. Interestingly, for reverse engineering of evidence materials, forensic experts are looking for features that allow for identification and classification, whereas for forensic comparison, the features should ideally emerge from random, uncontrolled processes that support discrimination. If these new leads result in the arrest and a court case against a suspect, forensic intelligence findings are often not presented as the primary evidence in court. This makes sense because as soon as a potential lead exists and a suspect is identified, options emerge to obtain reference material for a forensic comparison that could lead to a higher evidential value with respect to the involvement of the suspect. This is the case if at the crime scene biological traces or fingermarks of the unknown perpetrator have been found. However, also chemical profiling methods could be similarly applied if reference material is retrieved from a house search.

In this chapter, the use of forensic chemistry in forensic intelligence and the reconstruction of crime-related events through chemical analysis will be discussed. We will initially stay in the field of the forensic analysis of explosives to illustrate the concepts and application in case work. In doing so, this important forensic expertise area will also be discussed in a broader context. In addition, a new technique will be introduced and discussed in detail. This technique, **Isotope Ratio Mass Spectrometry** or in short **IRMS**, offers unique possibilities in forensic investigations, including the comparison of materials that are very pure and lack impurities at levels needed for chemical profiling. The feature that is forensically exploited with IRMS is the ratio of stable isotopes of elements present in the material of interest. Such ratios show slight variations that can be measured with great accuracy and precision with dedicated IRMS instruments. Additionally, the use of IRMS, a technique which was initially developed in the earth science/geology communities, applied to **human provenancing** will be discussed. It will be demonstrated how the assessment of elemental isotope ratios of human remains allows for the reconstruction of the residence and migration pattern of an unidentified victim.

After studying this chapter, readers are able to

- describe what types of crimes are associated with the abuse of explosives and how these explosives are chemically analyzed by forensic explosives experts
- understand why the chemical identification of energetic materials is challenging and requires a wide range of instrumental methods and strategies
- explain the concept of forensic intelligence and how chemical profiling in the absence of reference material can still be used to provide valuable information
- understand the concept of isotopes, list the major stable isotopes of light elements and calculate the standard atomic weight of an element based on the occurrence (mole fraction) of its stable isotopes
- understand the reasons why stable isotope ratios of elements can show slight variations, describe how these variations can be accurately measured with IRMS and explain why this requires dedicated equipment
- describe the extensive calibration and quality protocols required to accurately measure elemental isotope ratios with IRMS
- understand the use of IRMS in forensic material analysis and explain how elemental isotope profiles can be used to investigate potential links between explosives and precursors
- describe the principle of human provenancing and explain how IRMS analysis of different types of human remains can be used to reconstruct the origin and recent whereabouts of unidentified victims

In this chapter, you will learn more the about preparation and (ab)use of explosives.

NEVER TRY TO MAKE EXPLOSIVES, NEVER USE EXPLOSIVES ON YOUR OWN!

You are not only committing a crime, you are also endangering your own life and health and that of others.

7.2 Forensic explosives investigation

Trace and bulk explosives (also termed energetic materials) can be considered as chemical evidence material that is very directly linked to criminal activities. This is the reason why forensic institutes typically have a specialized team of forensic experts that conduct case investigations of crimes involving explosives and explosions. Compared to illicit drugs and ignitable liquids, some interesting commonalities and differences exist. In most countries, direct references to fire and explosions can be found in criminal law, in the Netherlands this is formulated as follows in Penal Law Code 175, which describes crimes that endanger the safety of individuals and goods:

Dutch Penal Law Code—Article 157

'He who intentionally starts a fire, creates an explosion or causes flooding, will be punished:

1. with an imprisonment of max 12 years or a fine of the fifth category if this could have resulted in serious damage to goods
2. with an imprisonment of max 15 years or a fine of the fifth category if this could have resulted in loss of life or serious injuries
3. with life imprisonment or an imprisonment of max 30 years or a fine of the fifth category if this has resulted in the loss of life'

Free translation from Dutch by the author.

So, in the Dutch criminal justice system, the abuse of ignitable liquids (to commit arson) and explosives (to cause explosions) is treated in the same penal law article, which in a way makes sense because of the potential mutual outcome of massive destruction of goods, infrastructural damage, and victims sustaining serious injuries or even loss of life. As we have already seen in **Chapter 1—An introduction to Forensic Analytical Chemistry,** the illicit drug law in the Netherlands (and in many other countries) is specifically linked to lists of banned substances (drugs of abuse classified either as "hard drugs" (List 1) or "soft drugs" (List 2)). This is not the case for explosives or ignitable liquids for which the legal focus is on the result of the abuse. Interestingly, the illicit drug law does not explicitly forbid the use of psychoactive substances, although this action requires possession which is defined as a crime. However, in principle somebody cannot be arrested for showing clear signs of drug consumption when he or she is not driving a vehicle and is not in possession of any listed substances. However, although criminal law does not explicitly list energetic materials typically, most countries do restrict individual access to explosives because of the risks involved. This of course does not apply to ignitable liquids, which are much more commonly used in society. A permit is not required to buy gasoline at a gas station or paint thinner at a DIY store because of the wide-spread legal use of such products. On the other hand, the legal use of explosives is restricted to specific activities in civil engineering (controlled demolition of infrastructure, construction of tunnels, avalanche prevention) or military operations. This enables a framework in which the possession of energetic materials is forbidden, but with exemptions for engineering and military use which also includes the companies and institutions involved in the supply, storage and safe use of explosives. In the Netherlands, this is specified in the special law on Firearms and Ammunition:

Law on Firearms and Ammunition—Article 2—second category

"Within the legal context the following objects as described in the various categories are considered as arms:

......

Category II

......

objects intended to target persons or goods by means of fire or explosives, with the exception of explosives for civil use when the use of these explosives has been granted in accordance with the law Explosives for Civil Use"

Free translation from Dutch by the author.

The law Explosives for Civil Use provides details of the exemption procedure.

The connection to firearms and ammunition for matters of possession is logical given the strict regulations of firearm ownership in the Netherlands. Furthermore, the propellants used in ammunition can also be considered an energetic material and can also be used to create **Improvised Explosive Device (IED)**, for instance the construction of a **pipe bomb** by filling a metal cylinder with **smokeless powder**. In countries with extensive private use and ownership of firearms (like the United States), smokeless powder is available and accessible in large quantities for re-use of ammunition. The main constituent of smokeless powder is **nitrocellulose**, an organic polymer explosive of the nitrate ester class (explosive classes will be discussed in more detail in this chapter). In so-called double base and triple base smokeless powders, respectively, **nitroglycerine** and **nitroguanidine** are added to the nitrocellulose propellant base. These compounds are also known as energetic materials.

Typically, crimes involving explosives are financially motivated or are intended to eliminate/injure an individual, damage an organization or even cripple an entire society. Threats involving explosives could also serve as a warning/deterrent or as a means to stop or force a certain action. Criminals use explosives to gain access to protected valuables and money. An example of this is the strong increase of ATM (Automated Teller Machine) raids in the Netherlands in the last decade. In an ATM raid, criminals use strong explosives to blow up an ATM in an attempt to get access to the money stored in the machine. Banks use protective measures including the application of security inks that severely stain the money bills when external force is exerted on the cassettes rendering them useless. Criminals sometimes attempt to clean the stained bills in an attempt to restore their value. In a similar fashion, explosives are sometimes used to crack open vaults and bank safes or to get access to protected, valuable jewelry. A mixed motive exists in extortion cases typically targeting retail chains or wealthy individuals. If the victims are willing to pay a substantial amount of money according to the instructions provided, the perpetrators will not execute their explosive attacks. However, in such cases motives can also be non-financial, for instance, when the perpetrators want to force a company to initiate or discontinue certain actions or are demanding certain statements to be publicly made (for instance, by acknowledging a certain wrong doing). In a targeted attack, explosives are used to eliminate an individual (assassination) or cause harm to an organization with no direct financial or societal motive. Perpetrators can act out of revenge or as part of gang violence. This type of abuse requires a modus operandi in which the explosives remain undetected and need to be concealed and covertly placed. The explosion is then initiated when the victim unknowingly triggers the device (*e.g.* a car or letter bomb), the perpetrator activates the bomb from a safe distance (*e.g.* a car bomb), or with the use of a timer (anticipating the target will be present near the explosive at a certain date and time). Without a doubt, the most destructive abuse of explosives in western countries has been related to terrorism. Terrorist attacks with firearms and explosives aim to destabilize societies by creating fear through destruction and loss of life on a large scale. Criminals are also involved in the theft and illegal production and trade of explosives. Military grade explosives are sometimes stolen or obtained from conflict areas to be sold to criminals or terrorists. In the Netherlands, a substantial black market exists for illegal fireworks. This is mostly related to New Year's Eve traditions in combination with the wish for more powerful fireworks than what is legally allowed. Such fireworks are often produced internationally for legal professional use, but sold illegally to individuals. The associated profits of illegal firework trade can be substantial and similar to criminal earnings from illicit drug production. Finally, explosive related crime also includes the clandestine, home-made preparation of energetic materials. So called **home-made explosives** or **HMEs** are produced out of a risky chemistry fascination (often by juvenile individuals) or as preparation for the criminal activities described above.

An explosion (see illustration above, based on a Wikipedia public domain picture) can be described as the sudden and rapid release of energy resulting in the production of heat and gas. The process is very rapid and the sudden onset of the explosion typically takes places in milliseconds. In this time range, an extensive amount of energy and gas is generated in the form of heat (often initiating a fire) and pressure (leading to a strong sound effect, the typical "*boom*" that humans associate with an exploding object). An explosion can be caused by both physical and chemical processes and can be accidental or deliberate in nature. Interestingly, in forensic institutions, forensic explosives experts tend to focus on cases involving chemistry-based, deliberate explosions, whereas physical and/or accidental explosive incidents typically are the domain of forensic engineers that look into industrial and traffic accidents and the collapse of structures like bridges and buildings. This can sometimes lead to confusion who should attend the crime scene to assist law enforcement because in the initial stages of the investigation it can be unclear what exactly caused the explosion and if this was an accidental or deliberate event. In general, physical explosion phenomena are related to accidents because the conditions required to trigger an explosion are difficult to control, prepare, and predict. Also dust explosions (a combined physical and chemical phenomenon triggered by the large relative surface area of small particles) and natural gas explosions in domestic homes are traditionally investigated by forensic engineers. This does not mean that establishing a physical cause or an accidental nature automatically ends the criminal investigation. Prosecution can still be warranted if there are indications of gross negligence or willful misconduct, for instance, due to the lack of safety protocols that according to industrial standards are mandatory for the operations leading to the incident. Additionally, gas explosions can also be deliberately staged, a theme that is frequently encountered in movies where victims unaware of the danger cause ignition in a gas-filled room by lighting a cigarette or flipping a light switch.

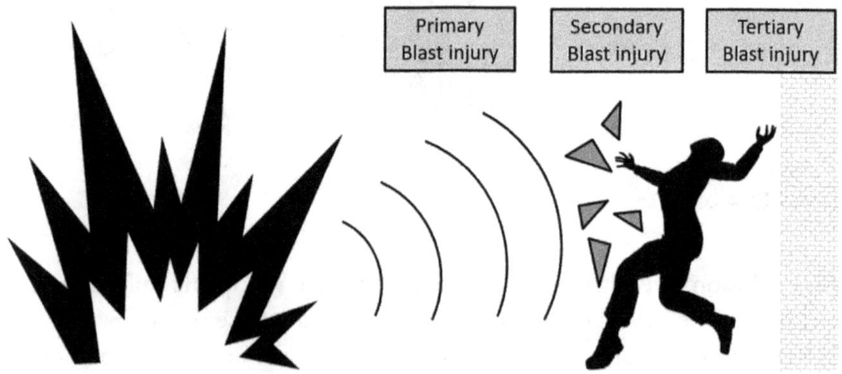

Illustration of the type of injuries of a victim in the vicinity of an explosion. *(Created with the use of Pixabay content)*

Victims that are in the vicinity of an explosion can sustain severe injuries or get killed instantly or at a later stage if the injuries cannot be treated adequately. The risks increase when individuals are situated close to the core of a very powerful explosion. The scale of the explosion directly depends on the amount of energetic material used in the device. Forensic explosives experts are frequently asked in court to reconstruct the explosion in terms of damage inflicted and to estimate the "*size of the bomb*". This is especially relevant for pre-explosion cases (where the bomb was not activated) or in post-explosion cases in the fortunate absence of targets, intended or accidental. To assess the severity of the alleged activities of the suspects, the legal professionals want to be informed on the potential outcome if an explosion had occurred or if individuals were situated at a certain distance from the blast. The magnitude of an (potential) explosion is typically expressed in "**TNT equivalents**" whereby the organic explosive TNT that was extensively discussed in the previous chapter (**Chapter 6—Chemical Profiling, Databases, and Evidential Value**) serves as a reference. The reason for this is the wealth of data that is available from decades of controlled test explosions with TNT. The equivalent of 1 kiloton TNT represents an explosion with an energy release of 4.184 GJ. This corresponds to 4184 J or 1000 calories when considering 1 gr of TNT (source: Wikipedia). On the basis of TNT testing, the amount of energy released during an explosion can be related to the damage inflicted to objects and infrastructure. Of course, such tests do not include humans which complicate the reconstruction with respect to injury potential. Additionally, the unpredictable nature of an explosion and the significant influence of environmental factors complicate the reconstruction (post-explosion) or prediction (pre-explosion). When considering the potential risk of (severe) injuries or loss of life, experts need to contemplate the several ways in which injuries can occur during an explosion. Typically, three categories are considered which are illustrated in the figure above. **Primary blast injuries** are the direct result of the blast wave hitting a victim. This wave can cause severe injuries to parts of the human body that are in direct contact with the atmosphere and are (partly) filled with gas such as the lungs, ears, and the gastrointestinal tract. Such injuries can be lethal, especially if a victim is exposed to a nearby **detonation** in which the explosion creates a **shock wave**, a pressure wave that exceeds the speed of sound in air. **Secondary blast injuries** are caused when a victim is hit by **shrapnel**, bomb fragments, or other hard objects that are propelled and dispersed by the explosion. Such objects can cause blunt force trauma or pierce the body causing severe bleeding (especially when an aorta is hit) or extensive damage to the internal organs. Explosive devices like hand grenades often include parts aimed at creating secondary blast injuries, for instance by embedding small metal balls in the construction or using a container that has been designed to undergo extensive fragmentation. Terrorists often include metal parts like screws and nails or marbles in IEDs to maximize the damage when activating the device in a big crowd. **Tertiary blast injuries** occur when the victim falls or is thrown against a wall or other object by the force of the explosion. This can result in both blunt and sharp force trauma similar to secondary blast injuries but this time with a mobile victim hitting a stationary object. Finally, one can also sustain **quaternary blast injuries** from indirect causes of being present near an explosion. For instance, burns from a fire that started after the explosion.

An essential aspect of the accurate reconstruction or prediction of the effects of a chemistry-based explosion is the identification of the energetic materials involved. The chemical identification of explosives also constitutes important evidence in indictments concerning illegal production and possession. This is why forensic explosives experts are also knowledgeable in the field of analytical chemistry and have analytical equipment available in the laboratory. As explained in **Chapter 2—Analytical Chemistry in the Forensic Laboratory**, the chemical analysis of explosives is not straight forward and requires an extensive portfolio of methods. When dealing with post-explosion traces obtained from crime scene swabs, trace analysis is conducted in complex and variable matrices. This typically requires a combination of a suitable separation technique and mass spectrometry. In pre-explosion cases, usually sufficient material is available for the forensic investigation unless the **Explosive Ordnance Disposal** squad (in the Netherlands as in many other countries the defusing and dismantling of bombs is conducted by military experts) had to inactivate the device via a controlled explosion out of safety considerations. With a sufficient quantity of intact material available for the investigation, experts prefer spectroscopic techniques for the identification because of non-invasive character, speed, and convenience of the analysis. These techniques include (N)IR and Raman spectroscopy, XRD, XRF (elemental composition), and if available NMR. In addition, colorimetric tests can be employed and the chemical reactivity of the material is tested by bringing a small amount in contact with a heated metal rod (the so-called hot needle test). This latter very simple and straight forward test is actually quite relevant to show the charge can indeed be ignited and will release the expected amount of energy. Sometimes IEDs are non-functional because the explosive material has not been prepared properly or has been desensitized through contamination or degradation. In addition to the pre- and post-explosion complexity, the chemical analysis of explosives is also challenged by the substantial diversity in the chemical composition of energetic materials. Although the desired effect to cause an explosion is the same, chemistry offers multiple ways to achieve this. Most explosive reactions are based on an oxidation reaction in which a carbon-rich fuel source reacts with an oxygen-rich oxidizer. After initiation, these reactions are typically strongly exothermic leading to the release of energy and often also yielding gaseous products that result in pressure build-up. When fuels such as candle-wax or wood react with the oxygen in the air, the energy release is not rapid enough to cause an explosion and the result is a fire in which the fuel is relatively slowly consumed and transformed in carbon dioxide and water. The fire stops when all fuel has been used up, the temperature is reduced or when the oxygen supply is stopped. Already in the 11th century in China the discovery of **black powder** demonstrated that oxidation reactions can proceed much more rapidly and violently when the oxygen for the oxidation can be provided from the material itself. Black powder is a so-called **pyrotechnic mixture** consisting of roughly 75 wt% of potassium nitrate (KNO_3, the oxidizer), 15 wt% of charcoal (pure carbon, C, the fuel), and 10 wt% of sulfur (S, also acting as a fuel). By creating a mixture of fine granulates of these three components, ignition with a flame causes a rapid burning or so-called **deflagration**. Because ambient oxygen is not required for the reaction to proceed, black powder can be ignited in a container that will explode when sufficient pressure is created through the formation of nitrogen and carbon dioxide. In the long history of human warfare, black powder has also been used in firearms and cannons for projectile propulsion. Although nowadays black powder is still produced on a large scale in China mainly for fireworks (firecrackers typically contain black powder), improved propellants have been developed over time such as the nitrocellulose based smokeless powders. More powerful pyrotechnic mixtures can be created by mixing potassium chlorate ($KClO_3$) or more preferably potassium perchlorate ($KClO_4$) with a fine metal powder such as aluminum (Al). A mixture of the latter two compounds is termed flash powder in which perchlorate (roughly 65 wt%) is the oxidizer and aluminum (roughly 35 wt%) is the fuel. The main reaction product is aluminum oxide (Al_2O_3) which is also slowly formed over time when aluminum surfaces are exposed to air and moisture. **Ammonium nitrate** (**AN**, NH_4NO_3) is a nitrate salt that is produced on a massive scale as it is essential as a fertilizer in the world food production. However, from industrial accidents in the past, it is known that AN is a very insensitive but also very powerful explosive under the right conditions. These characteristics also apply for **Urea Nitrate** (**UN**) which can be produced in a single step by reacting urea with nitric acid.

In civil engineering binary mixtures of AN (>90wt%) with fuel, abbreviated as **ANFO (Ammonium Nitrate Fuel Oil)**, are frequently applied for controlled explosions. ANFO is also used for controlled avalanche creation in winter resorts. The fuel component is mostly diesel but also kerosene can be used. ANFO is known for its excellent properties with respect to stability, ease of use, costs, and explosive power. ANFO is a so-called **high explosive** indicating that it can be detonated (as opposed to **low explosives** that deflagrate and need to be contained to cause an explosion) creating a powerful explosion that induces a shock wave. At the same time, ANFO is very insensitive and is categorized as a **tertiary explosive** (**blasting agent**) indicating that the explosion can only be initiated by a pre-explosion with a substantial booster charge. This is a very favorable combination of characteristics when considering safety and efficacy in construction projects. This is contrary to **secondary explosives** that only require a small amount of **primary explosive** to be activated. Primary explosives represent the most sensitive energetic materials which can detonate/deflagrate through the application of external force, friction/heat, or electrical discharge (a spark). Due to this sensitivity, primary explosives are the most dangerous class of materials to produce, handle, and store.

The pyrotechnic mixtures discussed above all contain inorganic constituents, which impacts the analysis methods that can be employed by the forensic explosives expert. In post-explosion trace analysis typically, ion chromatography (IC) or capillary electrophoresis (CE) has to be used to separate the ions of interest. With a regular LC system equipped with a reversed phase (RP) HPLC column, an ion-pair method could possibly be employed but most forensic laboratories prefer IC. Traditionally, IC is used in combination with conductivity detection but to increase the selectivity of the chemical identification recently the combination of IC with mass spectrometry is being explored. Counterintuitively, IC-MS is not straightforward as small ions tend to cluster in the ambient ionization source. However, pyrotechnic mixtures are not the only class of energetic materials that need to be considered when conducting a pre- or post-explosion forensic investigation. In more recent times, a whole range of very powerful organic explosives have been developed by mankind. These developments have mostly taken place in a military context and many of these organic explosives can be found in military weapons such as hand grenades, antitank grenades like rocket propelled grenades (RPGs), jet fighter missiles, and cannon ammunition. The continental wars in the 20th century fueled the development of new energetic materials with improved properties both in terms of storage stability, safety, and explosive power.

The organic explosives can be divided into several classes depending on the chemistry involved and the functional groups in the molecule. A majority of the organic explosives are based on the incorporation of nitrate groups onto a "carbon frame". This combined presence of the oxidizer (nitrate) and the fuel ("carbon frame") in a single molecule makes for a very efficient progression once the reaction is initiated. Instead of a powder mixture with separate granules, the two components needed for the oxidation reaction are present at the molecular level. This leads to a very rapid chain reaction in which the energy released by one reacting molecule triggers the reaction of the neighboring molecules. This exponential expansion leads to an almost instantaneous detonation once the main charge is sufficiently activated. The oldest class of organic nitro explosives is the so-called **nitro-aromatics** of which TNT is the most famous example. Forensic experts occasionally also encounter the nitro-aromatic explosives **tetryl** (2,4,6-trinitrophenylmethylnitramine) and **picric acid** (2,4,6-trinitrophenol, PA) in case work (Note: as the name suggests PA has acidic properties with a pK_a of 0.4 as the phenolate anion is resonance stabilized after loss of the hydrogen atom on the alcohol group). The nitro-aromatic explosives are produced by the nitration of the corresponding aromatic compound when exposed to concentrated nitric acid. The production requires multiple nitration and cleaning steps and leads to extensive and toxic chemical waste. Nitro aromatic explosives as encountered in case work are typically of military origin as they cannot easily be produced in a clandestine lab and are thus not considered as potential **home-made explosives** (**HMEs**). In more modern military devices, the nitroaromatic explosives have mostly been replaced by organic explosives from the nitro-amine and nitro-ester class. In **nitro-amines** explosives, the nitrate groups have been "anchored" on amine functional groups leading to $-N-NO_2$ bonds. Frequently encountered nitro-amines in forensic casework are **RDX** (cyclotrimethylene-trinitramine) and **HMX** (cyclotetramethylene-tetranitramine) which both can be produced by the nitration of hexamine. These compounds are typically from a military origin, although hexamine can be retrieved from camping fuel tablets allowing home-made production if one has access to fuming nitric acid. However, home-made nitro-amines are seldomly encountered by forensic experts. The well-known malleable explosive **C4** is based on RDX to which binders and plasticizers have been added. Nitro-esters are produced by the nitration of alcohols with nitric acid or a nitrate salt in the presence of sulfuric acid resulting in $-O-NO_2$ bonds. **Nitroglycerine** (**NG**), produced from glycerol, is one of the oldest and most famous members of the nitrate-ester class. In pure form NG is a colorless transparent oily liquid, which is highly impractical given its shock sensitivity. This drawback can be overcome by adsorbing NG on an inert carrier as was discovered by Swedish chemist Alfred Nobel. In 1867, he filed patents in England and Sweden on the invention that would make him famous and rich. By loading NG on diatomaceous earth, a silica-based natural material consisting of diatomic fossil remains, a very stable but powerful solid explosive mix is obtained which is known under the popular name **dynamite**. Nitrocellulose (NC) discussed previously as the main constituent of smokeless powder is also nitrate ester and is produced by reacting wood fibers, cotton cloth or paper with a mixture of nitric and sulfuric acid. NC predates NG as it was invented in 1832 while the production process was perfected in 1846. In general any carbohydrate molecule with a sufficient number of alcohol functional groups can serve as a raw material to produce nitrate ester explosives. A nitrate ester that is very frequently encountered in forensic casework is **PETN** (Pentaerythritol tetranitrate) which is produced from the polyol pentaerythritol. PETN was used on an extensive scale in World War II and is still used in military and civil explosive devices today. As a secondary explosive it strikes a good balance between stability and sensitivity and PETN is therefore primarily applied in boosters, detonators and detonation cords. PETN is also applied frequently in explosive mixtures such as **Semtex**. Semtex formulations contain PETN and RDX in variable ratios and a number of additives including plasticizers, binders, and antioxidants. Mixtures of PETN and TNT are known under the name **Pentolite**. In the Netherlands, PETN is the most frequently identified explosive in forensic case work although it is not often encountered as HME. In recent times, the nitrate ester **ETN** (**Erythritol tetranitrate**) has quickly gained popularity and it is nowadays frequently identified in forensic case work. This is the result of the widespread availability of its precursor erythritol, which is used on a large scale as a low-calorie sugar replacer. Production via enzymatic hydrolysis of starch followed by conversion of glucose to erythritol using genetically modified yeast has made this raw material available at large scale and at a very reasonable price. It can be purchased at a wide range of web shops for a price of roughly 10 euro per kg in crystalline, pure form. The reaction with either a mixture of sulfuric and nitric acid (the mixed acid route) or a combination of potassium nitrate and sulfuric acid (the nitrate salt route) is straightforward and leads via a recrystallization step to a very pure end product. ETN is a sensitive (primary) powerful high explosive that needs to be handled with care but can be melt-casted in desirable shapes via gentle heating because of its relatively low melting point of 61°C. All these characteristics have made ETN a popular HME. Several nitrate esters (NG, PETN, ETN) are also known as vasodilator medication to reduce blood pressure by widening the blood vessels. However, this dual use typically does not pose a problem in criminal investigations because the context for regular medical application will normally differ substantially from illegal use, for instance, with respect to the amount of material involved and its appearance (explosives are usually not prepared in tablet form). The peroxides **TATP** (**triacetone-triperoxide**) and **HMTD** (**Hexamethylene triperoxide diamine**) form a separate class of organic explosives that are not based on nitrate incorporation but on the reactivity of the -O-O- bond. The synthesis of TATP and HMTD is very straight forward and is based on the reaction of hydrogen peroxide with respectively acetone and hexamine in the presence of dilute sulfuric acid as a catalyst. TATP has been very popular among terrorists because of the general availability of the required raw materials (nail polisher for instance contains acetone, hydrogen peroxide is used in dilute form as a disinfectant or bleach and acids can be retrieved from several sources, *e.g.*, battery acid). However, this peroxide explosive is nicknamed "Mother of Satan" for a good reason, it is extremely sensitive. As a primary explosive, it can detonate from a small shock or slight friction. It is also a relatively volatile material having the highest vapor pressure of all organic explosives (excluding **nitromethane**). Therefore, TATP can also sublimate and recrystallize in high-risk areas such as screw lid ridges of containers. This makes the preparation of TATP-based IEDs very dangerous and several cases are known of explosions at terrorist safe houses where TATP detonated prematurely. An interesting aspect of the chemical analysis of TATP is that the molecule has two explicit conformer isomeric forms with relatively slow kinetics resulting in the separation and analysis of the individual conformers and strange "plateau" peak shapes in GC-MS. Additionally, TATP corresponds to the trimer but the reaction of acetone with hydrogen peroxide can also result in dimer, tetramer and even polymer formation.

This brief overview of the types of energetic materials and mixtures that forensic explosive experts encounter in both pre (bulk) and post (trace) explosion cases shows the chemical diversity and complexity that needs to be addressed. This also explains the broad range of analytical techniques that is employed to correctly identify the explosive material involved. For organic explosives, the methods of choice are LC-(HR)MS for trace and bulk analysis and spectroscopic techniques such as Raman, (N)IR, XRD, and NMR if intact material is available. This is also in combination with colorimetric assays and functionality testing. The chemical structures of all energetic compounds discussed are depicted below with their key characteristics described in two accompanying tables.

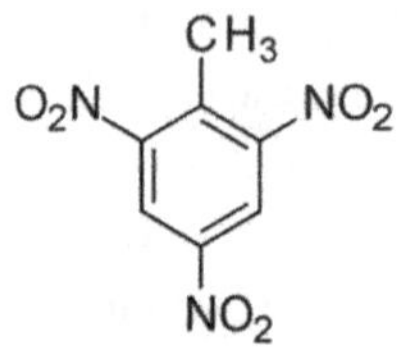
TNT

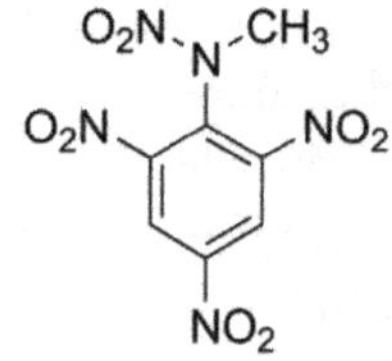
Tetryl

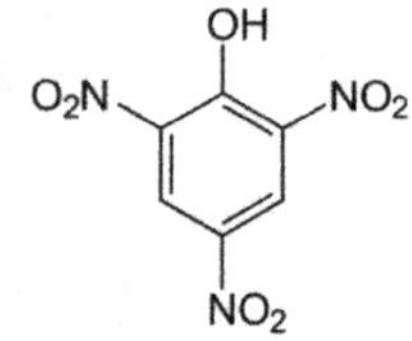
PA

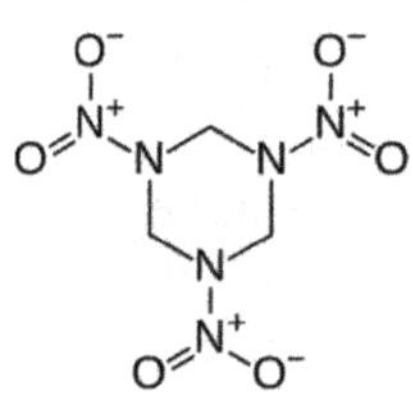
RDX

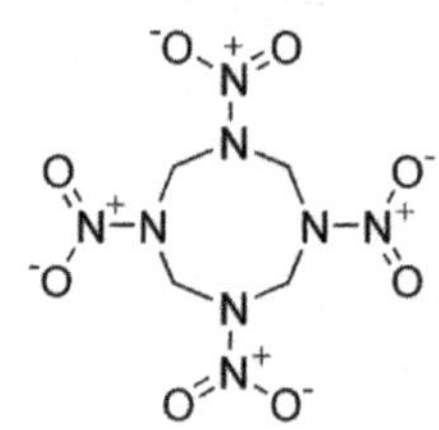
HMX

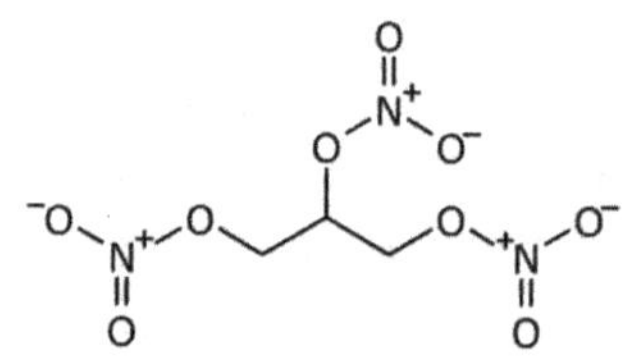
NG

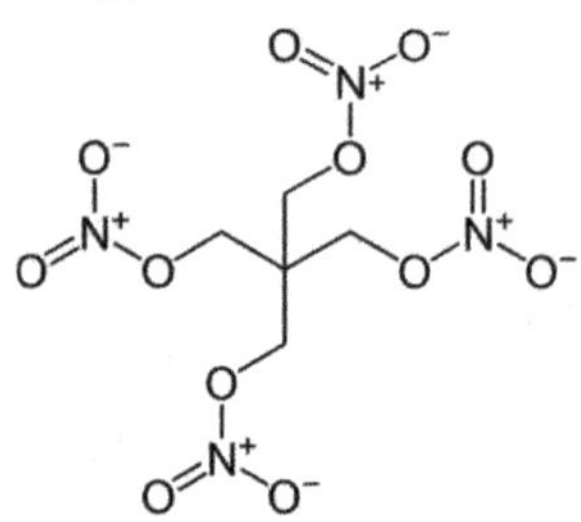
PETN

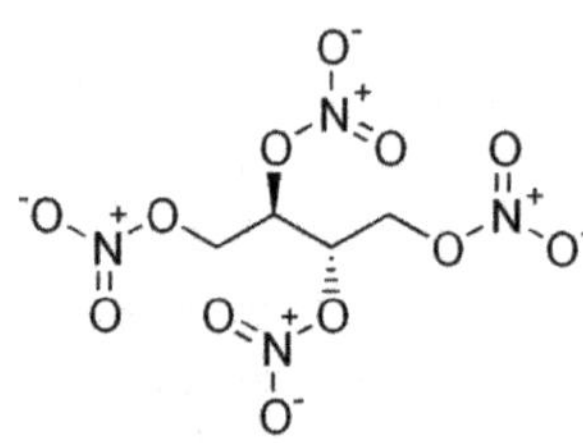
ETN

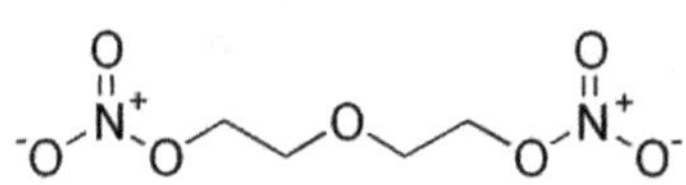
DEGDN

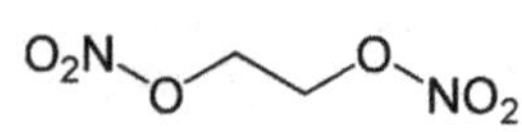
EGDN

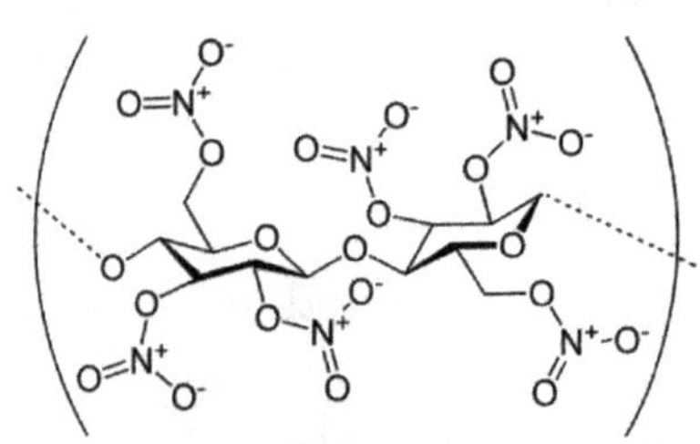
NC (monomer unit: glucose-trinitrate

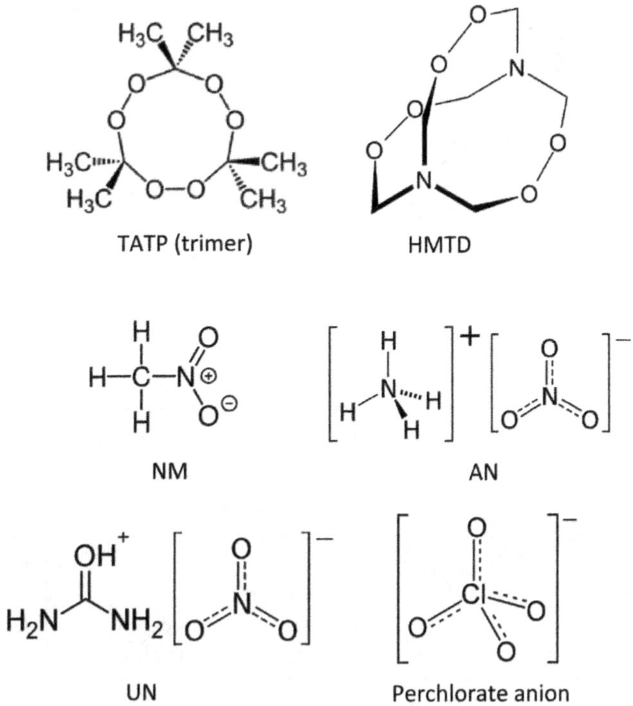

Molecular structures of energetic materials encountered in forensic explosive investigations. *(Wikipedia, free domain chemical structures, some were modified)*

Explosive	Mol formula	MW [Da]	Class/type	H/L	P/S/T	HME?
TNT	$C_7H_5N_3O_6$	227.1	Organic/nitro-aromat	H	S	No
Tetryl	$C_7H_5N_5O_8$	287.1	Organic/nitro-aromat/amine	H	S	No
PA	$C_6H_3N_3O_7$	229.1	Organic/nitro-aromat	H	S	No
RDX	$C_3H_6N_6O_6$	222.1	Organic/nitro-amine	H	S	No
HMX	$C_4H_8N_8O_8$	296.2	Organic/nitro-amine	H	S	No
NG	$C_3H_5N_3O_9$	227.1	Organic/nitrate-ester	H	P	No
PETN	$C_5H_8N_4O_{12}$	316.1	Organic/nitrate-ester	H	S	No
ETN	$C_4H_6N_4O_{12}$	302.1	Organic/nitrate-ester	H	P	Yes
DEGDN	$C_4H_8N_2O_7$	196.1	Organic/nitrate-ester	H	S	No
EGDN	$C_2H_4N_2O_6$	152.1	Organic/nitrate-ester	H	P	No
NC	$(C_6H_9(NO_2)_xO_5)_n$	–	Organic/nitrate-ester	L	S	No
TATP	$C_9H_{18}O_6$	222.2	Organic/peroxide	H	P	Yes
HMTD	$C_6H_{12}N_2O_6$	208.2	Organic/peroxide	H	P	Yes
NM	CH_3NO_2	61.0	Organic	H	T	No
AN	NH_4NO_3	80.0	Inorganic	H	T	No
UN	$CH_5N_3O_4$	123.1	Inorganic	H	T	No
FP	$Al + KClO_4$	–	Inorganic/pyrotechnic	L	P	No
BP	$C + S + KNO_3$	–	Inorganic/pyrotechnic	L	P	No

Overview of energetic materials encountered in forensic explosive investigations. Compound names: *BP*, black powder (mixture of carbon, potassium nitrate and sulfur); *DEGDN*, diethylene-glycol-dinitrate; *EGDN*, ethylene-glycol-dinitrate; *FP*, flash powder (mixture of aluminum and potassium perchlorate); *NM*, Nitromethane; *PA*, Picric Acid, other abbreviations are listed in the main text. In NC molecular formula x can be 1 (mononitrate), 2 (dinitrate) or 3 (trinitrate). Material characteristics: H(igh) versus L(ow) (detonation vs. deflagration), P(rimary)/S(econdary)/T(ertiary) (sensitivity) and HME? (is the compound encountered as home-made explosive in case work?). Information from NFI experts and Wikipedia

Energetic material	Energetic materials	Additives (F/A/B/P/S/T/D)	C/M
Dynamite	NG(>20%)	Diatomaceous earth (A)	C
C4/PE-4	RDX	Polyisobutylene (B), dioctyl sebacate or dioctyl adipate + mineral oil (P) (USA formulation) 2,3-dinitro-2,3-dimethylbutane (T)	M
Semtex	PETN + RDX	Styrene-butadiene (B), n-octyl phthalate, tributyl citrate (P), N-phenyl-2-naphthylamine (S), Sudan (D)	M
PEP 500	PETN	?	M
Pentolite	PETN(50%) + TNT(50%)	–	M/C
ANFO	AN	Diesel/kerosine (F)	C
ANNM	AN + NM	–	C
Amatol	AN + TNT(>60%)	–	M
Amatex	AN(51%) + TNT(40%) + RDX (9%)	–	M
Torpex	RDX(43%) + TNT(40%) + Al(17%)	–	M

Overview of professional civil (C) and military (M) energetic material formulations based on mixtures of explosives and the addition of fuel (F), adsorbents (A), binders (B), plasticizers (P), stabilizing agents (antioxidants) (S), taggants (T) and dyes (D). Information retrieved from NFI experts and Wikipedia

Given the complexity involved in forensic explosives analysis (in terms of chemistry and pre- and postexplosion setting) and the relatively modest case load (in terms of numbers, not in terms of expert capacity required) options for chemical profiling as described for TNT in the previous chapter (**Chapter 6—Chemical Profiling, Databases and Evidential Value**) have been limited for explosive materials in comparison with drugs of abuse. Another complicating factor is that several explosives are produced in very pure form through crystallization. TATP is an example of a home-made explosive that can be prepared in a straight forward single step reaction with broadly available raw materials:

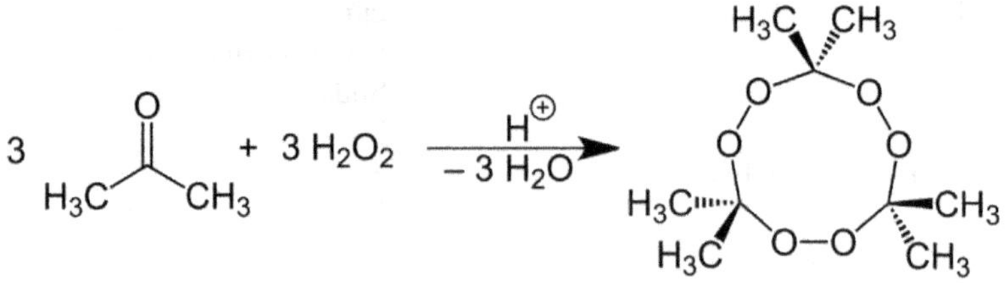

Reaction scheme of the production of TATP from acetone and hydrogen peroxide in the presence of an acid catalyst. *(Wikipedia, public domain)*

After addition of the reactants (acetone is typically added to an aqueous mixture of hydrogen peroxide and sulfuric acid), TATP crystals appear in the solution over time. Temperature control is not essential but to obtain yields of over 50% under ambient conditions, the reaction needs to be continued for several days. By filtering off the TATP crystals and rinsing the product with water a white crystalline TATP powder is obtained that is extremely pure. An impurity analysis with GC-MS or LC-MS will show that the TATP does not contain impurities at levels that allow chemical profiling. Hence, the approach discussed in the previous chapter would not be applicable if the perpetrator had used an IED with a main charge consisting of TATP. The use of home-made TATP in such a case would actually be more logical because access to TNT is not straight forward and requires special connections. So, for forensic explosives experts, it is important to have options for chemical profiling for explosives such as ETN and TATP, especially because reference material is likely to be present at the clandestine laboratory of the perpetrator. In the remainder of this chapter, we will discuss a very powerful MS based technique that has its roots in geology, but has been applied very successfully in forensic investigations. This technique is called **IRMS** and enables the forensic comparison at batch level of evidence and reference material on the basis of small variations of the isotope ratios of the elements present in the compounds. In the next paragraph, the fundamentals of IRMS will be explained in more detail. The measurement principle and the associated stringent quality requirements will be explained. Furthermore, it will be shown how this technique enables the study of the potential link between an explosive and its raw materials, allowing experts to address the question in a criminal investigation whether a raw material found with a suspect could have been used to produce the explosive charge retrieved from an IED at the crime scene. The application of IRMS in such a forensic chemical reconstruction will be demonstrated for the peroxide HME TATP and is based on a 2016 publication in the research group of the author. The corresponding reference has been included in the Further Reading section of this Chapter. In addition to forensic material analysis, IRMS can also be applied in the investigation of unidentified human remains. This forensic use of IRMS is termed **Human Provenancing** and can provide very valuable leads in cases where DNA profiles, fingermarks, or facial reconstruction have not resulted in the identification of an unknown victim. IRMS-based human provenancing is often explained by the popular phrase "*you are what you eat*" (and drink) as will be discussed further in paragraph 7.5. In forensic context, this refers to the fact that isotope ratios in body parts and tissues are related to the food stuffs that have been consumed by an individual and that many food stuffs carry a strong regional isotopic signature. Scholars or forensic experts that are planning to work with IRMS and want to gain detailed and comprehensive knowledge of the technique are recommended to read the book *Stable Isotope Forensics* written by prof Wolfram Meier-Augenstein (see Further Reading section for the full reference).

7.3 Isotope ratio mass spectrometry (IRMS)

To grasp the unique contribution of IRMS to forensic science, we must understand **stable isotopes**. A number of questions and answers will refresh your memory with respect to (or introduce you to) the elements of the periodic table, their stable isotopes, and the relative abundances with which we encounter them in our natural environment.

What does the name "isotope" mean and what attribute does it reflect?

The term isotope was introduced by English professor Frederick Soddy who received the Nobel prize in 1921 for his groundbreaking work on radiochemistry. The term was suggested to him by family friend Margaret Georgina Todd, a Scottish medical doctor and writer, after he explained his discovery that some radioactive elements possess different masses while exhibiting identical chemical properties. The name isotope is constructed from the two Greek words "isos" meaning "equal" and "topos" translating to "position." So isotope reads as "equal position" which refers to the periodic table of elements as introduced by Russian scientist Dimitri Mendeljeev. Isotopes of a given element have the same position in the periodic table indicating identical chemical behavior. At an atomic level isotopes share the same number of protons (and thus electrons in their neutral state) but differ in the number of neutrons and thus overall mass. This is indicated by the elemental atomic number Z (the number of protons), the number of neutrons N, and the mass number A according to the following standardized notation:

$$^{A}_{Z}\text{Element};\ N = A - Z$$

Pictures from left to right of Frederick Soddy (1921 Nobel Prize for his work on radiochemistry, 1877–1956), Margaret Georgina Todd (Scottish doctor and writer who suggested the term isotope to prof Soddy, 1859–1918), Dimitri Mendeljeev (who introduced the periodic table of elements in its current form, 1834–1907) and Harold Clayton Urey (1934 Nobel Prize for his work on isotopes and the discovery of Deuterium, 1893–1981). ***(Information and public domain pictures from Wikipedia)***

Most elements as listed in the period table have multiple isotopes, some of which are stable and some of which are radioactive. Radioactive isotopes (radionuclides, radiotopes) form into other radionuclides or stable elements over time through radioactive decay including the emission of radioactive radiation (γ radiation, i.e., high energy photons, α radiation, i.e., He nuclei, or β radiation, i.e., electrons). The rate of radioactive decay can vary from hours to thousands of years and is indicated by the half time ($t_{1/2}$), the time required to realize a 50% reduction of the isotope presence. Although valuable forensic information can be retrieved from the analysis of radionuclides (a very interesting example will be shown in paragraph 7.4), IRMS deals exclusively with stable isotopes. Stable isotopes do not undergo any elemental changes in the natural conditions encountered on earth and hence the isotope ratio, that is, the relative abundance of the isotopes of a given element, are in general considered to be constant.

Can you mention some important stable isotopes and list their relative abundance?

(If you do not know any, use a trusted Internet source such as Wikipedia).

Element	Isotope (%)	Isotope (%)	Isotope (%)	Isotope (%)
H	${}^{1}_{1}H$ (99.985%)	${}^{2}_{1}H$ (0.015%)		
C	${}^{12}_{6}C$ (98.89%)	${}^{13}_{6}C$ (1.11%)		
N	${}^{14}_{7}N$ (99.63%)	${}^{15}_{7}N$ (0.37%)		
O	${}^{16}_{8}O$ (99.76%)	${}^{18}_{8}O$ (0.20%)	${}^{17}_{8}O$ (0.04%)	
S	${}^{32}_{16}S$ (95.02%)	${}^{34}_{16}S$ (4.22%)	${}^{33}_{16}S$ (0.76%)	
Cl	${}^{35}_{17}Cl$ (75.77%)	${}^{37}_{17}Cl$ (24.23%)		
Br	${}^{79}_{35}Br$ (51.00%)	${}^{81}_{35}Br$ (49.00%)		
Sr	${}^{88}_{38}Sr$ (82.58%)	${}^{86}_{38}Sr$ (9.86%)	${}^{87}_{38}Sr$ (7.00%)	${}^{84}_{38}Sr$ (0.56%)
Pb	${}^{208}_{82}Pb$ (52.4%)	${}^{206}_{82}Pb$ (24.1%)	${}^{207}_{82}Pb$ (22.1%)	${}^{204}_{82}Pb$ (1.4%)

Overview of the isotopes and their abundance for some elements in the periodic table, the elements H, C, N, and O form the core of all organic compounds and their corresponding stable radionuclides are often termed "stable light isotopes" in IRMS studies

The **unified atomic mass unit** (indicated by **amu** or **Da** named after English scientist John Dalton who was the first to introduce the atomic theory) that is widely used in chemistry is defined at 1/12th of the mass of an unbound atom of ${}^{12}C$ in rest and in its electronic and nuclear ground state. The mass of 1 Da can be calculated by dividing the molar mass constant M_u by Avegadro's constant, N_A, the number of molecules in 1 mol. M_u is established from the experimentally determined molar mass of ${}^{12}C$ ($M({}^{12}C)$) leading to a mass of 1.66×10^{-27} kg for 1 Da:

$$1\ \text{Da} = \frac{M({}^{12}C)}{12.N_A}$$

Rather than using awkwardly small and complex numbers to express absolute atomic and molecular mass, the concept of unified atomic mass offers an elegant approach to express the mass of compounds relative to each other on the basis of a unified framework. The masses of an individual proton and neutron are almost identical (1.67×10^{-27} kg) but slightly higher than 1 amu (1.0073 and 1.0087 amu, respectively) while the mass of the much smaller electron equals 9.11×10^{-31} kg or 0.000549 amu.

Why is the mass of proton and a neutron higher than 1 amu?

That the mass of an individual proton or neutron is slightly higher than 1 Da is a direct consequence of the definition unified atomic mass based on the ^{12}C atom. The ^{12}C atom consists of six protons, six neutrons, and six electrons in a bound state. These atomic binding forces are very strong and according to Albert Einstein's most famous formula, the combination of these 18 individual subatomic particles to form a single carbon atom will result in the conversion of a small amount of mass in energy which is released in the formation process. This mass-to-energy conversion at atomic level is called the **mass defect**.

Can you determine the binding energy of one atom ^{12}C?

Use the following mass numbers:

One proton = 1.0073, one neutron = 1.0087 and one electron = 0.000549 amu.

One atom ^{12}C is made up of six protons, six neutrons, and six electrons. Combining these elements without any binding energy considerations leads to a unified atomic mass of 6×1.0073 amu (protons) + 6×1.0087 amu (neutrons) + 6×0.000549 amu (electrons) yielding a total of 12.0993 amu. Because 1 Da is defined on the basis of the ^{12}C atom, the mass defect equals 0.0993 amu or in other words the weight of one ^{12}C atom is 0.0993 amu less than anticipated on the basis of the masses of its subatomic parts. If we now consider Einstein's equation relating mass (m) and energy (E) through the speed of light constant ($c = 3.0e8$ m/s):

$$E = m \cdot c^2$$

the binding energy of one ^{12}C atom therefore corresponds to 1.4e-11 J or 92.5e6 eV. The energy unit electron volt (eV) relates to the kinetic energy gain of a single electron accelerated across a 1 V potential difference which this equals the charge of an electron (1.6×10^{-19} C).

The atomic mass in amu or Da is nothing more than a numeric ratio that expresses the mass of elements and molecules compared to the ^{12}C reference. As such, different stable isotopes of an element will exhibit variable masses on the basis of the varying number of neutrons, which is expressed in the **relative isotopic mass**. However, when performing a chemical reaction using raw materials that have not been isotopically fractionated (which is typically the case), the molecular mass of the raw materials and the reaction product will reflect the natural abundance of the stable isotopes as encountered in our natural world. This is expressed via the so-called **Standard Atomic Weight** ($A_{r,std}$). The atomic weight of a chemical element is determined as the average of the individual relative isotope masses weighed according to their abundance in the earth's environment. It is the most common and practical atomic weight used, e.g. to determine molar mass of molecules on the basis of elemental formulas. As such, it is listed as the elemental weight in the periodic table of elements but it is interesting to realize that it is a weighted average and does not generally reflect the mass of individual isotopes.

What is the Standard Atomic Weight of Pb in Da?

(Use the abundance and masses of the Pb isotopes listed in the table above).

Note that this weighted elemental mass is noted in the Periodic Table of Elements

Element	Isotope (%)	Mass	Abundance [%]	Mass fraction
Pb	$^{208}_{82}Pb$ (52.4%)	208	52.4	109.0
	$^{206}_{82}Pb$ (24.1%)	206	24.1	49.6
	$^{207}_{82}Pb$ (22.1%)	207	22.1	45.7
	$^{204}_{82}Pb$ (1.4%)	204	1.4	2.9
Sum			100	207.2

Calculation of the Standard Atomic Weight of Lead as the abundance weighed average of the molecular mass of its stable isotopes

For a long time, it was assumed that the natural abundance of stable isotopes was constant and was set as our planet was created in the solar system. As scientific insights developed and mass spectral instruments became more advanced, it became clear that small differences in isotope abundances exist due to chemical and physical processes in the environment. The mass differences induce isotope fractionation processes and have a small but significant effect on chemical reactivity. This means that the standard atomic weight, which is based on the abundance of the stable isotopes of an element, actually can vary depending on the origin of the material. In the latest versions of elemental tables generated by official international bodies like IUPAC (International Union of Pure and Applied Chemistry) and CIAAW (Commission on Isotopic Abundances and Atomic Weights), this observed variation is accounted for by providing an atomic weight range in addition to the average value as is listed in the table below for a number of elements.

Element	$A_{r,std}$	Min	Max	Range [%]
H	1.008	1.00784	1.00811	0.027
C	12.011	12.0096	12.0116	0.017
N	14.007	14.00643	14.00728	0.006
O	15.999	15.99903	15.99977	0.005
F	18.998	—	—	0
Ne	20.180	—	—	0

Average standard atomic weight and the range reported by CIAAW since 2009 for a selection of light elements

Why do Fluorine and Neon have a single standard atomic weight?

(Note: two different explanations exist for these two elements).

The halogen Fluorine (F) is one of the 26 mono-isotopic elements in the periodic table. With only one isotopic species (100% abundance) the standard atomic weight equals the relative isotopic mass and no range will be observed. However, for the noble gas Neon (Ne), stable isotopes are known (^{20}Ne (90.48%), ^{21}Ne (0.27%), ^{22}Ne (9.25%)) and another reason must exist for not observing any range in the $A_{r,std}$. This reason can be found in the fact that neon as a noble gas is highly unreactive and very diffusive. Hence, fractionation due to physical and chemical processes does not occur.

For the elements that do exhibit a range in the standard atomic weight, the numbers clearly illustrate that the observed variation in isotopic abundance is relatively very small compared to the average weight. So assuming the abundance of the stable isotopes to be constant is quite acceptable for many chemical activities and considerations. However, with forensic IRMS, we want to deliberately zoom in on this small difference to compare materials and reconstruct events.

Why are regular mass spectrometers, even the high resolution instruments, not capable of measuring isotope ratios?

The challenge does not relate to mass resolution, and modern MS instruments have ample resolution to measure each individual isotope on the basis of the different number of neutrons in the atomic nucleus. However, to distinguish materials on the basis of a minute difference in the abundance of the isotopes, the amount of each isotope in a sample needs to be measured with extreme precision and accuracy. The need in IRMS to measure very small differences in isotope abundance is illustrated in the figure below:

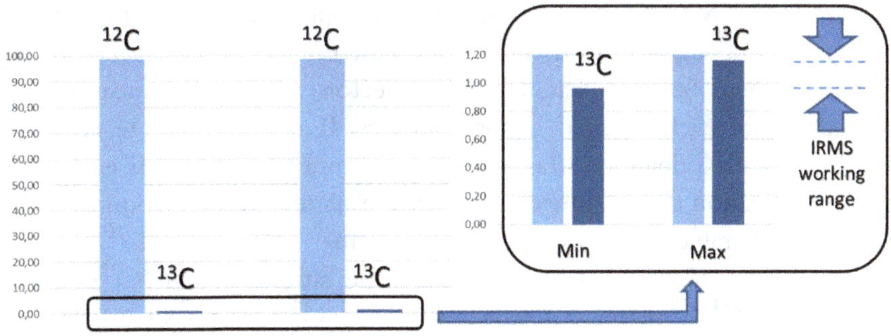

Illustration of the IRMS challenge: the working range, that is, the natural variation observed in the abundance of the ^{13}C isotope, is very small requiring a very accurate and precise measurement of the $^{12}C/^{13}C$ ratio in a material of forensic interest

For regular MS instruments, the variation in the signal intensity easily outweighs the difference in the isotope abundance. Therefore, instrument manufacturers have developed dedicated IRMS instruments to be able to measure isotope ratios of light elements with the desired accuracy and precision. The figures below sketch the working principle of IRMS instruments although actual configurations can differ between manufacturers. Isotope ratios for ^{13}C and ^{15}N can be assessed in a continuous flow elemental analyzer IRMS system (**CF-EA-IRMS**). To this end, a small amount of the material of interest (typically in the mg range) is placed in a tin or silver foil container and is flash combusted in the presence of oxygen. This process is carried out at a temperature of roughly 1000 °C in a reactor containing an oxidizing carrier material such as chromium oxide. This leads to the oxidation of all organic material into various gaseous compounds (N_2, NO_x, CO_2, SO_2, and H_2O). These gaseous reaction products are subsequently transported at a He flow rate of roughly 100 mL/min to a reduction reactor containing copper particles. This reduction step eliminates the presence of oxygen and quantitatively converts all NO_x to N_2. After removal of the water, the gaseous reaction products are separated isothermally on a packed GC column. The separated compounds are detected non-invasively using a thermal conductivity detector before entering the mass spectrometer. A helium make-up flow can be employed to optimize the signal intensity. The instrument can switch between the sample flow and reference gasses for quality control and calibration. Electron ionization using an electron beam is the preferred ionization method in line with GC-MS instrumentation. The mass spectrometer is typically set-up has a classic sector instrument in which the ions are deflected in a strong magnetic field. Separate Faraday cup detectors are used to measure the abundance of each charged isotopic species. IRMS systems are typically equipped with a detector array consisting of multiple units. For each cup, the amplification can be tuned to match the natural average abundance to create comparable ion currents. Monitored m/z values are 28 ($^{14}N_2$), 29 ($^{14}N^{15}N$) and 30 ($^{15}N_2$) for the ^{15}N isotope ratio and 44 ($^{12}C^{16}O_2$), 45 ($^{13}C^{16}O_2$, $^{12}C^{17}O^{16}O$) and 46 ($^{12}C^{17}O_2$, $^{13}C^{17}O^{16}O$, $^{12}C^{18}O^{16}O$) for the ^{13}C isotope ratio assessment. To account for the small ^{17}O contribution, a so-called ^{17}O correction algorithm can be applied. Isotope ratios are determined from the cup detector signal areas that are measured for the various isotopic species simultaneously as the gaseous products elute from the GC column and enter the mass spectrometer.

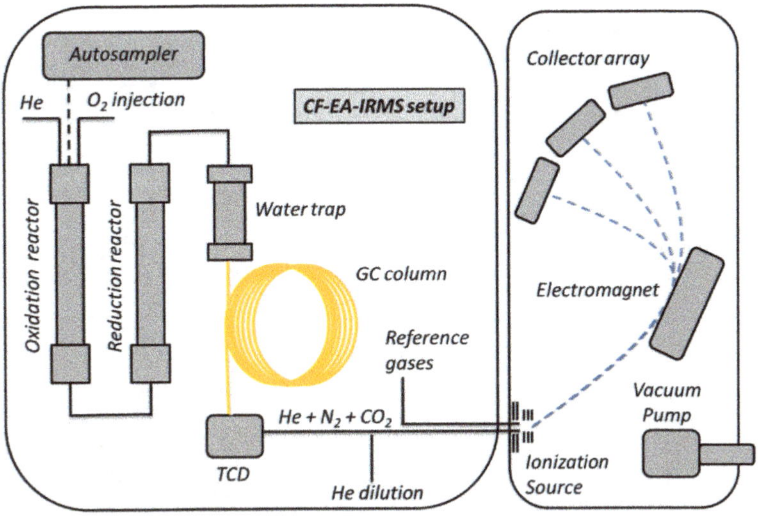

Schematic representation of the measurement of ^{13}C and ^{15}N isotope ratios with a continuous flow (CF)—elemental analyzer (EA)—IRMS system (based on Thermo Scientific schedules as depicted in the book Stable Isotope Forensics, see Further Reading section for full reference).

For the measurement of ^{2}H and ^{18}O isotope ratios, a different arrangement involving high-temperature pyrolysis is applied. Organic samples are quantitatively converted into H_2 and CO in a glassy carbon reactor operated at a temperatures exceeding 1400 °C. This setup is termed **CF-HTC/EA IRMS** (continuous flow-high temperature conversion/elemental analyzer IRMS). Monitored m/z values are 1 (^{1}H) and 2 (^{2}H, D) for the ^{2}H isotope ratio and 28 ($^{12}C^{16}O$), 29 ($^{13}C^{16}O$, $^{12}C^{17}O$) and 30 ($^{12}C^{18}O$, $^{13}C^{17}O$) for the ^{18}O isotope ratio assessment.

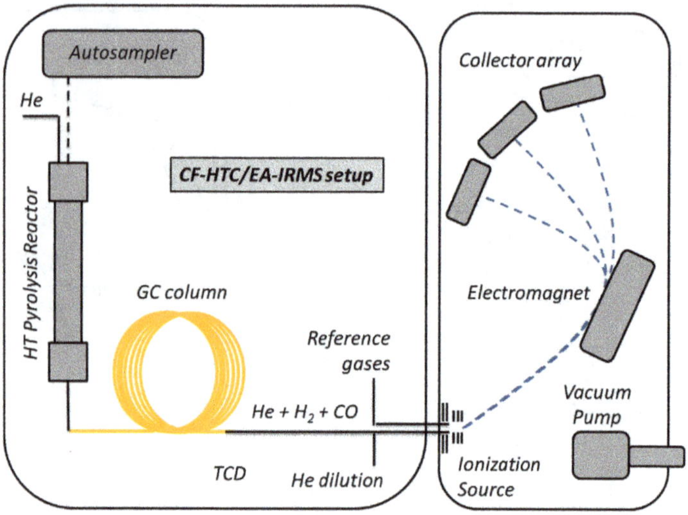

Schematic representation of the measurement of ^{2}H and ^{18}O isotope ratios with a CF—high-temperature conversion (HTC)/ EA—IRMS system (based on Thermo Scientific schedules as depicted in the book Stable Isotope Forensics, see Further Reading section for full reference).

For compounds containing nitrogen (many organic explosives are relatively nitrogen rich with exception of the peroxides) also N_2 is generated during the conversion and therefore the GC separation of the gaseous products prior to IRMS analysis is essential.

Why is a full separation of the gaseous products N_2 and CO required prior to the MS measurement of the ^{18}O isotope ratio?

IRMS systems have been carefully optimized to measure minute differences in the isotopic signal intensities to accurately establish the isotope ratios. The smallest interference will immediately lead to significant systematic errors. Therefore, it is necessary to achieve full separation before the compounds enter the mass spectrometer. However, in the ^{18}O analysis of N-containing compounds, this is also essential because of the **isobaric interference** of the coproduced nitrogen gas. For the measurement of both the ^{15}N (N_2) and ^{18}O (CO) isotope ratio, the ion signals at m/z 28, 29, and 30 are monitored. This is convenient with respect to the MS settings but also illustrates the need for the GC baseline separation of N_2 and CO.

Now with these dedicated MS sector instruments, isotope ratios for light elements (H, C, N, and O) can be measured with the desired precision and accuracy. However, work and quality control need to be very strict because of the very small isotopic ratio working range. To be able to compare and combine data from various labs and instruments and to ensure data consistency over time, suitable reference standards and internationally recognized protocols are required. To this end, an international community has been established of forensic experts that apply IRMS in case work. The FIRMS (Forensic Isotopic Ratio Mass Spectrometry) network plays a crucial role in setting international standards to ensure that IRMS findings are sufficiently robust to be used as evidence in criminal investigations. Readers that want to implement IRMS in their laboratory are advised to read the good practice guide and the guide on the interpretation of forensic IRMS data that have been published by the FIRMS community. These documents can be downloaded from the FIRMS website and full references can be found in the Further Reading section. The final part of this paragraph dealing with the measurement and quality protocols concerning IRMS is based on the information provided in these documents (in addition to the Stable Isotope Forensics book of prof dr Wolfram Meier-Augenstein).

Rather than reporting absolute values, isotope ratios for samples of interest are reported relative to internationally agreed references that act as "zero-anchors":

	Zero anchor	Explanation
$^{2}H:^{1}H$	VSMOW	**Vienna standard mean ocean water** *Distilled water standard prepared from ocean water. Ocean water is taken as a reference in the environmental water cycle in which ^{2}H isotope ratios are affected by water evaporation and condensation.*
$^{13}C:^{12}C$	VPDB	**Vienna pee dee belemnite** *A carbon reference based on marine fossil deposition (belemnite is an extinct squid like animal that lived in the late cretaceous period) found in deposits along the Pee Dee river in North and South Carolina in the USA). These deposits are known for very high $^{13}C:^{12}C$ ratios*
$^{15}N:^{14}N$	Air	**Atmospheric nitrogen** *Air consists for roughly 80% of N_2 which due to its unreactive nature and high volatility provides a consistent nitrogen isotope ratio reference point across the world.*
$^{18}O:^{16}O$	VSMOW	**Vienna standard mean ocean water** *Distilled water standard prepared from ocean water. Ocean water is taken as a reference in the environmental water cycle in which ^{18}O isotope ratios are affected by water evaporation and condensation.*
$^{34}S:^{32}S$	VCDT	**Vienna canyon diablo troilite** *An FeS (troilite) mineral sulfur reference from meteorite material found near Diablo Canyon in northern part of Carolina, USA. Meteorite mineral was chosen because of limited variability of S isotope ratios and because these values are believed to reflect initial conditions on earth.*

Internationally accepted light isotope reference materials used as 'zero anchors' in IRMS, the term Vienna refers to the International Atomic Energy Agency (IAEA) that defines and provides isotope reference materials. The IAEA houses in Vienna, Austria

The isotope ratio of light elements is expressed as a relative deviation from these anchor points. Because these deviations from the reference standards (Std) are very small, measured isotope ratios for samples of interest (S) are converted to **δ values** that express the relative shift in ‰ (per mil, 1/10th of a %):

$$\delta\left(^{13}C\right)_{S/Std} = \frac{\left(^{13}C/^{12}C\right)_S - \left(^{13}C/^{12}C\right)_{Std}}{\left(^{13}C/^{12}C\right)_{Std}} = \left(\frac{\left(^{13}C/^{12}C\right)_S}{\left(^{13}C/^{12}C\right)_{Std}} - 1\right)\cdot 1000[‰]$$

Metrology institutions like IAEA, NIST (National Institute for Standards and Technology), and the European IRMM (Institute for Reference Materials and Measurements) define and provide certified standards for IRMS analysis. As the original anchor materials deplete or are replaced based on new insights, a range of so-called secondary reference materials is commercially available. The certified isotope ratios of these materials are established from the primary anchor materials through interlaboratory exercises coordinated by metrology institutes. Another reason for having a wide range of accepted reference materials is to provide adequate calibration over the entire isotope ratio working range. Therefore, it is important to also have access to standards with isotope ratios that strongly deviate from the primary anchors. An overview of international IRMS reference standards is given in the table below. In addition to these official reference materials, also tertiary and in-house standards are employed in forensic IRMS laboratories. The δ values of these reference materials are not internationally recognized as they have been established with the use of secondary standards. Tertiary reference standards typically are established through interlaboratory comparison within IRMS networks. These materials play an important role as quality control samples to ensure the robustness of the IRMS instrumentation and data.

Primary IRMS standard	Code	Certified δ values [‰ versus anchors]			
		δ^2H	$\delta^{13}C$	$\delta^{15}N$	$\delta^{18}O$
TS limestone	NBS 19		+1.95		+28.65
Carbonatite	NBS 18		−5.01		+7.20
Lithium carbonate	LSVEC		−46.6		+3.69
Oil	NBS 22	−116.9	−30.03		
Sucrose	IAEA−CH−6		−10.45		
PE foil	IAEA−CH−7	−100.3	−32.15		
Caffeine	IAEA-600		−27.77	+0.91	−3.48
L-glutamic acid	USGS40		−26.39	−4.52	
L-glutamic acid	USGS41a		+36.55	+47.55	
Cellulose	IAEA−CH−3		−24.72		
Ammonium sulfate	IAEA-N-1			+0.43	
Ammonium sulfate	IAEA-N-1			+20.35	
Potassium nitrate	USGS32			+180.0	
Potassium nitrate	USGS34			−1.8	
Water	GISP	−189.7			−24.78
Water	GISP2	−253.8			−33.43
Water	SLAP	−428.0			−55.5
Water	SLAP2	−427.5			−55.5
Water	IAEA-604	+799.9			−5.86
Benzoic acid	IAEA-601		−28.81		+23.14
Benzoic acid	IAEA-602		−28.85		+71.28

Internationally available and accepted IRMS secondary reference materials (from the book Stable Isotope Forensics by prof dr Wolfram Meier-Augenstein, full reference given in Further Reading section)

Method validation and the use of control charts will be discussed in more detail in **Chapter 9—Quality and Chain of Custody**. The development of a reliable and robust IRMS method for forensic case work requires careful study and optimization of the following aspects:

- Sample homogeneity with respect to the isotope ratios and representative sampling and subsampling
- Sample stability during sample preparation and IRMS analysis (e.g., water adsorption of hygroscopic materials can lead to shifting d^2H and $d^{18}O$ values)
- Complete catalytic conversion and GC separation of the gaseous reaction products (e.g., incomplete conversion and N_2—CO peak overlap due to excessive N_2 formation for N-rich compounds such as explosives have been reported under standard conditions)
- Selection of appropriate reference materials for normalization of the δ values (ideally reference materials are matrix matched and cover the entire isotope ratio range encountered in the samples of interest)
- Sample sequences that ensure a sufficient number of sample repeats (typically 3—5) and include sufficient and regular analyses of reference materials and control samples
- Correction procedures for or elimination of blank signals, drift in δ values, detector non-linearity, and memory effects
- Calculation of the overall measurement uncertainty of the normalized δ values on the basis of the individual sources of variation (see illustration below from the FIRMS Good Practice Guide for IRMS, full reference provided in the Further Reading section)

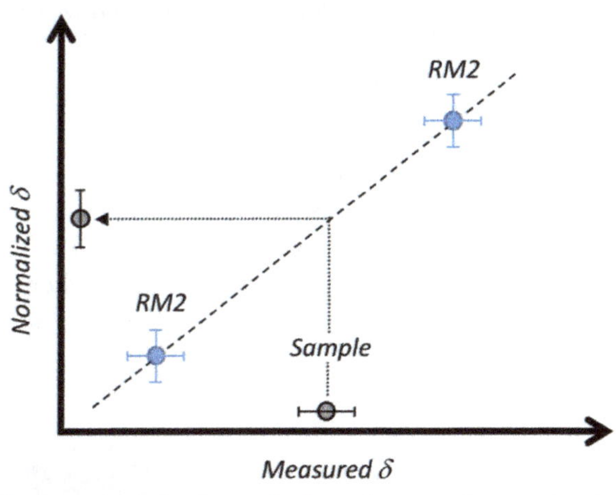

Illustration of the normalization of δ values and associated measurement uncertainty. *(Adapted from the FIRMS Good Practice Guide for IRMS, full reference given in Further Reading section.)*

By following the good practice guidelines, a robust IRMS method could be developed for the accurate and precise assessment of δ^2H, $\delta^{13}C$ and $\delta^{18}O$ values of TATP samples. This comprises all elements from which this energetic material is constructed as this organic peroxide explosive lacks the presence nitro groups and do not contain nitrogen. The opportunities this provides for chemical profiling and even synthesis reconstruction of high purity materials of forensic interest will be discussed in the next section.

7.4 Chemical profiling and synthesis reconstruction of TATP with IRMS

In **Section 7.2**, the overall reaction scheme for the production of TATP from acetone and hydrogen peroxide in the presence of an acidic catalyst is given. As discussed, TATP is a popular but very risky HME because of its shock and friction sensitivity as a primary explosive. The production is straight forward and yields a very pure end product without the need for specific equipment or expert chemistry knowledge. The required raw materials can be obtained with relative ease and without raising suspicion as this involves commodity chemicals that are used domestically and industrially on a large scale in a variety of products and applications. The discussion on the forensic use of the IRMS analysis of TATP is based on a study that was published in 2016 in the Journal of Forensic Sciences in a paper entitled "*The Potential of Isotope Ratio Mass Spectrometry (IRMS) and Gas Chromatography-IRMS Analysis of Triacetone Triperoxide in Forensic Explosives Investigations*" (reference can be found in Further Reading section). The illustrations shown in this section have been reprinted and adapted with permission of Wiley, the publisher of this forensic journal. Here we will primarily focus on the δ^2H and $\delta^{13}C$ results and the link between the raw material acetone and the end product TATP as the $\delta^{18}O$ findings were less convincing in the original study. As both acetone and hydrogen peroxide contain H, C and O atoms, the basic building blocks of organic peroxide explosives, it is important to understand the origin of these elements in the TATP molecule. This becomes clear when the reaction mechanism is studied in more detail:

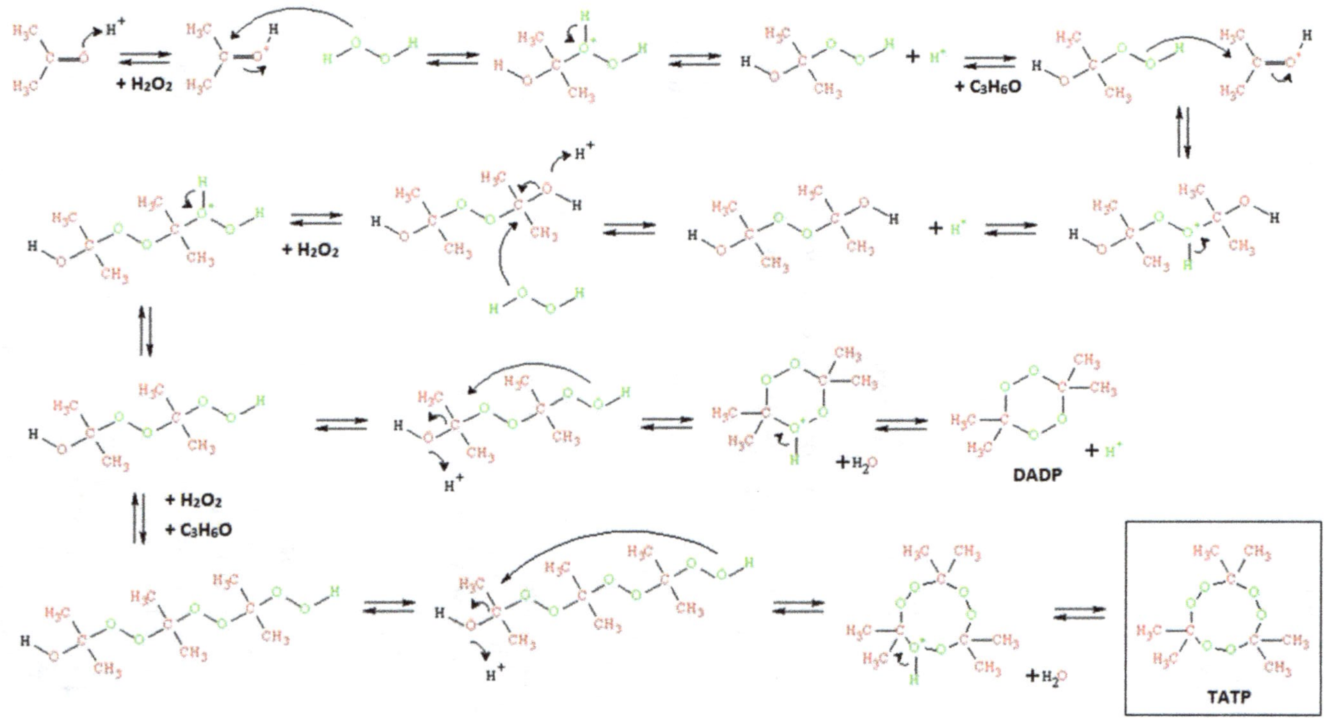

Reaction mechanism of TATP formation from acetone and hydrogen peroxide in the presence of an acid catalyst (elements originating from acetone are indicated in red and green represents hydrogen peroxide origin. *(Reprinted with permission from Bezemer et al. Journal of Forensic Sciences, 61 (2016) 1198–1207, https://doi.org/10.1111/1556-4029.13135)*

From the reaction mechanism, it becomes clear that in the TATP molecule, the C and the H atoms are originating from the raw material acetone, whereas the hydrogen peroxide is the source of the oxygen atoms. The O atom present in acetone as part of the ketone functionality (C=O bond) ends up in the water molecule that is formed during the final ring closure step yielding the cyclic TATP structure. So, to study the isotopic relation between acetone and TATP, we need to focus on the hydrogen and carbon isotope ratio while $\delta^{18}O$ data can provide insights how hydrogen peroxide affects the isotopic oxygen signature of TATP. However, two aspects complicate this latter comparison. First of all, hydrogen peroxide is not used in pure form but rather as a concentrated aqueous solution which means that the isotopic hydrogen and oxygen signature of H_2O_2 cannot be accurately assessed as the overall isotope ratio of the entire sample is measured with IRMS and thus will be a weighted average of the hydrogen peroxide and water in the sample. Furthermore, the working range of $\delta^{18}O$ values in the set of hydrogen peroxide samples studied was found to be quite limited in relation to the measurement uncertainty limiting the discrimination potential.

Similar to profiling methods based on chemical impurities as discussed in the previous chapter for TNT, the forensic potential of IRMS to discriminate and match materials is determined by the natural variation observed in the δ values. Ideally this range is much larger than the variation observed within a single source. Rather than producing a large sample set of TATP samples, the δ^2H and $\delta^{13}C$ working range can also be studied in a much easier and safer way by conducting a study involving the raw material acetone. Under the assumption that the isotopic H and C signature will be "translated" to the final product, the variation in isotope ratios for the precursor will also reflect the working range for TATP.

To that end, a substantial and forensically relevant set of samples was collected from acetone samples confiscated by the Dutch police in criminal investigations mostly related to illicit drug production. These 37 selected acetone samples were assumed to be unrelated on the basis of the available case information (although case related samples are forensically very relevant, the ground truth with respect to the origin of the material is often unknown and hence there is always a slight risk that some materials in datasets are unexpectedly linked). The subsequent IRMS analysis resulted in δ^2H values in the range of -210 to $-143‰$ with a standard deviation not exceeding 0.9‰ and $\delta^{13}C$ values in the range of −32 to −26‰ with a standard deviation below 0.1‰. The combination of H and C isotope ratio for each individual sample, depicted in the scatter plot below, provides a very positive outlook indeed. A substantial sample-to-sample variation is observed as the acetone samples are more or less randomly and evenly distributed within a considerable δ^2H and $\delta^{13}C$ working range. Assuming that the error bars (representing ±2 standard deviations observed for a three- to fivefold analysis for each sample) are indicative of the 95% confidence interval, it is clear that many (but not all) acetone samples can be distinguished on the basis of their bulk carbon and hydrogen isotope ratios. Because of the limited number of features considered (in this case the ^2H and ^{13}C isotope ratio) and the absence of any apparent correlation (as indicated by the scatter plot) in this case, there is no need to develop a score-based or feature-based model. Instead, the evidential value associated with matching δ^2H and $\delta^{13}C$ values of two samples can be estimated on the basis of the relative occurrence of such values in a representative database (the corresponding random match probability). This is not only determined by the isotopic working range but also by the distribution of the values for the reference samples in the database. Ideally, an even distribution is obtained of the samples over this working range. By binning the samples, this distribution is illustrated below for both the hydrogen and carbon isotope ratios for the acetone collection. With 37 samples, a total of seven and six bins were chosen over the δ^2H and $\delta^{13}C$ working range, respectively (rule of thumb as discussed in the previous chapter is to set the number of bins equal to the square root of the number of data points). The histograms show some variation and trends but for each bin the relative frequency is within 5%–30% indicating that acetone samples are distributed across the entire isotopic working range. Assuming that the hydrogen and carbon isotope ratios show now significant correlation (as indicated by the random distribution of data points in the scatter plot), the combined random match probability is 0.25%–8% depending on the actual δ^2H and $\delta^{13}C$ values. This indicates that corresponding hydrogen and carbon isotope ratios for two forensic samples yields LR values in the range of 10–400 on the basis of this limited acetone data set.

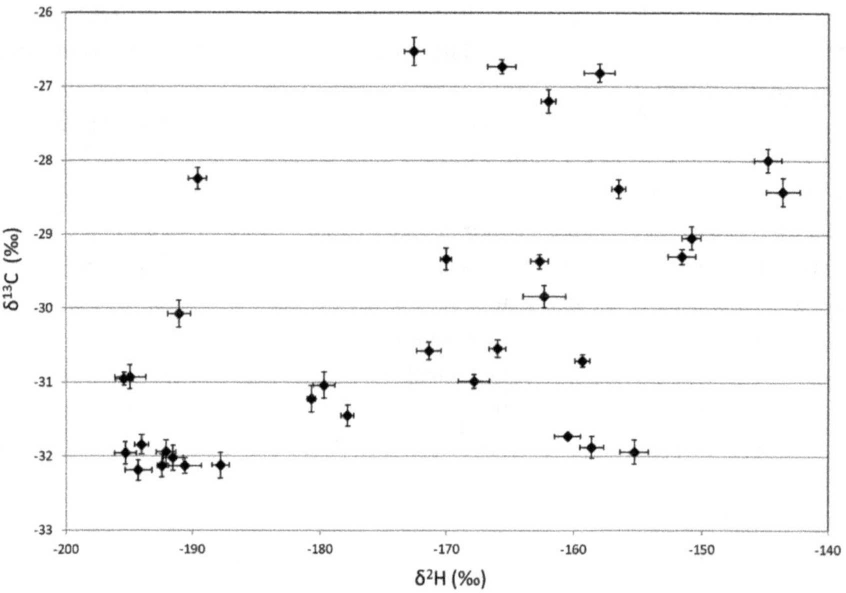

Scatter plot of δ^2H and $\delta^{13}C$ values (including 2SD error bars) for the individual samples of the acetone data set. *(Reprinted with permission from Bezemer et al. Journal of Forensic Sciences, 61 (2016) 1198–1207, https://doi.org/10.1111/1556-4029.13135)*

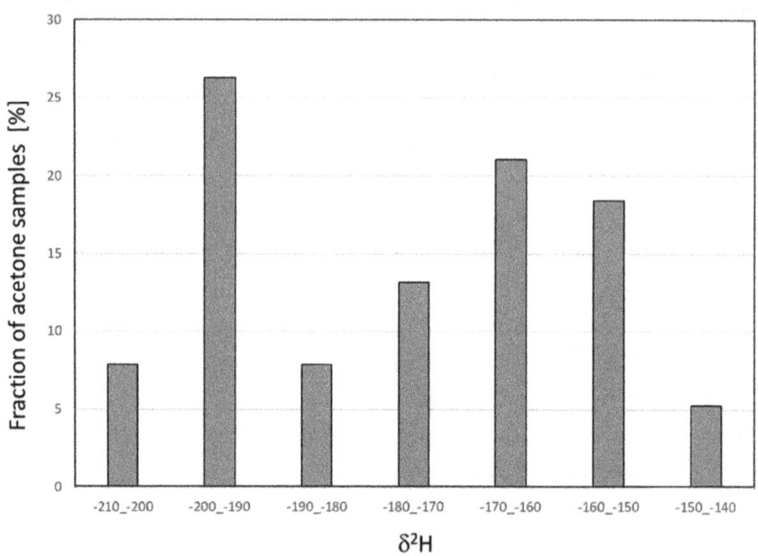

Distribution of δ^2H values for the acetone data set. *(Reprinted with permission from Bezemer et al. Journal of Forensic Sciences, 61 (2016) 1198–1207, https://doi.org/10.1111/1556-4029.13135)*

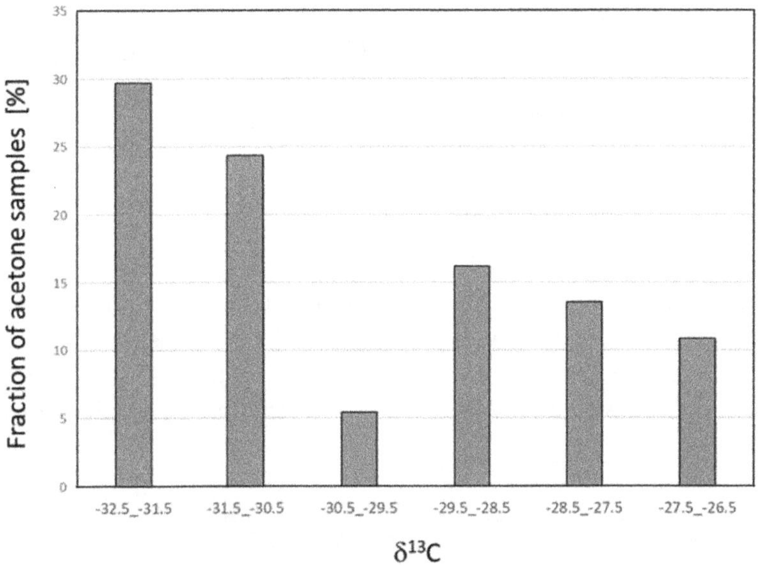

Distribution of $\delta^{13}C$ values for the acetone data set. *(Reprinted with permission from Bezemer et al. Journal of Forensic Sciences, 61 (2016) 1198–1207, https://doi.org/10.1111/1556-4029.13135)*

Another approach to illustrate the discriminating power (*DP*) of the developed IRMS methodology is to perform a pairwise comparison of all samples in the reference database and calculate the *DP* on the basis of the number of matching (m) pairs (unrelated samples with a $\Delta\delta$ within the measurement uncertainty) and the total number of samples (n) in the database:

$$DP(\%) = 100 \cdot \left[1 - \frac{2 \cdot m}{n \cdot (n-1)}\right]$$

For a number of k independent features, the combined DP_k can be calculated in a way similar to the *LR*:

$$DP_k(\%) = 100 \cdot \left[1 - \prod_{i=1}^{k} \left(\frac{2 \cdot m_i}{n_i \cdot (n_i - 1)}\right)\right]$$

For the acetone dataset, the $DP(\delta^2H)$ was 87% (87 matching pairs) and for $\delta^{13}C$ a *DP* of 84% (104 matching pairs) was obtained. Under the assumption of independence, the combination of the hydrogen and carbon isotope data should yield a discrimination power of 98%. On the basis of the scatter plot, a combined *DP* of 96% was determined experimentally (25 matching pairs), which is in good agreement and illustrates the absence of correlation between δ^2H and $\delta^{13}C$ for the acetone samples.

Next from this reference database, a number of acetone samples were selected with clearly distinguishable δ^2H and δ^{13}C values (samples that are located in different areas of the scatter plot). Under standardized conditions, small amounts of TATP were synthesized from these acetone sources to study the relation between the isotopic signatures of the raw material and the end product. A relatively short reaction time of 6 h was used to limit the TATP yield to keep the amount of explosive well below 2 gr for safety reasons. Sulfuric acid was used as the catalyst and was dosed at 1 mol% of the amount of acetone used in the reaction. The temperature of the aqueous reaction mixture was not controlled and the TATP crystals were rinsed with water. After drying, the light stable isotope ratios of the TATP samples were assessed with IRMS leading to the results depicted in the graph below:

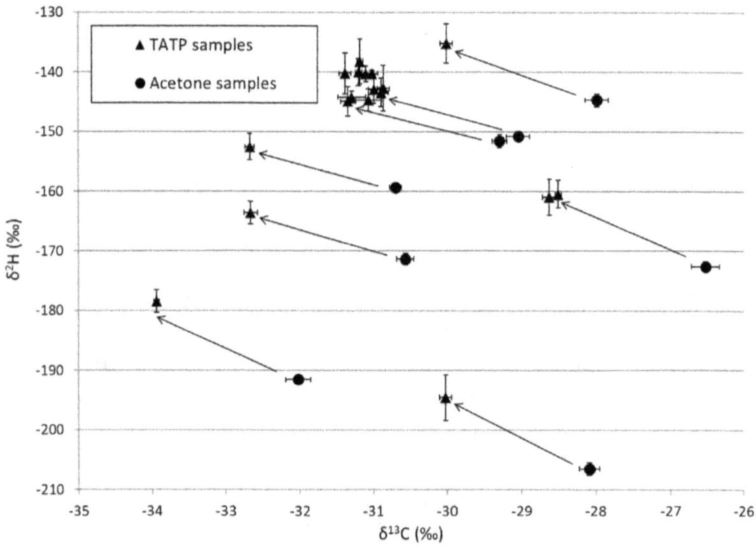

TATP δ^2H and δ^{13}C values produced under standardized conditions from various acetone sources. *(Reprinted with permission from Bezemer et al. Journal of Forensic Sciences, 61 (2016) 1198–1207, https://doi.org/10.1111/1556-4029.13135)*

The results show that although the measured isotope ratios for TATP significantly differ from those of the acetone precursors, the observed isotopic shifts are, however, very consistent. This means that the isotopic signature of the raw material is indeed "translated" (but not directly "copied") into the end product. That this translation is more or less repeatable under selected reaction conditions is shown below indicating that a potential link between acetone and TATP can be investigated by measuring the carbon and hydrogen isotope ratios for both samples. Interestingly, no correlation was found between the δ^{18}O precursor and end product which was expected on the basis of the reaction mechanism given earlier as the oxygen in the TATP molecule originates from hydrogen peroxide and not from acetone

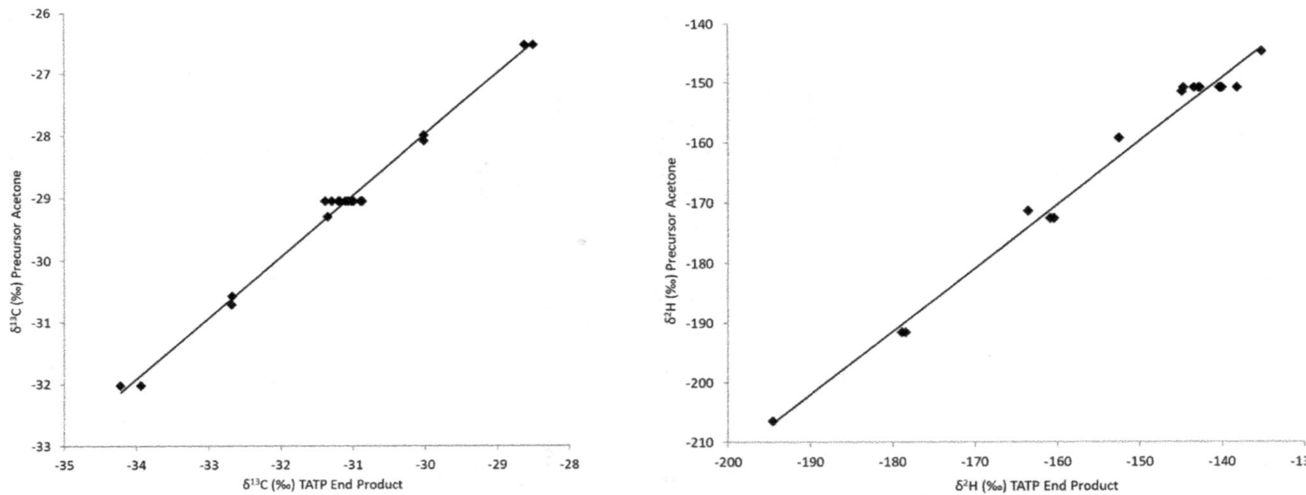

TATP δ^2H and $\delta^{13}C$ values produced under standardized conditions from various acetone sources. *(Reprinted with permission from Bezemer et al. Journal of Forensic Sciences, 61 (2016) 1198–1207, https://doi.org/10.1111/1556-4029.13135)*

For each reaction experiment, the δ^2H of TATP shifts to higher (i.e., less negative) values compared to the acetone source. This positive enrichment means that during the reaction relatively more deuterium is incorporated in the end product. For carbon, the opposite effect is observed as the TATP product exhibits a lower (*i.e.*, more negative) δ^{13}C value. A negative enrichment means that the end product is depleted from the heavier isotope (in this case ^{13}C) compared to the starting material. Isotopic fractionation is quite commonly observed during physical processes and chemical reactions. For physical processes like evaporation, diffusion, and crystallization, isotope fractionation is not governed by any chemical reaction but originate from the small but significant effect of the mass difference on physical parameters like vapor pressure and diffusion coefficient. Such processes are causing so-called **thermodynamic isotope effects**. Alternatively, **kinetic isotope effects** are the result of small but systematic differences in the rates of the individual isotopes in a chemical reaction. The heavier isotope usually exhibits slightly stronger bond energies which require a higher activation energy to initiate the reaction. This results in reaction rates being slightly lower for the heavier isotope and leads to an enrichment of the lighter isotope in the reaction product compared to the raw material. The degree of fractionation can be quantified by the **enrichment factor ε** that experimentally can be determined from the difference between the δ values of the precursor (RM) and the product (P) and is thus also expressed in ‰. Theoretically, ε is linked to the **fractionation factor α** (please note that the same symbol is used to indicate the odds in the formula of Bayes) which is defined as the ratio of the reaction rates of the heavier (k_H) and lighter isotope (k_L), respectively:

$$\varepsilon = \delta_{RM} - \delta_P = 1000 \cdot (\alpha - 1) = 1000 \cdot \left(\frac{k_H}{k_L} - 1 \right)$$

For the TATP batches produced under standard conditions, an average ε value of −2.10‰ was observed for δ^{13}C with a relative standard deviation of 6.5%. This is a relative shift of 7% when considering the average δ^{13}C value of the acetone precursors. For δ^2H, the average enrichment factor was 11.0‰ but with a high relative standard deviation of 32.5%. This shift also represents roughly 7% of the average δ^2H precursor value. The positive enrichment for hydrogen is difficult to explain as the hydrogen atoms in the acetone molecule are not directly involved in the chemical reaction that forms TATP. Possibly it is related to the crystallization of the TATP from the reaction solution.

Under standardized conditions, the yield of the TATP production was 2%–10% based on the amount of acetone in the reaction mixture. This low yield was chosen out of safety considerations as it limits the amount of TATP produced. However, such low yields are not very relevant and realistic when considering forensic case work. Typically, TATP is produced under less controlled conditions and with an aim to maximize the yield to create as much explosive material as possible for the main charge of the IED. To investigate the fractionation under variable conditions and higher yields, a second TATP synthesis experiment was conducted. In this series, all TATP was produced from one acetone precursor but with changing catalyst conditions (type and amount of acid) and sample clean-up and temperature control conditions and with variable reaction times in the range of 18–72 h. From all these variable conditions, only the reaction time significantly affected the observed fractionation. This could directly be linked to the yield which increased from 20 to over 60% with reaction time. When plotting the $\delta^{13}C$ value of the end product as function of the reaction yield, a similar linear correlation was observed as between TATP and its acetone precursor under standard conditions:

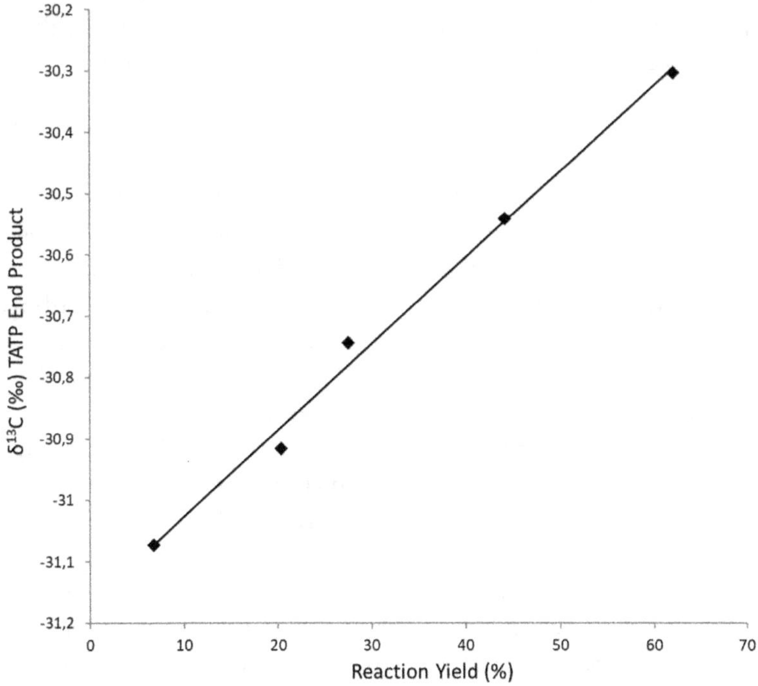

TATP $\delta^{13}C$ values as function of reaction yield for a selected acetone precursor from the dataset. *(Reprinted with permission from Bezemer et al. Journal of Forensic Sciences, 61 (2016) 1198–1207, https://doi.org/10.1111/1556-4029.13135)*

With increasing yield, the $\delta^{13}C$ value of the TATP became less negative indicating a reduced fractionation as the isotope ratio becomes more similar to the acetone from which the peroxide explosive was produced.

> Why does the $\delta^{13}C$ value increase with increasing reaction yield?
>
> What would the value be at 100% yield?

The explanation is quite straight forward if you consider the situation where the yield is 100%. If all acetone has reacted to form TATP, the kinetic isotope effect is nullified. As the reaction starts initially, there is a slightly higher reaction rate for the lighter isotope and thus the TATP formed will be depleted from ^{13}C at low yields. However, the remaining acetone in the mixture that will be slightly enriched in $\delta^{13}C$ will also be involved in the TATP production as the reaction reaches completion. In the end, one would expect that at 100% conversion the isotopic signature of the raw material is copied one-to-one in the end product. Indeed if the linear trend line in the figure above is extrapolated to 100% yield, the $\delta^{13}C$ value is almost identical to the value of the acetone precursor.

We have seen how IRMS can be used in chemical profiling and synthesis reconstruction of materials of forensic interest. This can be used in conjunction with methods that zoom in on chemical impurities such as the vacuum-outlet GC-MS method discussed in the previous chapter to quantify TNT trace impurities. A combination of IRMS and impurity profiling is often very powerful because these methods provide "orthogonal information." Orthogonal in this context means uncorrelated and combination of the findings typically leads to a much higher evidential values when in a forensic comparison two samples share the same isotope and impurity profile. In addition, the combination of IRMS and impurity findings also aids the synthesis reconstruction. As demonstrated in this section, the isotope signature of a material is related to its precursors, whereas it is typically less affected by the reaction conditions, especially for high yields. This means that IRMS provides insight in the question "*with what?*" and to some extent "*where?*" because raw materials can sometimes be traced to regions or certain factories on the basis of the isotope profile (see also the next section). In addition, impurity profiles provide information related to the "*how?*" question as impurities are linked to synthesis routes, the quality of the laboratory work and sample clean-up. Sometimes impurities can tell the forensic expert also something about "*when*" the material was made, provided that characteristic degradation markers can be identified. Of course, impurities can also originate from the raw materials and isotope fractionation can be a function of reaction conditions, so things are not completely black and white but the combination of IRMS and LC-MS/GC can be very valuable indeed in case work involving materials of forensic interest.

As demonstrated for TATP, isotope analysis can also offer chemical profiling opportunities for materials that are very pure and do not provide options for impurity analysis. As long as sufficient variation exists in isotope ratios of the raw materials involved, IRMS can be used to investigate the potential link between two forensic samples of the same, very pure compound. When we go back to the case discussed in **Chapter 6—Chemical Profiling, Databases, and Evidential Value** but now consider an IED with TATP instead of TNT, forensic experts would have to resort to IRMS to be able to provide evidential value that would assist in discriminating the two hypotheses considered in the case:

H_p (prosecution):	*The TATP from the IED is originating from the batch of TATP found at the home of the suspect*
H_d (defense):	*The TATP from the IED is originating from another source of TATP not related to the materials found at the home of the suspect*
E (evidence):	*The* δ^2H, δ^{13}*C and* δ^{18}*O values of the two TATP materials "match," that is, the difference between the values is within the measurement uncertainty for the isotope ratios of the three light elements*

Evidential value (*LR*) Finding an "identical" δ^2*H*, δ^{13}C and δ^{18}O profile of the two TATP samples is 100-1000 times more probable if the Prosecutor hypothesis is true then when the Defense hypothesis is true. This finding is incriminating for the suspect, not only are we dealing with the same type of explosive material, but it has now also been demonstrated that both samples have a very similar isotope ratio profile.

In addition, IRMS offers an option that typically is not available with impurity analysis and that is to investigate a potential link between a product and its raw materials in a way that has been demonstrated for TATP. In a forensic investigation, this could be of importance when at a suspect's home no reference batch is found for a direct comparison but rather the precursors that are needed to produce the material found at the crime scene. This is not an unrealistic scenario when considering explosives, the suspect might have produced one batch and used this entire batch in the construction of the IED. In case of TATP, an excess of acetone might then still be available for forensic investigation.

Can you define the hypothesis of the public prosecutor and the defense council in case of a potential link between the TATP found in the IED and a batch of acetone found at the home of the suspect.

H_p *(prosecution): The TATP from the IED has been produced with the acetone batch found at the home of the suspect*

H_d *(defense): The TATP from the IED has been produced by another source of acetone unrelated to the materials found at the home of the suspect*

When it comes to the evidence and the associated evidential strength, the situation is slightly more complex than the direct forensic comparison of two samples of the same composition. We now have to anticipate the effect of reaction yield on isotope fractionation (of which the forensic expert has no knowledge unless some other evidence, such as a laboratory notebook, is found detailing the reaction conditions). Basically, rather than reporting an *LR* value, the expert can interpret the degree of isotope fractionation of the TATP product in comparison with the potential acetone precursor. The basic question that is then addressed is whether the fractionation observed fits with the expected isotope shift when synthesizing TATP from the given acetone batch at a reasonable and realistic yield. If needed, background data for this assessment could be obtained by producing small amounts of TATP from the acetone for different reaction times and yields as previously demonstrated. This leads to a range of $\delta^{13}C$ values for which the TATP could have been produced with the acetone found in the home of the suspect. This range corresponds to an anticipated yield between 0% and 100% and hence within this range the evidential value will vary as TATP production at very low (excessive raw material consumption) and very high (excessing reaction times) yield are not very likely. However, outside this range, the conclusion can actually be quite absolute. In this case, the acetone related to the suspect could simply not have been used to synthesize the TATP in the IED because this either corresponds to a yield below 0% or in excess of 100%. This is visualized in the graph below for a TATP-acetone linkage study on the basis of carbon IRMS data:

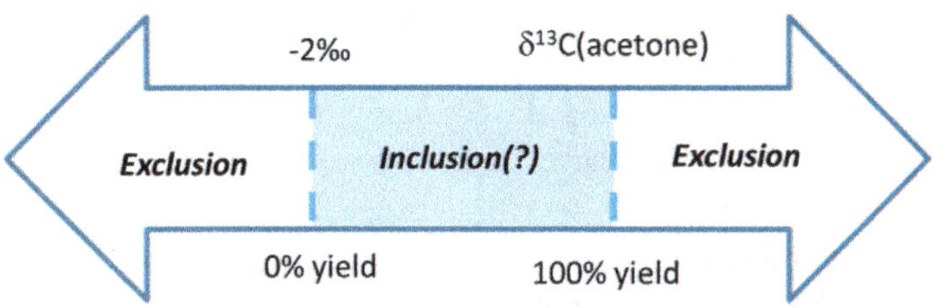

Visualization of the interpretation of $\delta^{13}C$ IRMS values when considering a potential link between a given acetone precursor and a batch of TATP using the data presented in this section

To conclude this part, we will now contemplate the applicability of IRMS when considering energetic material mixtures and postexplosion residues. Please, carefully read the following two questions and formulate your answer before continuing.

Semtex consists of a mixture of RDX and PETN with some additives.

Would it be possible to link a batch of Semtex to a source of RDX and PETN using IRMS?

IRMS is a "bulk method," this means that the isotope ratios measured for H, C, N, and O reflect the overall composition of the sample as all organic matter in the sample is fully converted in the oxidation/reduction reactors to the gases for which the isotope ratios are measured after separation. Hence, the reported δ values are a weighted average of the δ values of the individual components in the sample. Because these individual values are unknown, relating an individual component to the mixture, such as RDX to Semtex, is typically not possible when that component does not have an exclusive element that is not present in the other constituents. One way to tackle this problem is to separate the mixture in its individual components with a dedicated sample preparation procedure. If in the case of Semtex the RDX present could be isolated and characterized with IRMS a possible link with a batch of RDX could be investigated. However, sample preparation could be laborious and is quite an invasive process. One would also have to verify that the separated fractions are of sufficient purity. Another approach is to couple IRMS to GC and LC to establish compound specific isotope profiles. For more complex materials with constituents from different sources, a multicompound isotope profile will be highly characteristic thereby significantly enhancing the evidential value. In addition, it would be possible to investigate "partial links" for reconstruction purposes. The combination of compound specific isotope profile matches and non-matches could indicate that materials share a common main ingredient but that both batches were subsequently processed separately involving the use of different sources of additives. **GC-IRMS** and **LC-IRMS** requires high-end equipment and dedicated expertise as the separation techniques need to be carefully interfaced with the IRMS instrument. IRMS itself involves the separation of gases formed in the reactors, and this entire process needs to be done rapidly enough as compounds subsequently elute from the GC and LC column. GC and LC methods need to be carefully adjusted and optimized with respect to compound separation, compound mass, and mobile phase composition and flow to meet IRMS requirements. Although these techniques are very powerful, only a few forensic laboratories in the world use GC-IRMS and LC-IRMS regularly in casework because of the instrumental complexity and the expertise required.

Lets assume that in the forensic case at the hand the IED was successfully detonated.

Would it be possible to use IRMS to link TATP post-explosion residues to a batch of TATP found at the home of the suspect?

In addressing this question, we can again refer to IRMS as a "bulk method." In general, impurity and isotopic profiling is extremely hard if not impossible in a post-explosion scenario. After an explosion, typically only trace amounts can be retrieved and swabbing of surfaces or sampling of soil leads to a complex sample matrix including many unknown organic and inorganic compounds. In such a setting, the chemical identification of an energetic material and possibly its degradation products is an analytical challenge in itself. In addition, the violent chemical reaction creates impurity profiles related to the explosion that typically completely obscure the impurities that were originally present in the material. As discussed, chemical reactions also affect isotope ratios and therefore it is expected that an explosion will initiate significant isotope fractionation effects hampering a direct comparison of isotope profiles of post-explosion residue with intact reference material. When considering forensic explosives investigations, impurity analysis and the use of IRMS for chemical profiling and reconstruction purposes are in principle limited to cases where a sufficient amount of intact explosive is available to the expert. Possibly new forensic scientific insights and technological instrument developments will provide profiling options for pre- and post-explosion microtraces in the near future (how to innovate forensic science will be discussed in **Chapter 11—Multidimensional Innovation in Forensic Analytical Chemistry**).

7.5 Human provenancing: *you are what you eat and drink*

In this final section, we will discuss an interesting forensic application of IRMS which is termed **human provenancing**. Provenance is defined as the (historical) place of origin of an object, so human provenancing in forensic science indicates the use of forensic intelligence to reconstruct the distant and recent whereabouts (origin) of a person. This type of information is typically useful in cases involving unknown victims or perpetrators, especially in situations where fingermarks (when available) or DNA STR profiling of biological material have not resulted in a match and thus an identification. When in the Netherlands, unknown human remains are discovered and there are no obvious links to missing person cases, the authorities are often dealing with a crime involving human trafficking and a victim who has been residing in the Netherlands illegally or has just arrived in the country. With no official records and no prior arrests of the victim, biometric traces are unlikely to provide a match in missing person and forensic databases. A similar situation can exist with unidentified biological remains of suicide terrorists. After a successful attack, it is vital for law enforcement to find and arrest all participants in the terrorist cell. To this end, the identities of the deceased perpetrators need to be established as soon as possible. Of course personal documents, clothing, jewelry, dental status, body markings (*e.g.*, tattoos), and facial features (forensic face comparison) might provide important clues but these might not always be present or not characteristic enough to provide useful leads. If it was known from which region the deceased is originating, and how and when the deceased traveled to his/her final destination, this could yield new opportunities for identification. A team could for instance contact law enforcement colleagues in these regions to check missing person records. This in turn could lead to potential family members and the use of familial DNA techniques to link a name to the remains. Tactical information provided by the family member could subsequently result in new insights for the criminal investigation. Family members could for instance provide recent contacts and activities of the deceased individual.

By performing IRMS analysis of the human remains such geographic information can be obtained by making use of the vast body of knowledge that has been accumulated by geologists and IRMS experts over the last decades. Macro and local climatological and geological processes induce isotope fractionation leading to significant and consistent changes in δ values for several elements over geographical location and time. By collecting a substantial number of samples and building an isotope database, modeling the various environmental isotope fractionation effects and projecting the model in a GIS (Geographic Information System), so-called **isoscapes** can be constructed. Isoscapes are maps similar to isobaric weather maps but instead of indicating equal atmospheric pressure the contour lines now indicate a given δ value. Isoscape maps typically work with colors rather than lines to indicate a certain isotope ratio range. On the website www.waterisotopes.org data and models are provided by the University of Utah allowing researchers and forensic experts to construct water (precipitation) isoscapes for any region in the world. The website www.isomap.org offered by the University of Utah and the University of Purdue allows scientists to build their own models on the basis of available data.

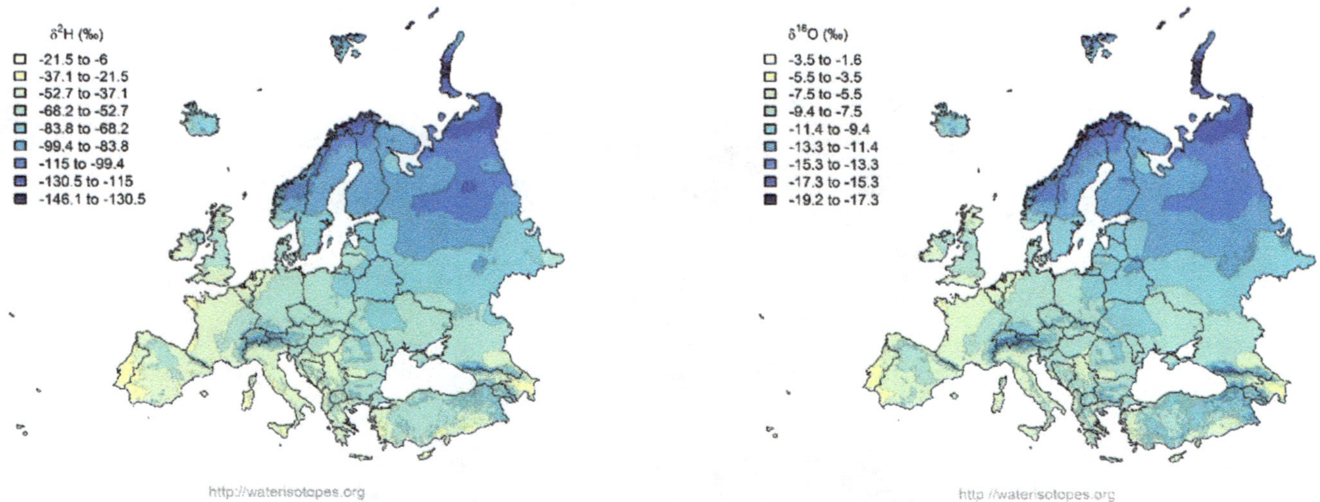

^{2}H and ^{18}O isoscapes of annual precipitation in Europe created on the website Waterisotopes.org (courtesy of Prof. Gabriel Bowen of the University of Utah) showing the strong correlation between hydrogen and oxygen fractionation

The isotopic composition of water precipitation typically has a strong influence on the hydrogen and oxygen isotopic signature in the local water supply and food chain. Plants, animals, and humans all require the consumption of large volumes of water on a daily basis. Tap water that is mostly consumed by humans in developed countries is typically created from river water and other fresh water reservoirs such as local lakes. As such, a very direct correlation between the isotope ratios for hydrogen and oxygen between precipitation and tap water is expected as was indeed demonstrated for USA tap water in a study by Bowen et al. published in Waters Resource Research in 2007 (https://doi.org/10.1029/2006WR005186).

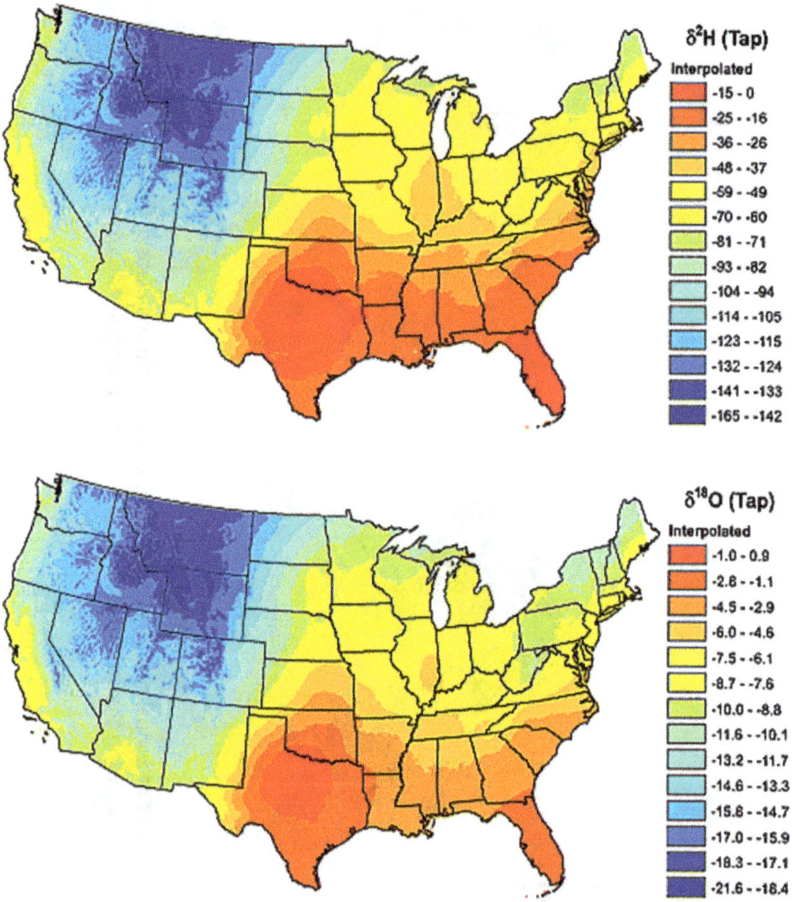

^{2}H and ^{18}O isoscapes of tap water in the USA created on the basis of precipitation data and a precipitation-tap water regression model. *(Reprinted from Bowen et al. Water Resource Research, 43, 2007, https://doi.org/10.1029/2006WR005186)*

Assuming that residents mostly consume local fresh water (also in their food), it can now be expected that this spatial isotopic distribution will be reflected in their tissues and organs. Complex fractionation processes will cause a significant shift in the actual δ values but if these processes are relatively consistent, the isoscope pattern can also be discerned when analyzing biosamples such as human hair, nail, teeth, and bone from donors from various regions with IRMS as has been demonstrated in several recent studies. To fully understand the forensic potential, we also need to take the temporal aspects of these biomatrices in consideration as is illustrated below. Basically, the isotopic signature of bone and teeth is related to the location where an (not too old) individual was born and raised, whereas continuously growing biomatrices such as hair and nail reflect more recent whereabouts. Especially for victims with long hair (assuming that the hair has not degraded yet and is of sufficient quality), segmented IRMS analysis can even provide a timeline demonstrating recent travel patterns. Temporal interpretation can be complex because of biodynamical remodeling processes during life (as occurring for instance in human bone) and the interaction of the human remains with the environment after death. A detailed discussion of these effects is beyond the scope of this book, and interested readers are referred to the work of prof Meier-Augenstein on Stable Isotope Forensics (reference given in Further Reading section). However, the forensic information that can be retrieved this way is quite clear. By measuring δ^2H and $\delta^{18}O$ of hair and teeth and interpretation of the data using isoscapes, the IRMS expert can deduce for instance that the unknown victim originates from the North-West of the United States but has been living for at least a year in Florida (for a case involving an unknown victim in this state). Alternatively, segmented hair analysis might indicate that the victim only very recently traveled to Florida. Such information can provide useful new insights and scenarios for the criminal investigation.

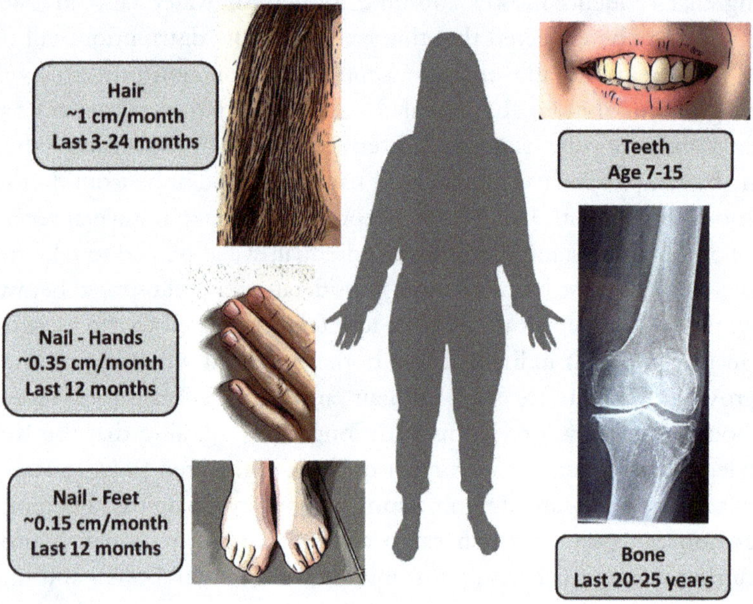

Illustration of human biological matrices that can be analyzed with IRMS and associated temporal information based on time of development, average growth rate, and remodeling characteristics

In addition to hydrogen and oxygen, also IRMS analysis of carbon and nitrogen can reveal important insights in human provenancing studies. These elements are strongly linked to the diet of an individual. At the start of the food chain, plants convert gaseous carbon dioxide and nitrogen to, respectively, hydrocarbons (sugar, starch) via photosynthesis and proteins via nitrogen fixation. These gaseous relatively unreactive compounds do not show significant fractionation, indeed ambient nitrogen is used as the scale anchor for ^{15}N. However, with every step in a complex food chain and the conversion of nitrogen to ammonia to amino acids to peptides and proteins, isotopic shifts will occur. The same holds for the conversion of carbon dioxide to sugar to carbohydrates such as starch and cellulose. A clear difference can be noticed for instance in the $\delta^{13}C$ and $\delta^{15}N$ values as measured in nail, hair, teeth, and bone of a vegetarian versus an individual with a more carnivorous diet. A diet that includes a lot of meat leads to relatively higher δ values.

A remarkable ^{13}C feature that provides interesting options for provenancing is based on the fact that different evolutionary paths of plant photosynthesis exist. Most cultivated crops consist of so-called C3 plants with the C3 label referring to the formation of a compound consisting of three carbon atoms in the first step of the photosynthesis. In contrast, C4 plants like corn and sugarcane exhibit a photosynthesis process that is more efficient in warm and dry climates and in which carbon dioxide is fixated through the formation of oxaloacetate, an intermediate with four carbon atoms. This leads to significant differences in $\delta^{13}C$ for C3 versus C4 plants. Because most processed foods in the USA contain a source of corn, the typical USA diet can be distinguished from the European diet on the basis of the carbon isotope ratio. This effect is so significant that the distributions of values that exist because of dietary variations across the American and European populations are almost fully separated as was demonstrated in a study by the Ehleringer group (see graph below).

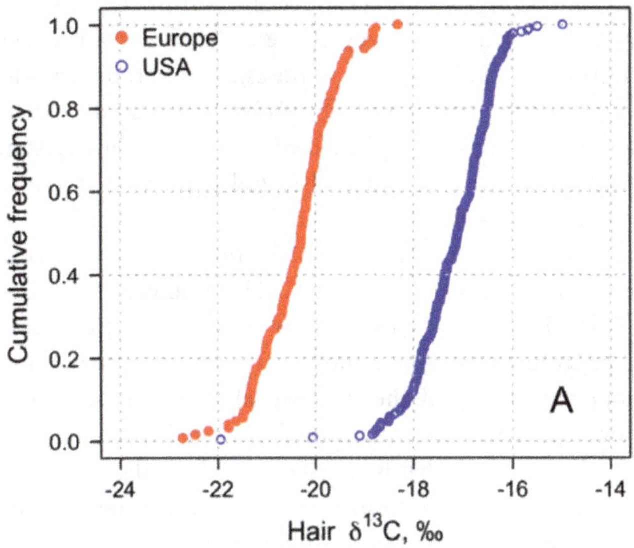

Cumulative frequency of $\delta^{13}C$ values in human hairs from European and United States residents showing the effect of a C3 versus C4 plants based diet. *(Reprinted from Valuenzuela et al. PLoS One, 7, 2012, e34234, https://doi.org/10.1371/journal.pone.0034234)*

The use of IRMS for forensic reconstruction and intelligence is not limited to human matrices. In principle, any biological sample with a characteristic isotope signature can reveal important information to support a criminal investigation. Typically, this involves biological specimens that are cultivated, smuggled, and sold illegally often in an international criminal setting. This includes poaching, ivory smuggle, and the illegal trade of endangered species. Within CITES (Convention on International Trade in Endangered Species of Wild Fauna and Flora), many countries collaborate to combat illegal trade, ensure the survival of endangered species and maintain biodiversity. IRMS analysis could provide clues whether animals have been bred domestically (as claimed by a suspect) or have been smuggled illegally. Isotope signatures could also indicate the region or country were the poaching occurred when encountering illegal animal products. The IRMS analysis of plant material can also be useful in investigations concerning the production, smuggling, and illegal trade of psychoactive substances. THC (the active ingredient of marijuana retrieved from the cannabis plant), cocaine (extracted from the leaves of the coca plant cultivated in Bolivia and neighboring South American countries), and heroine (produced from morphine harvested from poppy plants, cultivated extensively in Afghanistan) are examples of plant-based drugs of abuse. Although the multiple extraction, clean-up and chemical modification steps (*e.g.* the acetylation of morphine to create heroine) will induce significant changes in the isotope ratios, information concerning the geographical origin of the plant material can often still be retrieved. This is illustrated below with the special $\delta^{13}C$, $\delta^{15}N$ and $\delta^{2}H$ isoscapes for cocaine extracted from leaves of coca plants cultivated in various regions in Bolivia. In combination with the alkaloid impurity profile of the cocaine samples, this allowed the DEA (Drug Enforcement Administration of the United States) to pinpoint the origin of cocaine seizures, that is, to establish where the seized cocaine was originally cultivated. Such information can reveal smuggling and trading preferences allowing authorities to understand and successfully disrupt cocaine supply routes.

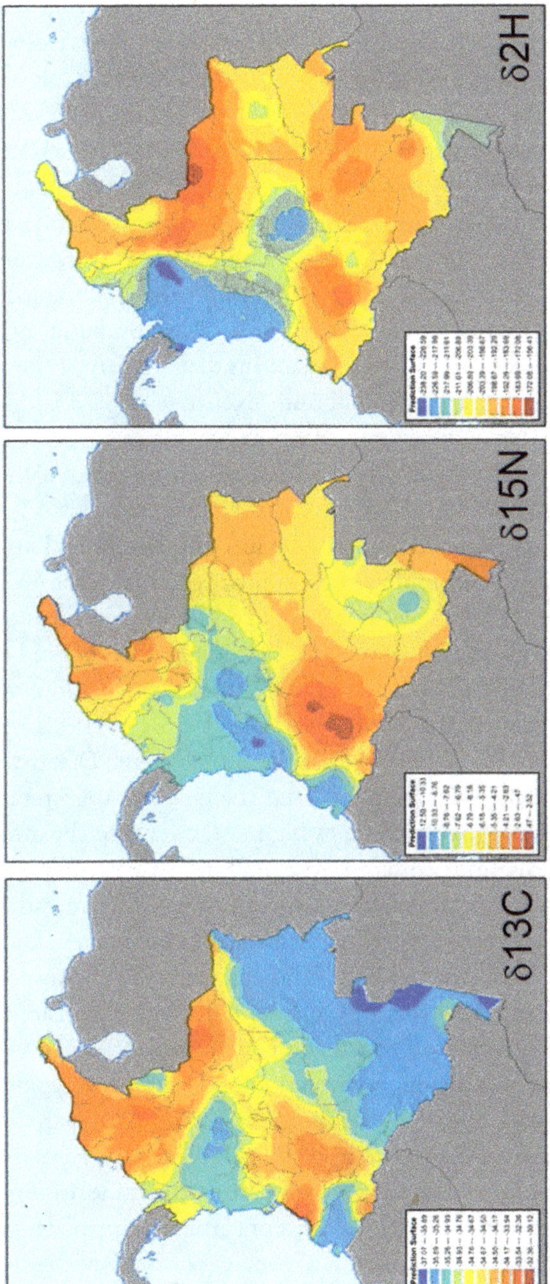

Cocaine isoscapes based on 336 authentic coca leaf samples and the use of ArcGIS Advanced software. *(Reprinted from Malette et al. Scientific Reports, 6,2016, https://doi.org/10.1038/srep23520)*

A final example of the use of isotope analysis for forensic intelligence purposes concerns the use of the ^{14}C isotope. This is not a stable carbon isotope but the radioactive decay of ^{14}C to ^{14}N proceeds very slowly and the reported half-life is 5730 years. For this reason, ^{14}C is an excellent feature to assess the age of very old archeological and paleontological objects in a well-known technique that is termed radiocarbon dating. In the earth's environment, a very low but relatively constant level of ^{14}C is encountered because this radioactive isotope is constantly being formed in the stratosphere from the interaction of nitrogen with cosmic radiation. This continuous formation in combination with the slow radioactive decay leads to a ppt (part per trillion) abundance of atmospheric $^{14}CO_2$ which subsequently translates to all living organisms that are part of the food chain. Once an organism dies, this equilibrium exchange of ^{14}C is disrupted and the original trace of radioactive carbon slowly starts to decrease through radioactive decay. The reduction of ^{14}C is thus a measure of the age of an object containing organic carbon and this concept was introduced by US Nobel laureate Prof. Willard Frank Libby in 1949. In criminal investigations, the dating of 10.000-year-old material is usually less relevant (except maybe for a forgery case where the authenticity of an ancient object is questioned). However, a man-made environmental phenomenon has offered unexpected contemporary dating opportunities. Extensive above-ground nuclear testing by several countries in the period between 1955 and 1980 resulted in a strong increase of ^{14}C abundance in the atmosphere. Due to the testing programs, a ^{14}C spike was observed across the globe in 1963 representing an abundance that was roughly twice as high as the level in the atmosphere in 1950. As concerns and criticism rose and nuclear testing was strongly reduced and primarily conducted underground, radioactive carbon levels in the atmosphere started to decrease again resulting in the radioactive ^{14}C "bomb pulse" illustrated below. This man-made phenomenon offers interesting contemporary dating opportunities for forensic intelligence purposes. Human teeth are formed at a certain age and do not continue to interact dynamically with the environment once the development is complete. Assuming that radioactive decay in these short time spans is negligible, the ^{14}C abundance as expressed in pMC (percent modern carbon) can be plotted in the bomb pulse curve to estimate the time of formation. In combination with medical knowledge concerning human dental development, this will lead to an estimation of the year of birth of the donor as was demonstrated in a brief communication in Nature in 2005 by the Swedish Karolinska Institut (www.nature.com/articles/437333a). When considering human biological matrices that continuously grow (hair, nails) or interact dynamically with the environment, the pMC value corresponds to the time of death. It should be noted that for correct interpretation, fractionation processes have to be taken into account and that ^{14}C measurements require special expertise, procedures, and equipment because of the low abundance. Where IRMS as discussed for light, stable elements is a technique that can be implemented within a forensic institute that has an interest to introduce isotope analysis in case work, work on radioactive ^{14}C will require a collaboration with a specialized laboratory. However, human provenancing is not an approach that is needed nor applied very frequently and in a large number of cases. Finding unidentified human remains is relatively rare in developed countries and in most cases the remains can be linked to a missing person report or quickly lead to an identification on the basis of biometric evidence and/or contextual information. The use of IRMS to indicate date and place of birth, recent whereabouts, and migration patterns and thus even time of death is limited to those high-profile cases were the remains cannot be identified by traditional tactical and technical means. In such rare cases, isotope analysis can provide the much-needed breakthrough to ultimately solve the case. A wonderful demonstration of how fundamental and profound scientific insights into the earth's geochemistry can assist the rule of law!

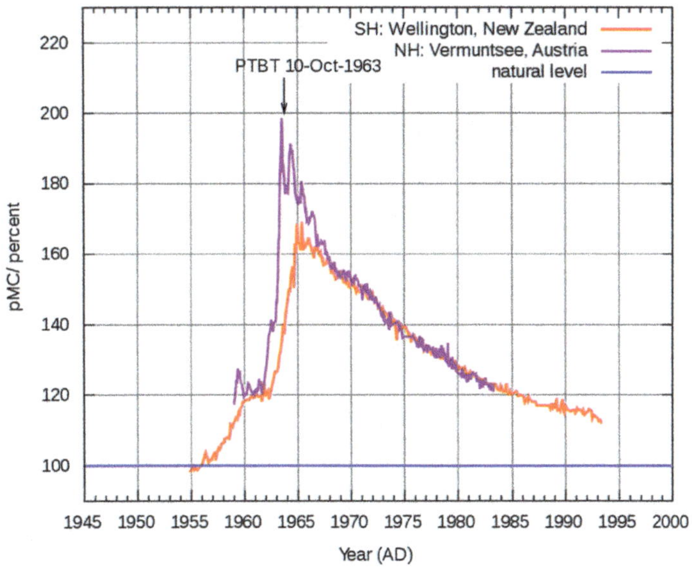

Relative atmospheric abundance of ^{14}C (pMC = percent modern carbon, a relative measure indicating the increase in ^{14}C after 1955) in the aftermath of above-ground nuclear testing in the period 1955–1980. *(Wikipedia)*

Further reading

Bezemer, K., Koeberg, M., van der Heijden, A., van Driel, C., Blaga, C., Bruinsma, J., van Asten, A., 2016. The potential of isotope ratio mass spectrometry (IRMS) and gas chromatography-IRMS analysis of triacetone triperoxide in forensic explosives investigations. J. Forensic Sci. 61, 1198–1207. https://doi.org/10.1111/1556-4029.13135.

Doyle, S.P.G., van der Peijl, G. (Eds.), 2020. FIRMS Guidance for the Forensic Interpretation of Isotope Ratio Data, first ed. FIRMS. 978-0-948926-37-2 (downloadable from the FIRMS website: https://www.forensic-isotopes.org/gfi-ird.html.

Dunn, P.J.H., Carter, J.F. (Eds.), 2018. Good Practice Guide for Isotope Ratio Mass Spectrometry, second ed. FIRMS. 978-0-948926-33-4 (downloadable from the FIRMS website: https://www.forensic-isotopes.org/gpg.html.

Meier-Augenstein, W., 2018. Stable Isotope Forensics – Methods and Forensic Applications of Stable Isotope Analysis. Wiley, Hoboken, NJ, USA, ISBN 978-1-119-08020-6.

CHAPTER 8

From data to forensic insight using chemometrics

Contents

8.1 What will you learn?

In **Chapter 1—An introduction to Forensic Analytical Chemistry**— the various aspects of chemical analysis were briefly discussed. In addition to the actual hands-on activities such as sample preparation and instrumental analysis, also the processing and interpretation of the recorded data constitutes an important and often crucial aspect. With the ongoing advancement of technology in the laboratory, instrumental datasets continue to increase both in size and complexity. When performing a single analysis of a sample with comprehensive 2D chromatography followed by a high-resolution mass spectrometric characterization of the eluting compounds, a single datafile of several gigabytes is not uncommon. Such a datafile contains a wealth of chemical information but also a lot of noise and not all aspects are of interest for the question at hand. Retrieving the useful, relevant information from the raw data and presenting this is in a meaningful way is not a simple

Chemical Analysis for Forensic Evidence
ISBN 978-0-12-820715-4
https://doi.org/10.1016/B978-0-12-820715-4.00012-2

matter. In analytical chemistry, the field of **chemometrics** is dedicated to the use of data science, mathematics and (multivariate) statistics to generate chemical information, insight, and understanding. In this chapter, you will be introduced to some basic chemometric methods that are also used successfully in forensic analytical chemistry. This includes a detailed introduction to **principal component aalysis** (**PCA**), a very frequently used, so-called **unsupervised method** to reduce the dimensionality of complex data sets to visualize underlying patterns and reveal "hidden information" in complex data. In addition also supervised methods such as **linear discriminant analysis** (**LDA**) will be discussed to generate powerful models from reference "**ground-truth**" data. Traditionally, chemometrics is used predominantly to valorize spectroscopic data. Techniques like FTIR, Raman, and NIR spectroscopy can be used to rapidly interrogate samples of forensic interest in a non-invasive manner. Unfortunately, the application of spectroscopy without chromatographic separation leads to complex, "mixed" spectra as each compound contributes to the overall spectrum. Furthermore, spectral datasets tend to be highly **correlated** as the resolution of the instrument exceeds the typical adsorption bands. Dimensionality reduction can, therefore, be very useful to retrieve valuable information from the spectral sample data. To demonstrate the potential of chemometrics in forensic chemistry, we will return to the issue when analyzing NPS isomers with GC-MS as discussed in **Chapter 4—Qualitative Analysis and the Selectivity Dilemma**. In that chapter, it was shown how using an additional technique like GC-IR can overcome the issue of near identical electron ionization (EI) mass spectra as encountered for some illicit drug ring isomers. But is it really true that these mass spectra do not contain any information that would allow a forensic chemist to determine which isomer is present in the sample? Visually identical does not necessarily mean that the spectra are the same from a data perspective. The human brain is typically not capable of discerning patterns from complex, multivariate datasets and therefore chemometric methods are required if we want to investigate the NPS mass spectra in more detail. Before applying advanced data analysis on multivariate mass spectra, we will also explore the options offered by **library match scores**. Such match scores generated by the MS instrument software indicate the degree of similarity between mass spectra and are used to create rank lists of potential library compounds for the identification of an unknown peak in the GC-MS chromatogram. These lists are checked by the forensic expert and based on visual inspection of the spectra a final decision is taken. The main question addressed in this chapter is whether the forensic expert can confidently identify NPS isomers on the basis of GC-MS findings through the use of new, data-driven, chemometric approaches. If successful, the benefits of using novel data science solutions are evident, instead of investing budget and capacity in instrumental solutions to solve the selectivity challenge, the problem could be tackled by simply taking a smarter and closer look to the data that is already available!

After studying this chapter, readers are able to

- explain the difference between Euclidean distance and cosine similarity as features to express the similarity of EI mass spectra of drug isomers
- develop a score-based model for drug isomer differentiation based on library match scores
- construct a ROC (Receiver Operating Characteristics) curve from library match score distributions
- explain the basics of PCA
- explain the basics of LDA and how *LR* values are estimated from LDA-based NPS isomer differentiation
- understand why LDA is often used in combination with PCA
- understand the difference between unsupervised and supervised multivariate data analysis methods
- list the main unsupervised and supervised chemometric methods and their use in forensic analytical chemistry
- discuss the potential benefits and challenges associated with the application of chemometrics in forensic case work

8.2 Library match scores and ROC curves

In **Chapter 4**, we have discussed the selectivity dilemma for regular illicit drug screening strategies based on **GC-MS** due to the continuous introduction of **NPS (New Psychoactive Substances)**. Even when all potential isomers are considered and reference **EI** mass spectra are available, identifying the correct molecular structure can be challenging because of the high similarity of the spectra. Possibly, the GC separation can assist the expert and a correct identification can be based on the observed retention time. However, volatile amphetamine-based NPS isomeric species can coelute especially when routine GC-MS methods are used that have been optimized to cover a broad spectrum of listed compounds in a short overall run time. This was illustrated with the example of the ring isomers **2-Fluor Amphetamine** (FA), **3-FA**, and **4-FA**. Assigning the right isomeric FA form by the expert is especially important in the Netherlands (and several other countries with similar legislation) because the NPS 4-FA was listed as a hard drug (list I) in Netherlands since 2017, whereas currently (2021) the other two isomers are not included in the banned substance lists of the Dutch illicit drug legislation. So, a misidentification in the forensic laboratory would almost automatically result in a wrongful conviction or acquittal. The corresponding EI-MS spectra provided in **Chapter 4** are shown once more because these data sets form the starting point for the data science strategies and chemometric methods discussed in this chapter.

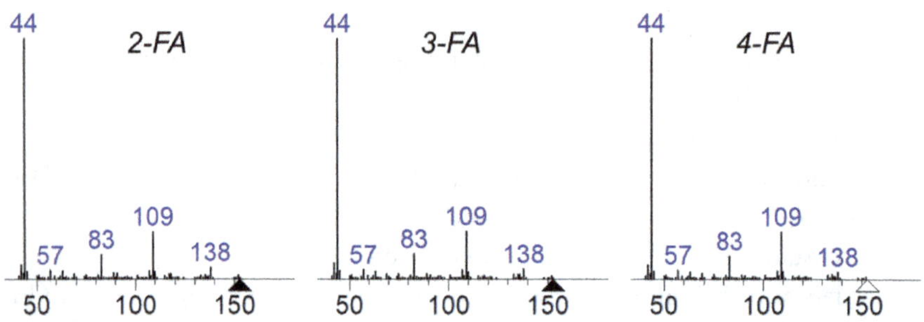

The 70 eV EI mass spectra of (from left to right) 2-, 3- and 4-fluoro-amphetamine analyzed as methanol extracts with GC-MS. The spectra are visually identical and the FA isomers typically co-elute when applying broad GC screening methods. *(Courtesy of Ruben Kranenbrug, Amsterdam Police Laboratory)*

Additionally, the forensic illicit drug expert must always consider the possibility of encountering a new compound for the first time in a case work sample. In such a situation, no reference material is available and consequently the spectrum is not yet available in the MS database. Consequently, it was concluded that additional analysis is required to meet the "**NPS challenge**" and to confidently present evidence in court with respect to the composition of suspected drug samples. It was demonstrated that the use of techniques like **GC-VUV** and **GC-IR** in addition to GC-MS can adequately address the selectivity dilemma, both in terms of distinguishing known isomers and warning the expert for the potential presence of a novel compound. At the end of the **Chapter 4**, we confidently stated "problem solved"! However, this solution comes at a price as it requires an investment in terms of budget and expert capacity. The forensic institute will need to invest in high-end, expensive laboratory equipment and will need to train its experts to work with the new instrumentation. Tailor-made methodology will have to be developed, optimized, tested, and validated (as will be extensively discussed in **Chapter 9—Quality and Chain of Custody**) and new databases will have to developed. The new equipment will regularly have to be serviced to ensure the quality of the analysis and occasionally will have to be repaired when it breaks down. Where GC-MS is extremely robust equipment after decades of technological developments, GC-IR and GC-VUV systems are relatively new and especially GC-IR systems are known to be less robust with respect to high volume operation. Additionally, a malfunctioning GC-MS instrument usually does not affect the case throughput as several instruments have been installed in the laboratory. But when a forensic institute invests in GC-IR or GC-VUV, a single instrument tends to be acquired and consequently an instrumental failure directly increases the back log. So extending the analysis scheme in high volume illicit drug screening will inevitably result in a higher cost per sample due to equipment investments, laboratory infrastructural, and operational costs, the use of extra consumables, and the additional expert time required to complete the investigation. This in most cases will also result in an increase in the lead time, *i.e.,* the time between receiving the physical evidence and providing the final report to the authorities that have issued the request. When considering smaller forensic laboratories that service local regions, an expansion of the analytical equipment might even be unfeasible because of lack of budget or expert capacity. This triggers the question whether we are really using the full potential of the GC-MS data and associated EI-MS spectra. Is the use of GC-IR or GC-VUV really necessary or could we somehow unlock subtle, hidden information that would allow the expert to identify the correct isomeric form solely on the basis of the measured mass spectrum?

To explore this we need to study the data in more detail. Zooming in on the baseline of the EI-MS spectrum illustrates that for the lower abundance m/z signals the highly similar mass spectra do show small differences for the various FA isomers:

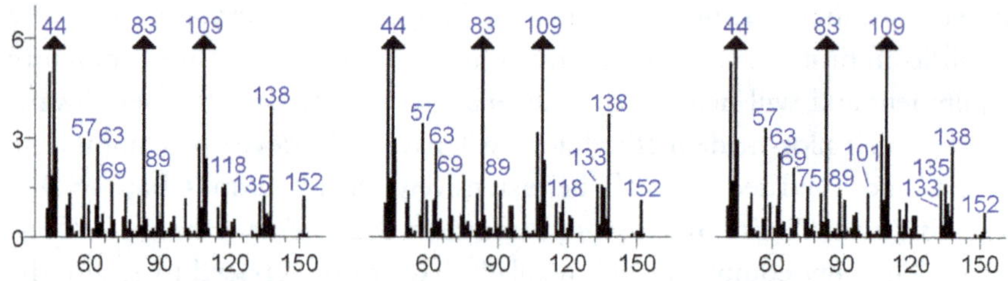

Baseline detail (0%–6% relative abundance) of the 70 eV EI mass spectra of (from left to right) 2-, 3-, and 4-fluoro-amphetamine to inspect small differences in the isomer spectra. *(Courtesy of Ruben Kranenburg, Amsterdam Police Laboratory)*

But are these differences significant and indeed related to small differences in the fragmentation chemistry of FA isomers? Especially for minimal fragment abundances the relative standard deviation can be large when repeating a GC-MS analysis of the same compound standard. Basically, what needs to be investigated is the "**between-isomer**" variation in relation to the "**within-isomer**" variation in a way that is very similar to what was discussed in **Section 6.5** of **Chapter 6—Chemical Profiling, Databases, and Evidential Value**. In addition, also the long term and instrument-to-instrument consistency needs to be inspected to ensure that minor details in the MS spectra are not subject to temporal drift or affected by minor differences in instrument configuration. However, such systematic deviations could be tackled by analyzing the reference isomer standards next to the case sample. Although this will require some additional lab work, in this way the unknown spectrum can be linked to the correct isomer for the given instrumental conditions. Of course, such an approach will not work when dealing with a novel substance.

But before considering isomer identification robustness, we need to take a step back and investigate whether any isomer-specific information is present in the EI-MS spectra and if so, find ways to extract such information and make it available in a scientific but convenient way to the experts. One could take a data science approach collecting a large reference, known "**ground truth**" dataset by repeated analysis of reference isomer standards. The collected mass spectra are then handled as a multivariate dataset used to create chemometric models as will be explained in more detail in **Sections 8.3 and 8.4**. But first we will take a more down to earth approach by exploring the **library match scores** calculated by the software of the MS instrument to create **rank lists** of candidate structures for the unknown compound. Depending on the instrument vendor typically a scale of 0—1, 0—100, or 0—1000 is used to express the degree of similarity between the measured and a library mass spectrum. A value of 1, 100 or 1000 is indicating truly identical spectra which is typically not achieved because small random and systematic variations prevent the "perfect match." The regular process involves "manual" inspection of the mass spectra of the top candidates in the rank list by the forensic expert. Usually, a certain match score threshold is applied based on experience, *e.g.*, only library spectra yielding a match score exceeding 0.9, 90, or 900 are inspected because the expert knows that below this value significant differences will exist and hence the case sample will not contain those reference compounds. The remaining candidate spectra are then carefully compared to the measured spectrum. The final identification is done by the expert making it a subjective process that could be sensitive to bias and human error. However, this expert decision is an essential step as was discussed in detail in **Chapter 4—Qualitative Analysis and the Selectivity Dilemma** as the software is incapable of considering the entire chemical context and MS library search algorithms are not flawless. Experienced GC-MS users know that depending on the complexity of the measured mass spectrum, the composition of the library and the existence of isomers and similar structures, the actual "ground truth" compound is not always be positioned as the top candidate in the rank list. Simply taking the highest rank score in the reference compound list can thus lead to an erroneous identification. In addition, such a simple approach would fail when for a new compound the reference spectrum is not available (yet) in the library.

How could we investigate whether MS library match scores contain FA isomer specific information that could be used to establish which FA isomer is present in a case sample? (Note: all the three FA isomer spectra are available in the MS library).

The easiest way to explore the potential of library match scores to discriminate between FA NPS isomers is to analyze standards and case samples of known composition, and inspect the reported library match scores against the three isomers. If discrimination is feasible, the correct isomer will consistently be placed at the top of rank list indicating a small but statistically significant difference in the match score. A typical result is illustrated below.

Reference spectrum in the library of	Match score for a sample containing		
	2-FA	**3-FA**	**4-FA**
2-FA	**964 ± 11**	923 ± 5	923 ± 5
3-FA	922 ± 4	**964 ± 11**	952 ± 6
4-FA	921 ± 5	946 ± 11	**964 ± 12**

Average GC-MS FA isomer library match scores and standard deviations for 30 runs of standards and samples containing 2-FA, 3-FA, and 4-FA on a single GC-MS instrument.
Copied from Kranenburg et al. Forensic Science International, 302 (2019) 109900, https://doi.org/10.1016/j.forsciint.2019.109900.

The table shows that the correct isomer (true positive) consistently yields the highest average match score. Does this mean that for the identification of FA isomers we can simply apply a match score threshold? If not how could we use these findings?

Although the average match scores reveal isomer-specific features in the mass spectra, applying a simple threshold will not work. The standard deviations indicate that a significant overlap in match scores will exist. The most striking example is 4-FA versus 3-FA, for a sample containing 4-FA the average match score against 3-FA (false-negative from a legal perspective) is only 12 points below the average score for the correct 4-FA reference spectrum (true-positive). In the reverse situation, a sample containing 3-FA will yield an average match score against 4-FA (a false positive from a legal perspective) that is 18 points below the average match score for the correct assignment.

So we need a different approach to establish the evidential value for isomer differentiation on the basis of EI-MS match scores. To this end, we first need to understand the way the vendor MS software finds potential candidate structures in the database and calculates match scores. It should be appreciated that such software has not been developed for isomer discrimination in forensic drug analysis, but rather to rapidly screen thousands of reference spectra in EI-MS databases to compile a rank list of potential candidates on the basis of spectrum similarity. Although forensic drug analysis laboratories usually have in-house compiled dedicated spectral libraries containing all known illicit drugs, NPS, frequently abused medicines, adulterants and raw materials, commercial EI-MS libraries can also be used for broad screening purposes based on extensive collections of reference spectra. The 2020 release of the NIST/EPA/NIH EI-MS library (https://chemdata.nist.gov/dokuwiki/doku.php?id=chemdata:start) that is frequently installed with GC-MS instrumentation contains over 300.000 unique compound spectra. If a comparison of a measured spectrum with a single library entry would only take 1 s, screening the entire database would still take 3.5 days! Fortunately, smart design of the search routines provides rank lists in real time as the analyst is interrogating the GC-MS data. In the Further Reading section, a 1994 reference can be found that describes various search algorithms for EI-MS libraries. The most frequently used method is based on the so-called **dot-product function**. If we consider an EI mass spectrum as a vector in an M dimensional space (where M is the mass range for which the ion abundances are measured with unity mass resolution between masses x and y), the dot-product of two mass spectra, $\boldsymbol{m_1}$ and $\boldsymbol{m_2}$ (vectors are generally represented in bold notation), is defined as:

$$\boldsymbol{m}_1 \cdot \boldsymbol{m}_2 = \sum_{i=x}^{y} (A_i)_1 \cdot (A_i)_2$$

where $(A_i)_1$ and $(A_i)_2$ are the (maximum normalized) abundances at mass-over-charge ratio m_i of mass spectra 1 and 2, respectively. If these factors are considered as rows or columns in a data matrix, the dot product can also be represented as a matrix multiplication with T indicating the **transpose of a matrix**, i.e., the conversion of a row to column or vice versa.

$$\boldsymbol{m}_1 \cdot \boldsymbol{m}_2 = \boldsymbol{m}_1 \cdot \boldsymbol{m}_2^T = \boldsymbol{m}_1^T \cdot \boldsymbol{m}_2$$
$$(A_x)_1 \cdot (A_x)_2 + (A_{x+1})_1 \cdot (A_{x+1})_2 + \ldots + (A_y)_1 \cdot (A_y)_2$$

The dot product of two vectors returns a scalar, a (one dimensional) number/quantity which in a **Cartesian coordinate system** equals the product of the vector lengths (indicated by the notation ||..||) and the cosine of the angle Θ between the two vectors.

$$\boldsymbol{m}_1 \cdot \boldsymbol{m}_2 = \|\boldsymbol{m}_1\| \cdot \|\boldsymbol{m}_1\| \cdot \cos\theta$$

The dot product can therefore be used to express the angular similarity of two vectors or in other words the similarity in the vector direction in the M dimensional space.

$$\cos\theta = \frac{\boldsymbol{m}_1 \cdot \boldsymbol{m}_2}{\|\boldsymbol{m}_1\| \cdot \|\boldsymbol{m}_1\|}$$

Similar to the **Euclidian distance** that was introduced in **Chapter 6—Chemical Profiling, Databases, and Evidential Value**, the angle Θ can also be used as a feature to express the similarity of two multivariate datasets such as chemical impurity profiles or compound mass spectra. The relation between this so-called **cosine similarity** and the Euclidian distance is visualized below with a simplified 2D representation of a mass spectrum for two compounds consisting of two different abundances for two identical m/z values.

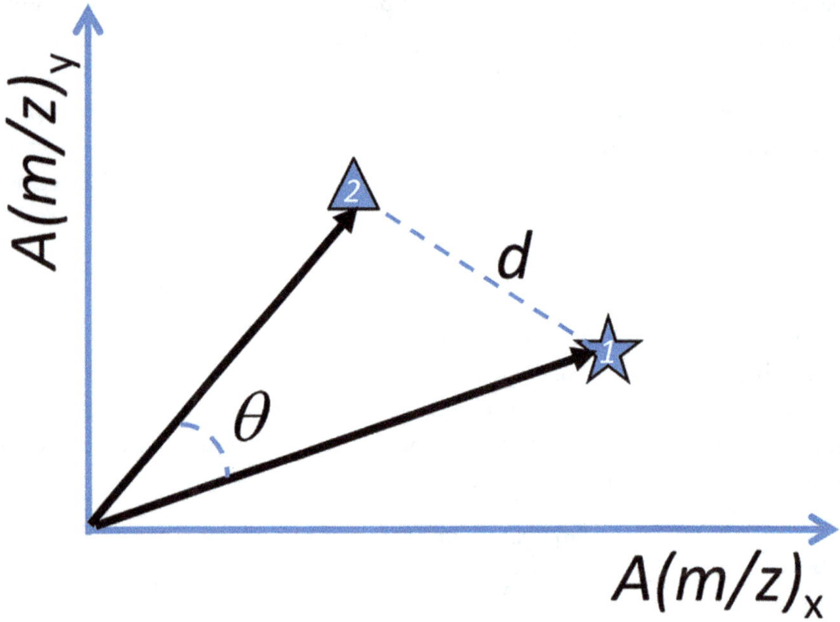

Shortest Euclidean distance d between and the vector angle Θ of two mass spectra depicted as vectors in a two-dimensional mass space, where A indicates abundance for the 2 m/z values considered (indicated by x and y) and where the subscripts 1 and 2 denote the two spectra, e.g., a measured spectrum of an unknown compound and a library spectrum of a reference compound

This figure also illustrates that like the Euclidean distance the vector length in an M dimensional space can be calculated using Pythagoras's theorem. This leads to the following formula to directly establish the cosine similarity from two tabulated mass spectra

$$\cos\theta = \frac{\sum_{i=x}^{y}((A_i)_1 \cdot (A_i)_2)}{\sqrt{\sum_{i=x}^{y}((A_i)_1)^2} \cdot \sqrt{\sum_{i=x}^{y}((A_i)_2)^2}}$$

$$\cos\theta = \frac{(A_x)_1 \cdot (A_x)_2 + (A_{x+1})_1 \cdot (A_{x+1})_2 + \ldots + (A_y)_1 \cdot (A_y)_2}{\sqrt{(A_x)_1^2 + (A_{x+1})_1^2 + \ldots + (A_y)_1^2} \cdot \sqrt{(A_x)_2^2 + (A_{x+1})_2^2 + \ldots + (A_y)_2^2}}$$

In addition to the Euclidean distance also the spectral angle can be used to express the degree of similarity between the multivariate profiles of two compounds or samples. What are the differences between both methods and in what situations will one method outperform the other? How can we tackle the potential weakness of the spectral angle approach?

An obvious difference between the Euclidean distance and the spectral angle is the fact that the cosine similarity yields values between 1 (identical vector angle, $\cos(0°) = 1$) and 0 (orthogonal vector angle, $\cos(90°) = 0$) while there is no maximum for the Euclidean distance. Another difference is the fact that in Euclidean distance a lower value means a higher degree of similarity while a good fit on the basis of spectral angle is indicated by a value near the maximum of 1 (please note that the EI-MS library match scores ranges of 0–100 or 0–1000 can simply be achieved by multiplying the cosine similarity with a factor of 100 or 1000, respectively). When using the so-called **cosine distance**, defined as 1-cosine similarity, this inequality is resolved as a cosine distance of 0 is now reflecting perfectly aligning spectral vectors. However, the most important consideration is of course that different similarity parameters are calculated. The cosine similarity only considers the direction of the vectors and not their length, whereas the Euclidean distance is sensitive to the absolute position of the sample points in the M dimensional space. This is why cosine similarity is often used for spectral data as in reflection spectroscopy the overall magnitude of a measured signal can vary significantly while the spectrum itself, the variations in response as function of wavelength or m/z vale are actually consistent. This data analysis method is also often used in hyperspectral imaging (HSI) and in this context often termed **Spectral Angle Mapping (SAM)**. Using the spectral angle method can be beneficial for datasets where the abundance ratios remain the same while the absolute signal might differ from analysis to analysis. To optimize results for Euclidian distance, it is important to normalize the data as much as possible and in principle this is the case for an EI-MS spectrum where the base peak is scaled to 100% (although signal saturation at a high abundance might negatively affect the outcome). Cosine similarity on the other hand can be more sensitive for relatively low ion signals near the threshold of the MS detector. In GC-MS, it is known that EI-MS spectra at trace compound concentrations can have a "binary appearance" as below a certain ion current no abundance is recorded for a given m/z ratio. This can lead to orthogonality in the data which results in a low cosine similarity score, whereas the Euclidean distance is actually indicating that the profiles are quite similar. This effect can be mitigated by applying a threshold to exclude low abundancies in the comparison. The differences between Euclidean Distance and Cosine similarity is illustrated in the figure below.

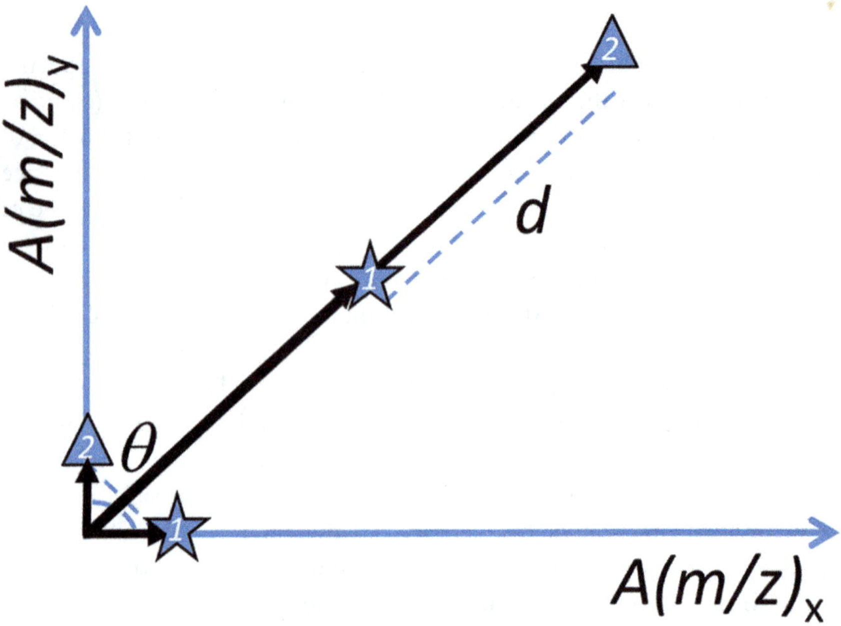

Illustration of the sensitivity of the Euclidean distance d for differences in absolute abundance values and the sensitivity of the vector angle Θ for low abundances near the detection limit of the mass spectrometer

Despite the discussed differences with the Euclidean distance metric, the match score as reported in many commercial systems to match an unknown EI-MS spectrum to library spectra of reference compounds can be regarded as a feature that quantitatively expresses the degree of similarity when it is based on the cosine similarity. What approach that was introduced in **Chapter 6** can we now use to investigate whether the measured EI-MS spectrum can assist in the FA isomer differentiation? What hypothesis pair would you define? And what is the main difference with the application discussed in **Chapter 6**?

As the library match score based on the cosine similarity is a feature that quantitatively describes the similarity of two EI-MS spectra, we can build a score-based model in line with the approach that was described for TNT impurity profiling in **Section 6.5**! First an appropriate hypothesis pair needs to be defined. An interesting difference with the model described in **Chapter 6** is that we are now dealing with a so-called **closed set comparison**. Assuming that the GC retention time and measured EI-MS spectrum is typical for fluor amphetamine (i.e., that no other compounds exist that elute at the same time and have a very similar mass spectrum), the challenge is to correctly determine which of the three isomers is present in the sample. The sample contains either 2-FA, 3-FA, or 4-FA, no other outcome exists, whereas in the case of the chemical impurity profile of TNT, the Defense hypothesis considers any potential source in an open and unknown population ("*The TNT from the IED is originating from another source of TNT not related to the materials found at the home of the suspect*"). This difference is reflected in the hypothesis pair. With a focus on the criminal activity, the Prosecutor hypothesis will focus on the isomer that is listed as a drug of abuse in the illicit drug law, which in this example is 4-FA:
H_p: "*The sample confiscated from the suspect contains the listed hard drug 4-FA*".
H_d: "*The sample confiscated from the suspect does not contain 4-FA but an unlisted FA isomer*".

As two FA isomeric forms are known that are unlisted, the forensic illicit drug experts needs to consider two comparisons: 4-FA versus 3-FA and 4-FA versus 2-FA. There is no need from a criminal investigation perspective to consider the 3-FA versus 2-FA option. The overall evidential value will be determined by the "*weakest link*," i.e., the comparison that yields the lowest *LR* value which will be obtained for the unlisted isomer with the EI-MS spectrum that has the highest similarity to the 4-FA reference spectrum.

In the closed set framework, the forensic expert is assuming that no other compound exists that yields a retention time and mass spectrum similar to the FA isomers. Given the high selectivity of GC-MS this is a reasonable assumption. However, can you think of another possibility that could complicate matters considerably? What additional assumption does the forensic expert need to make?

The forensic expert is assuming that the sample contains either 4-FA, 3-FA, or 2-FA but an additional option is that multiple FA isomers are present in the material! The measured EI-MS spectrum in this case would be a composite spectrum as the isomers coelute from the GC column. This will lead to library match scores that have not been modeled properly and will be compared to pure compound reference spectra. However, from casework it is known that NPS isomer mixtures are very rarely encountered. Although it is occasionally observed and analytical strategies are available based on GC-IR and GC-VUV to detect the presence of multiple isomers, the option of FA mixtures will not be considered further in this chapter. However, the forensic expert should always be aware of this possibility.

To build a robust score-based model and apply it for FA isomer differentiation in case work using GC-MS, the forensic expert now needs to take the following steps in line with the methodology introduced in **Chapter 6—Chemical Profiling, Databases, and Evidential Value**:

- Measure and establish the reference spectra by analyzing 2-FA, 3-FA, and 4-FA reference standards and storing these spectra in the EI-MS library
- Over a given time period, compile an extensive dataset for samples for which the ground truth is known with respect to FA isomer composition and which reflects typical analysis and sample variations and GC-MS instrument fluctuations
- Document the match scores for the 4-FA reference spectrum comparison for this dataset and ensure that the dataset contains a balanced number of true-positive (4-FA in the sample) and false-positive (3-FA or 2-FA in the sample) scores
- Bin the match scores for the 4-FA true-positive, the 3-FA false-positive, and 2-FA false-positive data sets in a histogram to establish the relative frequency of a given match score range (as a rule of thumb the bin size should be in the same order of magnitude as the square root of the number of data points)
- Convert the discrete frequency distributions to continuous probability density functions by fitting the data to a Kernel Density function (a weighed sum of normal distributions of different average scores)
- For a case sample containing an FA isomer, establish the EI-MS match score for the 4-FA reference comparison

- Use the continuous KDE distributions to estimate the probabilities of finding such a match score for a sample containing 2-FA, 3-FA, and 4-FA
- Establish the evidential value of the measured EI-MS spectrum and library match score for the 4-FA comparison by dividing the probabilities of 4-FA versus 3-FA and 4-FA versus 2-FA

The histograms and corresponding continuous probability density functions as constructed from a dataset of 30 GC-MS runs for each FA isomer is given below.

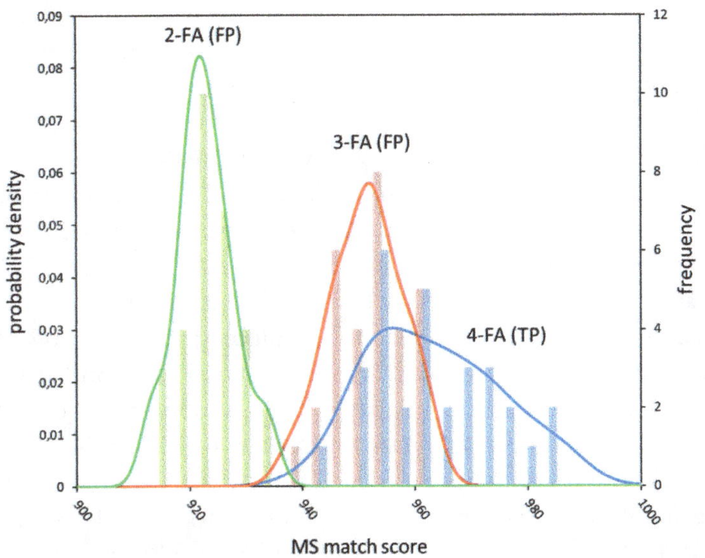

Histograms and KDE probability density functions of library match scores versus the 4-FA EI-MS reference spectrum of 30 GC-MS runs of standards and samples containing 2-FA (green, known false-positive), 3-FA (red, known false-positive), and 4-FA (blue, known true-positive). *(Reprinted with permission from Kranenburg et al. Forensic Science International, 302 (2019) 109900, https://doi.org/10.1016/j.forsciint.2019.109900)*

These findings are in-line with the conclusion from the tabulated match scores that the differentiation between 2-FA and 4-FA on the basis of the EI-MS mass spectra is relatively straight forward whereas the 4-FA/3-FA pair is more critical. The KDE probability functions of 2-FA and 4-FA are almost fully separated indicating high *LR* values for true-positives. Match scores exceeding 940 are rarely encountered for samples containing 2-FA, whereas only a small fraction of the 4-FA samples will yield a score below this threshold. When considering the distribution for 3-FA, a threshold needs to be set at 970 to prevent a false-positive 4-FA identification. However, because the 4-FA true-positive and 3-FA false-positive distributions show extensive overlap setting, such a threshold would result in a very significant false negative fraction which is unacceptable from a high-volume case work perspective. For a significant number of 4-FA containing samples, a match score below 960 (where both distributions intersect) would actually yield an *LR* value below one which misleadingly supports the defense hypothesis that the sample does not contain 4-FA but an FA isomer, most likely 3-FA.

An interesting alternative approach to illustrate the selectivity and sensitivity of a test is the construction of a **Receiver Operator Characteristic** or **ROC** curve. In such a curve the true positive rate (TPR, sensitivity) is plotted as function of the false positive rate (FPR, 1-selectivity) and this provides insight with respect to optimal decision thresholds settings for different applications and requirements. The ROC curve is applicable for a binary test, which in this case translates to the decision whether 4-FA is present or not. Because two isomers exist, two ROC curves, 4-FA versus 3-FA, and 4-FA versus 2-FA can be constructed. A ROC curve can be prepared by varying the threshold and establishing the TPR and FPR by determining the area of the probability distribution function between the set threshold and its maximum value. When starting with the maximum threshold, the perfect EI-MS library match score of 1000, it is clear that on the basis of both distributions both the TPR and FPR are zero. This is the starting point at the lower left of the ROC curve. As the threshold is lowered, the TPR will start to increase while the FPR remains (almost) zero as the area under the false-positive curves is negligible for match scores exceeding 970 (point A). The TPR continues to increase as the match score threshold is further lowered but now also the FPR starts to increase for 4-FA versus 3-FA. To reach a TPR of 1 (no false-negatives, i.e., no test results indicating the absence of 4-FA for 4-FA containing samples) the threshold needs to be set at 930 but this will also result in a significant number of false-positive results for 2-FA containing samples (point B) and a near 100% FPR for 3-FA containing samples. A perfect test can only be obtained for fully separated true-positive and false-positive probability density functions (please note that for probability functions the values can be very low but never exactly zero, so the terminology "separated distributions" is somewhat misleading). By setting a threshold between the two distributions, a square-shaped ROC curve is obtained indicating that a 100% TPR can be accompanied by a 0% FPR. The associated **area under the curve** (**AUC**) is one for a perfect test. For a test that adds no information to support the decision the ROC curve follows the diagonal, this indicates that for any decision threshold the TPR equals the FPR and hence that both probability density functions completely overlap. The test performs no better than a coin flip to decide whether 4-FA is present ("when it is heads the sample contains 4-FA and with tails I will report the presence of an unlisted FA isomer"). This diagonal is included as a point of reference in a ROC curve, the associated AUC is 0.5. The construction of the ROC curves for the 4-FA versus 3-FA and 4-FA versus 2-FA binary classifier is illustrated below. Depending on the test requirements, a match score decision threshold can be chosen to prevent false-positive outcomes (important in a forensic setting), prevent false-negative outcomes (important in a security setting) or provide optimal test results (maximum TPR, minimal FPR). The AUC value indicates the predictive value of the EI-MS match score to establish the presence of 4-FA in a sample. An AUC of 0.79 for the 4-FA/3-FA couple is unfortunately not sufficient in a forensic case work setting.

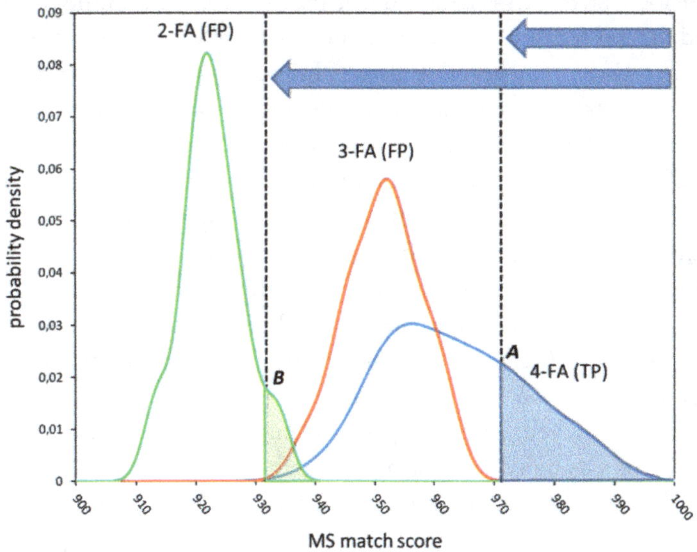

Illustration of the construction of a ROC curve from probability density functions. Starting from a maximum EI-MS library match score of 1000 against the 4-FA reference, the area under the 4-FA curve (TPR) and 3-FA and 2-FA curve (FPR) is determined as the match score is continuously lowered to construct the ROC curve. *(Adapted from previous figure)*

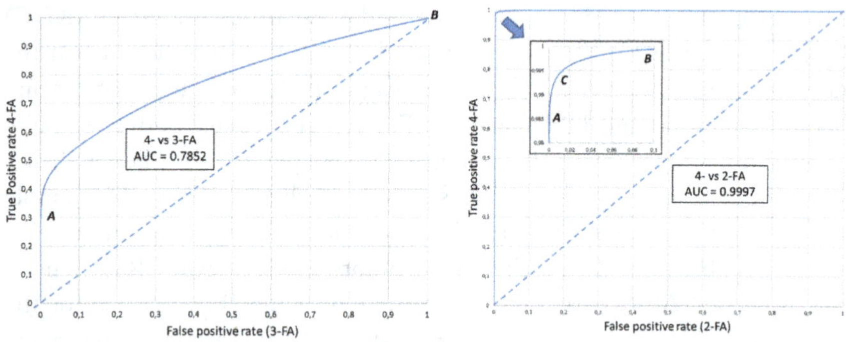

ROC curves for 4-FA versus 3-FA and 4-FA versus 2-FA, point A is indicating the TPR for which the FPR is negligible (no false-positives!) and point B is indicating a 100% TPR (no false-negatives!) and the associated FPR. The AUC for 4-FA versus 2-FA is almost 1 indicating a near perfect test. Optimal performance of the test is obtained at point C with a TPR of approximately 99.5% and an FPR of 1% at a match score threshold of 935. *(Reprinted and adapted with permission from Kranenburg et al. Forensic Science International, 302 (2019) 109900, https://doi.org/10.1016/j.forsciint.2019.109900)*

A closer look at the EI-MS spectra of the fluor amphetamine isomers has revealed the existence of subtle differences between the various ring isomers that at first glance are invisible to the human expert eye. However, our attempt to access and exploit this information through the use of library match scores against spectra for reference standards has only be partly successful. Clearly, the library match score provides a lead with respect to the actual FA isomer in the sample, but the associated evidential value is insufficient for a robust identification. This could either mean that the information "hidden" in the mass spectra is limited or that the methodology to extract this information is not optimal. Score-based models provide relatively modest *LR* values as the rarity of individual features is not considered. Therefore, the assumption that actually more information could still be retrieved from the spectra seems plausible. However, this will require more advanced data analysis methods and in the next section powerful chemometric techniques such **PCA** and **LDA** will be introduced.

8.3 Exploring NPS EI mass spectra with PCA

For the library match score study, we have created an extensive dataset of 30 GC-MS runs for each FA isomer. This equates to 90 EI-MS spectra which at a mass range of 41–154 at unit mass resolution leads to a 90 × 114 data matrix:

Sample Info		Abundance								
Analysis	FA isomer	m/z 41	m/z 42	m/z 43	m/z 44	..	m/z 151	m/z 152	m/z 153	m/z 154
1	2-FA	0.xxx	0.xxx	0.xxx	1	..	0.xxx	0.xxx	0.xxx	0.xxx
2	2-FA	0.xxx	0.xxx	0.xxx	1	..	0.xxx	0.xxx	0.xxx	0.xxx
3	2-FA	0.xxx	0.xxx	0.xxx	1	..	0.xxx	0.xxx	0.xxx	0.xxx
:	:	:	:	:	:	..	..	..	..	..
30	2-FA	0.xxx	0.xxx	0.xxx	1	..	0.xxx	0.xxx	0.xxx	0.xxx
31	3-FA	0.xxx	0.xxx	0.xxx	1	..	0.xxx	0.xxx	0.xxx	0.xxx
32	3-FA	0.xxx	0.xxx	0.xxx	1	..	0.xxx	0.xxx	0.xxx	0.xxx
33	3-FA	0.xxx	0.xxx	0.xxx	1	..	0.xxx	0.xxx	0.xxx	0.xxx
:	:	:	:	:	:	..	..	..	..	..
60	3-FA	0.xxx	0.xxx	0.xxx	1	..	0.xxx	0.xxx	0.xxx	0.xxx
61	4-FA	0.xxx	0.xxx	0.xxx	1	..	0.xxx	0.xxx	0.xxx	0.xxx
62	4-FA	0.xxx	0.xxx	0.xxx	1	..	0.xxx	0.xxx	0.xxx	0.xxx
63	4-FA	0.xxx	0.xxx	0.xxx	1	..	0.xxx	0.xxx	0.xxx	0.xxx
:	:	:	:	0.xxx	:	..	..	..	..	..
90	4-FA	0.xxx	0.xxx	0.xxx	1	..	0.xxx	0.xxx	0.xxx	0.xxx

FA isomer EI-MS data set represented as a 90 × 114 data matrix.

The mass range in GC-MS exceeds the range listed above for the FA dataset (41–154) and instrument specifications typically list a mass range of 2–1000. What is the reason that m/z values below 41 and higher than 154 are excluded from the dataset?

The molecular mass of fluor amphetamine is 153 Da and because in EI the molecular ion formed by electron removal has the highest m/z value (other ions are exclusively formed by fragmentation), there is little use of adding abundance data for higher m/z values. This will only add noise to the data set. The upper limit is set at 154 Da to also include the signal for the molecular ions that contain one ^{13}C isotope. The lower mass range often contains signals from traces of low molecular mass gaseous compounds. To exclude this source of variation, m/z values below 41 are excluded from the data set. This cut-off value is chosen because the most abundant trace compound in the Earth's atmosphere is the noble gas Argon with a typical concentration exceeding 0.9% (v/v). The main stable isotope of Argon has a molecular mass of 40 Da. Air leaks in a GC-MS instrument can be investigated by monitoring the abundances at m/z 28 (N_2), 32 (O_2), or 40 (Ar).

Why is the abundance at m/z 44 for all FA isomers and GC-MS runs equal to one and what does this mean for the dimensionality of the data?

The answer can be found by inspecting the EI-MS reference spectra that were depicted in the previous section. EI-MS spectra are typically normalized against the most abundant peak by setting the ion current at this m/z value at 1. For all FA isomers, the most dominant ion in the spectrum corresponds to the primary amine fragment in the molecule that splits off in a so-called α cleavage (structure is given in **Chapter 4**) resulting in a strong signal at m/z 44. By normalizing the most abundant ion, the dimensionality of the dataset is reduced to M-1 as the normalized m/z 44 signal can no longer be used to differentiate the NPS isomers. Effectively this column could be removed from the dataset. In the remainder of this section, we will see that sometimes it can be beneficial to start with the raw instrument read out instead of the normalized data. By using the actual ion currents possibly more effective ways of data preprocessing can be pursued.

The main challenge is now to effectively interrogate this dataset to find potential differences between the FA isomer spectra that could be exploited to confidently determine whether a forensic case sample contains 2-FA, 3-FA, or 4-FA (as indicated earlier we will not consider the option that a sample could contain a mixture of FA isomers). For a human forensic expert, it is very difficult to scan such a substantial dataset and then look for FA isomer discriminating leads while considering the large degree of correlation that could exist between the different signals in the mass spectrum. A strong limitation in human inspection of multivariate data is the fact that the world as we know and experience it is limited to three dimensions (four dimensions if we include time) and consequently our vision and brain has evolved to navigate and function in this 3D spatial environment making it impossible for humans to oversee and understand multidimensional datasets with $M > 4$ (3D projections that change over time as captured in a movie could possibly still be manageable). In this NPS data set, we now have to deal with 113 dimensions (*M*-1, excluding the normalized m/z 44 signal)! Even if this dimensionality is severely lowered by removing signals below a certain threshold, additional steps are required to allow effective human expert interrogation, interpretation, and application of the findings.

How would you as a forensic expert "manually" inspect the dataset in an attempt to find features that could be used to assign the correct the FA isomer?

Some main fragments (excluding m/z 44) in the mass spectra might exhibit small normalized abundance differences as function of the Fluorine position on the aromatic ring. This could be explored by plotting combinations of two ions (2D scatter plots) or three ions (3D scatter plots) and applying labels to discriminate between 2-FA, 3-FA, and 4-FA data points. If such differences exist and are significant, point clouds for the various isomers will be separated in the 2D or 3D m/z space. We could now exploit this by projecting a new point for a case sample in these plots. If the MS data for the case sample leads to a combination of m/z abundancies that are very similar to the 4-FA reference samples, this new point will be situated in the "4-FA point cloud." On the basis of this finding, the expert can now confidently state that the sample contains the listed hard drug 4-FA. The most prominent signals in the FA mass spectra are found at m/z 83, 109, and 138 and the corresponding ion structures have been explained in **Chapter 4—Qualitative Analysis and the Selectivity Dilemma**. The 3D and 2D scatter plots below show that this approach is surprisingly effective.

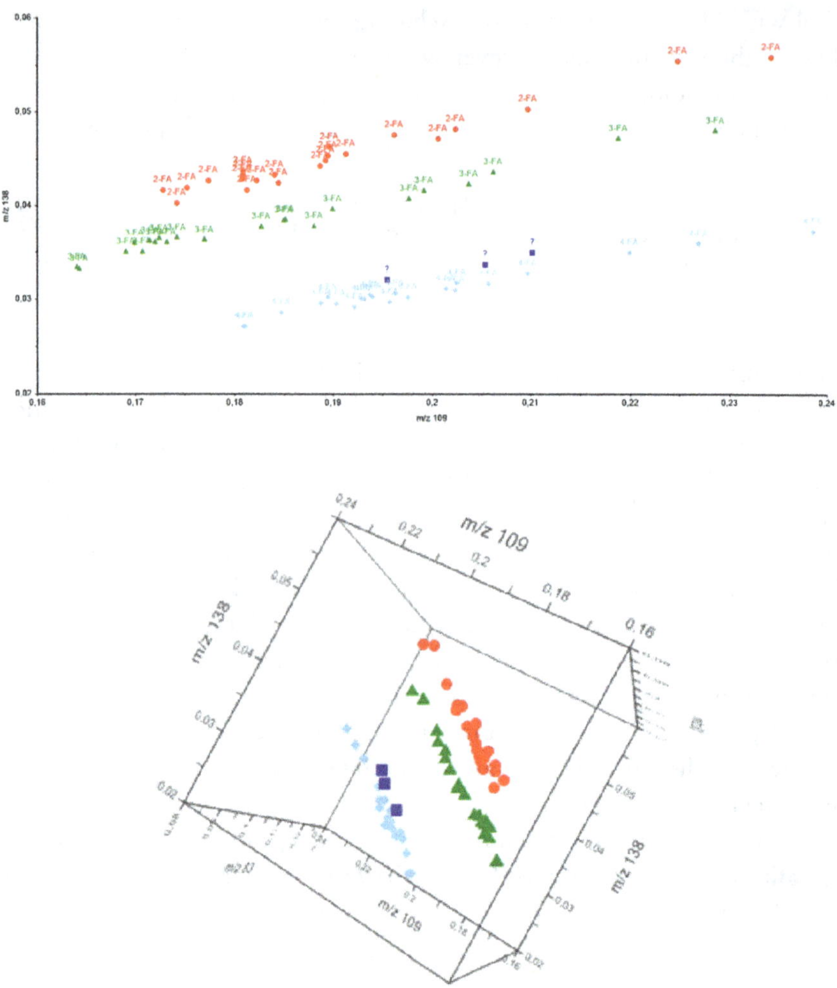

2D and 3D scatter plots of normalized abundance at m/z 83, 109, and 138 for 2-FA (red dots), 3-FA (green triangles), and 4-FA (light blue diamonds) and the projection of a case work sample containing an unknown FA isomer analyzed in triplicate (dark blue squares). The EI-MS spectrum for the unknown sample clearly projects in the light blue diamond point cloud suggesting the presence of 4-FA. *(Plots were created with Unscrambler version 10.5 from Camo Software, Aspentech)*

Interestingly, isomer discrimination is only possible when considering at least two ions. Projecting the data on the individual axis in the 2D and 3D plots will result in significant overlap of the data points hampering isomer discrimination. By exploiting the existing correlation for two or three ions, the data point clouds for the three isomers become fully separated in the 2D and 3D plot space, respectively. It seems that the variation in the normalized intensity is governed by a general phenomenon that is not related to the isomer species that is being analyzed. The 2D plot suggests that a linear relationship exists between the normalized m/z 109 and m/z 138 abundance with a similar slope but different off-set for the three NPS isomers. The variation could be instrumental by nature or could be related to the compound concentration and overall ion currents. This will be explored in more detail later.

Despite the success of the 2D and 3D FA isomer abundance plots, exploring a multivariate dataset in this way is quite inefficient and could lead to suboptimal outcomes. For more complex mass spectra a substantial number of 2D and 3D combinations will need to be investigated. Furthermore, substantial isomer specific information might be present in low abundance signals that are ignored in this approach. Finally, it would be desirable if trends and data variations that are not related to isomer-specific effects could be eliminated yielding more condensed and specific point clouds for each isomer. Generally, it would be highly desirable to prevent trial and error approaches but rather explore the full potential of the M-1 dimensions of this multivariate dataset to be able to process the dataset and optimize the model parameters to enable a robust FA isomer identification.

A very powerful way to constructively analyze multivariate data and visualize trends and correlations, and create feature models was introduced in 1901 by the English statistician and mathematician Karl Pearson (1857—1936), also known for the Pearson correlation coefficient that was discussed in **Chapter 6**. In the 1930s, this method was independently reinvented and developed by American statistician Harald Hotelling (1895—1973) who also introduced the name **PCA** (source: Wikipedia). Nowadays, **PCA** is a very popular data analysis method that is frequently used by scientists from a wide variety of disciplines. PCA is generally applicable irrespective of the nature and background of multivariate continuous data. Within analytical chemistry, PCA is an important data analysis tool and often used in combination with spectroscopic techniques such as NIR (near infrared) spectroscopy. NIR spectra which are easily obtained with affordable and portable instrumentation contain detailed chemical information but are difficult to interpret. For this reason, NIR spectroscopy is often combined with **chemometrics** to develop models that link spectral data to sample features of interest. A forensic example of NIR spectral data analysis will be shown in **Chapter 11—Innovating Forensic Analytical Chemistry**. Chemometrics is generally considered as a special branch of analytical chemistry aiming to "extract information from chemical systems" (source: Wikipedia). Chemometricians are analytical chemists that have specialized in data science or data scientists that have become experts in analyzing chemical data. Although the correct application of chemometric data analysis methods like PCA requires a good understanding, the availability of dedicated software enables analytical chemists with limited programming experience to perform multivariate data analysis. Frequently analytical equipment comes with software that includes the option of PCA analysis. In the remainder of this section, the readers will be introduced to PCA which will be applied to the fluor amphetamine isomer EI-MS data set. The aim is to acquaint the reader with the basic principles of PCA and the steps required to develop robust models. Important PCA model parameters will be discussed allowing students to perform their own PCA analysis using dedicated software packages. The mathematical details of PCA are considered beyond the scope of this book just as an in-depth explanation of the algorithms that are required to calculate the principal components (NIPALS, SVD). More information can be found in the Further Reading section of this chapter.

So, what is the main idea behind PCA? In PCA a new coordinate system is generated to project multivariate data in a way that reduces the dimensionality while capturing as much of the available information as possible. This new coordinate system to project the multivariate data points (in our example EI mass spectra) consists of so-called **principal components (PC)** that are made up of linear combinations of the original data variables, so in the case of the EI-MS dataset for the FA isomers, a PC is given by:

$$\mathrm{PC}_z = b_{z\ x} \cdot A_x + b_{z\ x+1} \cdot A_{z\ x+1} + \ldots + b_{z\ y} \cdot A_y$$

where z is 1,2,3,... for the first, second, and third principal component, respectively, and $b_{z\ x}$ represents the regression coefficient of the zth principal component for the (normalized) abundance A at mass channel x, $x+1$, ...,y. These regression coefficients, or so-called **PCA loadings,** are a very important source of information as they reveal which ions in the mass spectra are most influential for that given principal component. For the first principal component (PC_1), the loadings are obtained through an iterative process to maximize the variation of the projections of the data points on the PC axis. In this way, maximum information present in the dataset is captured in the first principal component. This process is repeated for the second principal component with the condition that the second axis is orthogonal to the first, i.e., that the dot product as discussed in the previous section is zero or in other words that the angle between the two vectors/axes in the M dimensional space is 90°. This process can be conducted M times until the number of principal components equals the number of m/z variables in the mass spectrum. However, to capture a near 100% of the data variance, typically, only a limited number (3–5) of principal components suffices. To determine how many principal components are useful to consider, the **explained variance** can be plotted as function of the number of PCs. As soon as there is no significant further increase in the variance including extra principal components will not add any value. All the available information is now condensed in a PCA dataset of strongly reduced dimensionality compared to the original data. This will allow the inspection of the multivariate data in 2D and 3D by calculating the so-called **PC scores** of all the mass spectra in the dataset and creating scatter plots of PC_1 versus PC_2, PC_1 versus PC_3, PC_2 versus PC_3, PC_1 versus PC_2 versus PC_3, etc. With the information fully condensed in the first few principal components, patterns will hopefully emerge that will link scores to sample features of interest. In our example, it is expected that if the EI-MS mass spectra contain isomer-specific leads the PC scores will group according to FA isomer. If these FA isomer clusters are completely separated in a 2D/3D PC score plot, the PCA model can be used to assign the correct isomer for case work samples of unknown composition by calculating the PC scores of the measured mass spectrum using the loadings as determined from the reference ground truth data set and projecting the new data point in the PC score plot. If this point is projected in the reference cloud for a given FA isomer, this will indicate that the sample will contain this fluor amphetamine species and not any of the other isomers. This approach is similar to the plots shown in this section for the three main m/z signals in the EI-MS spectra of the FA isomers but with one substantial and very important difference. In the PC score plots, the information of the entire mass spectrum is reflected thereby utilizing all available information in a consistent and efficient manner. If such PC score plots do not reveal any isomer specific information, this indicates that the EI-MS spectra of 2-FA, 3-FA, and 4-FA are simply too similar and that the additional techniques introduced in **Chapter 4** are essential to assign the correct isomer. However, this final conclusion can only be drawn when the data have been optimally prepared for the PCA and the correct PCA settings have been applied. This will be discussed in more detail next.

An essential step prior to PCA is to **mean-center** the data. Typically, PCA will not lead to meaningful results without mean-centering **(MC)** and dedicated PCA software suites perform this step automatically. The mean centering process is illustrated below and involves the calculation of the average signal for each individual variable in the dataset and subsequently subtracting this average from the signal of each individual entry. It is important to realize that the average value is calculated over the (variable) data columns and not over the (sample) data rows. So for each individual m/z variable, the average abundance for all samples in the dataset is calculated and this average is subtracted from each normalized abundance in the m/z column. This is fundamentally different than correcting on the basis of the row average which would mean that each sample is modified individually and only on the basis of the variable data for that sample. Row-based modifications are part of the data preprocessing prior to PCA and include the normalization on the basis of the most abundant signal at m/z 44 that was already conducted. Subtracting the average of the signal for all variables in one row (i.e., for one sample) is a form of baseline correction that is often done for spectroscopic data. A procedure named **SNV (Standard Normal Variate)** is used very frequently to correct for baseline offset and drift in NIR reflection and Raman spectra and involves the subtraction of the spectral average and division by the associated standard deviation. SNV is a data preprocessing step applied to data rows whereas mean centering is part of the PCA and is applied to data columns.

Data Preprocessing

Sample Info		Abundance								
Analysis	FA isomer	m/z 41	m/z 42	m/z 43	m/z 44	..	m/z 151	m/z 152	m/z 153	m/z 154
1	2-FA	0.xxx-0.yyy	0.xxx-0.yyy	0.xxx-0.yyy	0	..	0.xxx-0.yyy	0.xxx-0.yyy	0.xxx-0.yyy	0.xxx-0.yyy
2	2-FA	0.xxx-0.yyy	0.xxx-0.yyy	0.xxx-0.yyy	0	..	0.xxx-0.yyy	0.xxx-0.yyy	0.xxx-0.yyy	0.xxx-0.yyy
3	2-FA	0.xxx-0.yyy	0.xxx-0.yyy	0.xxx-0.yyy	0	..	0.xxx-0.yyy	0.xxx-0.yyy	0.xxx-0.yyy	0.xxx-0.yyy
:	:	:	:	:	:	..	..	..	..	..
30	2-FA	0.xxx-0.yyy	0.xxx-0.yyy	0.xxx-0.yyy	0	..	0.xxx-0.yyy	0.xxx-0.yyy	0.xxx-0.yyy	0.xxx-0.yyy
31	3-FA	0.xxx-0.yyy	0.xxx-0.yyy	0.xxx-0.yyy	0	..	0.xxx-0.yyy	0.xxx-0.yyy	0.xxx-0.yyy	0.xxx-0.yyy
32	3-FA	0.xxx-0.yyy	0.xxx-0.yyy	0.xxx-0.yyy	0	..	0.xxx-0.yyy	0.xxx-0.yyy	0.xxx-0.yyy	0.xxx-0.yyy
33	3-FA	0.xxx-0.yyy	0.xxx-0.yyy	0.xxx-0.yyy	0	..	0.xxx-0.yyy	0.xxx-0.yyy	0.xxx-0.yyy	0.xxx-0.yyy
:	:	:	:	:	:	..	..	..	..	..
60	3-FA	0.xxx-0.yyy	0.xxx-0.yyy	0.xxx-0.yyy	0	..	0.xxx-0.yyy	0.xxx-0.yyy	0.xxx-0.yyy	0.xxx-0.yyy
61	4-FA	0.xxx-0.yyy	0.xxx-0.yyy	0.xxx-0.yyy	0	..	0.xxx-0.yyy	0.xxx-0.yyy	0.xxx-0.yyy	0.xxx-0.yyy
62	4-FA	0.xxx-0.yyy	0.xxx-0.yyy	0.xxx-0.yyy	0	..	0.xxx-0.yyy	0.xxx-0.yyy	0.xxx-0.yyy	0.xxx-0.yyy
63	4-FA	0.xxx-0.yyy	0.xxx-0.yyy	0.xxx-0.yyy	0	..	0.xxx-0.yyy	0.xxx-0.yyy	0.xxx-0.yyy	0.xxx-0.yyy
:	:	:	:	:	:	..	..	..	..	..
90	4-FA	0.xxx-0.yyy	0.xxx-0.yyy	0.xxx-0.yyy	0	..	0.xxx-0.yyy	0.xxx-0.yyy	0.xxx-0.yyy	0.xxx-0.yyy
Average		0.yyy	0.yyy	0.yyy	1		0.yyy	0.yyy	0.yyy	0.yyy

PCA settings

Mean centering of the FA isomer EI-MS data indicating that the average value is calculated for the data columns and not for the data rows

Why is mean centering of the data so important to obtain meaningful PCA results?

After mean centering each variable (in our example each unit m/z in the range of the mass spectrum) will reflect the variation in the data set. Deviations from the variable average can be linked to features of interest associated with the given spectrum. Without mean centering the iterative process to find maximal variation will focus on the absolute mean abundance values as function of m/z and not on the variation in abundance. This is the reason that PCA without mean centering usually does not yield valuable insights and leads to poor model performance.

An additional step that can be considered in the PCA process is the so-called **auto-scaling (AS)** of the data, also termed data standardization. Auto-scaling in PCA is the "column equivalent" of the row-based SNV spectral data preprocessing step that was explained above. This means that in addition to mean centering each corrected abundance value in the m/z column is divided by the standard deviation as is illustrated below.

Sample Info		Abundance								
Analysis	FA isomer	m/z 41	m/z 42	m/z 43	m/z 44	..	m/z 151	m/z 152	m/z 153	m/z 154
1	2-FA	(0.xxx-0.yyy)/0.zzz	(0.xxx-0.yyy)/0.zzz	(0.xxx-0.yyy)/0.zzz	0/0	..	(0.xxx-0.yyy)/0.zzz	(0.xxx-0.yyy)/0.zzz	(0.xxx-0.yyy)/0.zzz	(0.xxx-0.yyy)/0.zzz
2	2-FA	(0.xxx-0.yyy)/0.zzz	(0.xxx-0.yyy)/0.zzz	(0.xxx-0.yyy)/0.zzz	0/0	..	(0.xxx-0.yyy)/0.zzz	(0.xxx-0.yyy)/0.zzz	(0.xxx-0.yyy)/0.zzz	(0.xxx-0.yyy)/0.zzz
3	2-FA	(0.xxx-0.yyy)/0.zzz	(0.xxx-0.yyy)/0.zzz	(0.xxx-0.yyy)/0.zzz	0/0	..	(0.xxx-0.yyy)/0.zzz	(0.xxx-0.yyy)/0.zzz	(0.xxx-0.yyy)/0.zzz	(0.xxx-0.yyy)/0.zzz
:	:	:	:	:	:	..	..	..	..	..
30	2-FA	(0.xxx-0.yyy)/0.zzz	(0.xxx-0.yyy)/0.zzz	(0.xxx-0.yyy)/0.zzz	0/0	..	(0.xxx-0.yyy)/0.zzz	(0.xxx-0.yyy)/0.zzz	(0.xxx-0.yyy)/0.zzz	(0.xxx-0.yyy)/0.zzz
31	3-FA	(0.xxx-0.yyy)/0.zzz	(0.xxx-0.yyy)/0.zzz	(0.xxx-0.yyy)/0.zzz	0/0	..	(0.xxx-0.yyy)/0.zzz	(0.xxx-0.yyy)/0.zzz	(0.xxx-0.yyy)/0.zzz	(0.xxx-0.yyy)/0.zzz
32	3-FA	(0.xxx-0.yyy)/0.zzz	(0.xxx-0.yyy)/0.zzz	(0.xxx-0.yyy)/0.zzz	0/0	..	(0.xxx-0.yyy)/0.zzz	(0.xxx-0.yyy)/0.zzz	(0.xxx-0.yyy)/0.zzz	(0.xxx-0.yyy)/0.zzz
33	3-FA	(0.xxx-0.yyy)/0.zzz	(0.xxx-0.yyy)/0.zzz	(0.xxx-0.yyy)/0.zzz	0/0	..	(0.xxx-0.yyy)/0.zzz	(0.xxx-0.yyy)/0.zzz	(0.xxx-0.yyy)/0.zzz	(0.xxx-0.yyy)/0.zzz
:	:	:	:	:	:	..	..	..	..	..
60	3-FA	(0.xxx-0.yyy)/0.zzz	(0.xxx-0.yyy)/0.zzz	(0.xxx-0.yyy)/0.zzz	0/0	..	(0.xxx-0.yyy)/0.zzz	(0.xxx-0.yyy)/0.zzz	(0.xxx-0.yyy)/0.zzz	(0.xxx-0.yyy)/0.zzz
61	4-FA	(0.xxx-0.yyy)/0.zzz	(0.xxx-0.yyy)/0.zzz	(0.xxx-0.yyy)/0.zzz	0/0	..	(0.xxx-0.yyy)/0.zzz	(0.xxx-0.yyy)/0.zzz	(0.xxx-0.yyy)/0.zzz	(0.xxx-0.yyy)/0.zzz
62	4-FA	(0.xxx-0.yyy)/0.zzz	(0.xxx-0.yyy)/0.zzz	(0.xxx-0.yyy)/0.zzz	0/0	..	(0.xxx-0.yyy)/0.zzz	(0.xxx-0.yyy)/0.zzz	(0.xxx-0.yyy)/0.zzz	(0.xxx-0.yyy)/0.zzz
63	4-FA	(0.xxx-0.yyy)/0.zzz	(0.xxx-0.yyy)/0.zzz	(0.xxx-0.yyy)/0.zzz	0/0	..	(0.xxx-0.yyy)/0.zzz	(0.xxx-0.yyy)/0.zzz	(0.xxx-0.yyy)/0.zzz	(0.xxx-0.yyy)/0.zzz
:	:	:	:	:	:	..	..	..	..	..
90	4-FA	(0.xxx-0.yyy)/0.zzz	(0.xxx-0.yyy)/0.zzz	(0.xxx-0.yyy)/0.zzz	0/0	..	(0.xxx-0.yyy)/0.zzz	(0.xxx-0.yyy)/0.zzz	(0.xxx-0.yyy)/0.zzz	(0.xxx-0.yyy)/0.zzz
Average		0.yyy	0.yyy	0.yyy	1		0.yyy	0.yyy	0.yyy	0.yyy
Standard Deviation		0.zzz	0.zzz	0.zzz	0		0.zzz	0.zzz	0.zzz	0.zzz

Mean-centering and auto-scaling of the FA isomer EI-MS data set

What is the effect of auto-scaling and when is this applied in PCA?

Would you recommend auto-scaling for the MS data set and why/why not?

Auto-scaling is often a sensible step when analyzing composite multivariate datasets, i.e., datasets that consist of variables that have been obtained from different sources and analysis methods. For instance, the PCA tutorial by Bro and Smilde, for which the reference can be found in **Section 8.6**, discusses a wine dataset that includes alcohol level, pH, sugar residues, and density. These variables have a different chemical/physical background and are measured by separate methods. In such a setting, variables with relatively high values can dominate the PCA because the high average signals are accompanied by substantial absolute measurement variations (at equal relative standard deviation). As a result nearly all the variance will be explained by the first principal component which is dominated by a few variables. With such a disbalance in variable magnitude, subtle variations as function of features of interest (such as the type of grape used to produce the wine, the country of origin or age) are not reflected in the PCA. Intuitively, one can appreciate that the unit chosen for a given variable should not affect the PCA outcome. In other words, whether the sugar residue is expressed in ppm (parts per million) or ppb (parts per billion) should not affect the PCA model. This can indeed be assured by auto-scaling the data as each variable is given an equal weight. However, making each variable equally important makes much less sense when considering multivariate data that is measured with a single method which is the case for our FA isomer dataset and also in general for spectroscopic data. A typical mass spectrum consists of a limited number of fragment ions and hence for many m/z variables the measured abundance will be near the threshold and will mostly represent chemical and/or instrumental noise. A process in which the impact of non-informative and "noisy" variables is willingly increased, will lead to a reduction in model performance. Indeed, also for the FA isomer EI-MS dataset the best results are obtained with mean centering only (also for NIR datasets auto-scaling is usually omitted). The PCA score plot of PC_1 versus PC_2 and the associated loadings are depicted below.

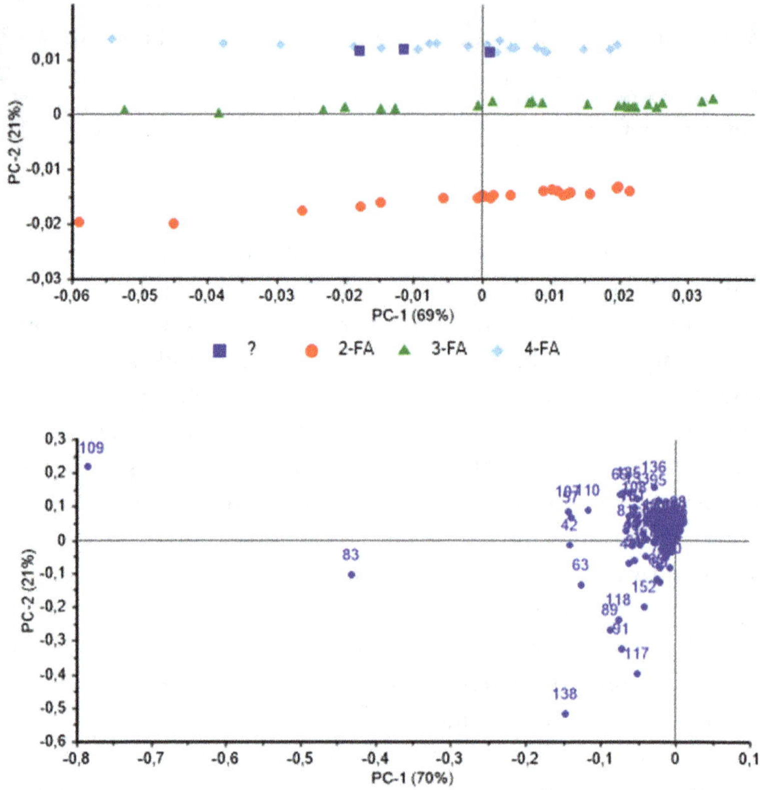

PCA score plot (top, PC_1 vs. PC_2) of the FA isomer EI-MS data and associated loadings plot (bottom) showing the regression coefficients for the various ions (m/z) in the mass spectrum. The three case work samples (dark blue squares) are projected in the score plot. *(Created with Unscrambler version 10.5 from Camo Software, Aspentech)*

Inspection of the PCA results leads to several interesting insights and questions:

1. The score plot shows an FA isomer pattern that looks very similar to the 3D scatter plot of the abundance at m/z 83, m/z 109, and m/z 138. Can you explain this similarity on the basis of the loading plot?
2. The first principal component (PC_1) explains 70% of the variance of the data but is this variance related to the FA isomer structure? If so explain the relation and if not what other effect could be causing this substantial variation?
3. The second principal component (PC_2) explains 21% of the variance of the data but is this variance related to the FA isomer structure? If so explain the relation and if not what other effect could be causing this substantial variation?
4. What measures could be taken to further improve the PCA model and achieve a more robust separation of the isomer cluster?

The PC score plot of PC_1 versus PC_2 is at first sight a bit disappointing, and the FA clusters are clearly separated but show a wide distribution in the PC_1 scores. This is an undesirable effect; on the basis of the first principal component, it is impossible to assign the isomer structure because the associated PC_1 scores for 2-FA, 3-FA, and 4-FA reference samples show a high degree of overlap. Still PC_1 accounts for 70% of the variation in the mean centered data. This must mean that this first principal component is zooming in on a dominant source of variation that is related to a generic phenomenon that is applicable to all EI-MS spectra irrespective of the FA isomer type in the sample. Because the spectra are normalized on the most abundant signal at m/z 44, this phenomenon is not simply a matter of sensitivity fluctuations. However, it could actually be the result of this normalization process when the m/z 44 abundance is too high which can lead to saturation and a non-linear response of the detector. This would cause the abundance of all ions of interest in the mass spectrum to rise or fall depending on the degree of saturation of the base peak in the mass spectrum. This hypothesis is supported by the PC_1 loadings, the absolute value of the loading is increasing in the order of m/z 138, m/z 83, and m/z 109 and this corresponds to the relative abundance of these main ions in the FA EI-MS spectrum with the highest peak in the mass spectrum at m/z 109. So, the higher the normalized abundance, the more significant the contribution of this m/z to the first principal component. This is expected for data that has not been auto-scaled, higher variable values are accompanied by a higher absolute standard deviation leading to a greater contribution to the variation in the dataset. This also explains why the PC score plot looks so similar to the 3D scatter plot that was obtained by plotting the abundance of these three ions. Although the second principal component accounts for "only" 21% of the data variance it is of direct relevance for the task at hand. The fact that the scatter plot shows a clear separation of the three FA isomer clusters can solely be attributed by PC_2. If all data points are projected on the PC_2 axis, condensed and totally separated ranges are obtained for 2-FA, 3-FA, and 4-FA. This is exactly what we need to confidently assign the correct isomer structure. The second principal component clearly has extracted isomer specific information from the mass spectra. When the PC_2 loadings are inspected we can see that in addition to the three main ions in the mass spectrum also additional m/z variables significantly contribute to the scores (*e.g.*, m/z 117, 91, and 89) revealing more complex ion chemistry processes that allow for FA isomer differentiation despite the fact that the EI-MS spectra are visually very similar. So, how could we further improve the PCA model to increase the robustness of the isomer assignment? The fact that the variation in the dataset is dominated by a generic and poorly understood phenomenon is not preferable. If this phenomenon starts to affect the performance over time, unknown case samples might not project into the FA isomer clusters leading to inconclusive results or possibly even erroneous conclusions. Possibly, it would require very frequent reruns of FA isomer standards which costs precious time and money. One way to address this problem is through further data preprocessing. In the original published study for which the reference is given in the Further Reading section of this chapter, several additional data preprocessing methods were explored by exporting the raw, un-normalized data from the MS instrument. Indeed it could be shown that the original hypothesis was correct and that the variation covered by PC_1 could be attributed to the very high m/z 44 abundance. By omitting this m/z column from the dataset and normalizing on the basis of total ion current (*i.e.*, dividing by the sum of all recorded abundances), the model was drastically improved as all principal components were now revealing isomer specific details in the mass spectra. Another approach that was very effective was to select the data in the m/z 100–200 mass range and apply total abundance normalization after selection. However, also for the current dataset improvements are possible by exploiting the fact that PC_1 has captured the uninformative part of the data variance. So basically we can use PC_1 as a filter and focus on PC_2 and PC_3. Indeed the explained variance plot is showing that also the third principal component still captures a significant amount of data variation (7.5%). In total, the first three principal components explain 98% of the data variation which can be further increased to 99% by including PC_4. Higher principal components add little information and can be ignored. This demonstrates the high degree of correlation in EI mass spectra which is not that surprising considering fragmentation pathways in EI. This demonstrates how the information present in a dataset consisting of 112 dimensions in the form of recorded normalized abundances for 112 m/z channels (excluding m/z 44 used for normalization) can be captured by three to four principal components in a PCA model.

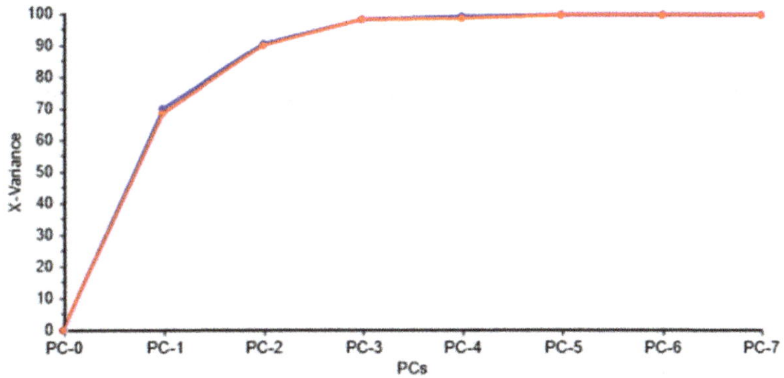

Explained variance as function of number of principal components in the PCA model of the FA isomer EI-MS data. The blue line represents the results for the entire data set, and the red line depicts the results for the validation data based on 20 segments of 3–4 random samples that are removed from the dataset and projected on the PCA model. *(Created with Unscrambler version 10.5 from Camo Software, Aspentech)*

Robust PCA models require sufficient data points to ensure that the effect of a single deviating sample or measurement has limited effect on the PCA model and the scores and loadings. Dedicated PCA software packages typically include automated validation protocols based on the "**leave-one-out**" principle. In the leave-on-out procedure, reference samples are removed one by one and projected in the resulting model. The projection is compared to the scores when that sample is included in the model. By combining the results of all leave-one-out validation experiments, the robustness of the PCA model can be tested. This dataset clearly contains sufficient samples and mass spectra for each isomer (a balanced distribution of data for the features of interest is also an important aspect to consider) as the explained variance for the validation data fully coincides with the results for the full data set. Before building a PCA model, it is also advisable to check the reference data set for outliers that could negatively affect model performance. Typically software packages also include outlier detection options. Illustrated below is the so-called influence plot which indicates how individual reference data points fit and affect the model. A detailed mathematical analysis is beyond the scope of this book, but the Hotelling's T^2 statistics is a measure of the influence of the sample on the model; high values mark measurements with a substantial impact on the model. On the y-axis of the plot, the F-Residuals indicate the distance of a data point to the PCA model. High values indicate that the model provides a poor description of that specific sample or analysis. Potential outliers are situated in the top right part of the influence plot and represent data points that do not fit the PCA model but nonetheless have a strong influence on the model parameters. Removing such samples from the reference data set typically results in very significant changes to the scores and loadings and improved model performance. On the basis of the influence plot, there was no reason to remove any EI-MS spectra from the reference data set.

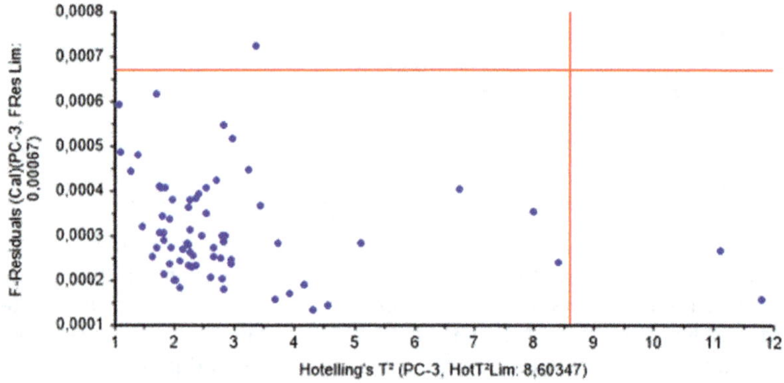

Influence plot for the FA isomer EI-MS data *(Created with Unscrambler version 10.5 from Camo Software, Aspentech)*

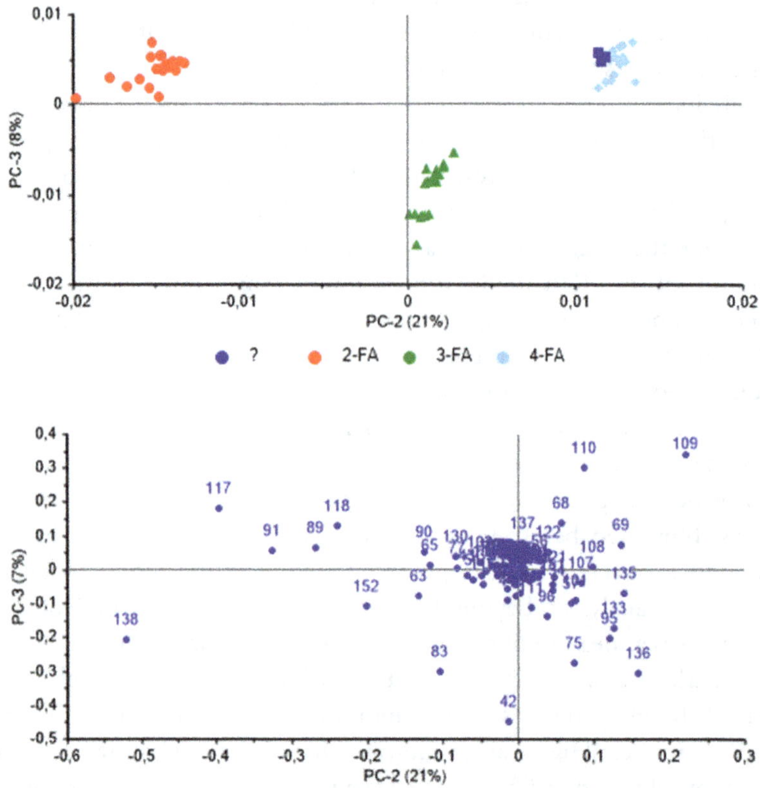

PCA score plot (top, PC_2 vs. PC_3) of the FA isomer EI-MS data and associated loadings plot (bottom) showing the regression coefficients for the various ions (m/z) in the mass spectrum. The three case work samples (dark blue squares) are projected in the score plot. *(Created with Unscrambler version 10.5 from Camo Software, Aspentech)*

With the robustness of the PCA model ensured and the reference data set checked for outliers, we can now focus on the retrieval and visualization of the isomer-specific information in the EI-MS dataset. The effect of the base peak abundance is captured by PC_1 and hence by ignoring this PC (which in a way is uncommon because the first principal component covers most of the data variation) hopefully the subsequent principal components will reveal isomer-specific grouping. The PC_2 versus PC_3 score plot given above indeed confirms these expectations. The unknown sample projects nicely in a condensed 4-FA cluster and because the isomer clusters are fully separated this provides convincing evidence that the case work sample contains the listed hard-drug 4-FA. The loading plot lists a substantial number of ions in the FA mass spectrum that contribute to the isomer specific differences in the PC_2-PC_3 space including m/z 136, 118, 110, and 109. However, the fragment ions exhibit a strong correlation as 99% of the variance can be explained by the first four principal components.

Overall, the results are truly remarkable! Where it was argued in **Chapter 4—Qualitative Analysis and the Selectivity Dilemma** that the identification of NPS isomers requires additional techniques like GC-IR and GC-VUV, we now discover that the application of a powerful multivariate data analysis like PCA enables isomer differentiation from GC-MS data. Of course, care has to be taken not to make general statements on the basis of a single example. Similar chemometric studies have to be conducted for more NPS (ring) isomer sets to fully appreciate the potential. Furthermore, we should not forget that to assign the isomer structure we have to build the PCA model first which takes time and effort and requires reference standards (or case work samples for which the isomer species has been confirmed by other analytical techniques) and a substantial data set ideally collected over a substantial period of time to include typical instrument fluctuations. Possibly, due to instrument drift, such a modeling exercise needs to be repeated for a given instrument over time. The applicability of a PCA model for a given instrument or set of instruments needs to be studied further, possibly data preprocessing can be optimized to further increase instrument span and model lifetime. Another interesting topic for further investigation is the model behavior when dealing with a mixture of isomers or when a new NPS isomer is encountered for the first time in case work. In the latter situation, the unknown EI-MS spectrum will be projected in a model that was created with a dataset that did not include this isomer species. Of course this will not result in a correct identification but hopefully the case work sample will now project outside the existing isomer clusters in the PCA score plot. This will provide a clear warning to the forensic expert that the sample under consideration contains a different compound. A similar outcome is also desirable for isomer mixtures. Possibly, a "mixed" EI mass spectrum will project spatially between the two PC score clusters of the individual isomers which will also act as an alarm and will signal that the case work sample has to be studied in more detail.

In addition to all these considerations, the forensic expert also needs to be able to express the evidential value of the isomer assignment. Ideally in the form of an *LR* value as was explained in **Chapter 6—Chemical Profiling, Databases, and Evidential Value** or at least by providing decision thresholds and associated false-positive and false-negative error rates. In the example shown, the mass spectra of the case work sample project quite clearly in the 4-FA cluster. But how would the expert interpret the results if the projection is close to (and much closer in comparison to the other isomer clusters) but not exactly in the cluster? What criteria need to be met to assign the isomer structure and when will the result be inconclusive? Such questions are very hard to answer with PCA, which is an exploratory multivariate data analysis technique that reveals trends by reducing data dimensionality. However, additional chemometric techniques are available to build on the PCA findings and address these important aspects as will be shown in the next section.

8.4 Differentiating NPS isomers with PCA-LDA of EI mass spectra

We will start this section with a challenging question to the reader:

> PCA is a so-called unsupervised data analysis method.
>
> What information in the EI-MS dataset is not used when building the PCA model?
>
> (Tip: Inspect the table in the previous section illustrating the FA isomer EI-MS 90 × 114 data matrix).

The PCA model is constructed purely from the EI-MS data and in this process the ground truth, i.e., the knowledge that a certain FA isomer is being analyzed, is not exploited. The information is included in the table to be able to label the datapoints after the PCA model has been constructed. This allows the expert to inspect if the model has extracted isomer-specific information from the multivariate data and to exploit score plot clustering to identify NPS isomers in unknown samples. It is a feature of **unsupervised** data analysis methods, like PCA, that sample labels are not used in the model creation. In PCA, the maximum variation in the entire data set is sought irrespective of the nature of the individual data entries. This leads to robust models that do not easily suffer from **overfitting** (insufficient data in relation to the model complexity and **extrapolation** (the analysis of samples that provide a model response that has not been covered by the reference data). However, unsupervised methods fail to make use of quite essential information that could be used to maximize the difference between the various sample classes. This is exactly what a **supervised** method like **LDA** does. LDA has been introduced by the famous British geneticist, statistician, and mathematician sir Ronald Fisher (1890–1962) who has been critical in the development of modern statistical science. He introduced ANOVA (analysis of variance) and the F-test (which is named after him) and formulated the principles of experimental design.

Pictures from left to right of Karl Pearson (1857–1936), Harald Hotelling (1895–1973) and sir Ronald Fisher (1890–1962). *(Information and free domain pictures from Wikipedia)*

LDA, like PCA, is a technique that reduces the dimensionality of a multivariate dataset by creating a new coordinate system by a linear combination of the original data variables. However, in PCA the first principal component is obtained by maximizing the projected overall data variance, whereas the LDA algorithm seeks to optimize the difference between classes while minimizing the data variation within a class. This difference is visualized below. For a detailed mathematical explanation of LDA, the reader is referred to academic chemometric course books or the tutorial in the free arXiv repository listed in **Section 8.6**. Dedicated data analysis software packages in addition to PCA typically offer a number of classification techniques including LDA. LDA is a supervised method for binary classification which typically can be very valuable in mass quality screening ("good" vs. "bad," "accept" vs. "reject," "in spec" vs. "out spec"). In forensic chemistry, binary classification is applied in chemical profiling where the main question is whether a crime scene sample and a given reference material share the same origin ("same batch" vs. "different batch"). In the current example, three possible FA isomers have to be considered. However, LDA can easily be expanded to include three or more classes in a so-called multiclass LDA approach. The framework for LDA-based NPS isomer differentiation is very similar to the use of PCA as described previously. The reference data set with 30 unit resolution EI-MS spectra for each known FA isomer class is used to create a multiclass LDA model. The dimensionality reduction is focused on maximizing the separation of the three FA isomer classes. This illustrates why LDA, unlike PCA, is a supervised data analysis method, the class labels in the data set are used to maximize the group separation. Spectra of unknown case samples are now analyzed by the LDA model and assigned to one of the three classes.

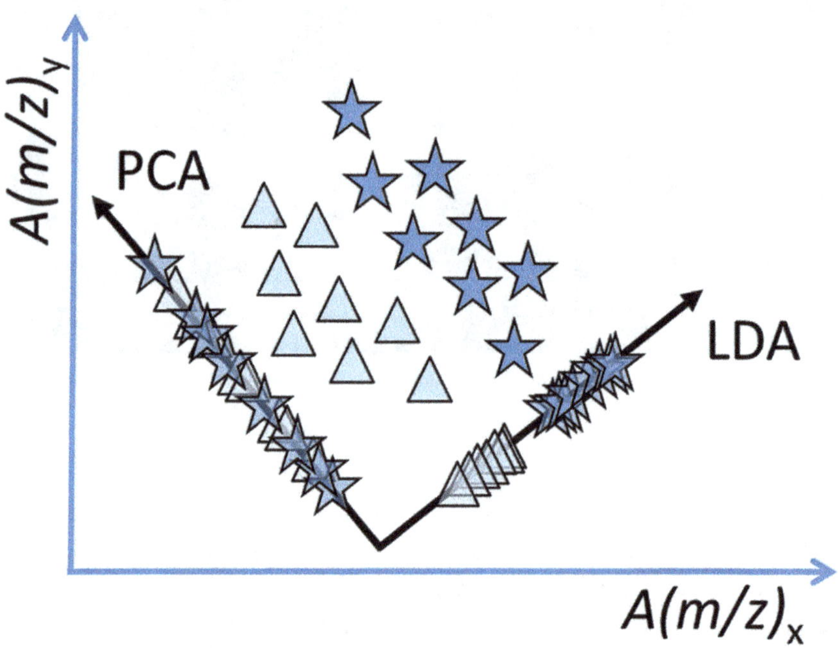

Illustration of the difference between PCA and LDA in a hypothetical 2D EI-MS dataset with two classes of interest (triangles and stars, respectively), the new coordinates are defined according to maximal variance of the data for PCA while in LDA the ratio of the between and within class variance is maximized

To be able to apply LDA correctly, the dataset needs to meet a number of model requirements. First of all, the variables for the various classes should be distributed normally. This means that in the current dataset the variation in the abundance for a given m/z channel should exhibit a distribution that is Gaussian and can be defined by its mean and standard deviation. In addition, this distribution should be similar for the various classes allowing the standard deviations to be pooled. The LDA model assumes that the data for the different classes exhibit the same degree of variation and only differ in the mean value. Considering the reference dataset creation and the data characteristics (normalized abundance for unit mass m/z channels), these assumptions seem reasonable. However, to be sure one could perform statistical tests on the m/z data columns. Normality can be checked with the **Lilliefors** test which is based on the **Kolmogorov–Smirnov** method. The equal variance assumption can be investigated by using the **Levene** or **Bartlett** test. Applying these statistical test methodologies showed that the LDA conditions, with a few exceptions, are met for the FA EI-MS reference data set. For the main ions in the mass spectrum, the normalized abundance for the 2-FA, 3-FA, and 4-FA isomers appears to be normally distributed with a similar standard deviation as is illustrated below. Please note that when class variability differs, discriminant analysis might still be feasible using a non-linear classifier such as a quadratic function (**QDA**) or the Mahalanobis distance (the scaled distance of a datapoint to the class center).

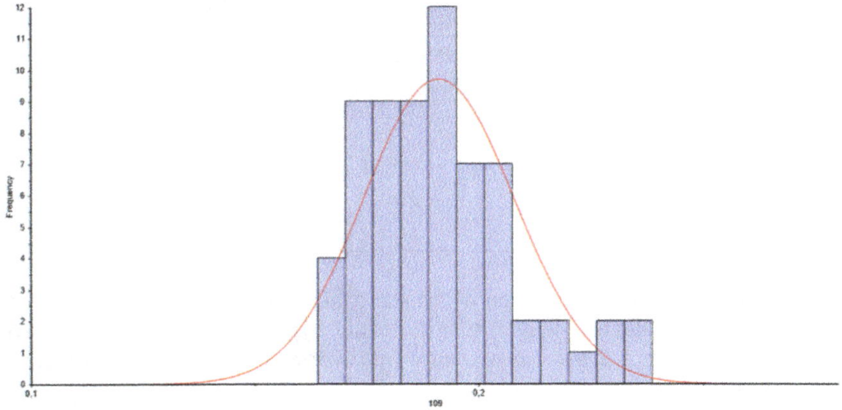

Histogram showing the abundance distribution for the m/z 109 data for the entire model data set (including all FA isomers) with in red the fitted Gaussian curve with a mean relative abundance of 0.19 and a standard deviation of 0.017. *(Plots were created with Unscrambler version 10.5 from Camo Software, Aspentech)*

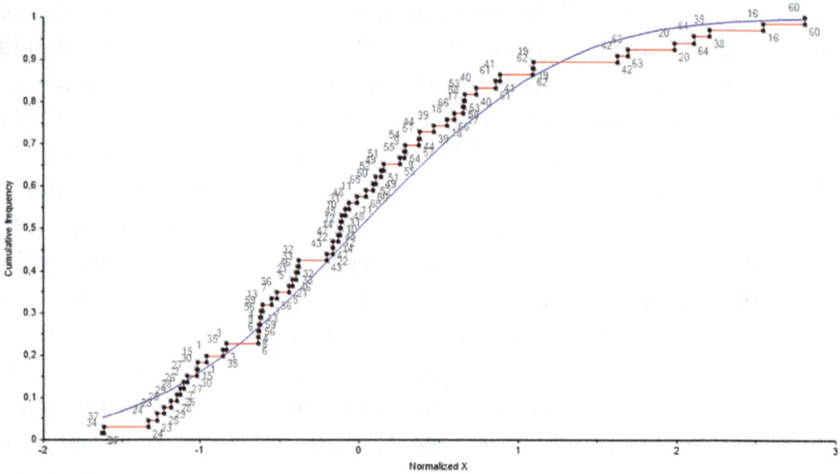

Kolmogorov–Smirnov (K–S) normality plot showing the results of the Liliefors normality test based on the average abundance and standard deviation from the m/z 109 model data. The data are plotted against the empirical cumulative distribution function, and the maximum difference is used as the parameter for the K–S statistical test. The null hypothesis that the data is normally distributed cannot be rejected. *(Plots were created with Unscrambler version 10.5 from Camo Software, Aspentech)*

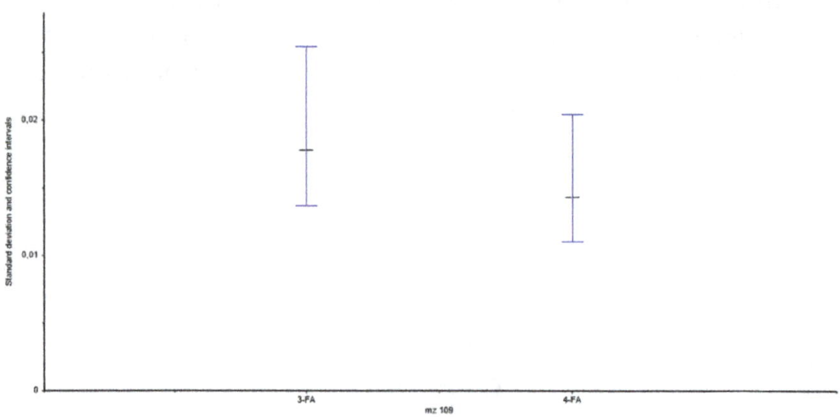

Bartlett's test shows the distribution of the standard deviation of the m/z 109 abundance data for the 3-FA and 4-FA standards. The null hypothesis that the variances are equal cannot be rejected. *(Plots were created with Unscrambler version 10.5 from Camo Software, Aspentech)*

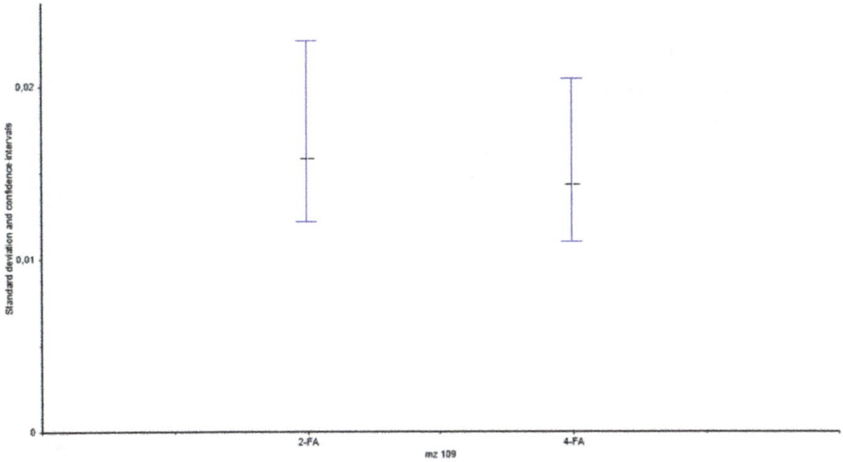

Levene's test shows the distribution of the standard deviation of the m/z 109 abundance data for the 2-FA and 4-FA standards. The null hypothesis that the variances are equal cannot be rejected. ***(Plots were created with Unscrambler version 10.5 from Camo Software, Aspentech)***

As a supervised data analysis method like LDA is typically very sensitive to overfitting phenomena. Therefore, robust LDA models require a substantial reference data set to ensure that the number of entries per class exceeds the number of variables. With the current 90 (entries) x 113 (m/z channels) data set we are clearly not meeting this requirement. The number of analyses per isomer class would have to be increased fourfold which would require a substantial additional laboratory effort. Such excessive data needs will hamper the implementation of LDA in forensic casework especially when reference data needs to be occasionally updated.

What strategy/methodology could be used enable robust LDA modeling without the need to increase the current, limited dataset?

What we need is a reduction in the dimensionality of the multivariate data set while maintaining maximum mass spectral information in order to distinguish the FA isomer forms. This is exactly what PCA does as described in the previous section! Therefore, LDA is often combined with PCA especially for spectroscopic datasets that typically combine a high dimensionality with a substantial degree of correlation. It should be noted that when data is easily acquired and of a limited dimensionality, the direct application of LDA can be preferable because as illustrated above, the first principal components do not necessarily contain class specific information. However, if sufficient principal components are considered that explain near 100% of the variance also the isomer-specific information is maintained and this will be utilized in the subsequent LDA model as is illustrated below.

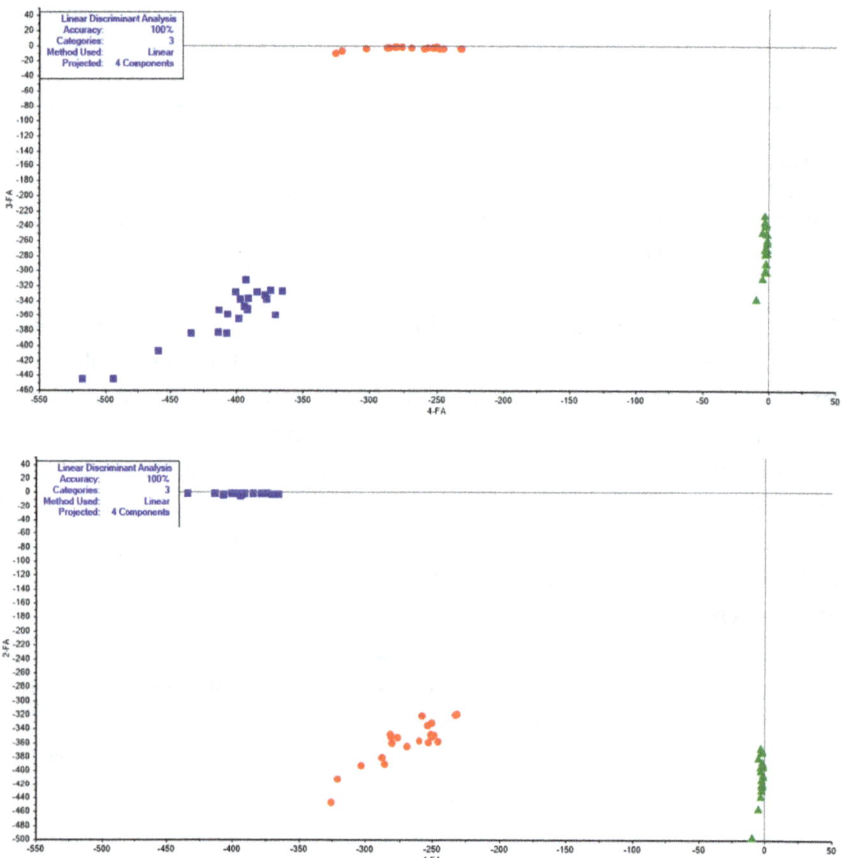

LDA 2D score plots of (top) 4-FA (green triangles) versus 3-FA (red dots) and (bottom) 4-FA (bottom) versus 2-FA (blue squares) of the FA isomer EI-MS model data. *(Created with Unscrambler version 10.5 from Camo Software, Aspentech)*

Using the first four principal components (explaining nearly 100% of all the variation in the dataset), a three class LDA model is constructed that perfectly assigns the correct FA isomer for validation exercises on the model data. The degree of class separation is more extensive in LDA compared to PCA which is to be expected because LDA uses the class labels of the model data to maximize this separation. However, the LDA plots as generated by Unscrambler need some further explanation. When applying the LDA model on the three case work samples the following result is provided by the Unscrambler interface:

Case work sample	Class assigned	2-FA	3-FA	4-FA
1	4-FA	−394	−245	−2,64
2	4-FA	−394	−256	−1,82
3	4-FA	−378	−246	−2,24

In line with the PCA projection results, the LDA model predicts that all three case samples contain the listed 4-FA isomer. This prediction is based on the data provided in the class columns. Although a detailed discussion of LDA theory is outside the scope of this book, it is important to know that LDA actually is based on Bayes rule! This makes the use of LDA well suited to estimate *LR* values or the evidential value associated with the isomer assignment. The discriminant values for the FA isomers given in the table actually correspond to natural logarithm of the posterior probabilities when assuming equal prior probabilities for each class considered prior to applying the LDA model.

Does it make sense to assume an equal prior probability for each FA isomer class?

Alternatively, prior probabilities could be estimated from recent case work, would that be a better alternative?

How could we establish the LR value for the 4-FA isomer assignment of the case work samples (Tip: the answer can be found in **Chapter 6**).

When considering the fluor amphetamine isomers, the 4-FA variant is most frequently encountered in Dutch forensic case work. However, because this compound is a listed hard drug since 2017, it might be that over the course of time this situation will change as drug producers shift to 3-FA and 2-FA as legal alternatives. Although it might seem plausible to use case work insights to estimate prior probabilities, it is actually advisable to consider equal prior probabilities to ensure that the classification is only based on the acquired mass spectral data and the use of FA model data. A simple thought experiment further supports this approach when we consider a situation where a forensic illicit drug analysis expert is faced with 3-FA or 2-FA in a case for the first time. As all prior cases showed the presence of 4-FA (assuming that the expert was able to differentiate the isomers), the estimated prior probabilities would be one for 4-FA and 0 for 3-FA and 2-FA. With a prior probability of zero, the posterior probability will remain zero irrespective of the outcome of the chemical analysis. In this way, the introduction of a new isomer will never be recognized! This is issue is resolved by assuming an equal prior probability for each isomer form. This is also in line with the purpose and function of the LDA model. We apply the model when we know from the analysis that the sample contains fluor amphetamine but so far have been unable to assess which of the three possible isomers we are dealing with. In such a situation, assuming an equal prior probability is a "fair starting point." It also makes "life easier" for estimating the evidential value of the isomer assignment because at equal prior probabilities the *LR* value equals the posterior odds if we remember Bayes formula (repeated from **Chapter 6**):

$$\frac{P(H_p|E)}{P(H_d|E)} = \frac{P(E|H_p)}{P(E|H_d)} \cdot \frac{P(H_p)}{P(H_d)}$$

Or in condensed form:

$$\alpha_{posterior} = LR \cdot \alpha_{prior}$$

which reads as "*the posterior odds equals the likelihood ratio times the prior odds*".

A complication which needs to be addressed is the fact that we have to consider three possible isomers, whereas the Bayesian framework is based on two (mutually exclusive) hypotheses. As was discussed in **Section 8.2**, the differentiation between 3-FA and 2-FA is less interesting from a criminal law perspective as the main question is whether the sample contains the listed hard drug 4-FA. However, two alternatives need to be considered (4-FA vs. 3-FA and 4-FA vs. 2-FA) and this leads to two different *LR* values from the LDA model assessment. This is also the reason that two LDA score plots are shown. Basically, the generic hypothesis pair given in 8.2 needs to be split up to be able to consider both FA isomer alternatives:

4-FA versus 3-FA

H_p: "*The sample confiscated from the suspect contains the listed hard drug 4-FA*".

H_d: "*The sample confiscated from the suspect contains the unlisted compound 3-FA*".

4-FA versus 2-FA

H_p: "*The sample confiscated from the suspect contains the listed hard drug 4-FA*".

H_d: "*The sample confiscated from the suspect contains the unlisted compound 2-FA*".

With this refinement, we are now able to convert the discrimination values (natural logarithm of the posterior probabilities) to *LR* values for the given isomer pair. Please note that an equal prior probability of 0.33 for all FA isomers translates to a prior probability of 0.5 when only considering two of the three isomers (*i.e.*, 0.33/0.67). In this way, the prior odds remain one which interestingly would not be the case when the more generalized defense hypothesis as given in **Section 8.2** is considered (in this case the prior odds would be equal to 0.5 with prior probabilities of 0.33 for 4-FA and 0.67 for the other two isomers combined). By using some basic mathematical rules concerning the logarithm function we can now calculate the evidential values for the identification of 4-FA in the case work samples:

$$\log(LR) = \log\left(\frac{P(H_p|E)}{P(H_d|E)}\right) = \log(P(H_p|E)) - \log(P(H_d|E))$$

Case work sample	Class assigned	Log(*LR*) 4-FA/2-FA	Log(*LR*) 4-FA/3-FA
1	4-FA	170	105
2	4-FA	170	110
3	4-FA	161	106

What would you conclude from these results?

It is clear that the identification of 4-FA using EI-MS data and the LDA reference model is extremely robust! The associated evidential value is enormous and seems to exceed *LR* values that are normally reported with DNA matches (log(*LR*) > 9). As discussed before, care has to be taken to report such excessively high *LR* values on the basis of a limited data set. The high *LR* values are the result of the fully separated isomer distributions which consequently leads to extremely low posterior probabilities for the defense hypothesis. Such low probabilities can only be estimated through severe extrapolation and are therefore characterized by a high degree of uncertainty. Although this requires proper *LR* calibration with significantly reduced adjusted *LR* values, the results nonetheless are very convincing. Fully separated isomer distributions clearly allow for a very confident assignment of the correct FA isomer on the basis of the measured EI-MS spectrum. Interestingly, the evidential value, although very high, is less for 4-FA versus 3-FA than for 4-FA versus 2-FA. This is in line with the match score findings described previously which showed completely separated score distributions for 4-FA (true-positive) and 2-FA (false-positive) but extensive overlap for 4-FA and 3-FA (false-positive). This indicates that the EI-MS spectra of 4-FA and 3-FA are more similar and therefore more difficult to differentiate. However, whereas a general similarity parameter such as a library match score based on cosine similarity provides limited discrimination, a PCA-LDA model specifically designed to extract isomer specific information is perfectly capable to assign the correct FA isomer. This shows the power of multivariate data analysis and how chemometrics can provide meaningful and relevant forensic information on the basis of small variations in analytical data (as long as these variations are robust and significant).

> Does this mean that we no longer need to invest in additional methods and instruments such as GC-IR and GC-VUV to resolve the NPS challenge? Imagine that you are a manager of a brand new forensic illicit drug analysis laboratory. Would you now only invest in GC-MS equipment and hire a data analysis expert to create the chemometric models to establish the correct drug isomers? What additional information would you need to take such an important decision?

The convincing results described above raise the question whether the complemental analysis involving GC-IR and GC-VUV as described in **Chapter 4** is still necessary for the robust identification of listed drugs and associated NPS isomers. However, before we return to GC-MS and the use of chemometric data analysis to tackle the NPS challenge, a substantial body of additional research will be required. First of all, the validity of this approach has been shown for just one class of NPS isomers. Possibly, the outcome will be less favorable for other compounds than fluor amphetamine. For successful implementation in high volume illicit drug case work, it is essential that the methodology can be broadly applied. Furthermore, the creation of a PCA-LDA model is time consuming as a substantial dataset needs to be established on the basis of reference standards or case samples for which the composition is known. In this respect, it is very important to understand the performance of the model over time to get a feel for its "expiration date." MS instruments need regular tuning to ensure mass accuracy and sensitivity and small differences in parameter settings might affect model performance when case samples are analyzed after a tune. At this stage, we also do not know whether a model created on one MS instrument can be successfully applied to case data generated on other instruments of the same or different model or from the same or different manufacturers. In the most unfavorable situation the PCA-LDA model is instrument specific and can only be used for a limited period of time. This would mean that whenever a fluor amphetamine isomer is encountered in case work a new reference data set would have to be acquired to create a valid PCA-LDA model. This is a capacity intensive and unattractive way of working and would possibly favor a more straight forward approach by acquiring a GC-IR or GC-VUV instrument. However, possibly a chemometric model can be created that is relatively insensitive to MS instrument settings and can be broadly applied for a long period of time. This would open the way to a range of software applications to establish the correct isomer species of numerous NPS in forensic drug analysis laboratories around the world using reference data sets that have been centrally generated. An additional interesting concept is that reference data are generated by various laboratories on a range of instruments to create a robust model that is created centrally and shared with all participants. The investment of each individual laboratory in terms of expert capacity and instrument analysis time remains limited while the model is based on an extensive dataset that encompasses all possible sources of variation. Such a joint international effort will also assist in the court acceptance of evidence generated with chemometric data analysis methods (aspects of validation, quality and admissibility requirements of forensic evidence will be discussed in detail in the next chapter). Future research by the international forensic community will have to point out whether such an approach is feasible. As a final critical note, it is important to understand that the use of the PCA-LDA model will always result in the selection of one of the three FA isomers even if the sample does not contain FA but a different NPS (*e.g.*, a fluor methamphetamine isomer). The model is limited to FA and establishes which isomer provides the best fit to the data even if all the fits are poor. Therefore, it is essential that the GC-MS analysis has unequivocally demonstrated the presence of fluor amphetamine in the sample. A clear indication might be provided by the model output in the form of relatively low posterior probabilities for all potential classes and hence the expert must always carefully inspect the discrimination values to make sure that the isomer form has been properly assigned. However, as discussed before, GC-MS is a very selective technique for compounds with different molecular mass or isobars and isomers with different functional groups. Fluor methamphetamine is for instance easily distinguished from fluor amphetamine on the basis of the mass difference of the most abundant fragment in the EI mass spectrum (m/z 58 vs. m/z 44).

8.5 The use of chemometric methods in forensic chemistry

By studying the previous sections, the reader has obtained a good general understanding of multivariate data analysis and more specifically two important chemometric methods, i.e., PCA and LDA. However, chemometrics is a quickly evolving field and new insights and methods are constantly emerging. Chemometrics is considered a separate scientific field at the interface of analytical chemistry and data analysis, and extensive study is required to confidently select and apply the correct methods for the question at hand. Readers that want to know more about chemometrics and want to understand the theory behind the methods are referred to academic course books and are advised to take a dedicated university course on multivariate data analysis. In this final part, an overview of the most frequently applied chemometric methods will be given. In addition to the basic principles, the rationale to apply a certain method will be explained. For each method, an example from recent literature will be shown where the method was successfully applied in a forensic setting. It is important to note that this section aims to provide a bird's-eye view and should not be considered as an exhaustive review of the use of chemometric techniques in forensic science. There are also no criteria associated with the forensic examples shown, other than that the author has read these articles and was inspired by the work presented. For every example shown, several other equally valuable forensic studies will be available. The aim is to provide the reader with an appreciation of the "chemometric tool box" and how the available methods can successfully be used in forensic science and can be implemented in case work.

Principal component (PCR) and partial least squares (PLS) regression

In **Chapter 5—Quantitative Analysis and the Legal Limit Dilemma**—we have discussed the relevance of quantitative analysis in forensic toxicology. The part that was not discussed in detail is the actual construction of the calibration curve using calibration standards and **linear regression** as it was assumed that the readers are acquainted with this process. Concentrations of the calibration standards are chosen to cover the range of interest (i.e., covering concentrations as typically encountered in case work samples) at equal intervals. In LC-triple quad MS, the peak height or (better) the peak area is measured for a selected fragment ion, typically in triplicate, for each of the calibration standards. For optimal accuracy and precision the standards should be "matrix-matched," i.e., in the case of the DIU analysis prepared in whole blood free of drugs of abuse, should undergo the same sample pretreatment and should be corrected for ionization suppression and precipitation losses by adding a deuterated internal standard (IS) at a given concentration and using the analyte/IS peak area ratio (A/A_{IS}). Assuming a linear response in the concentration range covered by the standards a calibration line is now constructed using linear regression. The slope and intercept are set to minimize the sum of squares of the difference between the measured and predicted peak area ratios of the calibration standards (The residual sum of squares or RSS). The resulting calibration line can now be used to convert any measured peak area ratio for a case work sample to a concentration of the illicit drug in whole blood. The coefficient of determination R^2 is often given to indicate the goodness of fit, i.e., the extent to which the calibration line (the model) is representative for the calibration data. The process is illustrated below:

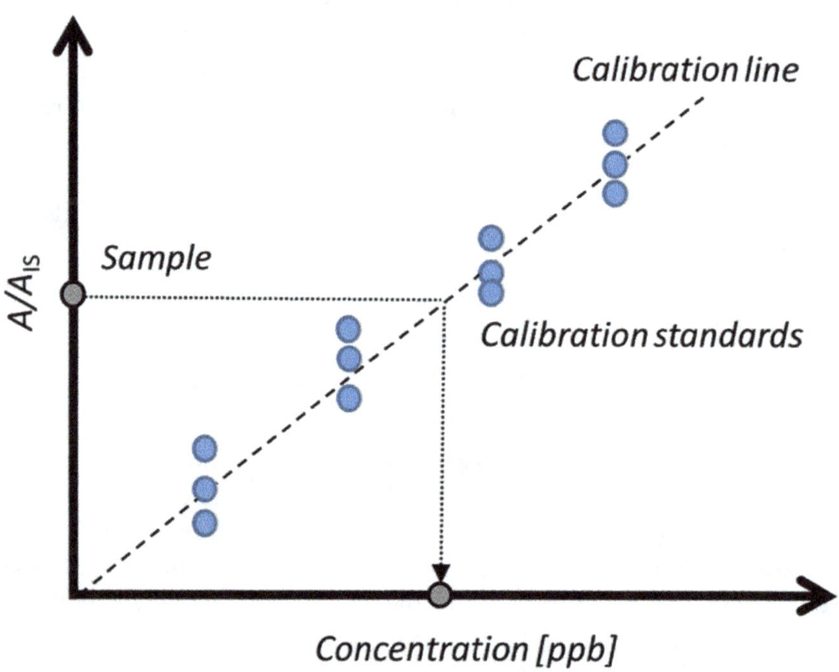

Construction of a calibration line through linear regression, the calibration line has a slope and intercept for which the lowest sum of squares between the predicted and measured peak area ratios for the calibration samples is obtained.

The calibration line is given by the intercept a and slope b for which the residual sum of squares (RSS) is minimal:

$$\left(\frac{A}{A_{\mathrm{IS}}}\right) = a + b \cdot C$$

The coefficient of determination (R^2) can be calculated from the RSS and the total sum of squares (TSS) which is related to the variance and is given by the sum of squares of the difference of the measured and average peak area ratio of the entire calibration data set:

$$R^2 = 1 - \frac{\mathrm{RSS}}{\mathrm{TSS}}$$

where

$$\mathrm{RSS} = \sum_{i=1}^{n} \left(\left(\frac{A}{A_{\mathrm{IS}}}\right)_i - (a + b \cdot C_i) \right)^2$$

$$\mathrm{TSS} = \sum_{i=1}^{n} \left(\left(\frac{A}{A_{\mathrm{IS}}}\right)_i - \overline{\left(\frac{A}{A_{\mathrm{IS}}}\right)} \right)^2$$

The "R squared" varies between 1 and 0 where a value of 1 indicates a perfect fit of the calibration line through the calibration data (RSS = 0).

The concentration of a case work sample can now easily be calculated

$$C_{\mathrm{sample}} = \frac{\left(\left(\frac{A}{A_{\mathrm{IS}}}\right)_{\mathrm{sample}} - a \right)}{b}$$

This scheme is used on a daily basis in forensic toxicological laboratories and is also named **simple linear regression** because we are considering the relation of two variables, analyte concentration versus analyte peak area (or peak area ratio when an internal standard is used). In the context of multivariate data analysis, we are considering a one-dimensional data set with only one data column containing the measured peak areas (or peak area ratios) as the concentration is known and used to (quantitatively) classify the samples. But how would we conduct a similar calibration scheme for quantitative analysis when we are dealing with a multivariate, correlated dataset. Such a situation occurs if we would measure a NIR spectrum for each of the calibration standards. A technique as NIR is not sensitive enough for trace analysis in a complex matrix as human whole blood and is typically not applied in forensic toxicology but could for instance be used to estimate the active ingredient level in illicit drug formulations. It could be quite valuable to use a rapid and noninvasive technique as NIR to quickly assess the level of cocaine in confiscated powders or the level of MDMA (XTC) in a large batch of tablets. But a typical NIR spectrum will not provide a clear signal that can be exclusively attributed to the analyte of interest. As the composition changes, the spectrum as a whole will show variations that could possibly be used to estimate active ingredient levels. The required multivariate models can be created with **Principal Component Regression** (**PCR**) or **Partial Least Squares** (**PLS**) regression. In a way, these chemometric methods can be regarded as the quantitative variants of PCA and LDA. PCR is an unsupervised technique that combines a dimensionality reduction by establishing the principal components and multiple linear regression (MLR). In MLR, a linear relationship is established between the concentration of the reference samples and a selected number of principal components that explain a majority of the data variance:

$$C = d_0 + d_1 \cdot \mathrm{PC}_1 + d_2 \cdot \mathrm{PC}_2 + d_3 \cdot \mathrm{PC}_3 + \ldots$$

In the same way as was illustrated for simple linear regression, the model can now be used to estimate the concentration of unknown samples by measuring the NIR spectrum, apply PCA dimensionality reduction and use the MLR model. PLS can regarded as a supervised variant of this approach as the dimensionality reduction is conducted to obtain principal components (which in PLS are called **Latent Variables**) for which the scores of the calibration measurements correlate with the concentration of the active ingredient. As a result typically less latent variables are required to develop the MLR model and therefore PLS is used more frequently than PCR also in forensic practice. However, some experts argue that using an unsupervised approach is scientifically more rigorous. For many applications, PLS and PCR will yield very similar results. A recent example of the use of a PLS model to estimate the cocaine level in street samples using handheld Raman spectroscopy is given below.

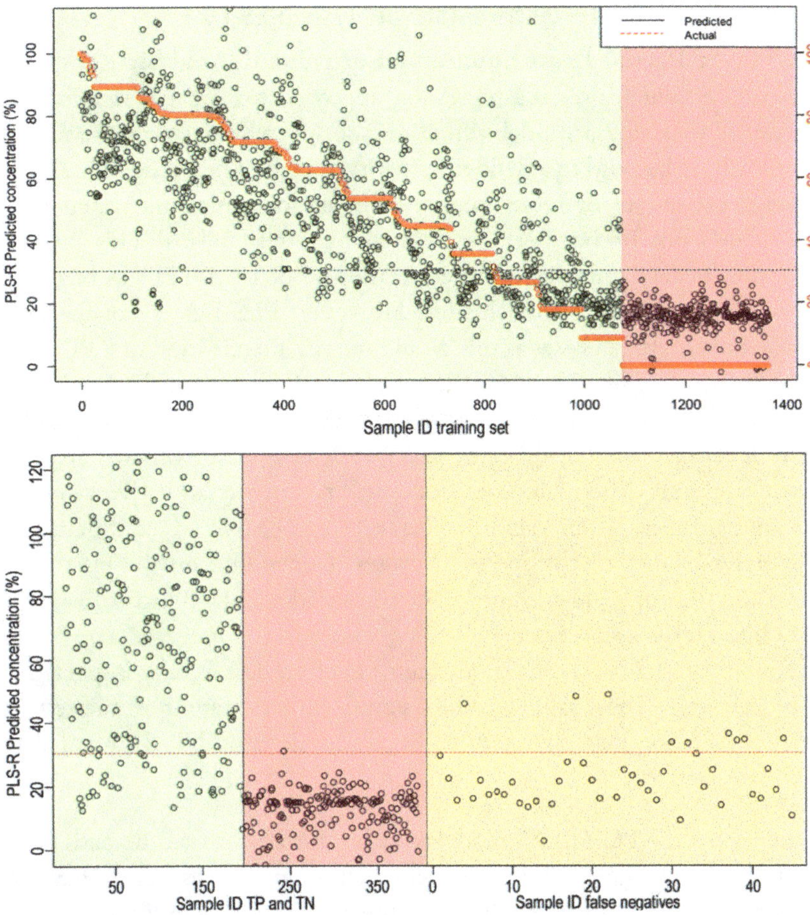

Cocaine concentrations predicted with PLS regression of handheld Raman spectra of (top) the training set (case samples and binary mixtures) and (bottom) a retrospective sample set consisting of case samples from 2015 to 2020. In red, concentrations as measured with GC-MS and in black the predicted concentrations with PLS-Raman are indicated. Green shading represents true-positive samples (cocaine containing), red shading indicates true-negative samples, and the yellow shaded area contains samples that gave false-negative outcomes due to the low cocaine level in the samples. The ground truth was established with GC-MS analysis. The dotted lines indicate the identification threshold set at 30 wt% cocaine for the PLS model. *(Reprinted from Kranenburg et al. Drug Testing and Analysis, 13 (2021) 1054–1067, https://doi.org/10.1002/dta.2993)*

Partial least squares—discriminant analysis (PLS-DA)

PLS-DA or **Partial Least Squares-Discriminant Analysis** is a special form of PLS which uses the regression approach for binary classification. The aim is to create a model which can decide whether a new and unknown sample belongs to a defined category A or B. The model is based on a reference data set of representative samples for which the category is known up front or has been established by other means. This type of application bears strong resemblance to the use of PCA-LDA as has been extensively discussed in this chapter. However, PLS-DA is not as easily extended to multiple classes as this would require the creation of a PLS-DA model for each pair. The MLR applied in PLS is "tuned" for binary classification by applying a fictive feature level of 1 for one class and a feature level of 0 for the other class. MLR is now based on two artificial calibration points. The regression model will return a quantitative value for unknown samples. If this value is close to 0 or to 1, we are confident that the samples belong to the specified classes. Depending on the class definitions and the application, appropriate thresholds can be set to convert the values into a class assignment. However, values in the range of 0.3–0.7 or values with substantial standard deviations need to be carefully inspected as this indicates that these samples might significantly deviate from the test set that was used to create the model. Such samples possibly do not fall in either category.

In addition to the PCA-LDA model to assign FA isomers on the basis of EI MS spectra, the figures and table below illustrate the use of a PLS-DA model to the same effect. In this case, the model has been limited to the 4-FA/3-FA pair as it was previously shown that the mass spectra of these two isomers are the most challenging to discriminate. Like PCA-LDA also the PLS-DA model confidently predicts that the three case work samples contain 4-FA. The predicted values in the ranges of 0.92–0.95 are very close to 1 which was assigned to the 4-FA reference samples.

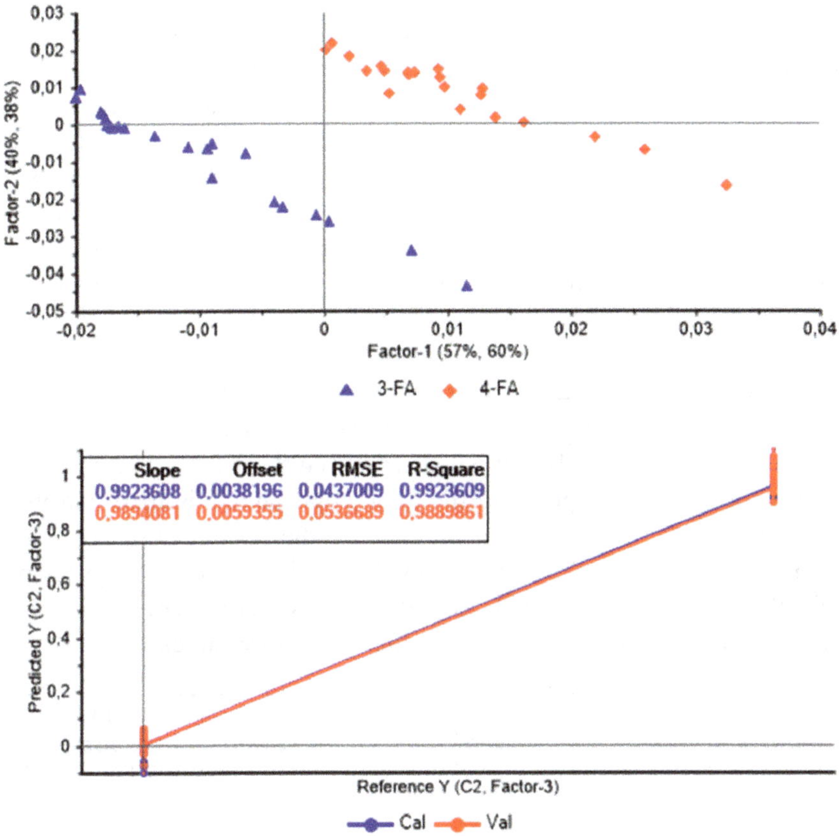

Case work sample	Predicted value	Standard deviation	Class assigned
1	0.92	0.05	4-FA
2	0.95	0.05	4-FA
3	0.93	0.05	4-FA

PLS-DA score plot (top) predicted versus reference plot (middle) of 4-FA (red diamonds) versus 3-FA (blue triangles) isomer EI-MS model data and table (bottom) with predicted values and isomer assignment for the three case work samples. In the PLS-DA model, the 4-FA reference standards were labeled with a 1 and the 3-FA reference standards were labeled with a 0. *(Created with Unscrambler version 10.5 from Camo Software, Aspentech)*

Soft independent modeling of class analogy (SIMCA)

SIMCA, an abbreviation for **Soft Independent Modeling of Class Analogy**, is a method that can also be used for classification on the basis of multivariate reference data. The term soft, as opposed to hard, refers to the fact that multiple (or no) classes can be assigned to a given sample. This characteristic of SIMCA is due to the fact that for each class a PCA model is established on the basis of the reference data for that class. An unknown sample is subsequently projected on each class model and on the basis of two distance metrics, the **object-to-model distance** and the **leverage** (distance to the model center), it is decided whether the sample is assigned to a given class. Thresholds can be set based on the variation observed in the reference data set. For this process to be meaningful, it is important that the different classes are sufficiently separated in a regular PCA model that includes all the reference data. It has already been shown that this condition applies for the EI MS spectra of the fluor amphetamine isomers. Creating three different PCA models on the basis of the 2-FA, 3-FA, and 4-FA reference data and applying the SIMCA classification method yields results that are fully in line with the other chemometric methods and also confirms the presence of 4-FA in all three case work samples:

Case work sample	2-FA class Assigned?	3-FA class Assigned?	4-FA class Assigned?
1	No	No	Yes
2	No	No	Yes
3	No	No	Yes

SIMCA can be regarded as a supervised technique even though it is based on PCA because the PCA models are class specific. The approach can become a bit cumbersome when a lot of difference classes have to be considered. Care also has to be taken that sufficient data is available for each class to create robust PCA models. A very interesting and powerful application of SIMCA is a situation where one class is of interest and well defined, whereas it is much more difficult to introduce additional classes because of the background of case samples, which is variable and unknown. For instance, in a setting where the main interest is to check whether a sample is authentic while a wide variety of counterfeits exists. In such a situation, SIMCA can be applied to investigate how unknown samples project on the "authentic" class. A decision is then made whether the unknown sample belongs to the class of interest, *e.g.*, whether the sample is authentic. If the decision thresholds are not met the sample is deemed to be a counterfeit/forgery without the need to link this to a specific type. Such an approach is not feasible in the previously discussed models which are based on the unsupervised or supervised differentiation of various pre-defined classes.

Support vector machine (SVM)

Support Vector Machine or in short **SVM** represents a powerful suite of multivariate data analysis methods that can be used for multivariate regression and classification. SVM allows for non-linear separation of classes (as opposed to LDA) by mapping the data in a new feature space. The so-called support vectors are the data points in the new feature space that mark the best separation between the classes (source: Unscrambler). A detailed discussion of SVM principles and theory is beyond the scope of this book but SVM is considered an interesting alternative when initial data analysis reveals the overlap of PCA class clusters. Additionally, the use of support vectors as critical reference samples for class differentiation also typically leads to less overfitting issues compared to other non-linear methods. An example of the use of SVM for the correct classification of semen, blood, or mixed crime scene stains on the basis of non-invasive Raman spectroscopy is given below.

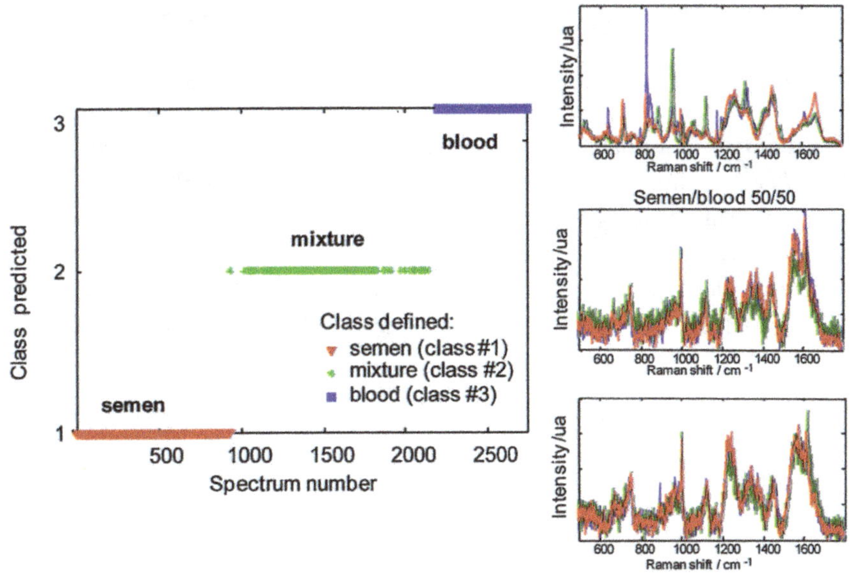

Combined use of SVM regression (for data selection) and SVM discriminant analysis (3 classes, pure blood, pure semen, or blood semen mixture) spectroscopic Raman data (spectra for each class displayed on the right) for the correct identification of blood, semen, and mixed crime scene stains. *(Reprinted with permission from Sikirzhytski et al. Forensic Science International, 222 (2012) 259–265, https://doi.org/10.1016/j.forsciint.2012.07.002)*

Hierarchical cluster analysis (HCA)

HCA stands for **Hierarchical Cluster Analysis**, a popular chemometric technique that can be used to inspect and visualize hidden structures in multivariate data sets. PCA is used in a similar fashion and also HCA is an unsupervised method because it does not use any sample labels as part of the cluster analysis. The HCA results are visualized in a so-called **dendrogram** (in Greek: tree-drawing). At the root of the HCA tree, we find the smallest clusters each consisting of the two samples in the dataset that are most similar. For multivariate data, this corresponds to two points with the shortest distance in the n-dimensional space. As discussed in detail in **Chapter 6**, this can be assessed by calculating the Euclidian distance of all possible pairs in the dataset. Other distance metrics can be employed such as Manhattan (city block), Canberra, Mahalanobis, Pearson's correlation, and cosine similarity. In the dendrogram, all closest pairs are sorted along the x-axis, whereas the relative distance (similarity) is indicated on the y-axis. In subsequent steps closest points and clusters are connected until in a final step the two remaining clusters are linked. For the connection of clusters, different approached can be followed:

- Single linkage—the shortest distance between points in two clusters
- Complete linkage—the largest distance between points in two clusters
- Average linkage—the average distance for all possible cluster point combinations
- Ward's linkage—the increase in the TSSs due to the joining of clusters

On the basis of available information, a relative distance can be set to yield the expected number of classes. All individual samples within each of the classes visually group together in the dendrogram while also substructures within a class are illustrated. An example of the use of HCA is given below where the dendrogram illustrates the substructure of 18 blue spray paints on the basis of transmission FTIR spectra measured using an IR microscope.

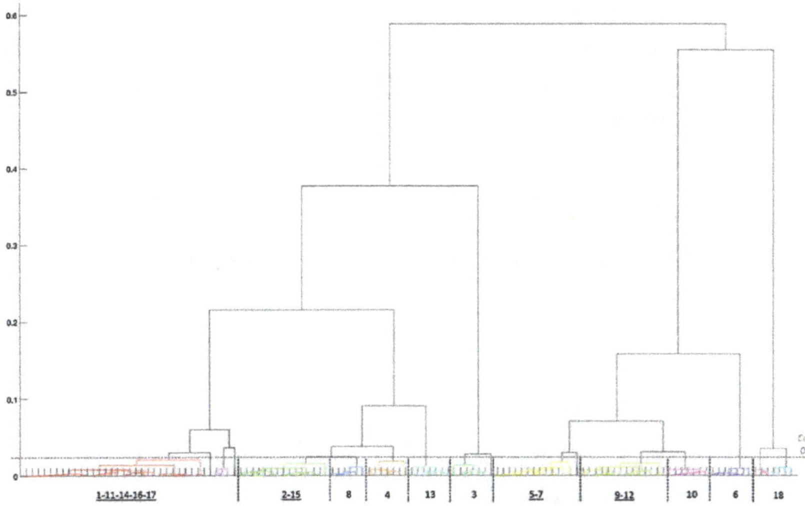

HCA dendrogram of 18 blue spray paint cans on the basis of transmission FTIR spectra. Pearson's correlation was used as the distance metric in combination with complete linkage. The cut-off value (0.021) for the number of clusters is based on optimal within and between source assessment as established from ROC curves. *(Reprinted with permission from Muehlethaler et al. Forensic Science International, 244 (2014) 170–178, https://doi.org/10.1016/j.forsciint.2014.08.038)*

k nearest neighbors (k-NN)

The final chemometric method discussed is the **k-NN** or **k-nearest neighbors algorithm**. This surprising simple and elegant approach is **non-parametric** (not considering probability distributions and associated parameters such as mean and variance and no modeling), can be used for regression and classification purposes, and is considered to be a **machine learning algorithm**. Given a representative, fully characterized, and sufficiently comprehensive reference data set, the class of an unknown sample is assigned as the class of the k reference samples that are most similar. Similarity in an n-dimensional data space requires a suitable distance metric such as Euclidean distance or any of the other options that were discussed. If the neighboring reference samples are from different classes, the class is decided by majority voting. For this reason usually odd values for k are chosen to ensure a "clean" decision. The number of neighbors considered is a very important parameter with respect to the performance of the algorithm. For "close decision" samples, the actual appointed class can change when the k value is adjusted as is illustrated below. Increasing the k value usually reduces classification noise but can also result in more diffuse class boundaries. For unbiased majority voting, it is important that the data set is balanced with respect to the number of entries per class. If this is not possible, votes from nearest neighbors can be weighed according to their distance to the unknown sample. For highly correlated multidimensional data, dimensionality reduction via PCA can improve the performance of the k-NN algorithm. When unknown samples can consist of components or compositions not accounted for in the reference data a distance metric threshold can be implemented to prevent misclassification. As a consequence strongly deviating samples are not classified because an insufficient number of nearest neighbors are located inside the distance threshold.

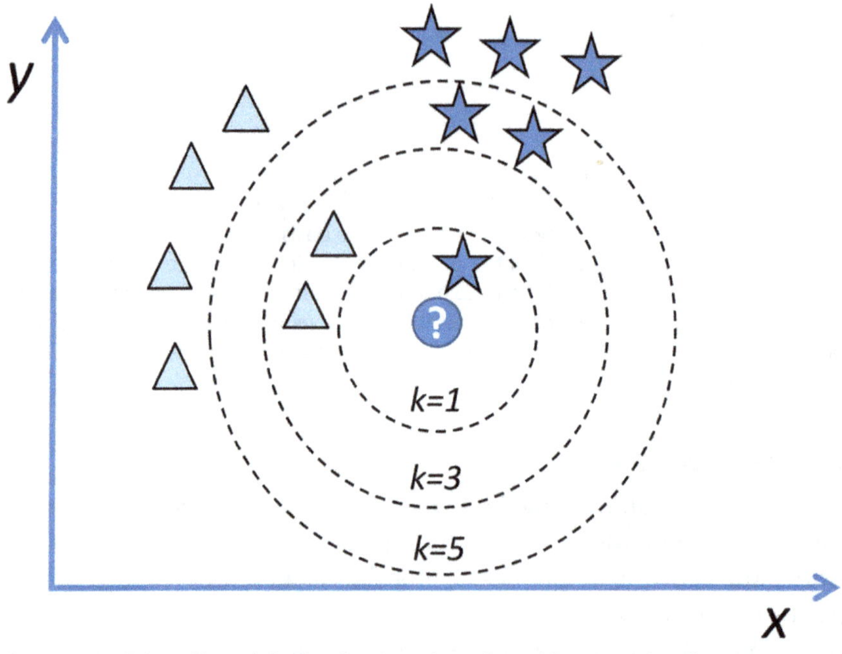

Illustration of the effect of the *k* value (number of neighbors) on the class assignment by majority voting of an unknown sample projected in a 2D dataset with features x and y. For $k = 1$ the star class applies, for $k = 3$ the class switches to triangle and for $k = 5$ the unknown sample is considered to belong to the star class again

In a recent study, the k-NN algorithm was successfully applied for the detection of cocaine with a low-cost, handheld NIR scanner with a limited wavelength range. As is illustrated below, the algorithm was used for classification (cocaine HCl, cocaine base or cocaine free) and subsequent estimation of the cocaine level for positive samples. This example illustrates the potential of k-NN, a very powerful yet elegant algorithm, provided that sufficient relevant reference data is available. This makes this methodology especially suitable in combination with rapid screening methods such as portable NIR but less favorable when acquiring data is time-consuming or when a substantial set of reference samples is difficult to obtain.

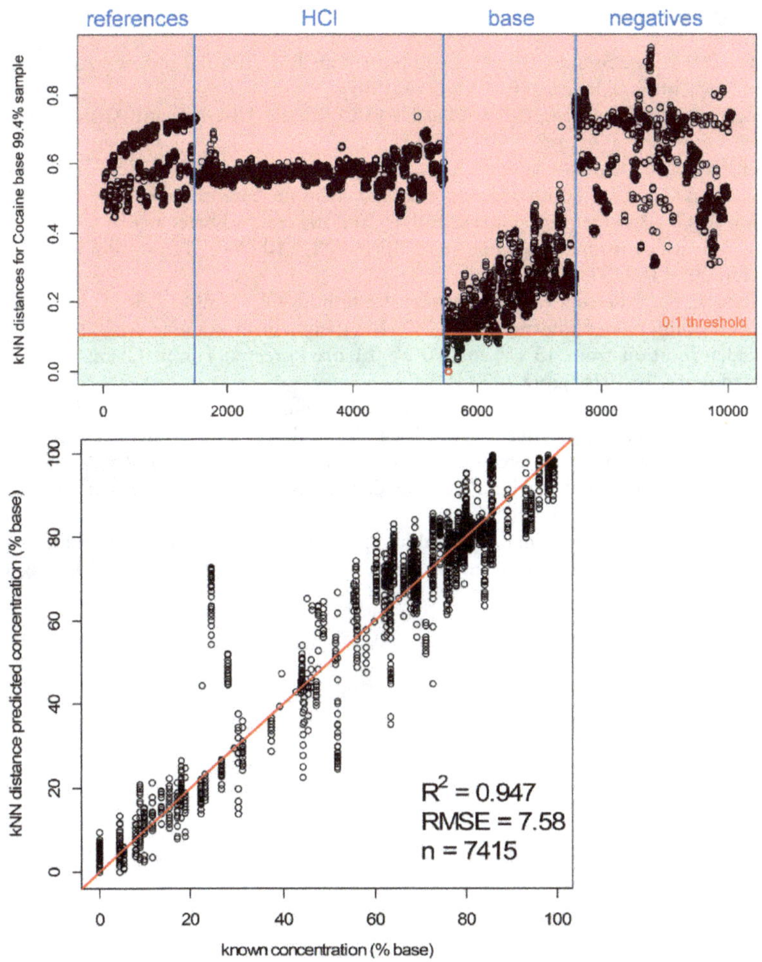

Top: Euclidean distance of NIR spectra of over 10.000 reference samples to the spectrum of a relatively pure cocaine base casework sample (red circle). The reference cocaine base samples are sorted from high to low concentration. The red line indicates the Euclidean distance threshold applied in the selection of nearest neighbors. Bottom: Predicted concentration as function of the known level of all cocaine base samples using the k-NN algorithm. *(Reprinted from Kranenburg et al. Drug Testing and Analysis, 12 (2020) 1404–1418, https://doi.org/10.1002/dta.2895)*

Further reading

Bro, R., 2014. Age Smilde. Principal component analysis (tutorial review). Anal. Meth. 6, 2812–2831. https://doi.org/10.1039/c3ay41907j.

Ghojogh, B., Crowley, M., 2019. Linear and Quadratic Discriminant Analysis: Tutorial. Cornell University arXiv.org, 1906.02590vol. 1.

Kranenburg, R.F., Garcia-Cicourel, A.R., Kukurin, C., Janssen, H.-G., Schoenmakers, P.J., van Asten, A.C., 2019. Distinguishing drug isomers in the forensic laboratory: GC-VUV in addition to GC-MS for orthogonal selectivity and the use of library match scores as a new source of information. Forensic Sci. Int. 302, 109900. https://doi.org/10.1016/j.forsciint.2019.109900.

Kranenburg, R.F., Peroni, D., Affourtit, S., Westerhuis, J.A., Smilde, A.K., C van Asten, A., 2020. Revealing hidden information in GC–MS spectra from isomeric drugs: chemometrics based identification from 15 eV and 70 eV EI mass spectra. Foren. Chem. 18, 100225. https://doi.org/10.1016/j.forc.2020.100225.

Kranenburg, R.F., Verduin, J., Weesepoel, Y., Alewijn, M., Heerschop, M., Koomen, G., Keizers, P., Bakker, F., Wallace, F., van Esch, A., Hulsbergen, A., van Asten, A.C., 2020. Rapid and robust on-scene detection of cocaine in street samples using a handheld near-infrared spectrometer and machine learning algorithms. Drug Test. Anal. 12, 1404–1418. https://doi.org/10.1002/dta.2895.

Optimization and testing of mass spectral library search algorithms for compound identification. Am. Soci. Mass Spectrom. 5, 1994, 859–866. https://doi.org/10.1016/1044-0305(94)87009-8.

CHAPTER 9

Quality and chain of custody

Contents

9.1 What will you learn?

The previous chapters have provided a comprehensive spectrum of investigative options. In doing so the most important forensic expertise areas that employ analytical chemistry have been introduced and discussed. This "matrix" of knowledge should enable you to assess the potential options for the chemical analysis of physical evidence in a given case and what instrumentation should be employed. The remaining chapters in this book will focus on the quality, reporting, and innovation aspects of forensic casework methods. In analytical chemistry, quality plays an important role as was already indicated in **Chapter 1—An introduction to Forensic**

Chemical Analysis for Forensic Evidence
ISBN 978-0-12-820715-4
https://doi.org/10.1016/B978-0-12-820715-4.00013-4

Analytical Chemistry. Based on the reported outcomes important decisions are taken and the more critical these decisions are, the more organizations tend to invest in quality systems to ensure the reliability of the analytical data. Reliability relates to trustworthiness; decision-makers should not have reasons to doubt the forensic chemical findings and the interpretations of the forensic experts. This means that the outcome should not be prone to errors. However, there is only one way to totally exclude errors and that is to not perform any forensic investigation at all! Where humans are at work errors will always be made, it is a fact of life. No matter how experienced a forensic expert is and no matter how many prior cases he/she has reported, making an error in an ongoing investigation is always a possibility. In a way, being an expert also comes with a certain vulnerability, using your experience makes you efficient and effective but can also create **bias** and **tunnel vision** and can make you insensitive to spotting weak points in a procedure or methodology. After handling so many similar cases, an expert might perform an investigation in an "auto-pilot mode" making him/her disregard certain warning signs. So, quality is not about excluding mistakes (which is impossible), it is about detecting, minimizing, and mitigating them. Quality is also about a vigilant and proactive attitude in an organization to learn from mistakes to constantly improve its performance. The ultimate decision in criminal investigations concerns the guilty/not-guilty verdict in the corresponding court case as decided by a judge (in the **inquisitorial judicial system** as applied in most Western European countries) or a jury (in the **adversarial judicial system** as maintained in the USA and the United Kingdom). Such a decision typically has an enormous impact on the lives of the suspects, victims, and their family members. Court rulings in high-profile cases can even severely affect an entire society. With the stakes that high, it is important that the forensic expert reports and testimonies can be trusted and do not lead to unnecessary controversy. Forensic investigations aim to strengthen the information position of the "**triers of fact**" to support the quality of their decisions, thus preventing wrongful convictions and acquittals as much as possible. These false positive and false negative outcomes of court rulings can severely reduce the trust in the judicial system and lead to a collective sense of injustice and insecurity. The fact that forensic investigations can play a vital role in the outcome of a criminal procedure warrants the substantial investment of forensic institutes and laboratories in their quality systems. A conviction "**beyond reasonable doubt**" also requires forensic findings "beyond reasonable doubt." In a legal framework, this also requires a fully documented trail of all forensic activities from the evidence collected at the crime scene until the expert testimony in court. The **chain of custody** regarding physical evidence and forensic traces retrieved from these items must be secured at all times. This allows experts to demonstrate the integrity of the presented evidence and to indicate, which organization had possession of and was responsible for the physical evidence at any stage of the investigation. Such information also allows accurate reconstruction when there is a suspicion of potential **contamination** during the laboratory work leading to a false positive outcome.

In this chapter, we will discuss all aspects of quality assurance in forensic investigations. This typically applies to all forensic expertise areas but of course, we will focus on the implementation of a suitable quality framework in a forensic chemical analysis setting. We will start with quality from a legal perspective and the measures taken to ensure that forensic expert testimony and the reported findings are **admissible** in a court of law. The requirements for forensic expert evidence as laid down in a code of criminal proceedings in a country provides a clear guideline of the quality measures that need to be implemented by the forensic institute. The need for accreditation will be discussed, the validation process for accredited methods will be illustrated and critical aspects with respect to the chain of custody, contamination prevention, and quality control during case work will be explained.

After studying this chapter, readers are able to

- understand the judicial definition of lawful evidence and describe the role and obligations of the forensic expert in criminal investigations as described in the Code of Criminal Proceedings
- explain the function of a register of court experts within a national criminal justice system
- explain the importance of a Code of Conduct for forensic experts and list the key elements of such a code
- list the Daubert criteria for admissibility of forensic evidence and assess whether (new) case work methods meet these criteria
- describe the ISO 17025 accreditation framework, explain the differences with the ISO 9001 quality standard and translate the ISO 17025 framework to forensic settings as described in the ILAC G19 guideline
- describe the types of contamination that can occur in a forensic case work laboratory and formulate appropriate measures to minimize contamination risks
- explain the Chain of Custody concept and indicate its importance for forensic investigations
- understand the need for method validation as part of the accreditation process
- describe the three levels of quality control in forensic case work and explain the differences between a proficiency test, a ring test, and a blind test
- construct a Shewart control chart for quantitative analysis in forensic investigations
- explain the importance of peer review in forensic investigations

9.2 Ensuring quality in forensic expertise

To understand the quality requirements of forensic expert reports and court testimonies we need to consider the role of forensic science in the criminal justice system. This is related to what was discussed in **Chapter 1—An introduction to Forensic Analytical Chemistry**, the activities of forensic experts are dictated by the criminal justice system in which they operate. This does not only apply to the questions that they answer, it also affects the quality measures they take while examining physical evidence. Before exploring this in more detail in this section, we first need to briefly address the way criminal justice has developed and is applied in civilized countries. The origin of contemporary criminal justice systems in most Western European and North American countries dates back to the 18th–19th century and is linked to the ideas of the French revolution (1789–99) and the declaration of independence of the United States of America (1776). Fundamental human rights such as religious freedom, freedom of speech, equality of man, the separation of church and state, and the balance of powers that form the basis of 21st century liberal democratic societies are strongly connected to these pivotal moments in history. These developments were in part inspired by the work of enlightened French writers and political philosophers Voltaire (1694–1778) and Montesquie (1689–1755). The work of Voltaire focuses on freedom of speech and religion and the "separation of church and state." Based on the study of the old Romans, French judge and political philosopher Montesquie publishes his ideas on the need for the separation of powers to achieve political liberty in his work *The Spirit of the Laws* (1748). The **Trias Politica** ensures a societal system of checks and balances and prevents the abuse of power by establishing three independent functions: the **legislative**, the **executive,** and the **judicial** branch. The legislative branch comprises the politicians that represent the constituents and can define, propose and pass laws. The government of a country, or the executive branch, is tasked with implementing and enforcing legislation. Finally, the judicial branch checks that the laws are correctly interpreted and assures a correct execution of legislation. The judiciary also investigates and rules in case of legal disputes and conflicts, either between parties (civil law) or between the state (representing society and victims) and a party (suspects) in criminal law.

Although forensic science also is frequently of value in civil cases, most forensic experts in governmental service are exclusively active in criminal law cases. For this reason, we will focus in this section on penal law and criminal investigations. Such an investigation typically starts when victims or witnesses report to the police. When law enforcement officers find clear indications of a crime, they file a police report and start a criminal investigation. Such an investigation is led by a **public prosecutor** (in the USA also called the **district attorney** often abbreviated to DA) who is appointed to the case. The public prosecution office investigates crimes, gathers evidence with the aid of police and forensic experts, instructs law enforcement to arrest and detain suspects, and ultimately prosecutes these suspects in court. It does so to ensure a safe and just society and to seek retribution on behalf of the victims. Instead of seeking vengeance and causing a vicious cycle of violence, the public prosecutor pleas on behalf of the victims of a crime. When the suspects are found guilty beyond a reasonable doubt in a court of law, the punishment in modern societies will consist of fines, incarceration (being locked up in prison), or a combination thereof depending on the severity of the events. For brutal crimes involving the loss of life, often of multiple victims, the perpetrator can also be sentenced to death in some countries. However, most societies do not support this "*an eye for an eye*" principle and enforce life imprisonment as the maximum sentence. Imprisonment of perpetrators thus not only serves retribution but also contributes to safety as it removes potential recidivists from society for the duration of the imprisonment. Safety is also served as the risk of imprisonment can potentially act as a deterrent, preventing an individual from actually committing a crime. In modern societies, the view also exists that convicted criminals can better their life and deserve a second chance. So, an additional function of imprisonment is to allow perpetrators to fulfill their debt to society and provide guidance to a better life.

While being a victim of a (violent) crime is extremely shocking and traumatic, becoming a suspect in a criminal investigation and being arrested by the police can be equally impactful (although for notorious repeat offenders this might be considered a risk of the profession). In popular lectures, I sometimes ask the audience what they find worse, a criminal who is incorrectly set free after actually committing a crime or the wrongful conviction of an innocent suspect. In my experience, a substantial majority will point to the latter situation as the biggest mistake a court can make. My explanation for this consistent outcome is that most people in my audience have never committed a crime, nor have ever been convicted and from that perspective, the idea of suddenly being arrested, put in detention, and ultimately being convicted is simply unbearable. The rights of the defendant in a court case must be carefully considered and protected, also because in a criminal investigation the public prosecutor represents society and has substantial support and means to conduct the investigations. Some very important measures and principles in criminal justice aim to support suspects. First of all, the public prosecutor, as a civil servant, should focus on truth-finding and not on relentlessly prosecuting suspects (although this is sometimes questioned for district attorneys who present themselves as hardened crime fighters). This means that when there is evidence that convincingly and clearly shows that a suspect is not involved in a crime, the public prosecutor must act accordingly and immediately warrant the release of the suspect and must publicly announce his/her innocence. The importance of forensic science to provide such undisputable exculpatory evidence is often not sufficiently emphasized but cannot be underestimated. In this respect, non-matching DNA profiles can be just as important as a "case solved by a DNA match."

Additionally, each suspect is entitled to professional legal counsel even when he/she cannot afford a lawyer. The defense lawyer possesses the required legal expertise to effectively defend the interests of the suspect and to question the facts and evidence as presented by the public prosecutor. In this capacity, the defense counsel can also criticize the forensic reports, present additional forensic expert reports, and testimonies, and request additional forensic investigations. This fundamental right to legal support is explained when someone is read his/her Miranda rights by a police officer when being arrested in the USA:

> *You have the right to remain silent. Anything you say can be used against you in court. You have the right to talk to a lawyer for advice before we ask you any questions. You have the right to have a lawyer with you during questioning. If you cannot afford a lawyer, one will be appointed for you before any questioning if you wish. If you decide to answer questions now without a lawyer present, you have the right to stop answering at any time.*
>
> Wikipedia

This phrasing can probably be recited by many because it is so often proclaimed in movies and TV series. The term Miranda refers to a USA 1966 Supreme court ruling in the case of Miranda versus the state of Arizona. The signed confession of the kidnapping and rape of a young woman was deemed inadmissible by the Supreme Court because the defendant was not explicitly informed of his rights beforehand. In addition to the right to have a lawyer, the Miranda statement also indicates that a suspect has the right to refuse to answer any questions or make any statements. This is linked to the fundamental right that a defendant cannot be forced to incriminate him or herself and that any statements made can be used as evidence in a court of law. So, after being explicitly and properly instructed of these rights, a confession after an arrest will constitute very incriminating evidence in the subsequent court case and will almost certainly lead to a conviction (unless other facts and evidence clearly indicate that the accused made a false confession).

As the district attorney prosecutes on behalf of society and the victims of a crime and the defense lawyer pleas for the innocence of the suspect, an independent, unbiased and uninvolved third body needs to arrive at a final conclusion regarding the guilt or innocence of the defendant. A guilty verdict with an associated sentence requires the absence of "**reasonable doubt**" on the basis of lawfully admissible evidence and facts. The so-called **triers of fact** consist of a judge or a team of judges in the **inquisitorial system** and of a jury of civilians in the **adversarial system**. In the latter system, which is found mostly in Anglo-Saxon countries, the judge has a referee function overseeing and safeguarding the legal procedures, ruling on the admissibility of evidence and determining the appropriate sentence while leaving the ultimate decision on the innocence or guilt of the suspect(s) to the jury. In the inquisitorial system as found in most main-land Western European countries, the judge has an active investigative role and needs to arrive at a personal conviction that it has been lawfully and convincingly demonstrated during the court proceedings that the defendant committed the crime. If these conditions are met, the judge (or team of judges) will find the defendant guilty and will set an appropriate sentence. In both systems, there is a fundamental right to an appeal before a higher court with the highest judicial body being the Supreme Court.

Lady Justice, a generic symbol of justice based on Justitia, is a Roman goddess (in Greek mythology named Dice or Dike) representing fair judgment and high moral standards. Her attributes indicate impartiality (blindfold or closed eyes), balanced assessment of the points brought forward by the disputing parties (scale), and the authority to convict and sentence (sword)

To establish that someone committed a crime, the associated activities must have been defined as such in the criminal law of the given jurisdiction. As was already discussed as part of the NPS challenge in **Chapter 4—Qualitative Analysis and the Selectivity Dilemma**, a very important legal principle is that one cannot commit a crime that has not been described and ratified as such in the law prior to the alleged activity. So, producing, transporting, and selling a substance for its psychoactive properties clearly has a criminal motive but nonetheless is not an illicit drug crime if that given substance is not listed yet. To that end, the public prosecutor in his/her **indictment** will clearly specify, which articles in the penal law are associated with the charges brought forward against the defendant. Some of those articles for the Dutch **Code of Criminal Law** (in Dutch Wetboek van Strafrecht) were already discussed in previous chapters. Basically, these articles set the societal "rules of engagement," the set of rules all citizens must accept and abide by to ensure that every individual can live in freedom without being harmed. A very general principle is for instance that one cannot just take someone's property without explicit consent. Theft is, therefore, punishable by law. In the Netherlands, this is described in article 310 of the Code of Criminal Law (free translation from Dutch by the author):

He who takes any goods that fully or partly belong to another with the intent to keep those goods will be found guilty of theft and will be punished by a maximum of four years imprisonment or a fine of the fourth category

Whereas the Code of Criminal Law describes the rules of societal engagement and defines, which human activities are considered a crime, the **Code of Criminal Proceedings** (in Dutch: "Wetboek van Strafvordering") sets the rules for investigating and prosecuting such crimes. In the Dutch criminal justice system, this Code of Criminal Proceedings defines the tasks, responsibilities, and rights of all of the actors involved in a criminal investigation, including police officers, public prosecutors, defense lawyers, judges, suspects, victims, witnesses, and (forensic) experts. In the following paragraphs, we will discuss a number of important articles in the Dutch Code of Criminal Proceedings that illustrate the role and the contribution of the forensic expert and what instruments are available to investigate and ensure that this contribution meets the necessary quality requirements.

Article 339 for instance describes what types of evidence can be used by the judge to arrive at a decision:

- Own observations (of the judge)
- Statements by the defendant
- Statements by a witness
- Statements by an expert
- Documentation (incl expert reports)

This means that in the Dutch criminal justice system, as in many other countries, the (forensic) expert can communicate his/her findings through testimony in person or through a forensic report. In the Inquisitorial System, it is much more common to accept the forensic expert report as an objective and scientifically sound piece of information and for that reason, the author will only need to come to court when that information is unclear or when additional questions exist. The list given above triggers an interesting (also from a quality perspective) and almost philosophical question: what makes someone an expert, or in other words, what differentiates a witness from an expert statement? This point is addressed in Articles 51i-51m of the Dutch Code of Criminal Proceedings:

- A forensic expert is appointed by the court to conduct investigations and provide information in an area in which the expert holds special knowledge and expertise
- The expert is obliged to report truthfully, fully, and "to the best of his/her knowledge"
- An appointed expert is obliged to provide the services requested by the court
- Rules can apply with respect to the qualifications of the experts and means to check these qualifications
- There is a national register of forensic experts that is maintained by a special body
- If an expert is appointed outside this register this needs to be motivated by the court

Some interesting observations can be made from these bullet points. First of all, being an expert in a criminal investigation means that you hold specific knowledge (as opposed to what can be considered common knowledge) that is of interest to the case at hand. Furthermore, being a court expert is not a voluntary thing, if you are appointed by the courts to serve as an expert you are required to provide the requested services and report the results to the best of your knowledge and ability. Not doing so without providing a proper reason (such as illness) is in itself punishable by law! Finally, the latter three points have a clear link to quality and court instruments to check whether an expert is "up for the task." The Netherlands Register for Court Experts (NRGD, https://english.nrgd.nl/) is an important body to ensure expert quality and such an instrument can be found in several other countries. In the Dutch criminal justice system registered court experts can directly be appointed in a given case that requires his/her expertise. The registration ensures that the experts meet the quality requirements and that no additional investigation into his/her qualifications is necessary. Allowing only registered experts to be appointed is too limiting because legal professionals sometimes require special and rare scientific expertise. In principle any scientist can be appointed by the courts, however, without registration it requires the courts to motivate the appointment and demonstrate that the appointee holds the required knowledge. In the Netherlands everyone who becomes a court-registered expert in one of the registered areas of expertise (these are the areas for which a high and frequent demand exists in criminal investigations) can send in an application file that initiates the NRGD registration procedure. A special advisory committee including established experts in the field will assess the application. If the application dossier meets the requirements, the candidate will be invited for an oral examination. The committee will then advise the Board of Court Experts whether the applicant has the required knowledge, expertise, and experience to be added to the register, either conditionally or unconditionally. This advice is based on the assessment of whether the candidate:

- has sufficient knowledge and experience in the field of expertise
- has sufficient knowledge and experience concerning the legal aspects related to the field of expertise and the role of a court expert in criminal investigations in general
- is able to clearly communicate with the legal parties involved regarding the appropriate formulation of the questions to be addressed through forensic investigation in the field of expertise
- is able to plan and execute a forensic investigation in the field of expertise to answer the questions in accordance with appropriate standards
- is able to apply state-of-the-art investigative methods in the field of expertise
- is able to collect, document, and interpret the findings of the investigation in the field of expertise in accordance with the appropriate standards
- is able to explain the findings both orally and in a written report in a manner that is accurate, clear, consistent, and comprehensible for legal experts
- is able to complete an assignment in a generally accepted time frame that fits the legal proceedings
- is able to conduct the investigation in an impartial, independent, competent and trustworthy manner

In addition to the registration process, the NRGD also maintains the **Code of Conduct** for court-registered experts in the Netherlands. All court experts that are in the register are expected to know this code of conduct and to act accordingly when conducting a forensic investigation. Such a code is defined by many national and international forensic institutions illustrating the importance of maintaining very high ethical standards. Given the fact that their forensic findings can have a direct impact on the verdict and with that the lives of victims and suspect, the work of the forensic experts should be morally impeccable.

> How would you define the work ethics of a forensic expert?
>
> Individually or in a student group formulate your own Code of Conduct!
>
> Please note there is no right or wrong in this respect!

Although Codes of Conduct can differ, typically the following elements are included:

- The investigation needs to be conducted in an independent, impartial and unbiased manner, a potential conflict of interest or involvement in the case must be reported immediately
- The expert needs to carefully consider and stay within his/her boundaries of expertise, if the questions cannot be answered within this body of expertise this must be reported immediately
- The expert needs to assure that the investigation is conducted according to the state-of-the-art in the field of expertise and meets all applicable quality standards
- The expert needs to report any internal and external attempts to hamper the investigation or alter the outcome of the findings and conclusions
- The expert must conduct the investigation and document the findings such that they can be subjected to peer review
- The work needs to be conducted in a manner that conserves the physical evidence and associated traces as much as possible to enable counter/second opinion investigations

- The expert report must be provided in an acceptable time frame and must be complete, logically organized, and understandable (for lay readers)
- The experts need to report any perceived miscarriage of justice due to the incorrect interpretation of the results of his/her forensic investigation

With these instruments in place, judges can trust that the forensic experts that they appoint have the required knowledge and experience to answer the questions by performing a forensic investigation in their field of expertise. Furthermore, registered experts are knowledgeable about legal proceedings and their role, obligations, and limitations and can confidently report their findings by providing a forensic report or court testimony. They are also very aware of the very high ethical standards that "come with the job" and will immediately flag any issues that affect their neutrality and objectivity. When relatively rare expertise is needed in a case, the court should ensure that the appointed expert has the required knowledge and experience and has been properly instructed with respect to the legal framework and the ethical standards. However, although a qualified forensic expert is a prerequisite for an admissible forensic investigation, this does not necessarily mean that the work itself meets the quality requirements. Even the most experienced experts can make a mistake, can produce a poor report, or can misunderstand the questions at hand. So assuring expert quality solves only "half of the quality equation," the instruments that are in place to also check and ensure the quality of the forensic investigation itself will be discussed in the next section.

9.3 Ensuring quality of the forensic investigation

To understand the current views on required quality measures for forensic case work in criminal investigations, we must first consider how the "rules of engagement" in criminal law as described in the Code of Criminal Proceedings actually come about.

> How do guidelines and norms evolve in criminal law when a certain aspect (such as the quality and admissibility of forensic investigations and expert testimony) is not clearly defined in the Code of Criminal Proceedings?

New insights follow from individual cases and rulings and associated motivation by judges. Such jurisprudence is especially important in appellate courts; verdicts from supreme courts often specifically address complex legal issues and associated rulings serve as legal guidelines for subsequent cases. In this manner, legal interpretation remains consistent across various cases and jurisdictions. If such new insights have a significant and fundamental impact on criminal investigations and court cases, they are incorporated into the Code of Criminal Proceedings. The guidelines have then been transformed into new legislation that needs to be adhered to.

The important aspects of the scientific rigor of the forensic methodology applied, the robustness of the reported findings and the validity of the expert interpretation have been addressed in many cases in courts around the world. **Jurisprudence** providing quality guidelines for the admissibility of forensic evidence exists in many countries. An especially important Supreme Court ruling in 1993 concerning the civil case of Daubert versus Merrell Dow Pharmaceuticals Inc (113 S.Ct.2786) in the USA resulted in an internationally accepted framework for quality standards for expert findings. This concerned the claim of the Daubert family that their son was born with serious birth defects due to the use of the antinausea medication Bendectin of Merrell Dow Pharmaceutical during pregnancy. The company presented data to show the absence of such an effect of their medication, which was disputed by the plaintiffs who presented their own expert evidence based on animal experiments, chemical structure analysis, and reinterpretation of published studies. Both the district court and court of appeals ruled in favor of Merrell Dow Pharmaceutical on the basis that the methods applied by the Daubert family were not generally accepted by the scientific community. The Supreme Court subsequently overturned these views indicating that the **Federal Rules of Evidence** supersede the so-called **Frye rule** (Frye v. the United States, 54 App. D. C. 46, 47, 293 F.1d 13, 1014). Together with rulings in two other cases this legal debate lead to the so-called **Daubert standard** and associated **Daubert criteria** on the admissibility of expert evidence. Because of their relevance, these outcomes quickly found widespread acceptance also outside the USA. As a result, every registered forensic expert is aware of the Daubert framework and is able to cite the associated criteria.

Daubert standard

- The responsibility of assessing whether expert evidence is based on sound **scientific knowledge** and is therefore admissible sits with the judge (this is true for both the adversarial and inquisitorial system). As the judge lacks the required scientific knowledge and expertise, additional information is required to enable a fair assessment
- For expert evidence to be admissible it must be both **reliable** and **relevant.** Again it is a task for the judge to determine whether the expert evidence is sufficiently robust and trustworthy and provides useful information when considering the questions that need to be addressed by the triers of fact. Expert testimony that is not relevant can be detrimental because it can cause distraction from the real issue at hand and can even lead to misinterpretation when it is incorrectly assumed to be relevant
- Scientific knowledge in this respect indicates that the findings and conclusions of the forensic expert have been obtained in line with the **scientific method.** Acquiring knowledge according to the scientific method entails the testing of hypotheses by performing empirical studies. As more data and findings provide support, the presented hypothesis/model gains scientific acceptance. Once general acceptance has been realized, new insights might emerge leading to new, refined hypotheses, which in turn are either supported or refuted on the basis of additional experimentation. In this continuous cycle, it is important that hypotheses are tested and confirmed (or rejected) by different scientific groups to prevent **confirmation bias** and **tunnel vision.** The general tendency to focus on findings that support one's conceptions and to ignore or nullify ("reason away") data that does not support the main hypothesis is part of human nature and thus also exists with scientists and content experts. General scientific acceptance also requires critical peer review and total transparency with respect to the experimental conditions and the data obtained. Scientific studies should be made public in such a manner that independent scientists are able to repeat the experiments and ratify or criticize the outcomes and conclusions

As judges decide on the admissibility of all the presented evidence, it makes sense that they also assess the forensic reports and testimonies. However, as a legal professional, a judge typically will find it difficult to determine whether the expert findings and conclusions are reliable and have been gathered in line with the scientific method (Judges will probably feel more comfortable in determining whether the evidence is relevant for the case although this can also include complex scientific considerations). The Daubert criteria consist of a number of questions that assist the judge in this challenging task. For the evidence to be admissible, the questions must all be positively answered when considering the presented expert evidence.

Daubert criteria

1. Has the theory or technique been subjected to peer review and publication?
2. Is the theory or technique widely accepted in a relevant scientific community?
3. Can and has the theory or technique been tested (under case-relevant conditions)?
4. Is the (potential) error rate of the theory or technique known and deemed acceptable (under case-relevant conditions)?
5. Is the application controlled through the use of protocols and standards?

Acceptance in the relevant scientific community is usually demonstrated through the publication of the methodology in a peer-reviewed, internationally recognized (forensic) science journal and subsequent (positive) citation of the published work by other groups. Peer review ensures that the published work is not seriously flawed, is scientifically sound, and that the conclusions are adequately supported by the data. Forensic journals like **Forensic Science International**, **Science and Justice**, **Journal of Forensic Sciences**, **Forensic Chemistry,** and **Drug Testing and Analysis** take the peer review process very seriously, and at least two independent content experts carefully and critically review the submitted work. Rejection rates for these journals are considerable. Also, presentations at international forensic science conferences contribute to the scientific debate on new insights and methods. Within the international forensic community, it is well known that the first step in the introduction of a new method in case work is the publication of the corresponding research during the development phase and presentation of the results to the forensic community.

Daubert criteria three to five address important aspects that are not necessarily covered in a scientific publication. A prerequisite for accepting a manuscript for publication is that the work presented is novel (*i.e.* not done before by others or already exhaustively described in previous work from the authors). However, it is not mandatory that the method is directly applicable in casework and is of actual value in criminal investigations. It is a quite natural and scientifically sound process for highly innovative concepts to be gradually developed into valuable methods as several studies from different groups appear in literature over time. However, expert evidence based on a "wild idea" or "a hunch" can of course not be admitted in court as part of a criminal investigation. This requires extensive testing to show that the method can be applied under realistic case conditions and for physical evidence as encountered at the crime scene. Furthermore, to demonstrate the validity of the results in the current investigation, rigid quality controls need to be implemented. However, errors can always occur and therefore the forensic expert must also provide insight into the risk of **false-positive** (i.e., an erroneous incriminating result) and **false negative** (i.e., an erroneous exculpatory result) outcomes by reporting corresponding **error rates** preferably based on a substantial number of assessments for reference samples under realistic conditions.

Although the Daubert framework provides a clear and consistent guideline for magistrates to assess the admissibility of forensic casework and expert testimony, it still remains a challenging task to assess whether all the criteria are confidently met. The courts will request the appointed expert to answer the Daubert questions and provide convincing data that corroborates the answers. However, the expert will typically be of the opinion that all quality criteria are met, the conclusions are therefore trustworthy and the supporting data robust. If in doubt, a capable expert will immediately report this to the court or public prosecutor while clearly indicating the quality issues associated with the forensic investigation. The expert is also expected to provide total transparency in the forensic report and therefore is obliged to include any quality issue that occurred during the investigation. However, irrespective of the degree of professionalism and self-criticism of the expert, his/her views on meeting the Daubert criteria can never be totally objective and neutral. However, the involvement of additional experts to independently assess whether the presented forensic evidence is fully in line with Daubert is very inefficient, costly, and time-consuming. It is not realistic to expect that such a check is consistently conducted for all forensic casework in all criminal investigations. It is much more efficient to agree to a generally accepted quality standard that needs to be adhered to when conducting and reporting forensic casework. A forensic institute would then be tasked with implementing and maintaining this quality standard for its forensic activities and with involving an independent quality assurance organization to regularly inspect and confirm the quality of the work. Regular inspections (or **audits**) ensure that forensic casework is meeting the quality criteria, that irregularities and **non-conformities** are actively sought, detected, and repaired, and that potential quality risks are discovered in an early stage and adequately mitigated via improvement programs. Such an authority would also oversee the introduction of new casework methods for which the forensic laboratory should provide a **validation dossier** including method optimization studies, test data demonstrating robustness, associated error rates, and a quality framework for daily operation.

In the Dutch Code of Criminal Proceedings, no direct references can be found on acceptable quality frameworks. Articles in a legal context tend to be more generically formulated to prevent detailed discussions on case and expertise domain-specific peculiarities. As a result, scientific details regarding certain types of investigations or laboratory quality requirements are typically not included. In the Netherlands, such details can be specified in so-called **Decrees**. For Human Biological Traces, a special decree exists entitled "**DNA investigations in Criminal Cases**" that provides very clear instruction for the required quality measures for donor identification on the basis of DNA STR profiling. This is the only example in the Netherlands of a decree that provides specific instructions for a given forensic expertise area. The need for such a detailed decree has to do with the fact that DNA STR profile-based identification also includes a mandatory buccal swab for suspects of crimes for which pre-detention/custody applies and the inclusion and search of STR profiles in a national DNA database. In addition, article seven in the decree describes the associated quality requirements (translated from Dutch by the author): '*DNA investigations are conducted in the laboratory of the institute that has been accredited to do so by the* ***Dutch Accreditation Council*** *according to the requirements as indicated in the* ***NEN-EN-ISO/IEC 17025***'. A lot of information on the accreditation process and the need for and aims of accreditation can be found on the website of the Dutch Accreditation Council (abbreviated in Dutch to **RvA**, for information in English visit https://www.rva.nl/en). According to the RvA accreditation ensures the quality of the provided services and products. The requirements to achieve accreditation are described in several ISO/NEN standards. The term ISO refers to international standards as set by the International Standardization Organization (https://www.iso.org/home.html). These standards are often translated and tuned to the domestic situation by local organizations, which in the Netherlands is the NEN Institute (https://www.nen.nl/en/), in Dutch the abbreviation for "Dutch Standards." A broad range of standards exists to cover numerous domains and activities; the ISO 17025 standard provides a quality framework for test and calibration laboratories. Maintaining an ISO 17025 accreditation is a requirement for such laboratories to ensure that their reported results and analysis certificates are accepted. This quality assurance is important as the results of the chemical analyses are often used as compositional information for calibration standards, to demonstrate that products meet specifications, or to show that environmental regulations are met. The ISO 17025 bears some similarities with the more extensively used ISO 9001 standard but is stricter in terms of the quality requirements and also includes scientific and laboratory operation aspects. ISO 9001 is broadly applicable to businesses and operations (broad scope) and focuses on process management, documentation, and customer service. An overarching ISO 9001 principle is that "*you do what you document and you document what you do*." The first part indicates that you work according to protocol and the second part that you carefully record the results. This also applies to ISO 17025 but with a narrower scope (only considering the activities of testing and calibration laboratories), and with the additional requirements that the services and products (the results of laboratory chemical analyses and tests) are conducted according to the state-of-the-art in the field and that the quality of the reported results is guaranteed. So, using an old-fashioned method according to documented protocols does not suffice in an ISO 17025 framework, the work needs to be state-of-the-art and the quality of the reported results needs to be demonstrated. To ensure these strict standards are met, regular audits are coordinated by the national accreditation council involving external experts as it requires content matter knowledge to assess whether the work is up to standard. When during these inspections non-conformities are found, these need to be swiftly and convincingly addressed by the host institute and the improvement actions need to be communicated to the accreditation council. Issues that are very severe or not adequately addressed could ultimately lead to the council suspending or terminating the accreditation. This often has dramatic consequences as it effectively prevents an organization to continue its operation. Consequently, accredited institutions continuously invest in quality and make sure to meet the requirements and quickly resolve issues. An organization that takes quality very seriously needs to have an open atmosphere when it comes to reporting flaws and incidents. Personnel should be encouraged to identify issues, make them known to management, and take initiative to resolve them. Such actions should be rewarded as opposed to enforcing quality by punishing those who make a mistake. As indicated before, mistakes can never be completely eliminated, a sign of quality is to proactively look for flaws and learn from them to prevent making the same mistakes in the future. A specific part of the ISO 17025 approach is that the impact of non-conformities is considered in the broadest possible context. Institutes need to consider potential implications for prior casework and similar flaws that can be conducted in other expertise domains. Although it makes sense for forensic laboratories to use the ISO 17025 framework for the chemical analysis of physical evidence (the required infrastructure, instrumentation, procedures, and methods are very comparable to those used in testing and calibration laboratories in other sectors), the standard itself is not focused on and does not mention forensic science. Therefore, a need exists for a "translation" of the ISO 17025 framework to forensic practice. Such a dedicated description of 17025 compliant quality control in a forensic setting has been provided by **ILAC**, "*the international organization for accreditation bodies*," in their **G19** guideline (see Further Reading section how to download the latest version of this standard aimed to "*provide guidance for forensic science units involved in examination and testing in the forensic science process by providing application of ISO/IEC 17025 and ISO/IEC 17020*").

> ISO 17020 is the quality standard for inspection bodies. Which part of the criminal/forensic investigation could make use of a quality framework based on this standard?

If we consider the activities of an inspection body, organizations that perform audits and check whether operations are in line with protocols and requirements, we can see similarities with a forensic crime scene investigation. Documenting the scene, reconstructing events, and the detection and securing of physical evidence that is subsequently transported to the forensic laboratory for further investigation. The main difference and therefore also the challenge when developing an ISO 17020-based CSI standard is that a crime scene is uncontrolled and therefore each scene can provide new challenges that require expert adaptation and flexibility. Inspectors have a reference, i.e., they know what they need to encounter for a positive assessment of the organizations that they inspect. Nevertheless, it is possible to define a quality framework for how to conduct a crime scene investigation. CSI experts working transparently according to generally accepted guidelines are very valuable as the crime scene investigation is a crucial first step that can "make or break" a case. New and experienced CSI colleagues being trained according to the guideline will also ensure consistency, i.e., irrespective of the experts conducting the investigation, the outcome (documentation, interpretation, secured traces) will be comparable. This is also a very important aspect as the CSI process cannot be automated and crucial decisions strongly depend on the level of experience and commitment of the experts. A level of subjectivity will therefore always be part of any crime scene investigation, making the process as reproducible and objective as possible will benefit the criminal investigation and court case.

In the remainder of this section, we will now discuss the various aspects (or modules) of an ISO 17025 compliant quality framework in a forensic setting. These modules are listed below covering aspects that need to be addressed by forensic laboratories that want to operate under ISO 17025 accreditation:

- Values
- Management
- Scope
- Personnel
- Infrastructure
- Validation, Quality Control, and Uncertainty
- Audits and Incident Reporting and Mitigation
- Complaint Procedures
- Reporting and Documentation
- Document and Information Management

How would you define a quality framework for forensic casework?

Individually or in a student group formulate your own QC guideline!

Discuss essential quality aspects for each of the modules.

Only continue reading after finishing this assignment.

Compare your views with the current views on quality in forensic science as described below. Did you include additional points, or did you exclude certain aspects?

Please note there is no right or wrong in this respect, although it of course helps to achieve accreditation if you follow accepted guidelines.

Values

It is important for an organization to have and express core values as part of its professional activities. These values serve not only external communication purposes but are also engrained in the work ethics of the personnel. Shared values create team spirit and an atmosphere where it is accepted and encouraged to provide collegial feedback when risks and flaws are spotted. From the ISO 17025 perspective (testing and calibration laboratories), these core values include **neutrality**, **objectivity**, **confidentiality,** and **traceability**. It reflects customer focus but without any sacrifice to the scientific findings. It can be argued that objective results are in the end also in the best interest of the customer. That these values are very important to uphold becomes clear when considering the somewhat vulnerable position of testing laboratories in relation to their customers. Critical customers that are not happy with the presented data (as opposed to unhappy customers because of poor communication, cost increase, or delayed reporting) could threaten to move their business to another competing laboratory. These values are also directly applicable in a forensic science setting. Forensic experts conducting casework should remain at a safe distance (in a literal and figural sense) from the criminal investigation to ensure impartiality. The expert should be committed to providing neutral, objective, unbiased information that contributes to the criminal investigation and assists the trier of facts. It is definitely not the task of the expert to "*find and convict the perp*" (etrator). These values are for instance clearly expressed in the vision of the

Netherlands Forensic Institute (translated from Dutch by the author): "***Focused on the Truth, Guided by Science, for a Safer Society***." Unfortunately, the popular CSI TV series often depicts a totally different attitude of the forensic expert.

Management

Just as important as shared values is the commitment of senior management of forensic institutes to the quality of the work and the ISO 17025 standards. The management should **lead by example**, communicating and showing the importance of quality to the personnel. How could directors expect forensic investigations in their institute to be of the highest standards if they pay little attention to quality and do not secure sufficient funds and capacity to maintain the quality framework? As senior management is also responsible for negotiating and keeping budgets and for ensuring sufficient capacity for and short lead times of casework, it is by no means obvious that sufficient resources are dedicated to quality control. For that reason, senior management should secure sufficient capacity in the organization dedicated to quality control and the ISO 17025 accreditation. Typically, a reasonably sized forensic institute will have a small central QC team of 2–3 fte dedicated to quality and accreditation. This team collaborates with quality officers in the various departments and forensic expertise areas to prepare for and conduct audits, handle quality-related issues and complaints, conduct validation studies for new methods, and maintain quality documentation (guidelines, procedures, methods). Within the ISO 17025 framework, an **annual management review** is mandatory. Senior management is expected to take notice of the outcomes of internal and external audits, especially of established non-conformities, complaints received, actions taken, ongoing quality issues and improvement programs, training activities, and overall means and capacity available for quality activities. On the basis of this review, senior institute management is obliged to annually formulate decisions and actions to maintain the ISO 17025 standard and further improve the quality of the forensic casework.

Scope

Within an organization seldomly all the work is conducted within the ISO 17025 framework. It is therefore important to clearly indicate, which activities are accredited (and which are not) and this is clarified through the scope of the accreditation. The scope of the NFI accreditation can directly be downloaded from the website of the Dutch Accreditation Council when following the instructions given in the Further Reading section. The corresponding document lists per expertise area the accredited procedures and methods, the materials involved (the physical evidence being investigated), and the associated quality document numbers. There is a reason that the list includes many forensic chemistry-based methods, chemical analysis is frequently conducted by test and calibration laboratories and hence the associated activities fit well within the ISO 17025 framework. It is for instance much more challenging (but not impossible) to accredit forensic methods that are based on human expert comparison and interpretation because of the subjective and often qualitative nature of the investigation. Although accreditation is awarded on an institute level (the NFI holds an ISO 17025 accreditation since 1994 under accreditation number L146), the scope specifies the accredited methods. As the list of methods is not exhaustive with respect to forensic expertise areas and investigative methods, casework is also conducted and reported outside of the accreditation framework. In addition, sometimes a forensic investigation can be conducted partially under the ISO 17025 standard. For instance, only the chemical analysis part in the laboratory but not the subsequent subjective expert interpretation of the data. In recent times, forensic institutes have invested in bringing expert interpretation under accreditation and the scope of the NFI for instance also includes the interpretation of DNA profiles at source level. There can be good reasons why certain types of investigation are not within the scope of the ISO 17025 accreditation. It also does not mean that corresponding findings are inadmissible in a court of law. It does require the expert to specify and demonstrate the quality assurance measures for the presented forensic casework and to explain why this methodology is not ISO 17025 accredited.

Personnel

Having trained and capable personnel is essential for achieving robust and reliable results. For this reason, the ISO 17025 standard contains an extensive module on personnel requirements. One cannot just hire a new expert and have this person performing and reporting critical casework the following week. An extensive introduction and training period is needed during which the new colleague is instructed in the various tasks by experienced team members. Typically, after observing colleagues doing the work, the new team member can cautiously do the activity under supervision to gain experience. In the first attempts, it is typically a good idea to train with realistic test materials or physical evidence that is abundantly available (e.g., a large batch of seized drugs). To be allowed to contribute to casework, new trainees have to show that they can conduct the tasks according to the documented procedures confidently and flawlessly and with the same results as the more experienced colleagues. Within the ISO 17025 framework, this all needs to be carefully documented. Each employee keeps his/her own personal training dossier, this applies not only to new but also to more experienced team members to ensure that their expertise is maintained and remains state-of-the-art. When novel methods are introduced in the forensic laboratory also experienced team members will have to be trained to use the new equipment. Managers confirm the education efforts by signing the training records. On a team level, the so-called **qualification matrix** indicates who is allowed to perform, what accredited activities at a certain point in time. As each step is documented in the case file and signed off by the operators, it can be demonstrated that only qualified and mandated personnel has performed the various stages of the investigation. This ensures that the investigation in a given case has been performed adequately and fully in line with the accreditation. It is quite common for various team members to contribute to a forensic investigation, forensic experts are quite often assisted by scientists, equipment specialists and/or laboratory assistants. When considering high-volume casework conducted at a department level, the various stages of the casework are sometimes also segmented for efficiency reasons. Co-workers are then dedicated to a certain aspect of the casework such as the sampling of forensic traces from the physical evidence. This workflow approach is for instance often encountered in forensic DNA investigations. However, the final product remains an expert report with the results of the forensic investigation. The registered court expert handling and overseeing the case will have final responsibility, signs the report, and will testify about the work in court if needed. Forensic institutes, therefore, offer dedicated training programs for their forensic expert positions. Such a program will include criminalistics and criminal law courses to explain the fundamental principles of forensic science and the role of the forensic expert in the criminal justice system. Furthermore, quality management instructions will be provided, and realistic courtroom exercises will be organized to become acquainted and comfortable with providing expert testimony. In addition to these exercises and courses, the junior forensic expert will acquire experience on-the-job under the close supervision of a more experienced colleague. As the practical skills are developed, the expert trainee will be allowed to provide a greater contribution to forensic casework. Under the supervision of a registered expert in the team, concept case reports will be prepared but not signed by the trainee. In this way, experience, knowledge and expertise are gradually developed as the junior forensic expert prepares for a final examination. Within the Netherlands Forensic Institute, the internal training program for forensic experts is very demanding and typically takes 2–4 years to complete. The program includes a courtroom test that is evaluated by "real-life" judges and public prosecutors and a final examination in which a legal professional and an unaffiliated, often international, expert are part of the exam committee. After successfully completing the exam, the NFI expert applies for admission in the Netherlands Register for Court Experts. Reexamination is required every 5 years to show that the forensic expert has maintained his/her knowledge and experience in the field and is capable of performing forensic casework according to the state-of-the-art. This is typically demonstrated by providing a list of reported cases and court appearances and by showing active involvement in innovation projects and teaching activities. Becoming and remaining a qualified, court-registered expert is a full-time job that requires a high degree of professionalism and work ethics and continuous investment in education and personal development. Again, a remarkable contrast to the image shown in the popular CSI TV series where the lead characters seemingly effortlessly cover a wide range of forensic domains and are able to conduct fingermark comparisons, generate DNA profiles and chemical identify explosives and drugs of abuse in a single episode. In real life, forensic experts seldomly are qualified to perform investigations in multiple expertise areas because of the dedication and effort that is required to assure the quality of the investigation in a single field.

Infrastructure

To be able to perform forensic investigations according to the strict ISO 17025 quality criteria, it is absolutely essential that the right infrastructure and organizational support are available. This corresponds to spacious, well illuminated, temperature and airflow controlled, state-of-the-art laboratories and sufficient support staff to ensure that the infrastructure consistently remains clean and well equipped. It also entails a smart use and design of the overall space available, including sufficient storage space under controlled climatized conditions (to maintain the integrity of the stored physical evidence) and an arrangement that supports the natural flow of the casework. Storage freezers and refrigerators are necessary for teams handling samples and extracts with limited stability at ambient conditions. When handling intrinsically dangerous materials such as potent drugs of abuse or explosives, special infrastructural measures are necessary to ensure the safety of the personnel. The Netherlands Forensic Institute for instance has an on-site bunker facility situated outside the main building where intact explosive material from casework is stored and sampled. Only small and harmless subsamples are subsequently analyzed in the laboratories in the main building. Just like any other laboratory handling dangerous materials, work with drugs of abuse or potentially dangerous materials secured from threat letters need to be handled in a fume hood while the analysts wear appropriate PPE (personal protective equipment). To adequately respond to incidents, laboratories need to be equipped with suitable means for impact mitigation such as fire extinguishers, showers, and eye wash stations. With respect to safety and security it is also important to acknowledge the sensitivity of work and the materials involved. Although luckily a rare event, targeted attacks on forensic institutes have occurred in the past in an attempt of criminals to destroy evidence, hamper the investigation, or steal drugs of abuse, explosives, or weapons. A shocking relatively recent example is the arson committed in August 2019 at the NICC, the national forensic Institute of Belgium. Extensive fire damage severely hampered the NICC operation for months and destroyed case evidence. This means that a forensic institute needs to invest in a tight security system, including 24/7 surveillance, a checkpoint for visitor admission, visitor registration, vaults to store sensitive items such as firearms, electronic access passes, restricted access areas, and external obstacles to prevent easy access to the main infrastructure. With more and more information processed and sent digitally, modern forensic institutes nowadays also require a state-of-the-art, well-guarded ICT infrastructure.

Photos of the Netherlands Forensic Institute located at the Laan van Ypenburg, the Hague, the Netherlands. This state-of-the-art infrastructure was realized in 2004 and has been specifically designed for forensic case work with dedicated laboratories, autopsy rooms, an indoor shooting range, an on-site bunker for the storage of explosive materials and controlled fire experiments, a visitor reception area, electronic access points and a front desk for receiving and storing physical evidence. *(Photos from and used with permission of the Netherlands Forensic Institute)*

The design and use of forensic infrastructure should be such that the risk of **contamination** is minimal. Although for many types of laboratories contamination is something that should be prevented, it is of crucial importance in a forensic setting. A false-positive result due to a contamination issue during casework can have devastating consequences for the suspects and victims in a criminal case and therefore needs to be prevented "at all cost." Although such risks can never be fully eliminated, they can be minimized through a strict quality regime. Also, measures can be taken to detect contamination when it occurs in order to minimize its impact. The risks involved are greatest for forensic microtrace analysis. Minor pollution occurring during the sample preparation for the GC-MS analysis of a relatively large batch of a seized drug will typically not affect the outcome of the chemical analysis. However, when considering ppb levels of the same drug in a whole blood sample, minor contamination can have much greater consequences. The same holds for forensic micro traces such as gunshot residues, glass fragments, and fibers or DNA contact traces. The smaller the object of forensic interest the bigger the chance and impact of contamination. Typically, in the forensic laboratory three types of contamination can occur during the transport, storage, and investigation of physical evidence:

1. **Primary transfer**—the spreading of case-related trace material on a single evidence item that alters the original trace pattern
2. **Cross-contamination (secondary/tertiary transfer, carry-over)**—the transfer of case-related trace material from one evidence item to another or from one prepared forensic sample to another
3. **External contamination**—the contamination of an evidence item, a forensic trace, or a subsequent sample with material originating from the forensic laboratory or the forensic investigator

What measures would you take to prevent contamination in the forensic laboratory?

Individually or in a student group formulate your own counter-contamination measures! Only continue reading after finishing this assignment.

The first and most important line of defense against contamination is related to the infrastructure. Within a forensic institute it is essential to separate those activities, which involve bulk and trace levels of the materials of forensic interest. The most striking examples are forensic toxicology versus illicit drug analysis and GSR analysis versus firearms examinations. Forensic experts that are involved in illicit drug analysis should not be involved in a forensic toxicological analysis. Additionally, equipment should never be shared for illicit drug identification and whole blood drug analysis. With the huge difference in concentration levels, the slightest carryover will immediately cause significant contamination for the trace analysis. Special precautions are also necessary for forensic firearms examiners as typically shooting incidents also require a gunshot residue investigation. A firearms examiner that has conducted a number of test firings at the indoor shooting range is completely covered with GSR particles. If such an examiner would subsequently join his/her GSR colleague to analyze GSR swabs on the SEM equipment, contamination is almost unavoidable given the fact that SEM allows for the detection and characterization of individual GSR particles, which are only a few microns in size. For this reason, firearm examiners at the Netherlands Forensic Institute that have worked at the shooting range are not allowed to visit the GSR section on the same day. Accurate reconstructions of shooting incidents do require a strong collaboration of both GSR experts and firearm examiners. So, rather than completely preventing any contact between these two expertise domains, strict protocols and smart use of infrastructure are preferred to minimize contamination risks. Within a given forensic domain substantial differences in levels and amounts can result in contamination risks. An example is the analysis of intact explosive material versus the trace analysis of the same compounds at trace level in post-explosion swabs. It is advisable that forensic explosive and explosions investigations are conducted in separate laboratories and equipment for pre and post-explosion cases. If this is not possible, care should be taken to prevent contamination of post-explosion samples with traces of intact explosives. For instance, through a smart order of casework analysis or by alternatively analyzing preexplosion and post-explosion samples on different days in combination with strict cleaning procedures and blank runs. For expertise areas that deal with both very high and trace levels as part of the investigation, providing an infrastructure that creates a logical workflow from "high to low" (or "bulk to trace") can also be effective, especially in combination with the regulation of air flows through the application of pressure differences. It is not uncommon for forensic laboratories to create a consistent overpressure in the trace sections to prevent contamination from incoming airflow. Special care should be taken that items related to the victim or the crime scene are never in the vicinity of items related to the suspect during the investigation. This needs to be considered for several expertise areas (e.g., GSR, glass, and fibers) but is especially important for forensic DNA investigations. Contamination of a clothing item of a sexual assault victim with forensic biological reference material from a suspect in the forensic laboratory can have dramatic consequences and could result in a wrongful conviction. For this reason, forensic DNA laboratories typically have different investigation rooms for the victim and perpetrator-related material. Care should also be taken that associated evidence items are not stored in close proximity and handled and transported simultaneously.

The second level of contamination control entails the use of working procedures and methods that limit the risks involved. A recommended way of working is to investigate one evidence item at a time (the "*one exhibit on the table*" principle) and to collect and process all the relevant forensic traces from this item before moving to the next exhibit. During the investigation of different items, a rigorous cleaning protocol should be employed including the use of reactive agents such as oxidizers and UV light to actively eliminate any remaining traces from the previous investigation. Ideally, this elimination is also confirmed by taking and analyzing blank swaps before switching to a different case or item. The final line of defense is the use of protective clothing and packaging/containers to prevent contamination. Unlike what you see in the CSI TV series, the use of protective clothing already starts at the crime scene and is continued in the forensic laboratory when examining the collected exhibits. The use of Tyvek suits, hair nets, latex gloves, shoe covers, and mouth masks makes the investigation uncomfortable and challenging but is absolutely essential to prevent external contamination. Without such protection, there is a serious risk that the investigators will contaminate the exhibits with their own biological cell material, fingermarks, hairs, clothing fibers and any other micro traces that they carry with them unknowingly. If this transfer goes unnoticed and is not recognized as external contamination, the traces could be detected during the investigation and could be interpreted as crime-related. This will typically not lead to a wrongful conviction but can however have serious implications for the admissibility of the evidence, could steer the investigation in the wrong direction, and ultimately lead to wrongful convictions. If for instance, a forensic expert would leave his biological cell material accidently on a clothing item of a victim and his retrieved DNA profile is subsequently considered to be of the unknown male perpetrator, this could lead to the release of a suspect in the case on the basis of tactical information and other forensic evidence. For this reason, it is nowadays standard procedure to have a **DNA elimination database** with the STR DNA profiles of all the employees that enter the forensic laboratories or handle physical evidence. In this way, the external contamination is detected early on and although this could still have serious consequences with respect to the admissibility of the evidence, it will be immediately clear that the trace is not case- but rather examination-related (a complicated situation arises when an employee of a forensic institute is actually involved in both the crime and the subsequent forensic investigation but luckily this is an extremely rare event). Spotting external contamination can be more difficult for traces of a nonbiometric nature such as clothing fibers. Investigators also need to be vigilant for external contamination from unexpected sources. Materials that are used during the investigation such as swabs, cloths, solvents, and tapes need to be totally free of contaminants and this needs to be checked and certified by the suppliers. However, sometimes when changing to a new supplier or when the supplier changes its production processes, unexpected external contamination may occur. One of the most infamous external contamination cases in the history of forensic science involved the police search in Germany, Austria, and France for a female serial killer based on an STR DNA profile that suddenly occurred in several criminal and murder investigations in the period 1993–2009. This "*Phantom of Heilbronn*" also named the "*Woman Without a Face*" turned out to be a female factory worker causing contamination of cotton swabs that were used for the sampling of forensic biological traces for DNA typing (source: Wikipedia, https://en.wikipedia.org/wiki/Phantom_of_Heilbronn). This example shows the importance of blank runs involving all materials except for the physical evidence and the importance of keeping an open and critical mind when obtaining forensic evidence that does not make sense from a tactical point of view.

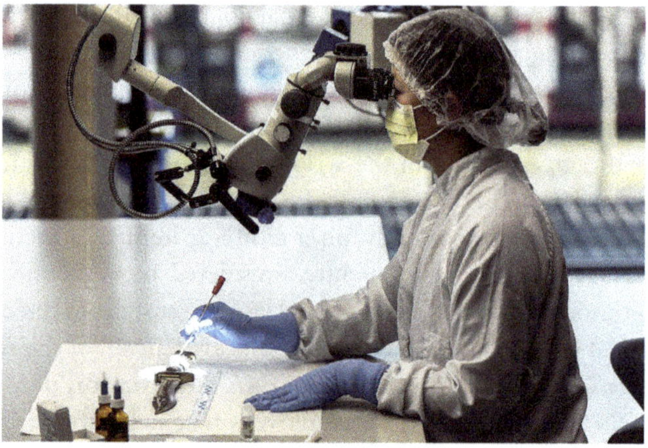

Investigation of an evidence item for the presence of human biological traces at the Netherlands Forensic Institute illustrating the "three lines of defense" to prevent contamination: a dedicated investigation room, strict protocols (one exhibit on the table), and the use of protective clothing. *(Photos from and used with permission of the Netherlands Forensic Institute)*

Chain of custody

Another critical quality aspect is associated with the physical evidence and the sensitive nature of a criminal investigation. The so-called **chain of custody** is quite specific to forensic science and covers infrastructural, organizational, and documentation aspects to maintain the integrity of the physical evidence at all times. In a criminal investigation, there is no room for doubt with respect to the origin and state of the exhibits found at the crime scene and the forensic traces secured from these items. An adequate chain of custody ensures that at every stage of the investigation ownership of the physical evidence items is clear and documented. Each custodian ensures the traceability of the evidence items, storage conditions that preserve their state, an accurate and complete recording of the handling and examination of the items. To demonstrate that a custody chain has not been compromised, each organization needs to carefully document the receiving and returning of the evidence items. During each step in which the custody is transferred to another partner in the criminal justice system, it is also very important to document the state of the items including detailed photographs according to forensic standards. Transport of evidence items from the crime scene to the forensic laboratory needs to be secured, documented, and conducted according to established protocols. Evidence items should not get lost or damaged during the logistic operations and new items should not suddenly appear "out of nowhere." In the Netherlands, the transport of evidence items from and to the forensic laboratory is arranged by the logistic services of the police using specially sealed crates. The crates are sealed at the crime scene/police station and the seal is only broken when logistic specialists of the front office of the Netherlands Forensic Institute process the evidence items and associated forensic requests. By matching the requests with the items, the forensic investigations are lined up in the correct order after consultation with the forensic experts. For evidence items that pose safety risks, such as firearms (that could still be loaded), first inspections are usually conducted by forensic experts. Any deviation from the agreed packaging protocols is immediately documented and reported, the same holds for items that might have been damaged during transport. This all serves to provide total transparency with respect to the fate of the evidence items during the investigation.

Preferably, forensic investigations are non-invasive thus preserving the state of the evidence and allowing for follow-up and counter investigations. However, many investigations are intrinsically destructive and alter the state of the evidence but still are essential for an accurate reconstruction of the events. Such an alteration can be relatively minimal, for instance by taking a small sample of a drug seizure to dissolve in an organic solvent for chemical identification with GC-MS. However, several types of investigation are quite invasive and significantly and irreversibly alter the state of the evidence. It is very important that such experiments are carefully prepared and successfully executed as "you only get one chance." In interdisciplinary casework involving various expertise areas, the correct order of the investigation is crucial in this respect. First, the investigations need to be conducted that are non-invasive and pose minimal contamination risks. Typically, biological and micro traces need to be secured first. When considering a firearm possibly used in a shooting incident, the firearms expert is the first to inspect the item and render it safe but also the last in line to perform the forensic investigation. A firearms examiner will conduct test firings at the shooting range to create reference bullets and cartridge cases for comparison against crime scene items. This process is very invasive and for instance, removes and replaces the original GSR traces. Therefore, all microtrace investigations (GSR, fingermarks, fibers, DNA) need to be conducted first. Before performing an invasive experiment, it is also important that all associated parties have been informed, fully understand the destructive nature of the test and agree to it. Tactical and legal considerations might outweigh the potential insights that are created with the destructive test.

Who owns the evidence?

Who decides if destructive analyses can be performed?

In the Netherlands, the public prosecutor assigned to the case is leading the criminal investigation and is therefore also responsible for the physical evidence. Typically, the Dutch police stores all the evidence including long-term storage if necessary. Storage of evidence is mandatory as long as legal procedures are ongoing and appeal options are available. Evidence items cannot be stored indefinitely and after the case has been solved and all legal options have been exhausted, exhibits are usually destroyed. The Netherlands Forensic Institute only temporarily receives the evidence items for the duration of the forensic investigation. Once the casework is finalized the items are returned to the police. An exception to this is samples that require special storage conditions only available at the forensic institute (for instance DNA extracts) or items that make for valuable additions to forensic reference databases. (for instance firearms). If the forensic experts want to conduct a destructive examination, the public prosecutor needs to grant permission. This sometimes requires the public prosecutor to seek approval for the examination with the investigative judge in the case.

Maintaining evidence integrity and assuring the chain of custody requires expertise and dedicated capacity, infrastructure, and ICT systems. At the so-called **Front Desk** of Netherlands Forensic Institute over 30 professionals work full time to process over 150 requests for a forensic investigation that are received on an average day. Annually, over 100.000 evidence items are processed, matched to the requests, and distributed over the various forensic expertise teams. From these items, multiple forensic traces can be retrieved by the forensic experts that also require registration. This results in numerous forensic reports describing the results and expert interpretation linked to the case, the request, and the exhibits. It is essential that the correct investigation is conducted for the given request and evidence item. At a total annual caseload of 60.000, a chain of custody-associated error rate as low as 0.1% seems impressive but would still result in 60 serious quality incidents per year. One instrument to prevent errors is to apply consistent item numbering throughout the criminal justice system. Typically, organizations have their own quality systems and have set up their own system for the administration of evidence items and associated forensic traces. Consequently, during the course of the forensic investigation, new numbers and codes emerge as more organizations get involved in the investigation. This is a potential source of confusion (in addition to increased administrative tasks) and therefore in the Netherlands a unique, single code is provided for each evidence item and this code is used throughout the investigation. This so-called SIN, translated from Dutch as **Trace Identification Number**, code is integrated into a **Track & Trace** system at the NFI using **RFID** (Radio Frequency Identification) tags, scanners, and detection poles. In this way, evidence items can be located in the institute and their movement in the building can be accurately monitored and reconstructed.

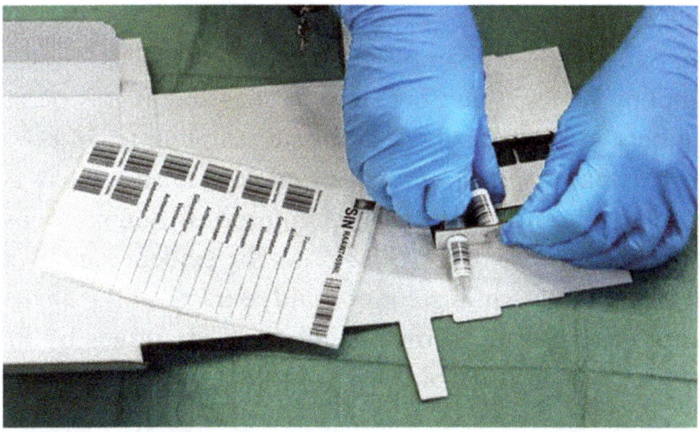

Application of Dutch SIN (Trace Identification Number) barcodes to forensic casework samples. *(Photos from and used with permission of the Netherlands Forensic Institute)*

Validation, quality control and measurement uncertainty

New methods and insights are constantly created and published in forensic science journals. Many forensic institutes also maintain substantial research and innovation programs. The most successful and valuable innovations are candidates for implementation in casework. As discussed, before the starting point is broad acceptance by the international scientific community, preferably by means of peer-reviewed scientific publications. However, a publication as such provides no guarantees that all Daubert criteria are met and that the presented methodology is fully in line with the ISO 17025 quality framework. Therefore, the next step toward transforming scientific discovery into a valuable case work method involves comprehensive **method validation** and presenting a **validation dossier** (together with the final method description) to the accreditation committee during an external audit. After a positive evaluation by the committee, the new method can be added to the ISO 17025 scope of the institute. If nonconformities are found these first need to be addressed in a way that meets the approval of the committee prior to implementation. The intention to add a new method to the scope has to be announced beforehand and the validation documents need to be provided to the committee prior to the audit. Typically, the committee will pay extra attention to the new method during the audit to make sure that the forensic experts covered all necessary quality aspects. It is not uncommon that during an audit forensic experts give a short presentation to explain the background and the details of the validation study. The ISO17025 standard and the ILAC G19 guideline provide clear instructions on what a method validation should entail. The following aspects should be addressed in the validation dossier:

- Scope
  - Purpose and limitations (questions that can and cannot be addressed with the methodology need to be specified)
- Storage, Sampling, and Sample Preparation
  - Sampling strategy: number of units analyzed in case of large sample numbers and statistically appropriate sampling schemes
  - Sample homogeneity: optimal sample size to adequately average out potential inhomogeneities while minimizing sample "consumption"
  - Sample stability: storage conditions for evidence items and subsequent samples such as solutions and extracts
  - Sample preparation: showing sufficient robustness by limiting the impact of small, unavoidable variations in the sample preparation process

- Analysis/Test
  - **Sensitivity** (in a binary test): assessing the **True Positive Rate (TPR)** of a test, the proportion of outcomes that correctly detect the presence of a compound, this also provides information on the **False Negative Rate (FNR)** as TPR + FNR = 1, Sensitivity = TPR/(TPR + FNR)
  - **Specificity** (in a binary test): assessing the **True Negative Rate (TNR)**, this also provides information on the **False Positive Rate (FPR)** as TNR + FPR = 1, Specificity = TNR/(TNR + FPR)
  - **Limit of Detection (LoD)**: establishing the minimal amount or concentration that can still be detected by the method, presence below this limit would result in a false negative outcome
  - **Limit of Quantification (LoQ)***: establishing the minimal amount or concentration that can still be quantified with an acceptable measurement uncertainty
  - **Linear range***: establishing the upper level and thus the concentration range for which the analytical concentration can be measured without significant systematic errors due to detector signal saturation. Above this level additional dilution of the sample is required
  - **Accuracy***: establishing the systematic error, the average deviation from the true value of realistic samples by analyzing a reference standard for which the actual level is known/specified or by comparing the outcomes with an accredited/accepted test method based on a different measurement principle
  - **Precision***: establishing the random error, the statistical variation observed due to small, random fluctuations when analyzing realistic samples
  - **Measurement uncertainty***: reporting the overall standard deviation and confidence interval (at setting boundaries, typically 95% or 99%) for the analysis of casework samples, might make use of pooled data from reference samples
  - **Repeatability**: the consistency of the test or analysis when repeated over time on the same instrument/setup, in the same laboratory, and by the same researchers
  - **Reproducibility**: the consistency of the test or analysis when repeated by different researchers from other laboratories using the same protocol and setup but with different instrumentation

- Interpretation and Conclusions
  - **Background levels**: demonstrating that the compounds of interest are not present at concentrations exceeding the detection limit in typical blank sample matrices (sometimes this is possible from a dual-use perspective or from prior exposures, if this is the case, the validation dossier must explain how this is handled and interpreted)
  - **Contamination**: optimizing work protocols and showing that the integrated cleaning steps effectively prevent contamination through true negative results for blank control measurements
  - **Robustness**: detailed statistical analysis for instance by applying ANOVA (analysis of variances) to understand the factors causing random and systematic variation and limit their impact through method optimization
  - **Hypotheses**: specifying the hypotheses pair that is related to the question and that determines the evidential value for a given result of the analysis/test
  - **Reference Databases**: constructing a relevant and sufficiently large set of reference data that for instance for chemical profiling methods can be used to estimate the RMP (random match probability) for a false positive result
  - **Evidential Strength/LR values**: providing LR values or LR ranges or associated verbal statements for a given analysis/test result and demonstrating the validity possibly through the use of LR calibration (as will be discussed in **Chapter 10—Reporting in the Criminal Justice System**)
  - **Conclusions**: providing the reporting framework and the final concluding statements for a given analysis/test result should be applied consistently in casework (as will be discussed in **Chapter 10— Reporting in the Criminal Justice System**)

Components indicated by an * are only relevant for quantitative methods, i.e., methods that measure a numerical value of a feature such as the concentration of a compound in the case sample. Qualitative methods determine the presence of a compound or provide some form of classification without quantitation. Diagnostic (binary) tests are in this context also considered qualitative methods.

In addition to the validation report, the audit dossier will also include the details of the final method to be applied on a regular basis in casework. The method will be based on the optimal parameters with respect to method robustness and performance as assessed during the validation. The committee will also investigate whether a satisfactory (i.e., in line with the ISO 17025 framework) quality regime has been included. The method should contain a number of quality checks as part of the procedure to ensure that the data obtained in a given case is meeting the expected performance parameters as documented in the validation dossier. These quality checks will provide indications of whether the equipment is still operating within specifications and will indicate when equipment maintenance and part replacement are necessary, *e.g.* when a capillary GC-column should be replaced in a GC-MS set-up because of poor peak shapes due to gradual contamination or thermal degradation of the stationary phase. Typically, three levels of quality control are considered for accredited methods. The first "line of defense" or the so-called first level QC involves the use of a control sample of known composition. Every time casework samples are analyzed this control sample is included in the sequence. In quantitative analysis, the measured concentration of this control sample can be compared against the known level and the random and systematic deviation can be checked against expected values as established during method validation. When using a **Shewhart control chart** (named after Walter Andrew Shewhart, an American physicist, engineer, and statistician, 1891–1968, source: Wikipedia), the reference level of the control sample is plotted over time. When deviations gradually increase and certain threshold values are exceeded this triggers the analysts to revise the system and only resume the case work analysis when the data for the control sample are within the specified limits. For the **Upper Control Limit** (**UCL**) and **Lower Control Limit** (**LCL**) typically ± three times the standard deviation from the mean value of a given control sample data set is taken. For normally distributed data excluding any significant deviations over time, 99.74% of all the data points should fall within the UCL and LCL range. Specific actions can be defined when levels are reported outside this range. When subsequent measurements show a systematic deviation resulting in reported values below the LCL or above the UCL, casework should be stopped and actions should be undertaken to revise or repair the equipment.

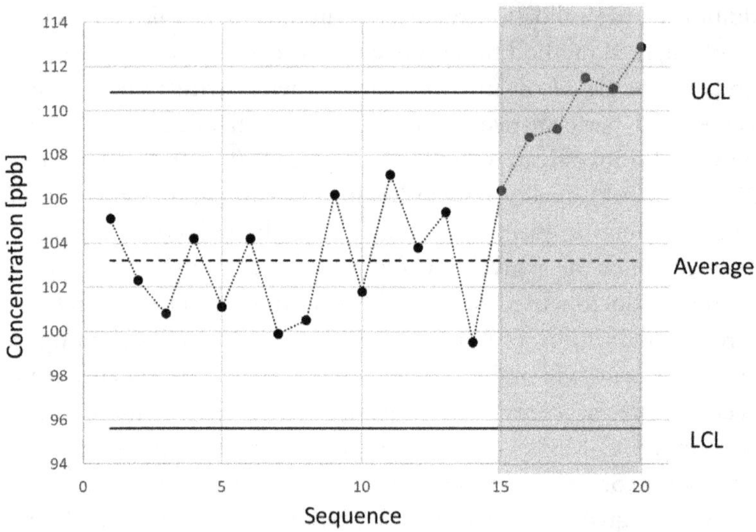

Shewhart control chart showing 20 consecutive data points for the control sample analyzed with each sequence, the first 15 points form the basis for the determination of the average value, UCL and LCL. The final 5 points illustrate the loss of process control. The control sample contains the analyte of interest at a concentration of approximately 103 ppb

In addition, occasionally a second level QC sample is included in the sample sequence. This sample is prepared beforehand with a known concentration (that differs from that of the control sample) by a team member not involved in the actual analysis. Only when the analyst has completed the measurements the true value is revealed and compared to the measured level of the analyte of interest. If the true value is within the measured range as determined by the confidence interval then this is another confirmation that the method will produce trustworthy data for the case samples (excluding specific effects associated with the nature of the casework samples, therefore QC samples should match the characteristics of the casework samples as much as possible).

Assuming that the control sample is representative of the actual samples from casework, why would we include the occasional analysis of a "known-unkown" (second level quality control) in our method?

For the control samples, the investigator/analyst knows all the details and hence knows what the outcome should be. This means that intentionally or unintentionally (cognitive bias) the data processing could be tweaked in favor of the expected outcome.

It would be even better if the control sample would originate from outside the forensic institute to prevent any bias in the QC preparation or access of the analyst to the sample information. To this end, institutes participate in so-called **proficiency tests** or **ring tests** to check the performance of their accredited methods, known as the third level of quality control. Proficiency tests are typically offered by commercial companies that have developed a range of tests for the most common accredited forensic casework methods. These tests can be ordered at any moment and the ground truth of the test samples is only revealed to the laboratories after the test results have been returned and inspected. A ring test is typically organized by an international network of forensic science institutes (see also **Section 9.4**) and involves the creation of a realistic casework sample set for which the ground truth is known by the organizers but not by the experts involved in the analysis. These samples are studied by all the teams participating in the ring test and the corresponding reports are collected by the organizers and an overview of the results is provided (and often shared with the community by means of a scientific paper in a forensic journal). The performance of each team is reported confidentially and an anonymous overview is created (e.g., citing the results of Institute A, Institute B, Institute C, etc). This overview is very valuable as it provides insights into the robustness of the methodology and associated error rates for the forensic expertise area and the accredited methods as a whole (i.e., across various forensic institutes). Providing individual results in these overviews anonymously prevents the "naming and shaming" of teams that for whatever reason scored below average and lowers the reluctance to participate.

What would you prefer, a proficiency test or a ring test?

And why?

Proficiency tests are straightforward to initiate and have been tailor-made to support the accreditation process. However, companies providing such tests have also been criticized for making them too easy and not representative of actual case work. Thus reducing such tests to a simple means to "check the quality box" for an upcoming audit. This adds little value and is not representative of the quality of the actual case work. In that respect, ring tests are much more valuable as samples are chosen and created by forensic experts to reflect the actual complexity encountered in casework. However, ring tests are laborious, time-consuming, and resource-intensive to organize.

The most realistic test environment is achieved with a **blind test**. In a blind test, a **fake case** is meticulously prepared without the experts being aware of the actual nature of the exercise. The fake case is processed with the other incoming cases and investigated and reported as such. The recipient of the forensic report is often a public prosecutor or a police officer not to raise any suspicion with the experts that they are actually being tested (forensic experts tend to have a special gift to sense something is out of the ordinary and that the case at hand could actually be a blind test, it is essential to the nature of the test that the fake case is not revealed as such).

What is the difference between a blind test/fake case and a proficiency/ring test?

And why is it important to introduce fake cases in addition to these tests?

Although proficiency and ring tests are valuable quality instruments they fail to test the day-to-day routine casework conditions. When lab workers know that they are being tested they will pay extra attention to the analysis and will double-check all the results. The most realistic condition is created when there is no awareness of an ongoing test, only then the results will be truly representative.

Audits, incident handling and complaint procedures

Typically external audits are conducted annually by the accreditation council. However, only a selection of accredited methods can be inspected with each visit in addition to any new methods, which have been submitted for accreditation by the host institute. A schedule is maintained to ensure that each method within the ISO17025 scope is externally audited at least once every 4 years. The invitation of external experts to join the audit committee is matched to the list of expertise areas and methods that are "up for inspection". The presence of external experts is essential to judge whether the methodology applied is in line with what is considered state-of-the-art in a given domain. In addition, QC specialists will inspect whether quality management, protocols, and methods are in line with the ISO 17025 framework. A good way to check whether accredited methods are correctly executed in day-to-day forensic casework is to conduct a dossier inspection. The auditors will then randomly select a number of recent case files and conduct a detailed inspection of the contents. The way the results have been obtained and how they are registered and documented in the case file should be fully compliant with the method description. If for instance a certain form has been created to document an investigation then one would expect to find such a form, fully filled out and properly signed in the dossier. If instead a simple paper note is found describing the findings, the auditor will register a non-conformity as the work in this particular case was not executed according to the applicable quality standards. In addition, also organizational and managerial aspects will be inspected to ensure that proper attention is given to quality within the forensic institute and that the overall efforts with respect to management reviews, employee training, and the physical and digital infrastructure are ISO 17025 compliant. After an audit process that typically takes 1–2 days, the overall findings are summarized by the audit team to senior management and the host institute is able to respond to the points raised and bring forward any objections. Although the audit team may use this feedback to adjust their recommendations, this is not a negotiation process. On the basis of the information provided and the conducted inspections, the auditors will provide a final report of the findings including a list of registered non-conformities (if any) that or minor or major concerns. A minor concern indicates that the deviation or omission is not critical to the outcome of the test/analysis and that the operation can be continued if the non-conformity is properly resolved in a given time period (typically 30 days). A major non-conformity indicates that the audit team has discovered a serious issue of non-ISO17025 compliance that is directly affecting the quality of the final results or the accreditation of the institute. This means that the methods and processes involved must be halted immediately and can only be continued when the issues are resolved in a satisfactory manner. Whether both minor and major non-conformities are adequately resolved needs to be decided by Accreditation Council on the basis of a report of the forensic institute. This report needs to include a thorough cause-effect analysis of the origin of the non-conformity and the actions taken to resolve the issue. The non-conformity analysis needs to consider the following four critical aspects:

- **Cause:** what is the reason for the non-conformity to exist, is it the consequence of a number of related issues, recent modifications, or of more general underlying issues such as improper training of personnel?
- **Magnitude:** has the non-conformity affected past investigations, does the non-conformity also appear in other methods and other teams? What actions have been undertaken to address the wider issue?[1]
- **Solution:** what actions have been undertaken to effectively resolve the non-conformity in a consistent manner, i.e., a structural solution in line with the quality framework as opposed to a "quick temporary fix" that is not sustainable?
- **Operationality:** does the solution work in practice, can it be shown/demonstrated that the non-conformity no longer occurs?

An accredited organization does not only rely on annual external audits to ensure that its activities are ISO 17025 compliant. Systems should be in place that ensures a continuous focus on and a dedication to quality. In addition, employees should "live and breathe" quality, be proud of the ISO 17025 status of their work, be vigilant in correcting and preventing mistakes, be creative in spotting options for improvement, and be proactive when it comes to implementation. In creating such a mindset, it is important that co-workers are rewarded (and not punished) for internally reporting incidents and non-conformities due to their own actions and

[1] The discovery of a non-conformity that has a significant impact on the outcome of a forensic investigation or raises doubts concerning the quality and trustworthiness of the findings can have very significant legal consequences when the errors apply to past casework. Forensic institutes are then obliged to contact the recipients of the former casework reports and inform them of the issue. In a worst-case scenario, *e.g.* when the forensic evidence was critical to the conviction of the suspect, this could result in a new trial, the acquittal of jailed suspects, and lawsuits of defendants for wrongful convictions.

those of their colleagues. There are two important instruments to facilitate these internal processes to maintain and raise the quality standards. First, ISO 17025 compliant organizations need to conduct regular internal audits and include a system of reporting and resolving issues in line with the external audit procedure. To make sure that the audits are meaningful and independent, QC contacts of the various teams perform internal audits of other expertise areas. This also assures that the best practices of one team are shared with and implemented in other parts of the organization. Results of the most recent internal audits need to be reported to the accreditation council as part of the annual audit. Internal audit rounds are typically organized within an institute on a half-year basis according to a schedule that assures that each accredited expertise area is internally audited at least once every 4 years. In addition, accredited institutions should also have a protocol and system for co-workers to report incidents and non-conformities. This allows decisive action to be taken after a quality incident has occurred. The employees involved can immediately register the issue and provide the details of the incident. The manner can then be resolved in a way similar to non-conformity handling in internal and external audits. It is important to electronically register the incident and provide supporting information when the issue has been resolved and the file is closed. External audit teams might request overviews of the incidents registered internally in the past year along with proposed and implemented solutions.

Commercial laboratories have a strong customer focus and aim for customer satisfaction in every aspect of the service except for the science involved and the outcomes of the test/analysis. Such customer orientation is somewhat less apparent in forensic institutes, especially as they are part of the local or federal government. In a way, this is a bit peculiar as a forensic investigation is a small part of a much bigger criminal investigation or court case. Public prosecutors, police officers, and judges can all be considered clients who make use of forensic reports to raise the quality of their work and decisions. And such clients can also discover flaws and errors when reading these reports, be dissatisfied because of delays or be disappointed with the investigation in relation to agreed actions. As part of the ISO 17025 framework, a **customer complaint protocol** must be implemented enabling complaint registration and archiving. Each registration should be timely investigated, resolved, or at least mitigated and discussed with the filer of the complaint. Partners in the criminal justice system should be made aware of the complaint procedure and should be encouraged to submit complaints if they are dissatisfied with the forensic services provided. Ideally, the process to file a complaint is straightforward and time-efficient for instance by providing a form on the institute's website.

Peer review

An essential instrument to detect and mitigate errors in forensic investigations is **peer review**. The "four eye principle" is applied by having a colleague with the required knowledge and skills thoroughly check the critical parts of a forensic investigation. Irrespective of the degree of expertise, everybody can develop "blind spots" for obvious mistakes when reviewing their own work. In ISO 17025, peer review is recommended for all critical findings, expert interpretation, the dossier, and the final report. For subjective expert assessments as seen in forensic firearms investigations, the actual comparison is repeated and individually evaluated by a fellow expert. At the NFI, the procedure even involves a third colleague when the two firearm examiners have dissenting opinions regarding the degree of similarity of the observed striation patterns. In chemical analysis, subjective interpretation of findings is usually less apparent but not completely absent, e.g., in fire debris analysis and the classification of ignitable liquids traces through expert interpretation of the HS-GC-MS data. In addition, chemical analysis typically generates a lot of instrumental data and the compiled dossiers and digital archives should be checked for completeness. For the final report, it is also important to meticulously check practical details such as correct case and evidence item numbers. In addition, the reviewer will comment on the forensic interpretation of the data and the readability of the text in the report. Aspects included in the peer-review are scientific rigor (are the statements made and conclusions drawn scientifically sound and supported by the body of knowledge in the field?) and textual clarity for lay persons (are the contents understandable for legal professionals without a scientific background, are they able to arrive at the correct conclusions independently when reading the report, will they grasp the limitation of the findings?). For testing laboratories, the final report describing the samples received, the work done, the data obtained, the data processing and interpretation, and final conclusions drawn, is the primary product that is sent to the customers. Therefore, detailed ISO 17025 standards exist for reporting and this will be discussed in more detail in the next chapter entitled **Reporting in the Criminal Justice System**. To ensure transparency initial versions of the report and comments from the peer reviewer should be archived in the case file.

Quality documentation and information management

To demonstrate that an institute is operating in line with ISO 17025 requirements, having a quality document and information management system is mandatory. To this end, forensic institutes typically use tailor-made, commercially available software packages from companies specialized in ICT solutions for ISO compliance. The ISO 17025 standard provides clear instructions on what should be included in such a system:

- **Quality Management Handbook** describing the quality policy of the institute and associated goals and implementation framework (organization, infrastructure, training, improvement programs, audits, procedures)
- **Document Management** of all procedures, methods, and other processes relevant to the ISO 17025 accreditation. In addition to actual methods, the archive should also contain previous versions providing transparency of the quality improvement process. Documents should have specific identifications codes or numbers so it can be demonstrated that work is being conducted according to the latest instructions
- **Incident Reporting** and **Improvement Project** modules to allow for continuous improvement with respect to Quality performance. Employees should be able to report quality incidents in an easy and swift manner. The system should signal open incidents and should archive, which actions have been taken to resolve the issue. Incident files should only be closed when the issue has been adequately addressed. Employees should also be encouraged to start quality-related improvement projects, the corresponding archives should demonstrate the work done and results realized before project files are archived
- **Internal Audit** module used for the planning, registration, and reporting of internal audits conducted according to described procedures. An institute-wide approach is required where QC points of contact who has been properly trained conduct audits at different teams and departments. In this respect, it is important that an auditor does not visit the same team in every audit round and does not audit work that is part of their own activities and expertise domain. Outcomes, as reported by the auditors need to be archived and subsequent improvement actions need to be agreed upon, initiated, realized, and reported. Established non-conformities during internal audits should be resolved with the same intensity as findings from the annual external committee
- **Management Reviews** to regularly evaluate the state of the quality system versus the ISO 17025 standards on the basis of goal and target realization, observations during internal and external audits, number and severity of quality issues and incidents, the success and outcomes of improvement programs and projects and the number and nature of customer complaints

Electronic Quality Management Systems should be easy to use for all employees, should include automatic backup functionality, should provide for accurate registration of quality-related activities, should automatically assign tasks and responsibilities according to defined roles, should provide alerts for open action points and associated deadlines and should allow for the easy creation of QC overviews for reviews and audits. Maybe this seems obvious to the reader but actually striking the right balance between ease-of-use and automation on the one hand and making sure you are meeting all the very strict ISO 17025 standards on the other can actually be quite challenging!

9.4 Quality through forensic networks, the importance of ENFSI and OSAC

In the process of getting a method accredited, specific challenges might exist for the different areas of forensic expertise. For each method, either instrumental or based on human assessment, the experts need to "translate" and implement the ISO 17025 standards. This should be based on the outcomes of the validation study and the construction of a robust quality scheme for routine operation as has been discussed previously. However, there are several ways to do this, there is no one single approach that is mandatory. Auditors will "only" check whether all quality requirements are met and implemented correctly for a methodology that can be considered state-of-the-art. From a legal perspective, it can be valuable for forensic laboratories to share best practices and follow a common approach when "translating" quality standards for a given type of forensic investigation. An example of such a shared approach was already given in **Chapter 4—Qualitative Analysis and the Selectivity Dilemma** when discussing the internationally recognized recommendations of the **Scientific Working Group for the Analysis of Seized Drugs** (**SWGDRUG**). Interestingly, these recommendations take into account that laboratories can have different preferences and budgetary means with respect to the methods used by introducing the scheme of Category A, B and C techniques. But by providing clear guidelines with respect to the combination of methods for robust identification of drugs of abuse, a laboratory can demonstrate to the courts and the national accreditation council that their instrumental choices meet international standards of forensic illicit drug experts. This demonstrates the importance of international networks of forensic institutes and experts in the realization of a generally acknowledged quality framework for forensic methods. Two organizations of specific importance in this respect are the **European Network for Forensic Science Institutes (ENFSI)** and the **Organization of Scientific Area Committees for Forensic Science (OSAC)**.

ENFSI (www.enfsi.eu) was officially established in 1995 and currently (June 2021) has 73 members from 39 European countries. Members, mostly national forensic institutes and laboratories, exchange information, share best practices, and conduct quality improvement programs to raise the quality of forensic science throughout Europe. ENFSI also hosts the European Academy for Forensic Sciences (EAFS), the major European conference on forensic science that is organized every 3 years. In addition, ENFSI facilitates information sharing and collaboration at the forensic expert level through 17 **Scientific Working Groups (SWGs)**. SWGs organize regular meetings where forensic experts from different European countries meet and discuss their field. Especially for relatively small expert teams handling a limited caseload, having good international contacts is essential for maintaining the knowledge base. Experts participating in an SWG initiate quality improvement projects such as the construction of European guidelines or the organization of an international ring test. Important European forensic quality documents can be found on the ENFSI website (see Further Reading section for details) including Best Practice Manuals for specific forensic expertise areas and Forensic Guidelines that are of a more general nature.

The USA counterpart OSAC has been initiated in 2014 and is coordinated by **NIST**, the **National Institute for Science and Technology**. This world-renowned US metrology institute is part of the Department of Commerce and produces a broad range of certified standards including for the forensic science community. In addition, NIST has been a critical force in improving scientific and quality standards in forensic investigations in the USA after the publication of two very critical reports on the limited scientific foundation and poor quality of frequently used forensic casework methods resulting in serious miscarriages of justice in the United States. In 2009 the **National Research Council (NRC)** published a much-cited report entitled *Strengthening Forensic Science in the United States: A Path Forward*. In 2016, this was followed by a report of the **President's Council of Advisors on Science and Technology (PCAST)** to president Obama entitled *Forensic Science in Criminal Courts: Ensuring Scientific Validity of*

Feature-Comparison Methods. The goal of OSAC is to *"strengthen the nation's use of forensic science by facilitating the development of technically sound standards and guidelines and encouraging their use throughout the forensic science community"* (source: https://www.nist.gov/osac). The OSAC organization consists of a Forensic Science Standards Board, seven scientific areas committees, and 22 expertise area subcommittees (June 2021). These committees have produced well over 50 standards that can be found in the OSAC registry. According to OSAC these forensic standards "*define minimum requirements, best practices, standard protocols, and other guidance to help ensure that the results of forensic analysis are valid, reliable and reproducible*" (source: https://www.nist.gov/osac/osac-registry).

The work done within ENFSI and OSAC is not only extremely important for raising the quality of forensic casework in criminal investigations around the world, but it is also providing a great source of information for forensic trainees and experts that want to implement a quality framework in their laboratory. In the next chapter, we will discuss the specific aspects of the forensic report, the final product of a forensic investigation.

Further reading

Code of Conduct, NRGD, Netherlands Register of Court Experts, 2017 downloadable from. https://english.nrgd.nl/.

ENFSI. ENFSI Codes, Guidelines and Best Practice Manuals. https://enfsi.eu/documents/.

I133. ISO/IEC 17020 Scope of the Netherlands Forensic Institute, the Latest Version of the Scope (In English) for Crime Scene Investigation. https://www.rva.nl/en/scopes/details/I333.

ISO/IEC 17025, 2017. General Requirements for the Competence of Testing and Calibration Laboratories, ISO a free information document can be downloaded atThe latest version of the actual standard is not freely available but can be ordered at ISO. https://www.iso.org/publication/PUB100424.html.

L146. ISO/IEC 17025 Scope of the Netherlands Forensic Institute, the Latest Version of the Scope (In English) for Crime Scene Investigation. https://www.rva.nl/en/scopes/details/L146.

Modules in a Forensic Science Process, ILAC-G19:08/2014, 2014. ILAC freely downloadable at. https://ilac.org/latest_ilac_news/ilac-g19082014-published/.

National Research Council, 2009. Strengthening Forensic Science in the United States: A Path Forward. The National Academies Press, Washington, DC. https://doi.org/10.17226/12589.

OSAC. OSAC Documents under the section Governing Documents and OSAC Registry. https://www.nist.gov/osac.

President's Council of Advisors on Science and Technology (PCAST), 2016. Report to the President, Forensic Science in Criminal Courts: Ensuring Scientific Validity of Feature-Comparison Methods. https://obamawhitehouse.archives.gov/sites/default/files/microsites/ostp/PCAST/pcast_forensic_science_report_final.pdf.

CHAPTER 10

Reporting in the criminal justice system

Contents

10.1 What will you learn?

With the quality instruments discussed in the previous chapter in place, a judge in a criminal case can now be confident that the assigned forensic expert indeed has the required skills, knowledge and expertise. In addition, the legal professionals can be ensured that the forensic findings meet the highest quality criteria due to the ISO 17025 accreditation of the applied methodology. Please note that this not mean that the forensic results are always free of mistakes, even in the most strictest of quality regimes errors can and will occur. However, this quality framework provides clarity of the forensic investigation and robustness of the results as several quality "checks and balances" are part of the protocol. Overall, this means that the generated forensic information is trustworthy as it has been gathered by genuine experts using state-of-the-art, internationally acknowledged methods under very strict quality control. As long as there no specific legal

Chemical Analysis for Forensic Evidence
ISBN 978-0-12-820715-4
https://doi.org/10.1016/B978-0-12-820715-4.00014-6

issues or other peculiarities associated with the presented work, the judges could safely and swiftly deem the presented forensic evidence admissible in the case at hand. This prevents lengthy discussions on the admissibility of the forensic evidence and allows the legal professionals to focus on the matter at hand, i.e., whether or not it has been proven beyond reasonable doubt that the suspect committed the criminal acts as described in the charges brought forward by the public prosecution office. However, to make optimal use of the information provided by the forensic expert when addressing this **ultimate issue**, it is important that the legal professionals and triers of fact understand the reports and testimonies of the experts. This is not a straightforward matter at all because it requires scientists to communicate in a way that remains scientifically sound, but at the same time is understandable for lay persons with a legal rather than a scientific background. As we will see in this chapter, considerable efforts are needed from forensic institutes to ensure that the work of their experts it correctly interpreted and valued by law enforcement and in the court room. In this respect, it is important to appreciate that a forensic investigation that yields critical information is literally worthless if the legal professionals and triers of fact do not understand anything the expert has written in the report or stated in court during the testimony. Much of what will be discussed in this chapter is of a generic nature, i.e., is broadly applicable irrespective of the forensic expertise area. However, there are some specific aspects to reporting when considering forensic analytical chemistry and of course this will be discussed in detail.

After studying this chapter, readers are able to

- list the ISO 17025 quality standards for reporting
- understand how forensic institutes raise the awareness and understanding of forensic expertise areas and associated reports with their partners in the criminal justice system
- explain the differences between a forensic scientific paper and a case work report

- construct a verbal conclusion to indicate the strength of the evidence when comparing two hypotheses according to the Bayesian framework
- explain the Prosector's Fallacy and the Defense Fallacy, two frequently occurring evidence interpretation errors in the court room
- understand why in forensic analytical chemistry different ways of reporting exist including the use of absolute statements and frequentist statistics in addition to the Bayesian framework
- write/construct a realistic forensic case work report

10.2 ISO 17025 reporting standards

A report, either sent electronically or as a paper document, is the primary way of communicating the results of a test laboratory to its customers and thus it comes as no surprise that the ISO 17025 quality standard contains various reporting requirements. Although nowadays reporting can be done in various ways to suit the needs of these customers, it is important that the reports meet the desired quality standards as part of the overall trustworthiness of the presented results. A simple example is that a report should be sent in a certain format that does not allow an individual to easily change the data undetected. An economic motive to "tweak the numbers" could exist when for instance a customer receives analytical data that indicates that a certain production batch does not meet the necessary specifications. In the end the test laboratory, the company receiving the analysis and society in general all lose when such foul play occurs. One can imagine that this will result in a criminal investigation when as a result a health or safety risk has occurred. Additionally, the underlying raw data and lab records of the analysis should always be easily accessible and traceable to the results described in a report. Such a paper and data trail is very important and should be checked during internal and external audits. This prevents mischievous actions where laboratories or its employees deliberately create fake reports for which no actual laboratory analyses have been conducted. Luckily, such deliberate malicious acts are very rare, but they have occurred in the past. Economic motives could exist for instance to provide companies with false reports that are then used to mislead the authorities with respect to the content of products such as highly toxic waste streams. Examples unfortunately also exist in the forensic domain, not so much for financial reasons but as a final resolution to deal with high caseloads and related personal stress or when forensic experts get too involved in the work of law enforcement and desperately want to help "catch the bad guys". In such a setting, it has occurred in the past that forensic lab results could be ordered on request by police officers to help them "solve the case". In the USA, this has resulted in several convictions of forensic lab workers and numerous miscarriages of justice. It is clear that such actions go against any forensic code of conduct and bring tremendous damage to the suspects, victims and societal trust in the criminal justice system.

The most important ISO 17025 reporting requirements are summarized below:

- Any report must be checked and approved through a peer-review process prior to release
- The results described in the report should be clear, accurate, unambiguous and objective
- The report should contain all information as agreed with the customer, as required by the accredited method and as needed for interpretation and final conclusions
- Each report should in principle at least contain
  - a title
  - the name and address of the test laboratory
  - the laboratory location where the analyses were conducted
  - the name and address of the customer (the requesting organization)
  - information provided by the customer concerning the request
  - a unique indicator/reference of the method(s) applied
  - a unique indicator/reference of the material(s) received
  - a short description of the materials received and their condition
  - the date of material/request receival
  - the date/period of laboratory analysis
  - the results of the analysis
  - if relevant: laboratory conditions and instrumental parameters during the analysis (when such information is not available in the method description)

- if applicable: the associated measurement uncertainty (when this is relevant for the interpretation of the results and key conclusions of the investigation)
- if applicable: a reference to a sampling plan/strategy or a sample preparation procedure
- if applicable: deviations from the regular, accredited methodology due to specific case circumstances
- the date of the official release of the report
- the names of the experts involved in the investigation
- the name and signature of the employee responsible for the report and the results on behalf of the test laboratory
- if applicable: statements regarding corrected versions
- if applicable: general disclaimers or confidentially statements.

In addition, the report should clearly distinguish what information has been retrieved from the customer. These parts should be directly copied and fall under the responsibility of the requesting body. The test laboratory is responsible for all other information in the report including the results of the analysis/test and the final conclusions. If a report contains expert interpretation and opinions (subjective statements), this must be clearly labeled as such and must be separately reported from the test/analysis results (objective data). ISO 17025 guidelines provide quite some flexibility with respect to report format allowing for electronic and paper versions of either extensive reports, simplified shortened reports or calibration certificates as long as this is agreed with the customer and key quality information is provided or easily retrievable.

In a forensic setting, requests for forensic investigation are typically received from the police under the leadership of the public prosecution office. However, requests can also come from the courts and in rare events directly from the defense council. In the legal system in the Netherlands, the NFI cannot directly work for defense lawyers because NFI investigations should always be used in a court of law and this could go against the obligation of the defense council to always serve the best interest of their clients. This means that exculpatory forensic evidence will be used by the defense lawyer to support the client but that unexpected incriminating findings will not be admitted as evidence in court. To circumvent this issue, the defense council typically approaches the public prosecutor or the court to request additional forensic investigations to be conducted by the NFI. This ensures that the new forensic results will be used in the case at hand but also means that the defense lawyer is taking a risk that this course of action could turn out to be detrimental to the client. If such a risk is unacceptable, the defense needs to obtain services from a commercial/independent forensic service provider. Another important principle in the forensic domain is transparency, the report and all associated information should be made available to all legal parties including the defense council. A typical report will not contain all the methodological details, and it will rather provide the associated codes under which these methods have been documented in the quality information management system. This means that at the request of the legal parties, details of these methods must be made available. Especially for the defense council such information is very important to assess and scrutinize the forensic work. Irrespective whether such criticism is scientifically fair, the obligation of the defense council is to raise doubts concerning the robustness and applicability of the forensic methodology. However, the legal demand for transparency comes at a considerable cost for the efficacy of forensic investigations. As details of the forensic methods also become known to the suspects, such information quickly spreads in the criminal community. As a result, perpetrators might modify their ways of working to nullify the forensic investigation. For this reason, forensic institutes together with law enforcement and the public prosecution office sometimes request for critical methodological details to be left out of the report and supporting documentation. Sometimes some information is then provided by expert testimony in closed chambers but nonetheless these two opposing interests—a fair trial versus a successful fight against (organized) crime—cause a strong judicial dilemma and affect forensic reporting.

10.3 Ways to raise forensic understanding in the criminal justice system

In the Netherlands the NFI, an agency of the Ministry of Justice and Security, has three official tasks as described in the "Staatscourant" (an official publication format through which the Dutch government announces new laws and other governmental affairs). Because no specific law exists in the Netherlands concerning forensic investigations, official matters regarding the national forensic institute are periodically announced through this format. In the most recent version from 2012, the main tasks of the NFI are defined as (translated by the author):

1. To conduct independent forensic case work of a technical, medical-biological and scientific nature and to report the findings thereof
2. To develop and implement new forensic investigative methods and techniques to increase forensic scientific knowledge
3. To be an (inter)national center of forensic knowledge and expertise

The fact that the NFI is such an internationally acknowledged and recognized forensic institute is directly related to the governmental and societal recognition that the institute should not only do forensic case work, but can also invest in future methods by maintaining a sizable forensic science innovation program in collaboration with academic institutions and commercial companies. The importance of ongoing forensic innovation and **Research and Development (R&D)** programs to improve and create new forensic methods for case work will be discussed in more detail in **Chapter 11—Innovating Forensic Analytical Chemistry**. Additionally, it is acknowledged that the NFI also has a third primary task to be a center of forensic knowledge and expertise and that as such the institute is expected to share its unique forensic experience and teach forensic science both on a national and international level. On an international level, this has resulted in the NFI providing advice and support to developing countries that are seeking to strengthen their rule of law by investing in forensic facilities and capabilities. On a national level, this translates to offering tailor-made forensic courses to partners in the criminal justice system. By investing in the forensic knowledge base of its primary customers, the NFI raises the impact and value of its reports in court. If police officers, public prosecutors, lawyers, and judges clearly understand the options and limitations of forensic methods, they will be able to correctly interpret and use the forensic findings in a given case. This effort in teaching forensic science and transferring forensic knowledge is considerable and an extensive course list is offered within the Dutch criminal justice system as published on the website (https://www.forensischinstituut.nl/trainingen-cursussen, in Dutch). Some of these courses are also provided internationally upon special request and after approval of the Dutch authorities (https://www.forensicinstitute.nl/training-and-expertise/training-and-courses).

The international forensic courses offered by the Netherlands Forensic Institute in 2021 as indicated on their English website version (https://www.forensicinstitute.nl/)

Although these teaching efforts are highly valued, raising the impact and understanding of forensic reports by educating the reader is frankly a somewhat indirect, cumbersome and not the most straight forward solution. It puts the burden with the customer to invest time and effort to learn forensic science, which is neither their main priority nor task. Furthermore, not everybody will be able or willing to follow courses in forensic science and hence investing in such programs will in the end only be partially effective. Therefore, it makes much more sense to invest in increasing the readability of forensic reports, making them fit-for-purpose, and providing general information regarding the methodology applied in an easily accessible manner. This puts the burden with the forensic institute and its experts to invest in reporting and science communication. This includes training experts how to communicate science to non-scientists (educating the writer instead of the reader!) and initiating discussions with the partners in the criminal justice system to better understand their perception and use of the forensic reports and to valorize this feedback to improve readability and efficacy. Of course, this needs to be established without any sacrifices to the scientific principles, soundness of the forensic interpretation, and validity of the conclusions. From the perspective of a police officer, public prosecutor or judge, a question whether the suspect fired a gun and killed the victim should ideally be answered with a simple yes or no. Having experts formulating a quite complex answer that only partially addresses the main question can easily lead to disbelief, dissatisfaction and criticism that the report is needlessly complex. However, it is often scientifically not possible to provide absolute answers and experts need to stay within the boundaries of their knowledge area. This is something that needs to be explained clearly but also requires understanding, appreciation, and acceptance of the basic scientific and quality principles by the receiving party. They should be adamant to uphold and defend these ground rules that make forensic science such a valuable, objective instrument and resist the pressure to "close or solve the case".

Criminal investigations tend to lead to substantial archives. For big cases it is not uncommon for legal professionals to carry numerous cardboard boxes completely filled with documentation into the court room. In such a setting of "heavy paper load" it is common sense to keep the forensic reports as short and concise as possible. Writing a 50+ page exhaustive forensic case report might not be read in detail and therefore important considerations might get lost when readers rapidly scan documents and only pay attention to the conclusions "on the final page". Therefore, applied methods are simply referred to by providing the document numbers and for high volume case work special arrangements are made for condensed reporting. These shortened report forms are accepted in the ISO 17025 framework and can be very effective when no extensive forensic interpretation is required, *e.g.*, for the chemical identification of drugs of abuse or forensic toxicological whole blood screening against legal DUI limits. In combination with electronic signing and digital reports sent over secured network, large case work volumes can be processed without serious backlogs and with lead times in hours instead of weeks. In such optimized processes the sample logistics (i.e., getting the samples to the laboratory) typically becomes the limiting step. Acknowledging that recipients of such reports should have a good understanding of the forensic expertise area and the methodologies applied, an expert team should also provide more generic information. For this reason, the NFI develops so-called **Technical Annexes** (in Dutch "Vakbijlages") that are sent together with the forensic case reports. These annexes provide information on the type of investigations conducted in a given forensic expertise area, and on the instruments and methods used, explain what type of questions can be addressed (and what are the limitations) and provide instructions how to read the report and interpret the results and conclusions. Of course, such information can also be provided using more modern ways of communication, for instance through instruction videos shot in the laboratory showing the actual instrumentation and experts demonstrating the investigation. Although such videos can also be used to educate and entertain the general public, forensic institutes have to take care not to disclose information that can be misused by criminals to obstruct the forensic investigation.

Vakbijlage

Identificatie van verdovende middelen

Inhoudsopgave

1. De vakbijlage algemeen

Het Nederlands Forensisch Instituut (NFI) kent een groot aantal typen onderzoek. Eén daarvan is de identificatie van verdovende middelen. Onderzoeksrapporten van het NFI worden vaak vergezeld van een vakbijlage zoals deze, hetdient als toelichting op het onderzoek en heeft een zuiver informatief karakter. De vakbijlage geeft weer volgens welke methoden en met welke technieken en hulpmiddelen een dergelijk onderzoek in het algemeen plaatsvindt, inclusief de manier van concluderen en rapporteren. De informatie die van toepassing is op een specifieke zaak staat altijd in het onderzoeksrapport vermeld. In rapporten die in het kader van identificatie worden opgesteld, is een verwijzing opgenomen naar de NFI-website waar deze vakbijlage is te vinden. Aan het einde van deze vakbijlage zijn een verklarende woordenlijst en een literatuurverwijzing opgenomen.

Front page of the Technical Annex (in Dutch) describing the forensic identification of drugs of abuse at the Netherlands Forensic Institute (downloadable from https://www.forensischinstituut.nl/over-het-nfi/vakbijlagen-en-informatiebladen)

10.4 Bayes, verbal conclusions, "popular" fallacies, and the hierarchy of propositions

In the next section we will discuss the forensic report and show a possible template for a report that is logically organized and therefore easy to read, but still comprehensive with respect to the case context, evidence material, results, interpretation and conclusions. The conclusions of the expert as an answer to the questions posed, form the most vital part of any forensic report and in the end, determines the added value to the criminal investigation and court proceedings. In **Chapter 6—Chemical Profiling, Databases, and Evidential Value**—the Bayesian framework was introduced as a scientifically robust way to express the evidential value in chemical profiling studies. However, this framework has a much wider impact when considering the role of the forensic expert in the criminal justice system and the formulation of conclusions from case work. Readers interested in a more detailed discussion of the role of the forensic expert and forensic science in the criminal justice system is referred to the book *Interpreting Evidence: Evaluating Forensic Science in the Courtroom* from Robertson, Vignaux and Berger (details are given in the Further Reading section). Given its importance, the "**Formula of Bayes**" in its full and condensed form is provided once more (the author feels that forensic scientists and experts should know this formula by heart):

$$\frac{P(H_p|E)}{P(H_d|E)} = \frac{P(E|H_p)}{P(E|H_d)} \cdot \frac{P(H_p)}{P(H_d)}$$

$$\alpha_{posterior} = LR \cdot \alpha_{prior}$$

Verbally, this indicates that the **posterior odds** ($\alpha_{posterior}$) equals the **likelihood ratio** (the ***LR*** **value**) multiplied by the **prior odds** (α_{prior}). The posterior odds are given by the ratio of the probability of the Prosecution hypothesis - H_d - being true/correct and the probability of the Defense hypothesis - H_d - being true/correct given all information including the forensic evidence. The prior odds are calculated through the ratio of the probabilities of the Prosecution - H_d - and Defense - H_d - hypotheses at the start of the forensic investigation when the forensic evidence is not available yet. The *LR* value is reported by the forensic expert and expresses the ratio of the probability of observing the evidence (not the hypotheses, it is important to understand and appreciate this subtle but crucial difference!) when considering the Prosecution and Defense hypotheses to be true, respectively. In this context, the Prosecution hypothesis is based on a scenario in which the suspect is committing the criminal activity of which he/she is charged in the indictment. The defense lawyer typically formulates an alternative scenario leading to a Defense hypothesis that exonerates the suspect. The forensic evidence, as expressed through the *LR* value is based on objective findings as obtained and interpreted by the forensic expert (who is neutral with respect to guilt or innocence), indicates whether the forensic investigation is more probable when considering H_p or H_d to be true. As discussed in **Chapter 6**, an *LR* value in the range of 0−1 represents exonerating evidence, i.e., forensic findings that support the Defense proposition, whereas *LR* values exceeding one indicate incriminating evidence as the findings become more probably when considering the Prosecution proposition. When the *LR* value is exactly 1, the evidence is equally probable under either proposition and the forensic investigation does not provide any assistance to the triers of fact.

- The Bayesian framework not only enables the determination of the evidential strength in light of the hypotheses considered, it also illustrates the role, responsibilities and professional boundaries of the forensic expert and provides a guideline for reporting the findings of a forensic investigation:

- The forensic expert is not responsible for the hypotheses, the propositions are derived from the scenarios and questions considered by the public prosecutor and the defense council. The forensic expert does often advise with respect to the formulation of the hypotheses to allow for a meaningful contribution of the forensic investigation
- The forensic expert should only report on the probability of the evidence given the hypotheses (the *LR* value). Because the expert is not aware of all the information and evidence in a case, he/she is not able to assess the prior odds and with that the posterior odds. It is also not the task of the expert to determine posterior probabilities, this should be left to the triers of fact (judges and juries)
- The evidential strength (*LR* value) depends on the hypothesis pair considered, the same findings can lead to very substantial differences in the *LR* value for different sets of hypotheses. When hypotheses are not properly defined and considered, forensic findings cannot properly "land" (i.e., be evaluated). If the defense lawyer refuses to provide an explanation, the most logical defense hypothesis has to be assumed in order to assess the evidential strength
- An absolute statement on the basis of a forensic investigation is in principle only possible when $LR = 0$ or in other words when the probability of the findings is zero when considering the hypothesis of the public prosecutor. In this situation, the posterior odds are also zero irrespective of the prior odds and the probability of the evidence under the defense hypothesis (assuming that this probability has a non-zero value). Very high *LR* values, indicating strongly incriminating evidence, lead to near unity posterior probabilities for the public prosecution hypothesis irrespective of the prior odds. However, even for astronomically high numbers this probability will never be exactly one and hence no absolute statement can be provided by the expert

It should be noted that these principles follow from logic and sound criminalistic reasoning and hence cannot be "sacrificed" for the sake of lay understanding. Although formulations can be adapted to improve readability, experts often cannot provide an absolute answer to a question in order to maintain scientific validity of the forensic report.

In **Chapter 6—Chemical Profiling, Databases, and Evidential Value**—it was shown how through a score-based, Euclidian distance model and within and between score distributions obtained from reference samples, *LR* values could be estimated. However, reporting *LR* values, i.e., a quantitatively expressing the evidential strength, in casework is relatively rare in forensic chemistry. First of all, it requires a thorough validation including the calibration of the *LR* value to bring numbers in line with the available (reference) data (a discussion of *LR* calibration and validation is beyond the scope of this book but some references have been included in the Further Reading section of **Chapter 6** for the interested reader). Additionally, for the type of material and question considered, the required reference data might not be available or the reference collection is not substantial enough. Finally, the nature of the findings might not easily be modeled and translated in probability density functions. This applies for instance when forensic experts visually perform comparisons of complex patterns or chromatograms and arrive at subjective conclusions with respect to the degree of similarity. As indicated in the ENFSI guideline for evaluative reporting in forensic science (downloadable from the ENFSI website, for details check Further Reading section) this does not mean that forensic comparisons cannot be used or reported when accurate LR numbers cannot be provided. Such a conservative approach would actually mean that many valuable forensic investigations should not be conducted because of their semiqualitative or partly subjective nature. As the ENFSI guideline states, it is, however, important that the experts are able to estimate the probabilities of the evidence under the considered hypotheses to arrive at a meaningful and sound assessment of the evidential strength ("*The conclusion should express the degree of support provided by the forensic findings for one proposition versus the specified alternative(s) depending upon the magnitude of the likelihood ratio (LR)*"). This should either be done on the basis of limited reference data available (sometimes generated as part of the investigation to create reference data for a specific case) or through so-called **expert elicitation** in which the knowledge and experience of the expert is translated in probability estimations. This is a complex process as estimating experience-based probabilities is far from easy, but one can imagine how a random match probability can be estimated on the basis of how frequently an expert encounters a certain feature or pattern during case work. It is clear that this does not justify the reporting of a single *LR* value, rather, this will lead to an *LR* range as an estimation of the evidential strength. As one should never overestimate the evidential strength for incriminating evidence (*i.e.*, *LR* values greater than 1), the lower limit of such a range should then be considered by the triers of fact. The readability of the forensic report is usually improved if such *LR* estimations are expressed by means of verbal statements. The ENFSI guideline states that "*a likelihood ratio may be expressed by a verbal equivalent according to a scale of conclusions* ....". Indeed many forensic institutes apply a verbal framework linking terms or codes to various degrees of evidential strength. An example of such a verbal scale is illustrated below:

The findings of the investigation are

- ***approximately equally probably***
- ***slightly more probably***
- ***more probable***
- ***appreciably more probable***
- ***much more probable***
- ***far more probable***
- ***extremely more probable***

if hypothesis one is correct, than if hypothesis two is correct.

where $H_1 = H_p$ for incriminating evidence, $H_1 = H_d$ for exonerating evidence and the following (log) *LR* ranges are linked to the verbal terms:

Verbal term	***LR* range**	**Log(*LR*) range**
Approximately equally probably	1–2	0
Slightly more probably	2–10	0–1
More probable	10–100	1–2
Appreciably more probable	100–1.000	2–3
Much more probable	1.000–10.000	3–4
Far more probable	10.000–1.000.000	4–6
Extremely more probable	>1.000.000	6+

Why is it important to link *LR* ranges to the verbal terms?

When the verbal terms are not "anchored", their interpretation and use will diverge. One individual or expert might use and interpret the term "far more probable" differently than another expert. Consistency is especially challenging between different expertise areas and this would result in forensic reports where the same verbal term would actually reflect a different strength of evidence for various types of investigation. This is highly undesirable and would lead to confusion with the triers of fact.

In addition, *LR* range calibration, forces the expert to properly substantiate the conclusions. When choosing a term of the verbal scale the expert must be able to convincingly demonstrate why a given *LR* range applies. In the evaluation process the expert must first estimate the evidential strength and then use the correct verbal term in the conclusion. Reversing this order would basically indicate that the applied methodology is lacking a sound scientific basis and sufficient reference data.

The reporting guidelines detailed above are scientifically very robust and ensure that even in the absence of sufficient reference data forensic experts can evaluate the findings and estimate the evidential strength in a correct manner. However, the phrasing is often perceived as complex and does not automatically enhance the readability especially for legal professionals with limited scientific background knowledge. For this reason, other phrasing is sometimes preferred such as the relatively popular "*support formulation*". By expressing that the forensic evidence supports hypothesis *X*, the requirement that the forensic expert only makes statements concerning the probability of the evidence is still maintained. Typically, such terminology is perceived as easier to read and grasp, however it can also lead to scientific and criminalistic issues as is illustrated below.

Consider the following statement in a forensic report:

"*The findings of the forensic investigation strongly support the hypothesis that the TNT retrieved from the IED originate from the TNT batch found at the home of the suspect*".

What is the main scientific concern associated with such phrasing?

For a forensic expert to be able to assess the evidential strength, a pair of hypotheses needs to be considered. As discussed previously, the *LR* value is the ratio of the probability of the evidence under H_p and H_d, respectively. The same results can lead to significantly different *LR* values for different defense scenarios (different H_d). Estimating the evidential strength requires two mutual exclusive propositions. A support formulation can be used but it must then also clearly specify the alternative hypothesis.

The correct use of the support statement is given in the ENFSI guideline for evaluative reporting in forensic science:

The findings of the investigation

- ***do not support*** *one proposition over the other*
- *provide* ***weak support*** *for the first proposition relative to the alternative*
- *provide* **moderate support** *for the first proposition rather than the alternative*
- *provide* **moderately strong support** *for the first proposition rather than the alternative*
- *provide* ***strong support*** *for the first proposition rather than the alternative*
- *provide* ***very strong support*** *for the first proposition rather than the alternative*
- *provide* ***extremely strong support*** *for the first proposition rather than the alternative*

For these verbal expressions the same *LR*/Log(*LR*) ranges apply as for the respective probability-based statements listed in the table above. Although lay readers might prefer the expression of evidential strength in terms of support of one hypothesis over the other, a correct understanding of the nature and meaning of logically correct expert conclusions remains difficult, not only for legal professionals but often even so for the forensic scientists themselves. This is generally attributed to the fact that the human mind has difficulty to apply probabilistic principles to matters that lie in the past, i.e., that have already occurred. We are more used to making decisions by estimating the probability that a certain event will happen in the future. Consequently, scientifically sound expert conclusions are often misinterpreted and incorrectly applied and explained in the courtroom. Two very infamous fallacies are the **Prosecutor's Fallacy** and the **Defense Fallacy**, and they will be discussed in more detail next. The Prosecutor's

Fallacy is also known as the error of "**transposing the conditional**" indicating that the probability of the evidence given the hypothesis is misinterpreted as the probability of the hypothesis given the evidence. In case of a DNA match this means that the random match probability is explained as the probability that the suspect is not the donor of the trace:

A partial DNA profile retrieved from a biological touch trace at the crime scene matches with the reference profile of the suspect in the case. The *LR* value of the partial profile match is roughly 1000. The forensic DNA expert reports a random match probability of 0.001 through the following statement: "*The probability that a randomly selected male individual not related to the suspect will yield a match with the partial DNA profile is one in a 1000*". After reading the forensic expert report, the public prosecutor concludes and states in court that the probability that the suspect is the actual donor of the biological touch trace from the crime scene is 99.9% (1 minus the random match probability times 100).

Why is this an incorrect interpretation of the forensic evidence?

What condition needs to apply for this prosecutor statement to be correct?

(Tip: this can be derived from the formula of Bayes).

The probability that the suspect is the donor of the biological touch trace is also affected by the prior odds. The public prosecutor is mistaking the *LR* value for the posterior odds. When we look at the formula of Bayes, it is easy to see that the likelihood ratio only equals the posterior odds when the prior odds are 1. This condition applies for equal prior probabilities of both propositions, or in other words that prior to the forensic investigation both propositions are equally probably. This "coin flip" situation is however far from obvious.

To explain the Defense Fallacy, we will stick with the same case example now switching to the view of the Defense council:

After reading the forensic expert report, the defense lawyer reasons as follows: "*The random match probability is 0.001. Of the Dutch population of approximately 17 million people roughly 8,5 million are male. This means that on average in addition to my client another 8499 Dutch males will yield a match. The probability that my client is the donor of the biological touch trace is therefore not 99.9% but rather 0.01% (1 divided through the number of potential donors times 100)*"

Why is this also an incorrect interpretation of the forensic evidence?

The defense lawyer is disregarding the fact that the suspect has been accused for a reason. The context and information concerning the case has led to a suspicion against his client. However, the defense council is arguing that every male in the Netherlands should be considered as the potential perpetrator. However, this includes many persons that could simply not have committed the crime because of their age (young kids and elderly people), condition (e.g., disabled people) or because they have a very solid alibi.

For the final part of this section, we will focus our attention on the hypothesis. As the hypothesis pair considered can strongly affect the evidential value, correctly defining them is of great importance. However, formulating appropriate hypotheses on the basis of the questions in the criminal investigation and the position of the public prosecutor and the defense council is certainly not straight-forward and requires experience and criminalistic skills. In the initial phase of the investigation, a suspect might not be available and hence a generic alternative hypothesis will have to be defined that is logical and favors a future suspect.

Consider the following four hypotheses pairs defined for a forensic glass comparison of glass particles retrieved from a suspect's clothing and reference glass from a window that was smashed in a burglary case:

Set 1:

H_p (prosecution) : *"The glass particles on the clothing of the suspect originate from the smashed window at the crime scene"*

H_d (defense) : *"The glass particles on the clothing of the suspect do not originate from the smashed window at the crime scene"*

Set 2:

H_p (prosecution) : *"The glass particles on the clothing of the suspect originate from the smashed window at the crime scene"*

H_d (defense) : *"The glass particles on the clothing of the suspect originate from another glass source"*

Set 3:

H_p (prosecution) : *"The glass particles on the clothing of the suspect originate from the crime scene"*

H_d (defense) : *"The glass particles on the clothing of the suspect do not originate from the smashed window at the crime scene"*

Set 4:

H_p (prosecution) : *"The elemental profile of the glass particles on the clothing of the suspect matches the profile of the smashed window"*

H_d (defense) : *"The elemental profile of the glass particles on the clothing of the suspect does not match the profile of the smashed window"*

Which of the hypotheses pairs do you prefer and why?

What concerns do you have with the other three hypotheses pairs?

The hypotheses given in set two are to be preferred:

H_p (prosecution) : "*The glass particles on the clothing of the suspect originate from the smashed window at the crime scene*"
H_d (defense) : "*The glass particles on the clothing of the suspect originate from another glass source*"

These hypotheses are mutually exclusive (they cannot both be true) and in this case also exhaustive ($H_p + H_d = 1$, they cannot both be untrue). Furthermore, the Defense hypothesis is specific, it is generally not advised to formulate the alternative simply as the inverse of the incriminating hypothesis, *e.g.*, "originate" versus "does not originate", as is the case in hypothesis set 1. This is very difficult to test. The issue with set three is that the hypothesis of the prosecution is not source specific, multiple glass sources can exist at the crimes scene. The hypothesis pair is also not mutually exclusive, both will be true if the glass fragments originate from another glass source from the crime scene, *e.g.*, another window or a drinking glass. The hypotheses described in set four are incorrect because they reflect findings of the forensic investigation. The evidence should not be mixed with the propositions because otherwise the probability of the findings given the hypotheses cannot be established.

When considering the hypothesis pairs above it is clear that they are affected/limited by the forensic methodology applied. The main question of the team conducting the criminal investigation would normally not relate to miniscule glass particles entrapped in the clothing of the suspect. Having glass particles on your clothing from a window smashed on a crime scene in itself does not constitute a crime. It could be the result of a criminal activity and although this seems logical other plausible explanations might exist. The main issue in the case is whether the apprehended suspect committed the burglary by smashing the window, entering the property and stealing a number of valuable goods. Criminal activities can often not be directly assessed by the forensic experts as results in an investigation are at so-called **source level**. At source level the probability of the evidence is considered for hypotheses relating to the origin of the (trace) material, i.e., source level hypotheses. Throughout the previous chapters we have seen several examples of such source-level investigations, *e.g.*, is the biological trace originating from male donor X or is the explosive material retrieved from the IED originating from the batch seized at the house of suspect Y? Whether such a link (in case the probability of the evidence is much higher when considering the prosecution hypothesis) is strongly incriminating depends on the context of the case and other evidence. The triers of fact must very carefully consider these case conditions and should not automatically assume that strongly incrimination source evidence is equally incriminating at **activity level** or **offense level** (where the offense level represents a criminal activity). In the ENFSI guideline for evaluative reporting in forensic science, it is recommended that experts specifically mention this when reporting forensic evidence at source level. Three very relevant studies were published in the period 1998–2000 by forensic scientists of the former Forensic Science Service (FSS) of the UK on **case assessment and interpretation** (CAI) and what was introduced as "**the hierarchy of propositions**" (full references are given in the Further Reading section). While propositions at offense level are considered by judges and juries in court, evaluating forensic findings at this level is typically not possible because it requires additional data, new methods, and case context information that is not always available. We will remain with the burglary case example to illustrate this.

Consider the preferred source level hypothesis set for a forensic glass comparison of glass particles retrieved from a suspect's clothing and reference glass from a window that was smashed in a burglary case:

Source Level Propositions:

H_p (prosecution) : "*The glass particles on the clothing of the suspect originate from the broken window at the crime scene*"
H_d (defense) : "*The glass particles on the clothing of the suspect originate from another glass source*"

Now define related Prosecution and Defense hypotheses at activity and offense level.

Try to come up with Defense hypotheses that represent plausible activities that could lead to the same evidence but in a scenario in which the suspect is not involved or has no criminal intent.

Source Level:

H_p (prosecution) : *"The glass particles on the clothing of the suspect originate from the broken window at the crime scene"*
H_d (defense) : *"The glass particles on the clothing of the suspect originate from another glass source"*

Activity Level:

H_p (prosecution) : *"The suspect smashed the window at the crime scene"*
H_d (defense) : *"The suspect walked by when another individual smashed the window at the crime scene"*

Offense Level:

H_p (prosecution) : *"The suspect smashed the window to access the shop and steal the valuables"*
H_d (defense) : *"The suspect smashed the window to assist someone in need at the shop, bystanders than stole the valuables"*

Please note that for all four propositions at activity and offense level, one would expect the suspect's clothing to contain glass fragments originating from the smashed window. Hence, the elemental match that is strongly supporting the Prosecution hypothesis at source level will not discriminate between the higher level propositions (i.e., $LR = 1$). However, the expert could possibly use other forensic findings such as the number and size of the glass particles retrieved from the clothing of the suspect. One would expect less particles to be present when the suspect did not smash the window himself. However, this would have to be tested by performing experiments, *e.g.*, by reconstructing the events as indicated by the suspect as accurately as possible and experimentally assess the glass transfer and the variability of the process. Typically, the evidential value of such investigations at activity level is much lower than matching source level profiles. This is especially true when considering trace transfer and persistence (TTP) phenomena of human biological traces and chemical microtraces such as glass, fibers and gunshot residues (GSR).

Given these complications, it might seem at first sight that the added value of the majority of source level investigations in forensic science is rather limited. However, in the entire context of the case and other information and evidence, providing strong links at source level can be of decisive importance. First of all, it typically forces the suspect to provide an alternative scenario because remaining silent (a fundamental legal right as was discussed in **Chapter 9—Quality and Chain of Custody**) will lead to generalized Defense propositions formulated by the investigative team. In addition, the context and facts concerning a case might make innocence propositions to explain the evidence highly improbable. When we consider the burglary case propositions, the defense hypothesis at activity level could be tested when there are eye-witness statements. At offense level, the idea of bystanders might be unrealistic when the valuables were stolen in the middle of the night, but the Defense proposition would be strongly supported when the scenario is confirmed by the victim that received assistance from the defendant. As indicated before, such aspects and details need to be carefully considered by the legal experts and the triers of fact, the forensic expert is only involved in providing evidence on the basis of forensic analysis and interpretation in his/her field of expertise. New scenarios can lead to new propositions that could be tested by performing forensic investigations as proposed by the forensic experts. However, ultimately there are limitations to what science can contribute to reconstruct events of interest in a criminal court case and this should be clearly and explicitly communicated and reported by the forensic expert.

10.5 Reporting forensic analytical chemistry investigations

In this section, we will turn our attention to forensic reporting in a specific forensic analytical chemistry context. So far, reporting has been discussed in a generic criminalistic framework that applies to every forensic expertise area, irrespective of the associated scientific domain. However, when considering chemical analysis in a forensic laboratory some specific reporting aspects apply as is illustrated in the question below.

> Although the Bayesian framework has been adopted by many forensic laboratories across the world and is generally preferred by forensic scientists, typically a majority of forensic analytical chemistry case reports does not contain Bayesian conclusions. Can you guess why?

If we go back to **Chapter 1—An Introduction Forensic Analytical Chemistry** and consider once more the definition of forensic analytical chemistry (the study of the separation, identification and quantification of chemical components of natural and man-made materials to answer questions of interest to a legal system), the following types of chemical analyses are conducted to address these questions:

- Qualitative Chemical Analysis: Chemical Identification (e.g., Drugs, Explosives)
- Quantitative Chemical Analysis: Quantitation (e.g., Toxicology)
- Chemical Profiling and Forensic Reconstruction

These activities have been discussed extensively in the book, with **Chapter 4** assessing chemical identification with a focus on high volume illicit drug case work, **Chapter 5** addressing quantitative analysis in forensic toxicology in relation to DIU legislation and **Chapters 6** and **7** dealing with chemical profiling and forensic reconstruction mainly focusing on forensic explosives. The Bayesian framework is ideally suited for chemical profiling and intelligence methods as has been demonstrated in **Chapter 6** with a score-based model for TNT impurity profiling with vacuum-outlet GC-MS. Working with profile similarities, distance scores and the characteristic nature of features (in feature-based models) bears resemblance to fingermark or DNA STR profile comparison, although in chemical profiling links are typically established at batch level, which means that multiple subsamples can exist that could generate the sample profile. The associated evidential strength is typically less for "matching" (i.e., very similar) profiles compared to biometric methods and more difficult to establish and validate because constructing representative reference collections is more challenging. However, the majority of the forensic analytical chemical investigations actually relate to high volume case work and this is associated primarily with chemical identification and quantitative analysis. For this type of work, applying a Bayesian framework for evidence interpretation and reporting of findings is less evident and much less frequently used as will be discussed in more detail next.

In a forensic institute, typically a majority of the chemical identification-based work is related to drugs of abuse. Other experts with a need to establish molecular structures are active in the field of forensic explosives investigation, fire debris analysis and **CBRN** (identification of chemical warfare and other threat agents). At the NFI, of all forensic chemistry related casework over 50% deals with illicit drug analysis. This translates to roughly two out of three of all chemical analyses being conducted by the forensic illicit drug experts of the institute. As was discussed in **Chapters 1, 2,** and **7**, the legal framework in many countries does not require a quantitative analysis and as a result high-volume GC-MS-based screening methods have been developed aimed at the rapid identification of listed substances. The entire process is optimized for speed and efficiency (of course at the required scientific and quality standards and under strict ISO17025 accreditation) and often includes automated reporting using standardized, shortened electronic report templates. Within the criminal justice system, such special arrangements need to be approved to ensure that that reports can be used as forensic evidence in a court of law. In many countries, innovation projects are aimed at enabling chemical identification of drugs of abuse "at the police station" and "on the street" and thus "out of lab" and "on the scene." This creates whole new demands and challenges as will be discussed in more detail in the next and final chapter (**Chapter 11—Innovating Forensic Analytical Chemistry**). But the chemical identification of listed substances "on the spot" further optimizes efficiency and improves the actions and decisions by law enforcement agencies. These actions and decisions relate to arresting perpetrators but equally to rapidly releasing or (even better!) not detaining innocent civilians that happen to be in the possession of suspicious materials that turn out to be fully legal. In this mode of operation, forensic reports typically do not contain an elaborate Bayesian conclusion to express the evidential value. Rather, the short report template will simply state the listed substances (if any) that were found in the confiscated sample. A straightforward and absolute expert statement that of course is easy to understand for every reader with limited chemistry knowledge. However, the question is whether such an approach is entirely scientifically sound and correct. To investigate this further, we first need to understand the scientific principles of chemical identification in forensic case work.

How can the process of chemical identification in forensic science be characterized?

Which step does it involve according to the framework of Paul Kirk ("*Criminalistics is the Science of Individualization*", **Chapter 1**).

In **Chapter 2—Analytical Chemistry in the Forensic Laboratory**—it was discussed that in forensic science, chemical identification is something fundamentally different than biometric identification (*i.e.*, the identification of an individual based on biometric features such as a fingermark or an STR DNA profile). In biometric identification, a set of very characteristic features is exploited to link a single individual from an entire population to a crime, location or a victim. In chemical identification we want to establish the chemical structure of a given compound present in a sample. Each chemical compound represents a specific arrangement of elements leading to specific physical and chemical characteristics such as volatility, reactivity, and biological activity. Chemical identification in forensic casework amounts to answering "the what question" at a molecular level. In the three steps discussed in **Chapter 1** this corresponds to the first step of Detection/Identification and not to Individualization. The main point of the discussion now pertains to uncertainty and evidential value in the chemical identification process.

> Can a forensic chemist make absolute statements regarding the chemical identity of an unknown compound?

According to the author (maybe to the surprise of the reader given the previous chapters), this intriguing question can be positively answered when considering the classical approach of structure elucidation in organic chemistry. This will be illustrated with a simple, somewhat hypothetical, example.

> A chemist is asked to identify the chemical composition of a very pure liquid sample. Using high-resolution mass spectrometry, the molecular formula has been determined as C_2H_6O (the measurement uncertainty is much lower than the elemental mass differences hence the compound must have this elemental composition). According to the laws of chemistry and chemical bonding, which stable molecular structures can be derived from this molecular formula?

For small molecules, the number of stable elemental arrangements is still limited. In this case, only two possible stable structures can exist: ethanol (alcohol) or dimethylether (ether). Both molecules have a molecular formula of C_2H_6O and hence cannot be distinguished with mass spectrometry no matter how high the resolution of the instrument (unless the different chemical functionality is exploited in the ionization mechanism).

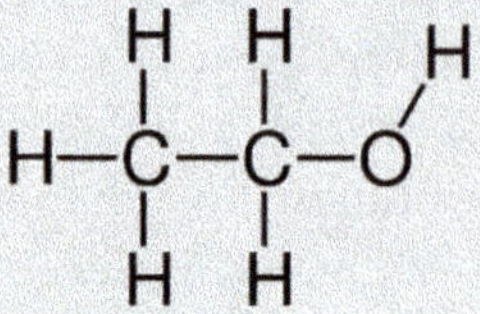

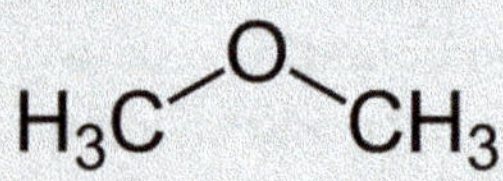

The task of the chemist has now been reduced to determine whether the very pure liquid consists of ethanol or dimethylether. What would be the next step in the structure elucidation/ chemical identification process?

The easiest way to determine whether the liquid consists of ethanol or dimethylether would be to smell it. These compounds have a distinctively different smell. However, this rather unscientific approach is not recommended because of health concerns. In structure elucidation, the chemist uses spectroscopy-based techniques to unravel the chemical structure. To this end, techniques like ^{1}H NMR (Nuclear Magnetic Resonance) and IR (Infrared) spectroscopy can be used. From reference data it can quickly be deduced which structure we are dealing with.

Using ^{1}H NMR (top spectrum) and IR Spectroscopy (bottom spectrum) the results depicted below are obtained. The chemist now arrives at a final conclusion concerning the chemical compound making up the pure liquid. What liquid are we dealing with? How does the chemist report his/her results, what statements should be made regarding uncertainty or evidential value?

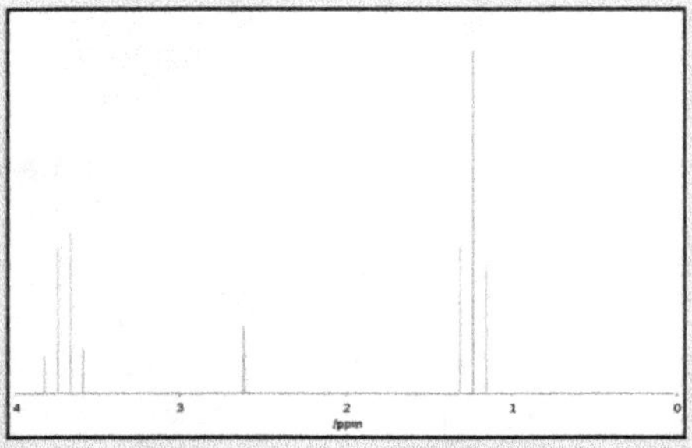

1H NMR spectrum *(Wikimedia Commons)*

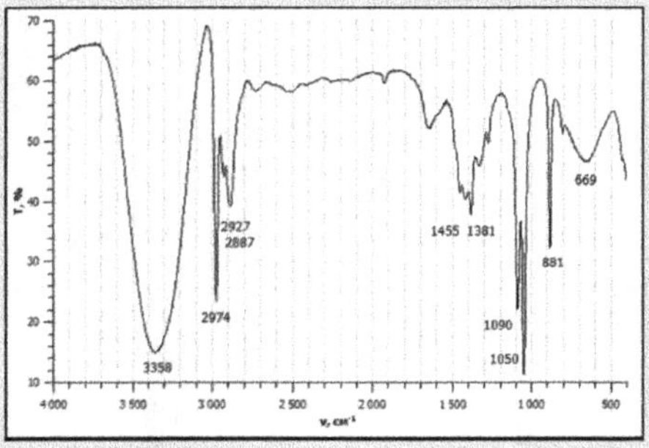

IR transmission spectrum *(Wikimedia Commons)*

These spectra can be attributed to ethanol. Although this could be deduced by comparing the spectra of the unknown liquid with reference spectra in the library, it is not required. In structure elucidation the chemist can use his/her scientific knowledge to deduce the chemical structure on the basis of the spectral information. This for instance follows from the peak positions and patterns in the NMR spectrum (the triplet and quartet peak pattern is indicative of a CH_3-CH_2 element in the molecule) and the broad peak seen in the IR spectrum which corresponds to the presence of an OH group in the molecule.

Given these results (HR-MS, NMR, and IR), there is no uncertainty in the final conclusion! On the basis of shared standards and guidelines in the chemical sciences, this liquid is made up of pure ethanol (alcohol). The HR-MS results limit the number of options to two compounds and the NMR and IR spectra allow the chemist to select the correct molecular structure out of these two options. The assignment is absolute and also does not require any contextual information, as the chemical identification is solely based on the findings and interpretation of the chemist.

Can we on the basis of this example now conclude that chemical identification in the forensic laboratory is absolute and hence that it is scientifically sound to report without uncertainty or without providing evidential values? Provide argumentation to explain your answer.

This example oversimplifies the task of the forensic expert to highlight the fundamental scientific point of the absolute nature of structure elucidation in chemistry. The reality in the forensic laboratory is too complex to state that chemical identification in forensic case work is also absolute. With increasing molecular mass, the number of stable arrangements of the associated elements increases exponentially and differentiating between all plausible options can become a daunting task even when using the most powerful spectroscopic techniques. In addition, forensic samples are seldomly pure and hence NMR and IR spectra will actually provide signals from various compounds in the sample. Although with NMR options exist to extract compound specific information from mixtures, typically separation methods are required to clarify the picture and to enable the identification of the individual compounds in the sample. However, there is also a very practical reason why the example is not representative for the way chemical identification in case work is conducted. Given the case load, reporting deadlines, and limited budgets, the forensic expert needs to efficiently conduct the investigation, finish the work in an acceptable time frame while ensuring a reasonable price per case. This does not allow for a tailor-made approach on a case-by-case basis using a diverse range of very expensive techniques until the expert feels that there is sufficient scientific evidence to confirm the molecular structure. In contrast, high-volume qualitative case work involves the use of screening techniques, accredited methods and the use of reference databases. This of course was already discussed extensively in **Chapter 4—Qualitative Analysis and the Selectivity Dilemma**. This chapter explains a chemical identification strategy based on the SWGDRUG guidelines by using a suitable combination of Category A, B, and C techniques to ensure sufficient selectivity. However, validation and accreditation of the preferred identification strategy (*e.g.*, colorimetric testing in combination with GC-MS analysis using both the GC retention time and the electron ionization mass spectrum match vs. a library spectrum to identify the presence of a listed substance) does not entail a false positive check against every compound known to man. This is also unrealistic as currently open-source chemical databases such as **Chemspider** (http://www.chemspider.com/) and **Pubchem** (https://pubchem.ncbi.nlm.nih.gov/) contain around 110 million chemical compounds and structures and this number is increasing every day as new natural and man-made compounds are continuously being discovered. With this many compounds the selectivity of the applied methodology must be very high indeed. However, of all these recorded compounds a majority will have a priori a very small probability of being present in a forensic sample. Should a forensic expert for instance consider a potential false positive compound that only has been synthesized in an academic laboratory once many years ago and since then has not seen any application or large-scale production? On the other hand, potential, still unlisted NPS isomers, known drug additives and adulterants, impurities, and degradation products and common chemicals such as food stuff additives, medicines, and functional chemicals must of course carefully be investigated when developing and validating screening methods for chemical identification in forensic case work. In this respect, attempting to quantify the method selectivity through a **random chemical match probability** (i.e., the probability that a random compound from a generic compound population would express the same features used to identify a listed substance) could even be misleading. The selectivity offered by GC-MS or LC-triple quad-MS techniques is known to be very high corresponding to a very high evidential strength when the retention time and mass spectra match that of a listed substance in combination with a positive colorimetric test. The mass spectrometer alone can generate a selectivity that corresponds to an *LR* value in the million range, although this is compound specific as some mass spectra and product ions are more specific than others. However, as was shown in **Chapter 4**, the forensic expert should not solely rely on the intrinsic selectivity of the applied methodology. In light of the NPS developments, new isomers can always emerge, and such isomers could be very difficult to distinguish from listed substances. As was previously discussed, the legal implications of such a false-positive outcome could be very significant, potentially leading to wrongful convictions. That the isomer was produced, sold and used with the same intentions is irrelevant as long as such compounds are not specifically included in the illicit substance lists. Given these complications, applying a Bayesian framework to express an evidential strength in chemical identification does not seem very useful. An interesting aspect in all of this is the "chemical nature of the complications." Although the case context can provide important information, even a priori considerations typically focus on chemical compounds such as raw materials, impurities, degradation and waste products. All these aspects are within the scientific domain of the forensic drug expert and are difficult to interpret for legal professionals. To infer that a certain drug was synthesized in a clandestine drug production facility, the forensic scientists will use his/her experience based on prior case work and training. If the raw materials and production routes are new to law enforcement, the forensic expert will use his/her scientific knowledge to unravel and reconstruct the chemistry involved. In light of this discussion, this author recommends to "simply" report the identified structures while ensuring that the applied methods are extensively and continuously screened for potential false positives on the basis of chemical expertise available. The report should also contain information with respect to the selectivity of the features used for identification and the potential existence of structural isomers not yet encountered in criminal investigations. To remain vigilant, forensic experts could randomly select case samples for an extensive chemical analysis using advanced spectroscopic techniques. In this way, the suitability of the applied methodology for given illicit substances can be put to the test and potential false-positives can be discovered.

Chapter 5—Quantitative Analysis and the Legal Limit Dilemma— discusses the fundamental nature of measurement uncertainty and the complications this causes when comparing an outcome to a fixed and absolute legal limit. Typically, uncertainty in quantitative analysis is described using frequentist statistics (i.e., establishing mean, standard deviation and confidence interval) and statistical hypothesis testing as is illustrated in section 5.4. In statistical hypothesis testing, the null hypothesis can either be rejected (the suspect is found guilty) or accepted (the suspect is considered to be innocent) and this can either be correct (true positive or true negative, respectively) or incorrect (false positive or false negative, respectively). When the triers of fact would only use the information provided by the forensic toxicologist to reach a decision, the test result would directly reflect an erroneous outcome when a Type I error or a Type II error occurs. In other words, a false-positive result would then automatically lead to a wrongful conviction while a false-negative outcome would result in a wrongful acquittal. In DUI cases where drivers have been subjected to regular roadside testing (as opposed to cases were potentially intoxicated drivers have caused a serious accident or have been stopped by the police because of erratic driving) it does not seem unrealistic that the verdict will be solely based on the analysis of the whole blood sample obtained from the suspect. A statistical approach that does not focus on the evidential strength of the forensic findings but rather accepts of rejects a hypothesis while reporting the associated error rates seems somehow appropriate. However, if we consider the whole blood analysis as a diagnostic test (considering two possible outcomes; the level of listed substance X is either below/equal to or exceeding the legal limit), a direct relationship exists between the *LR* value (diagnostic value) and Type I and Type II errors in statistical hypothesis test exists:

ground truth / ***verdict***	***H_0* is true suspect is innocent**	***H_1* is true suspect is guilty**
***H_0* is accepted suspect is acquitted**	No error Correct acquittal True Negative (1-α)	Type II error Wrongful acquittal False Negative (β)
***H_0* is rejected suspect is convicted**	Type I error Wrongful conviction False Positive (α)	No error Correct conviction True Positive (1-β)

$$LR = \frac{P(E|H_p)}{P(E|H_d)} = \frac{P(E|H_1)}{P(E|H_0)} = \frac{(1-\beta)}{\alpha}$$

where α is the probability of a type I error and β is the probability of a type II error.

Giving the example discussed in **Chapter 5** this results in a very high *LR* value. For the hypothetical case an average concentration of 65.4 μg/L (92%−101%, $\alpha = 0.01$, 19 d_f) was measured with LC-MS in the blood sample of the suspect. This corresponds to a lower level of the confidence interval of 60.4 ppb, well above the legal limit of 50 ppb. If we would adjust the α-value until this lower level of the confidence interval coincides with the legal limit, the corresponding *LR* value would roughly be 10^9.

How would the toxicological expert formulate the conclusion in a DUI case when applying the verbal probability terms and the Bayesian framework?

H_p (prosecution)	:	"?"
H_d (defense)	:	"?"
E (evidence)	:	"?"
Expert statement	:	"?"

H_p (prosecution)	:	*"When the suspect drove his car, the amount of cocaine in his blood exceeded the legal limit of 50 ppb"*
H_d (defense)	:	*"When my client drove his car, the amount of cocaine in his blood was below or equal to the legal limit of 50 ppb"*
E (evidence)	:	*A cocaine concentration of 65.4 ppb (92%−101%, a = 0.01, 19 d_f) was measured with LC-MS in the blood sample of the suspect*
Expert statement	:	*"The evidence is extremely more probable when the hypothesis of the prosecution is correct than when the hypothesis of the defense council is correct"*

The example shows that quantitative data can be evaluated and reported in the Bayesian framework without any issues and in a transparent manner. However, in most countries the more traditional frequentist approach to hypothesis testing is still preferred often in combination with predefined and agreed legal decision schemes.

10.6 A template for a forensic case work report

In this final section, a generic template for reporting forensic case work is presented. This template is inspired by the way NFI experts report their results within the Dutch criminal justice system. For each paragraph and section, specific aspects are discussed and an example is given. This example revolves around a burglary case and confiscated clothing of a suspect that contains glass particles that might originate from broken glass sources at the crime scene.

Forensic Case Work Report

Title

Police Investigation : POL-2022-239

Case Number : FS-2022-014

Responsible Expert : *John Doe*

Signature : John Doe

Date of request : *Month Day, Year*

Date of report submission : *Month Day, Year*

Forensic Services Amsterdam
Science Park 904, 1098XH Amsterdam,
The Netherlands
Telephone : 020-5257656
Email : clhc-science@uva.nl

Case information

In this part of the report, the context of the case is briefly described based on the information provided by law enforcement. This information should be neutral, punctual, and factual and should not strongly incriminate any suspect or support leading police scenarios. This could result in bias when the forensic expert is conducting the investigation. The context could include the time and date of the incident, alleged crime, location, and brief description of the crime scene and if available the names of the suspects. This part should also contain the contact details of the police officials and legal professionals that are leading the criminal investigation and are requesting the forensic services. Finally, the official date of the request should be included.

Requested forensic investigation

In this part, the original request from the public prosecutor, police force, or other parties involved in the criminal investigation must be cited. It is important to exactly quote the questions to the forensic expert as this is the starting point of the work conducted and the hypotheses considered (even when during subsequent discussions it becomes clear that the questions need to be adjusted). In court, the original "assignment" could play an important role when discussing the forensic findings and thus all parties involved need to be informed.

Forensic experts are used to a wide variety of formulations. From very specific questions related to the case conditions and the criminal investigation, to very brief statements requesting a certain product code (this is often the case for high-volume case work, e.g., illicit drug screening). Regularly, such requests are incomplete or erroneous and need to be adjusted after contact between the expert and the team coordinating the investigation. For clarity and transparency, it is important that these processes and the original requests are documented in the forensic report. Only reporting the adjusted request withholds potential important information to the triers of fact and the defense council. Some examples of typical request formulations are given below including a short explanation of the complications involved:

Do the fire debris samples taken from the store contain traces of ignitable liquids?

Are glass fragments present on the clothing of the suspect?

If the answers to these incomplete questions are affirmative, follow-up questions typically emerge requesting to specify the ignitable liquid class (e.g., gasoline) or to compare the glass fragments to a potential source at the crime scene.

Drug screening according to method FS01Drugs

Request for FS05GSR

FS01TOX, FS02TOX and FS03TOX

Typical examples of a high-volume screening requests from officials that coordinate and request many forensic investigations. The actual questions that are considered by the forensic expert are described in the method/protocol for the type of investigation that can be found in the quality system of the forensic institute. For readability and transparency, these underlying questions should be included in the report.

Did the suspect shoot the victim with the firearm we found on the crime scene?

Can you help us proving that the suspect stabbed the victim?

Examples of questions that cannot be answered by the expert directly (the presence of GSR particles does not necessarily mean that somebody fired a gun, the expert should not address the ultimate issue) or that reveal tunnel vision and bias. For the Defense council, such information can be very important.

Hypotheses

For the expert to be able to interpret the findings and report the evidential strength, the research questions need to be translated to suitable hypothesis pairs. Typically, these hypotheses are constructed in a case assessment and interpretation discussion between the experts and the team requesting the investigation. Although responsibility for the hypothesis pairs ultimately lies with the public prosecutor leading the criminal investigation, the forensic expert provides important input because the hypotheses will need to align with the possibilities offered by the forensic methodology. It is for instance in general not very useful to formulate activity level propositions when the investigation only provides source level insights. For forensic investigations that are conducted in high volume, the hypothesis pairs might also be standardized and provided in the method documentation in the quality system.

Sometimes it can be useful to define a hypothesis framework when multiple traces have to be compared against a potential crime scene source. This for instance is the case for forensic glass comparison in burglary cases:

Are glass fragments present on the clothing of the suspect and if so, do any of the glass fragments originate from the broken window of the jewelry shop?

Beforehand the forensic experts do not know whether they will find any glass fragments on the clothing of the suspect that has been sent in as evidence item. But if glass particles are found, this can lead to a relatively large number of forensic comparisons, one for each glass fragment. The results can be efficiently tabulated when considering the following generalized hypothesis pair:

Prosecution hypothesis

"*Glass particle **A** obtained from clothing item **B** originates from crime scene source **C***"

Defense hypothesis

"*Glass particle **A** obtained from clothing item **B** originates from another source not related to crime scene source **C***"

where **A** represents the evidence code for a retrieved glass particle, **B** represents the clothing item of the suspect (maybe multiple clothing items have been confiscated), and **C** represents the crime scene source (maybe multiple broken glass sources have to be considered as potential source).

Evidence items received

This part of the forensic report describes in detail the physical evidence items received. This includes the unique identifiers, *e.g.*, physical evidence codes, as provided by both the team leading the criminal investigation (these codes are typically generated within a police quality system) and the forensic institute receiving the physical evidence (these codes are typically generated in the internal quality system of the forensic institute). These evidence item codes are linked to the case through what should be unique case numbers. This link is essential as part of the chain of custody as this demonstrates that the evidence items investigated indeed originate from the crime scene. When considering numerous items and individual codes for each organization involved, the total number of codes increases rapidly, and several "unique" codes may exist for the same item. This increases the risk of an administrative error and confusion in court. For this reason, in the Netherlands the criminal justice system uses a national coding system for forensic evidence items. The concept of a single, national Trace Identification Number (in Dutch abbreviated to SIN number) has been discussed in more detail in **Chapter 9—Quality and Chain of Custody**. In the example given below, the first two columns providing the institution codes could be combined in one column when using SIN numbers.

Original Crime Scene Code	Forensic Institute Evidence item code	Description of the evidence item
POL-239-001	FS-014-001	Black coat (brand, model)[a] of the suspect
POL-239-002	FS-014-002	Red scarf (brand, model)[a] of the suspect
POL-239-003	FS-014-003	Reference glass sample from broken front window[a]
POL-239-004	FS-014-004	Reference glass sample from broken vitrine[a]

[a]The evidence description must be sufficiently detailed to prevent mix-up with other evidence items (although differentiation is also possible through the evidence item codes).

Evidence items to be investigated

This part could also be included in the previous section, but it could be useful to separately state the actual samples that are analyzed according the described forensic methodology. Quite often this will constitute forensic traces that have been found on or in the evidence items described in the previous section. Such traces are typically detected and secured at the forensic institute (although in the Netherlands such work is also conducted by forensic specialists of the Dutch police), which means that new evidence items are created that should be individually coded and linked to the original evidence item. In the example of the burglary case this refers to glass particles that have been found on the clothing evidence items and that will be individually analyzed and for which the findings will be compared to the reference glass sources from the crime scene.

Forensic Institute Evidence item code	Description of the forensic trace
FS-014-001-#01	Glass particle 1 retrieved from the black coat
FS-014-001-#02	Glass particle 2 retrieved from the black coat
FS-014-002-#01	Glass particle 1 retrieved from the red scarf
FS-014-002-#02	Glass particle 2 retrieved from the red scarf
FS-014-002-#03	Glass particle 3 retrieved from the red scarf

Forensic investigation

In this part of the report, the forensic expert assigned to the case concisely describes the analytical chemistry methods that have been applied to analyze the evidence items. This includes sampling protocols and sample preparation steps. Some general information can be included describing the methodology and instrumentation used and referring to scientific publications that form the basis of the analysis. Such information could also be provided in the form of a Technical Annex that is added as a separate document to the forensic report. Essential methodological details and parameters have to be included according to the ISO17025 reporting guidelines, but this is not necessary when such information is available in the method description. The case report can then simply refer to the associated document number in the quality system. All methodological details of the investigation should be carefully documented in the case file, but the reporting officer can decide which of these details are critical and should be included in the report. Of course, any significant deviations from the documented methodology due to case specific circumstances have to be reported in detail.

For the forensic glass comparison in the example of the burglary case, the NFI would typically employ LA-ICP-MS (Laser Ablation—Inductively Coupled Plasma—Mass Spectrometry) to create elemental impurity signatures for the glass traces secured from the clothing items (more traditional methods for glass comparison are based on refractive index measurements of elemental profiling with mXRF). An example of the description of the forensic investigation is given below:

*"The glass fragments from the clothing items and the reference glass from the crime scene were analyzed with Laser Ablation—Inductively Coupled Plasma—Mass Spectrometry (LA-ICP-MS). With LA-ICP-MS, trace elements can be accurately and precisely quantified using a reference glass standard doped with elements. The instrument and methods used have been described in detail in scientific literature: Andrew van Es, Wim Wiarda, Maarten Hordijk, Ivo Alberink, Peter Vergeer, Implementation and assessment of a likelihood ratio approach for the evaluation of LA-ICP-MS evidence in forensic glass analysis, Science and Justice, 57 (**2017**) 181–192. Signal intensities have been normalized using the Si signal (abundant element in glass) and concentrations have been determined as mg/kg (ppm) for 18 elements using the glass reference standard in a single-point calibration. All measurements were done in triplicate using point ablation on different areas of the glass particles. Using a dataset of 107 reference pane glass sources, a two-level feature-based was created. This model was used to compare the elemental profiles from the glass particles recovered from the clothing items of the suspect with those of the reference glass particles retrieved from the crime scene. Details of the element selection, model creation, and glass elemental profile comparison can be found in the appendix and case file."*

Findings

Findings should be regarded as factual information, i.e., observations or the recorded output of instrumental analysis. It is recommended that in case reports the findings, often in the form of analytical data, and the interpretation of these findings are provided in separate sections (as is the case in this template). In this way, it is very clear to the reader what can be considered as data (objective, undisputable) and what as interpretation (subjective, debatable). Unfortunately, in forensic practice such a line is not easily drawn as matters are not as "black and white" but rather that "many shades of gray exist". When we again take the burglary case and the glass analysis as example, the elemental concentrations measured with LA-ICP-MS can be considered as the "hard analytical data." However, the instrument output is an array of MS intensities (abundances) for various m/z values representing the ionized fractions of the 18 elements in the ablated samples. The abundance is normalized using the Si signal to account for instrumental variations. Next, the measured intensities of the reference glass standard are used to determine the concentrations of the 18 elements on the basis of the levels provided for the standard. Using a single calibration standard will lead to significant measurement uncertainty. All glass samples are analyzed in triplicate and the final result, the elemental impurity profiles for all glass samples in the case, will consist of average concentrations and the associated standard deviation/confidence interval. Although measurement uncertainty should be reported, ISO 17025 guidelines indicate that this is only necessary when it directly affects the interpretation and final conclusions. In this case, one could argue that the measurement uncertainty in the elemental concentrations is less relevant because the average profiles are compared using a feature-based model resulting in an evidential strength expressed as LR value. Using graphs and tables to illustrate the findings provide insight to the reader. Below the elemental profiles of the reference crime scene glass sources are provided as an example. In general, visualization in forensic reporting is recommended as means to put the results in perspective.

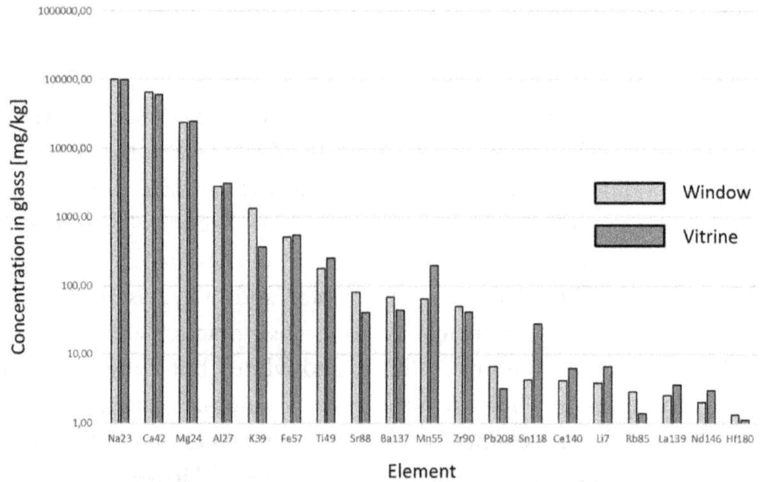

Example of the visualization of findings: elemental profile of reference window and vitrine glass recovered from the crime scene. The bar graph illustrates the substantial difference in concentration of the various elements in glass (mind the log scale!) and the key elements that differentiate the two reference glass sources (K, Mn, Pb, Sn, Rb)

Interpretation of the findings

In the interpretation section, the expert arrives at an estimation of the evidential strength of the findings in light of the propositions considered. This involves either a subjective or a more objective assessment depending on the type of investigation conducted. As discussed in the previous part, many forensic reports dealing with qualitative (chemical identification) and quantitative analyses do not interpret findings according to the Bayesian framework. Evidential strength can then still be qualitatively discussed considering the sensitivity and selectivity and the associated risk of respectively false-negative and false-positive outcomes. For quantitative analysis confidence intervals can be used to interpret the findings with respect to the legal limit thresholds. For the glass comparison on the basis of elemental impurity profiles, the evidential value can be expressed quantitatively because of the well validated and calibrated feature-based model built from an extensive dataset. The details of the modeling should be discussed in this section allowing readers to understand the various steps involved and the associated considerations and limitations. The outcome in this case is not only determined by the profiles measured for the case samples but also by the reference data used to create the model. A fair question is for instance how representative this data is for the type of glass considered in the case. Can a reference collection of pane glass samples for instance be used in the comparison of vitrine glass? If different types of glass have different elemental levels to optimize the glass properties for the intended use, the vitrine glass composition will be relatively rare leading to low random match probabilities for individual features (element concentration in glass) and thus relatively high *LR* values for "matching" profiles. The aspect of database relevance is often a point of discussion in chemical profiling. In addition to the previous point, too high evidential strengths can also be the outcome of applying global databases to a local case. In a world-wide population, features can be relatively rare compared to a local glass collection that takes into account the local use of glass panes in construction. Typically, local markets are dominated by a limited number of construction companies that obtain glass from a limited number of suppliers. So, information must also be provided on the modeling process and the reference data used (or an appropriate reference needs to be provided). This will allow readers and other experts to critically assess the work done and the validity of the reported evidential strength. For multiple comparisons as in our case example where several glass particles were found on the clothing of the suspect, *LR* values can be reported in a convenient table form:

Particle (A)	Clothing item (B)	*LR* (window) (C)	*LR* (vitrine) (C)
FS-014-001-#01	Black Coat	1250	~0
FS-014-001-#02	Black Coat	~0	5345
FS-014-002-#01	Red Scarf	3337	~0
FS-014-002-#02	Red Scarf	~0	3445
FS-014-002-#03	Red Scarf	1872	~0

The capital letter codes refer to the generic formulation of the hypotheses. The tabulated results indicate that there is strong evidence to support the prosecution hypothesis that the glass fragments originate from the broken window and vitrine at the crime scene. The findings are roughly 1000-5000 more likely under the same source than under the different source proposition and there are glass particles that match both crime reference glass sources

Conclusions

Readability of forensic case reports is considerably increased when the original research questions are taken as starting point for the conclusions. This "connects the dots" in the mind of the reader and stimulates the expert to formulate the conclusions as direct answers to the questions that initiated the forensic investigation. The conclusions should be as short and concise as possible, all relevant aspects covering the investigation should preferably be discussed in the Findings and Interpretation sections. This enables the forensic expert to convey the "main message" to the stake holders. Assuming that the questions are essential to the ongoing criminal investigation and court case, the expert answer in the form of the evidential value of his/her findings in light of the propositions represents the added value of the entire investigation. For the de burglary case and the LA-ICP-MS analysis of the glass fragments, the conclusions could be condensed to:

Are glass fragments present on the clothing of the suspect?

"*Yes, in total 15 glass particles were found on both garments but 10 glass fragments were not suited for LA-ICP-MS analysis. Two glass particles retrieved from the coat of the suspect and three glass particles retrieved from the scarf of the suspect could be analyzed with LA-ICP-MS.*"

If so, do any of the glass fragments originating from the broken window of the jewelry shop?

"All five particles analyzed with LA-ICP-MS yield elemental profiles that strongly resemble the profiles of either the broken window or the broken vitrine at the crime scene. Glass fragments matching the two crime scene sources were found on both the black coat and the red scarf of the suspect. Using the calibrated verbal scale as detailed in the appendix, finding this degree of similarity in the elemental profiles is ***much more probable*** *when the particles are originating from the broken window or the broken vitrine at the shop than when the particles were originating from another glass source."*

Appendix

In the appendix additional and supplemental information can be provided such as an explanation of the verbal scale used to express the evidential value:

"Forensics Services Amsterdam expresses the strength of the evidence according to the Bayesian framework using the following verbal scale:"

"The findings of the investigation are"
- *"approximately equally probable [as]"*
- *"slightly more probable"*
- *"more probable"*
- *"appreciably more probable"*
- *"much more probable"*
- *"far more probable"*
- *"extremely more probable"*

"If hypothesis one is correct, than if hypothesis two is correct."

"In this setting hypothesis one and hypothesis two indicate either the proposition of the public prosecutor or the defense council."

*"To promote transparency for the reader and uniformity among the different expert areas, Forensic Services Amsterdam has implemented a numerical calibration of the verbal terms. These definitions are expressed in orders of magnitude and are listed in the right column in the table below. e.g., the term "**slightly more probable**" means that the probability of observing the results of the investigation is considered 2 to 10 times larger when hypothesis one is true than when hypothesis two is true."*

Verbal term	*LR* range	Log(*LR*) range
Approximately equally probably	1−2	0
Slightly more probably	2−10	0−1
More probable	10−100	1−2
Appreciably more probable	100−1.000	2−3
Much more probable	1.000−10.000	3−4
Far more probable	10.000−1.000.000	4−6
Extremely more probable	>1.000.000	6+

The conclusion expresses the evidential strength of the results considering the hypotheses. The conclusion does not represent the probability that a particular hypothesis is true. Such a probability depends on other evidence and information and falls outside the forensic expertise domain and the scope of this report.

Further reading

Cook, R., Evett, I.W., Jackson, G., Jones, P.J., Lambert, J.A., 1998. A model for case assessment and interpretation. Sci. Justice 38, 151−156.

Cook, R., Evett, I.W., Jackson, G., Jones, P.J., Lambert, J.A., 1998. A hierarchy of propositions: deciding which level to address in casework. Sci. Justice 38, 231−239.

Evett, I.W., Jackson, G., Lambert, J.A., 2000. More on the hierarchy of propositions exploring the distinction between explanations and propositions. Sci. Justice 40, 3−10.

ENFSI, 2016. ENFSI Guidelines for Evaluative Reporting in Forensic Science, Approved Version 3.0. ENFSI downloadable via. https://enfsi.eu/wp-content/uploads/2016/09/m1_guideline.pdf.

Regeling, 2012. Regeling Taken NFI (Regulation Tasks NFI, in Dutch). Ministry of Justice and Security of the Netherlands. https://wetten.overheid.nl/BWBR0031558/2012-05-19/.

Vignaux, G.A., Berger, C.E.H., Robertson, B., 2016. Interpreting Evidence: Evaluating Forensic Science in the Courtroom, second ed. Wiley, Hoboken, New Jersey, USA. 978-1-118-49243-7.

CHAPTER 11

Innovating forensic analytical chemistry

Contents

11.1 What will you learn?

This final chapter is dedicated to innovation in forensic science in general and forensic analytical chemistry specifically. Forensic institutes usually struggle to find sufficient time for R&D projects to introduce new options for case work or to improve existing methods. High caseloads, backlog, and new urgent requests typically preoccupy forensic experts on a daily basis, and this can frustrate science and innovation efforts. For this reason, small-scale forensic laboratories sometimes exclusively focus on case work. However, innovation and fundamental scientific studies are essential to ensure and sustain the added value of forensic investigations. Society, crime, and law and order are dynamic and as new criminal activities, evidence materials and laws emerge, forensic institutes have to adapt to ensure that their services remain relevant. Additionally, progressing scientific insights and technological breakthroughs continuously offer new opportunities to assist criminal investigations and solve crime. However, the process of developing a novel idea into a robust forensic casework method requires a lot of time and effort from the forensic experts and scientists. This requires a clear understanding of the potential of new methods to make sound

Chemical Analysis for Forensic Evidence
ISBN 978-0-12-820715-4
https://doi.org/10.1016/B978-0-12-820715-4.00019-5

decisions on the investment of capacity and equipment budgets. This chapter will allow the reader to understand the why and how of innovation in forensic science. These insights will be used to address three innovation areas in forensic analytical chemistry, the detailed chemical analysis of microtraces, the interpretation of chemical evidence at activity level and rapid, on-scene chemical analysis of forensic evidence materials.

After studying this chapter, readers are able to

- explain why innovation is essential to **forensic science** and thus also for **forensic analytical chemistry** and describe the five main innovation incentives
- understand how **innovation** can be organized and embedded in a sustainable manner in forensic institutes
- formulate ideas and R&D projects and select techniques and methods to develop chemical profiling strategies for **microtraces**
- formulate ideas and R&D projects and select techniques and methods to enable forensic chemical analysis at **activity level**
- formulate ideas and R&D projects and select techniques and methods to realize the robust **chemical identification** of explosives and drugs of abuse directly at the scene using **portable equipment**

11.2 Five reasons to innovate

To be able to successfully improve, innovate, and introduce forensic methods, we first must understand why such methods are employed in the first place. Only if we truly grasp the added value of a case work method, will we be able to think how to increase this added value even further. This book started in **Chapter 1—An Introduction to Forensic Analytical Chemistry** with the formulation of the following definition:

> *"Forensic Analytical Chemistry entails the separation, identification and quantification of chemical compounds in natural and man-made materials to answer questions of interest to a legal system".*

Interestingly, this definition offers an elegant description but provides no clues when we want to consider the added value of a forensic analytical chemical investigation. To provide a framework, a more detailed analysis is required.

What is the main "product" of a forensic investigation? What do forensic institutes provide when they conduct and report case work and what do the partners in the criminal justice system seek when they submit a request?

Information!

The forensic report hopefully provides clear answers to the questions emerging from the criminal investigation.

When an institute reports the findings of a forensic investigation, this hopefully answers relevant questions that assist the criminal investigation and supports the criminal court proceedings. Although the investigation itself is typically quite physical and deals with physical evidence (with the exception of digital forensic science), the output and with that the main product is **information**. This information is based on scientific principles, has been obtained within a strict quality framework, and has been gathered by experts that are impartial and conduct their work according to the highest integrity standards. Hence forensic information normally is very neutral, trustworthy, and robust and can therefore be of vital importance to arrive at the right conclusion (*i.e.*, verdict) by the triers of fact. When we talk about improving forensic services through innovation, the aim is to increase the added value of the information.

If the main outcome of a forensic investigation is scientific information, what factors would now determine the added value of that information given the question addressed within an ongoing criminal investigation?

Typically, partners in the criminal justice systems that are conducting a criminal investigation are not that interested in the scientific principles of the applied forensic methodology. They just want a useful answer to their question which will help them to successfully solve the crime and bring the perpetrators involved to justice (they should be equally motivated to release innocent suspects and prevent wrongful convictions at all times).

So, what defines a useful answer to a question formulated in the request form submitted to a forensic institute? Basically, there are three elements that determine this:

1. The **strength** of the answer
2. The **quality** (in terms of robustness, trustworthiness, reliability, scientific validity) of the answer
3. The **relevance** of the answer and with that the question to the case at hand

It is interesting to explore a bit further why and how these aspects add value. The strength relates to how explicitly an expert can answer a question. In verbal terms, one might appreciate that the reply "*maybe*" is less convincing than a straight "*yes*" or "*no*." Even more disappointing would be the answer "*I do not know*". Forensic experts have a scientifically sound but somewhat elaborate way of answering case related questions. As discussed in detail in **Chapters 6** and **10**, a forensic expert that is using the Bayesian framework will report the evidential value (*LR* value) that indicates to what extent the evidence is supporting one hypothesis over the other (or more accurately: how much more probable the evidence is when one hypothesis is true than when the other hypothesis is true). The hypothesis pair is defined on the basis of the question that is formulated as part of the criminal investigation and is listed in the request. The higher the evidential value, the more probable the evidence becomes under the hypothesis of the prosecutor. If the *LR* value becomes very high (e.g., exceeding 1.000.000), the posterior probability of the prosecution hypothesis being true is approaching (but never equals) 1 (an interesting exception is the combination of a very high *LR* value with a very low prior probability, this can occur in cases of a DNA profile match in a database while there is no clear contextual link of the donor to the crime). This indicates that very high *LR* values provide very convincing expert answers that leave little room for doubt and uncertainty. The evidential value obtained depends on the context of the case (and with that the question asked), the nature and state of the evidence material but also on the methodology that is used to analyze and characterize this material.

The importance of the quality of the answer is easier to understand and explain. There is no added value to a very explicit answer ("*yes*") of a forensic expert when in court it turns out that this statement is lacking sufficient scientific support. Even worse, such a situation could do a lot of damage especially when it leads to wrongful convictions and could ultimately undermine the general trust in forensic science. A forensic finding is more trustworthy when it is obtained within a strict quality regime as discussed in **Chapter 9—Quality and Chain of Custody** and the applied procedures include several checks and balances. This adds more value as there is less reason for any party involved in the criminal investigation to doubt the outcome. This does not mean that in an appropriate quality framework, a forensic investigation will be free of errors, any claims in that respect should be pursued with caution and reservation. However, a good quality system does include measures to minimize errors and their impact as much as possible, and procedures to monitor, test, and report potential error risks. Quality and robustness can also be considered from a scientific angle. It can be argued that *LR* based methods, i.e., methods with which experts are able to report numerical *LR* values on the basis of representative reference databases, add more value than subjective methods involving an expert judgment based on experience and knowledge of the field. In addition, measurement uncertainty also affects the added value of a forensic investigation. A discussed in **Chapter 5—Quantitative analysis and the Legal Limit Dilemma**, the confidence interval in relation to the legal limit will determine the false-positive (type I error, wrongful conviction) and false-negative (type II error, wrongful acquittal) error rates. Although measurement uncertainty is typically seen as part of the quality characteristics of a method, this is usually also expressed in the formulation of the conclusion.

The relevance of the answer and with that of the question seems at first sight an odd aspect when considering the added value of a forensic investigation. Why would a question be asked in the first place when it is of limited relevance? The reason is that the questions that really matter in a criminal investigation often cannot directly be answered by the forensic expert. Either because this involves context information and knowledge outside of the scope of the expert but often also because state-of-the-art forensic methods force questions to be rephrased and "downsized" because otherwise no meaningful answer can be provided. In this respect, it is important to understand that relevant questions typically relate to actions of individuals (usually suspects), whereas the chemical analysis of evidence materials normally provides information at source level. The so-called hierarchy of propositions was addressed in detail in **Chapter 10—Reporting in the Criminal Justice System**. In addition to the example of the detection and analysis of glass particles in a burglary case discussed in that chapter, another insightful example concerns the analysis of Gun Shot Residues (GSR) in shooting incident investigations. In such cases, a suspect is stubbed, and these stubs are subsequently screened with SEM (Scanning Electron Microscopy) for the presence of particles with specific primer element combinations (i.e., Pb, Ba, and Sb) that are highly selective to ammunition and thus the discharge of a firearm. Of course, the main question in the criminal investigation is whether the suspect fired a gun during the incident. If several witnesses declared that they saw the suspect shooting the victim, the presence of GSR particles on the hands and clothing of the suspect is in line with those statements. However, the statements are irrelevant to the forensic investigation which should only be based on the analysis of the trace evidence. On the basis of the findings (the detection of numerous particles containing Pb, Ba, and Sb on the suspect stubs with SEM), the GSR expert would, however, indicate that he/she cannot make any statements regarding the actions of the victim. The reason for this is that the expert knows that when a gun is fired, substantial amounts of GSR particles can also be deposited on bystanders. If the suspect would state that he was standing next to the actual shooter and that the shooter ran off after shooting the victim, this could also explain the evidence. For that reason the question has to be rephrased for the expert to be able to provide a meaningful answer. NFI experts state that the presence of GSR particles can only indicate that the suspect was involved in a shooting incident without specifying an action (excluding the possibility of secondary transfer on the basis of the number of particles). At an even lower level, the question could be specifically targeted at the forensic investigation by simply asking the experts whether the stubs taken from the suspect contain any particles that are specific for discharging a firearm. Of course, such objective forensic scientific findings are of value to the case but in the end the triers must combine this with other information to arrive at the conclusion that indeed the suspect was the shooter. Please note that finding no GSR particles does not necessarily mean that the suspect did not fire the gun, this depends on the context and more importantly the time between the shooting and the sampling and whether the suspect was able to clean-up and change clothes undetected.

By identifying the three main factors that determine the added value of forensic information (evidential value, quality, and relevance), we now also understand how we can improve forensic methods through scientific research. However, before exploring this in more detail, we need to consider two important practical aspects concerning a forensic investigation that also proof to be strong drivers for innovation.

Let's assume that you are in need of a forensic service that can only be provided by a special contract laboratory. In your arrangement with this laboratory the forensic investigation has been agreed in detail including the case questions, the evidence material and associated transport, the laboratory methods that are going to be applied and the way the results will be reported.

What two other essential aspects will you have to discuss and agree on?

There is nothing specifically forensic about these practical aspects. For every product or service that is offered by a provider to a customer, both parties need to agree on the associated **cost** and **delivery time** (how much do you have to pay for your new furniture and when will it be delivered?). Because forensic science is considered to be a governmental task in many countries, costs are covered by society ("the tax payers") as forensic institutes claim part of the national budget. An exception, with a remarkable history, in Western Europe is the United Kingdom were forensic services are provided by commercial contract laboratories on the basis of regular tenders and agreed **Service Level Agreements** or so-called **SLAs**. But also in a more commercial setting, the costs for the forensic services are in the end paid for by the government. So, the general overarching question is how much a society is willing to invest in forensic science. At the level of the institute, an important follow-up question is how these funds should be distributed over the various expertise areas. Typically, institute directors always need to manage "shortage" and have to make difficult decisions how to spend the budget wisely and were to invest in expert capacity or expensive new laboratory equipment (and even more painful were to divest if no additional budget is provided for these investments). That forensic services are always short on capacity follows from the natural tendency to investigate all the potentially interesting evidence material in a criminal investigation. The team involved wants to solve the crime while not making any critical mistakes in the process. As forensic services can typically be requested without any budget considerations (it is very uncommon to agree on a budget for a given case beforehand), it is easy to understand how additional forensic investigations are requested that are not critical but "*nice to have, just to be sure.*" Although there are no direct financial consequences of such a "*better safe than sorry*" strategy, there is another price to pay in the form of increasing **lead times**. If all crimes are investigated with an excess of forensic requests, the forensic experts at the institute are faced with an increasing number of incoming cases that ultimately exceed the expert and instrument capacity. As a result cases get shelved and a case backlog starts to build up and grow as long as the capacity is insufficient compared to the number of requests that are received every month. This can lead to excessive delivery times of several months or even worse and initiates processes in which teams try to give their request a high-priority status. However, allowing exceptions to a fair **First In-First Out** (**FIFO**) principle will increase complexity and will increase lead times even further for the unprioritized cases. This case backlog "*runaway*" can only be stopped and reversed by increasing the expert capacity or by canceling a number of redundant requests. Request get redundant when lead times exceed the time frame of the criminal investigation and court session dates. This clearly illustrates how the cost of a forensic investigation and the lead time of a forensic report directly affect the added value. This has nothing to do with the scientific content, there is simply no added value when the report is provided after a judicial decision has already been made. This results in a total waste of forensic resources and should be prevented at all costs. In a situation in which a potentially valuable forensic investigation is extremely expensive or would take ages to finish, it could be a wise decision not to proceed but rather reserve the capacity for other cases. Such a decision is always very difficult to make because in the criminal justice system professionals typically work on a case-by-case basis and are very committed to an ongoing investigation. This makes it very difficult to see "*the bigger picture*" and to take a decision that is unfavorable to the case but is preferable from an overall case load perspective. An exception to this could be a high profile or cold case for which a specialized forensic investigation, such as a DNA mass screening, is the final resort to solve the crime. Under such special circumstances, typically dedicated funding and expertise is allocated.

The importance of the time needed and cost of a forensic investigation as illustrated above indicates that innovation efforts directed at making forensic case work cheaper and faster can be extremely valuable. Basically, improvements related to the evidential strength, quality, and relevance increase the **Forensic Information Value Added** (**FIVA**), while cost and lead time reductions improve the **Forensic Information Value Efficiency** (**FIVE**), an indicator of how much effort it takes to obtain the information. We now have identified **FIVE** lines along which forensic case work methods can be improved. Dedicated innovation programs and R&D projects aim to introduce new or improve existing forensic methods to:

1. increase the evidential strength to provide more convincing answers
2. improve the quality of the investigation to make the answers more robust
3. provide new investigative options to enable more relevant answers
4. speed up the investigation to reduce lead times and provide answers quicker
5. reduce the cost of the investigation to increase case work capacity and make answers less expensive

Which of these FIVE factors are most urgent to pursue in R&D projects depends on the forensic expertise area considered, the current state of the art of the methodology and the needs of the partners in the criminal justice system. It makes little sense to heavily invest in further improving the evidential value of an investigation when quantitative *LR* values in excess of 1.000.000 can already be reported and police and legal professionals are indicating that they really need forensic information to distinguish activity-based scenarios. When a valuable innovation strategy is established in collaboration with the key stakeholders, forensic experts will also have to establish what scientific and technological options exist to meet the needs. Is the analytical instrumentation currently available able to generate the desired information and how much time, effort, and budget would it require to implement a new instrument and forensic case work method in the laboratory? It is not uncommon that an incentive to innovate is triggered by a new and powerful analytical instrument that is introduced by a laboratory equipment manufacturer. Such instrumentation is seldomly brought to market for forensic use only (the forensic market is way too small and specific for that). However, when forensic experts "*have done their homework*" in the form of a clear innovation strategy, they are able to quickly recognize the forensic potential of a new instrument (e.g., LA-ICP-MS for detailed elemental profiling of microtraces such as glass fragments,

Orbitrap high-resolution MS for chemical identification of explosive traces in postexplosion residues, chemical profiling of ignitable liquids with comprehensive gas chromatography with mass spectrometric detection—GCxGC-MS). The figure below illustrates that valuable innovation projects and programs can be established when experts are aware of the FIVE factors to increase added value, customer needs are known and the right scientific knowledge and technological options are available:

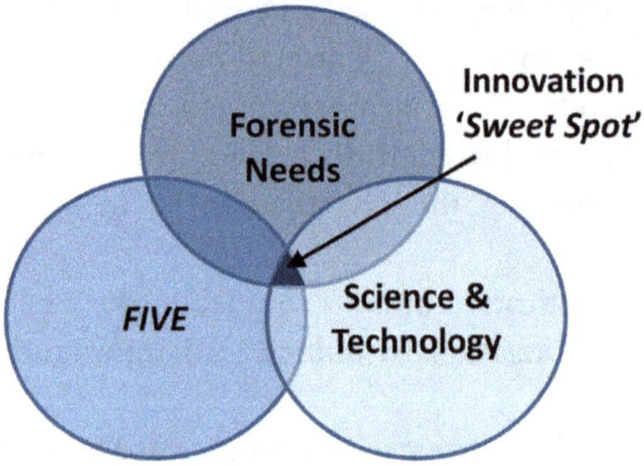

Schematic illustration how relevant forensic innovation programs can be established by considering the FIVE concept, the needs in a forensic expertise area, and new scientific insights and technological advancements *(Adapted from A.C. van Asten, Science and Justice, 54 (2014) 170–179, https://doi.org/10.1016/j.scijus.2013.09.003)*

Consider the two forensic methods described below. What meaningful innovation projects could you define to improve these methods? Which of the FIVE factors are key to increase the added value of the forensic information in criminal investigations? Provide argumentation for your choices (this will be a good exercise if you as a forensic expert need to convince senior management to invest in your R&D plans!)

(1) The chemical identification (qualitative) analysis of drugs of abuse (**Chapter 4**).
(2) The chemical profiling of explosives (**Chapters 6** and **7**).

The chemical identification (qualitative) analysis of drugs of abuse (Chapter 4)

A matter of concern that was addressed in **Chapter 4** was a selectivity issue due to the rapid introduction of NPS (New Psychoactive Substances) in the drugs of abuse market. This is an example where R&D efforts are required to meet an acute challenge that is related to societal developments. The investment in GC-IR or GC-VUV adequately resolves the issue and such improvements can be considered as falling under category number 1: increasing the evidential strength such that the expert feels confident to make an absolute statement with respect to the chemical structure.

Taking a step back, the chemical identification of listed substances is typically a large volume activity, handling several thousands of requests annually is not uncommon. This requires a very substantial logistic effort as all confiscated samples need to be sent to the forensic laboratory for further analysis. For large-scale forensic services, the reduction of lead time and cost of the analysis (categories 4 and 5) is usually very worthwhile and therefore also a target for innovation. In this respect, enabling police officers to conduct the chemical identification of drugs on the scene with easy-to-use, hand-held, automated equipment is of great potential value. However, selectivity is usually limited making confirmatory lab analysis currently still necessary.

The chemical profiling of explosives (Chapters 6 and 7)

Whereas the chemical profiling of drugs of abuse has a long history and can rely on substantial reference databases because of the high case load, the situation is less favorable for explosives for which a lack of profiling options and reference data exists. When we consider the chemical profiling method for TNT that was introduced in **Chapter 6**, a clear proof of principle was shown leading to a limited evidential strength (LR values in the range of 10–100 for samples with matching impurity profiles). To make sure that these methods can be used with confidence in case work, R&D efforts should focus on the addition of features (e.g., inorganic and organic impurities) and the expansion of the database to make sure that reference probability distributions can be assessed more robustly. By conducting such category one and two projects, chemical profiling of intact explosives can be implemented in forensic practice and can assist in criminal investigations.

11.3 How to stimulate and organize forensic science and innovation

In the previous section, we have seen how meaningful R&D projects can be defined that increase the value of the forensic information in case work. This ensures that scarce capacity and budgets are used effectively. Forensic science is an applied science and although in academic environments valuable fundamental work is conducted, any innovation, scientific insight, or new technology in principle always aims to assist in criminal investigations and solve crime. Of course, not every method described in scientific literature will ultimately be applied in case work, however, this does not mean that such efforts are wasted. A healthy process involves the introduction of a new idea, followed by more academic studies preferably from various research groups to test its validity, map its scope, and prepare its introduction in forensic practice. In such a setting, a body of literature ultimately supports a new case work method which is also part of the Daubert criteria as discussed in **Chapter 9—Quality and Chain of Custody**. In addition, science is exploratory and curiosity driven which means that not every good project or promising equipment investment will be successful. Results can be very promising but can also reveal unforeseen complications and challenges that are difficult to resolve and ultimately require the management to terminate a project without the desired improvement in casework.

The introduction above indicates that being involved in science and innovation programs takes considerable time, effort, and budget for a forensic institute. Such means are scarce and urgent case work understandably has a higher priority with the forensic experts. These conditions can make it challenging to consistently invest in the latest insights and methods. In this section, we will discuss how a successful forensic R&D program can be established and maintained alongside casework. A clear distinction should be made between the R (Research) and the D (Development) in R&D programs. Development typically involves the introduction of new equipment in the laboratory and ongoing work to ensure that the forensic methods remain state of the art and can handle new evidence materials or new forms of criminal activities. Development projects are therefore usually closely linked to case work and need to be conducted by the forensic experts and professionals in the forensic laboratory. Normally, development activities are conducted in-house possibly in collaboration with an instrument supplier but are not part of innovation programs with a consortium of external partners. Such extensive collaborations are usually part of the research and innovation activities that aim for improvements in forensic practice on a longer time scale (e.g., in the next 3–5 years). The question is how such programs can be realized in a sustainable innovation ecosystem without draining the limited resources of a national forensic institute. A convincing solution to this challenge can be found by seeking the right partners to create a platform for collaboration in forensic science and education. Countries like the USA, Canada, Australia, UK, Switzerland, and the Netherlands have established powerful national forensic networks. In such networks, special academic centers, schools, or institutes that specialize in forensic science and education often play a central role. In the academic year 2020–21, in total 14 of such academic institutes specializing in forensic science organized a unique international lecture series named **Crossing Forensic Borders** to stimulate knowledge exchange and collaboration. An academic environment usually stimulates interdisciplinary collaboration and provides access to the latest scientific insights and technological developments. This means that when a connection to academy exists, the forensic potential of scientific breakthroughs can quickly be explored. It definitely helps that most scientists find the idea that their invention could assist in solving crime very interesting and therefore they typically are quite eager to collaborate.

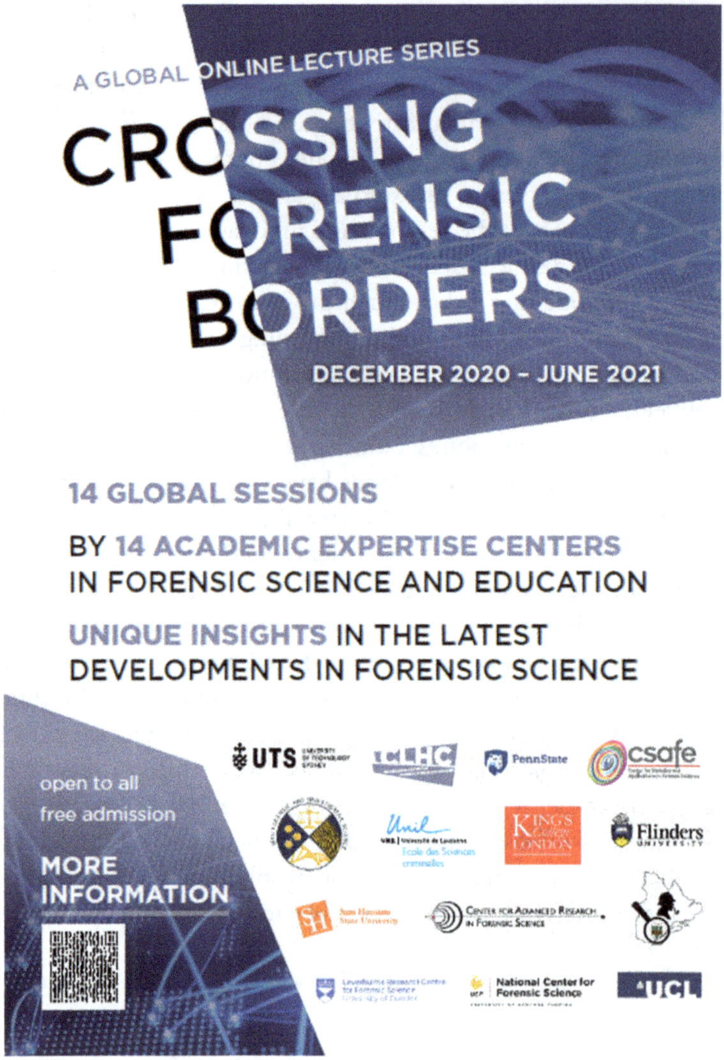

Crossing Forensic Borders poster announcement introducing the 14 academic forensic centers from the USA, Australia, Canada, UK, Switzerland, and the Netherlands contributing to the international lecture series

An additional reason for forensic institutes and law enforcement agencies to collaborate with universities is the access to research capacity. It is much more efficient for a forensic expert to supervise (Ph.D.) students in a joint project than to conduct all the research him/herself. Of course, supervision and guidance takes time and effort and is essential to ensure the quality of the work done. However, this investment will pay off once the students are capable of independently performing forensic scientific studies and reporting the results in the form of a presentation or academic paper. Depending on the topic such (Ph.D.) student projects can be conducted in university laboratories but because forensic institutes usually have dedicated and specialized instruments, it could also be useful to enable students to work directly at the forensic institute as part of an internship. This of course requires some precautions to ensure that the students have no access to case work and do not have direct access to physical evidence in ongoing criminal investigations. It could also require planning with respect to the access to instruments that are frequently used for regular case work. However, inviting talented students to conduct research in a forensic environment can be very valuable for several reasons. First of all, conducting academic research in a forensic laboratory fully involves the experts and this typically ensures that the work is representative and relevant for case work. This improves the odds that the results are actually implemented and expand the forensic toolbox of the forensic expertise area. Additionally, a forensic institute is introduced to young scientific talent and might consider offering jobs to these students if openings emerge. On the basis of an internship of several months or even for the duration of a complete Ph.D. project, the qualifications of a student are clearly demonstrated. On the basis of this experience, forensic experts can safely conclude whether a student would be a great addition to the team. In this way, supervising students and contributing to forensic academic education provides a return of investment as the future of forensic science is ensured by preparing new generations for a career as forensic expert or scientist.

However, to transform a novel idea into a powerful new case work method, two essential elements are still missing. First of all, tech companies and instrument manufacturers need to be involved to be able to develop robust technology from novel scientific insights. This is not necessary when a specific forensic methodology is developed on existing equipment using generally available consumables (such as the chemical profiling of TNT with vacuum-outlet GC-MS as discussed in **Chapter 6**). However, when innovative methods require dedicated or modified instrumentation, new (single use) sampling tools or reference standards, the quality required in case work can only be achieved when such items are produced on a larger scale under tight specification by specialized companies. As part of the casework, new instrumentation can be acquired, and the required tools and materials can be ordered as the new method becomes available to the forensic community. Performing case work on instrumentation that has been developed in an academic environment as part of a research project is typically not sustainable. The final element that needs to be added to the formula is financial means, the budget to conduct the research and develop new technology. Typically, universities require research funding to appoint temporary research staff such as Ph.D. students, whereas forensic institutes and other partners in the criminal justice system operate with tight budgets focused on criminal investigations. Reserving a substantial amount of the annual budget for exploratory research with external partners is usually not an option nor is the use of such budgets to directly appoint Ph.D. students. Companies can make decisions to invest in projects that show a good revenue prospect with the introduction of new products and services. However, the forensic market is relatively small and highly specialized and hence for most companies developing forensic technology is considered to be a spin-off activity with a modest market potential (it must be noted that there are companies that exclusively produce equipment, consumables, and standards for forensic casework but this is rare and such companies are relatively small). This results in companies being interested to participate in forensic projects but reluctant to provide the budget for the activities. With an intended use to support criminal investigations and share the findings with the forensic community without any restrictions, the exclusive financing of a forensic science project by a commercial partner is also not optimal. To resolve the budget challenge, consortium partners could apply for external funding by submitting their forensic science research proposals in suitable programs. Academia is used to finance their research in this manner and scientific staff often spend a significant amount of their time on grant acquisition. When forensic science proposals are submitted within generic funding schemes, forensic projects will have to compete with a broad range of science and application areas. This usually limits the success rate and to really boost forensic science dedicated national funding programs are required. In this respect, the substantial Horizon 2020 EU funding schemes dedicated to security (Secure Societies work program) can offer interesting options for advancing forensic science in a European setting.

As illustrated below, we now have all the elements for a successful forensic innovation program: a research consortium that is linked to forensic practice through representatives from the criminal justice system and forensic experts that want to expand their options to conduct case work. The scientific basis and a suitable academic research infrastructure and culture are provided by the university, whereas commercial and industrial partners have the technological capability to design and produce new instrumentation and associated prototypes. The budget for these activities is sourced externally by a national or international funding agency after a successful submission of the proposed research by the consortium. Possibly the funding requires the partners to contribute cash and/or in-kind but these contributions typically represent a fraction of the overall costs to conduct the research.

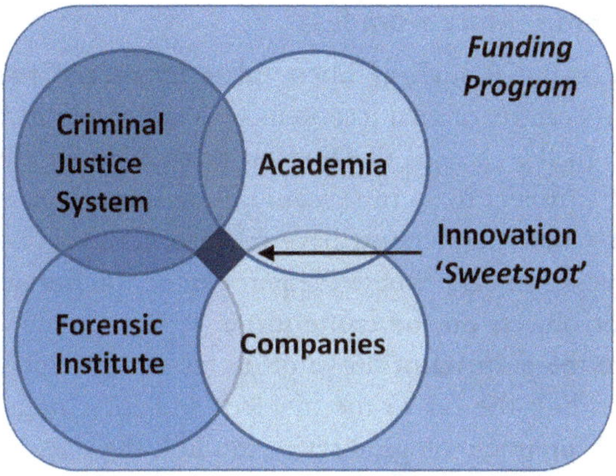

Schematic illustration how relevant forensic innovation programs can be realized by establishing a consortium representing the criminal justice system, academia and industry, and acquiring external funding *(Adapted from A.C. van Asten, Science and Justice, 54 (2014) 170–179, https://doi.org/10.1016/j.scijus.2013.09.003)*

11.4 Advancing forensic analytical chemistry

In discussing ways to successfully organize R&D in forensic science, the previous two sections were quite generic. The innovation strategies and models can be applied to every forensic science domain and are not specific to forensic analytical chemistry which is the topic of this book. Therefore, in this final section, we will focus our attention on promising innovation directions for forensic chemistry. Three themes will be explained and discussed in more detail and promising scientific and technological alternatives to create new investigative options will be presented. The theme "**More from Less**" focuses on trace evidence and the options for chemical profiling and detailed chemical characterization when there is very little material to work with. The theme "**From source to activity**" explores the options to interpret forensic chemical findings at activity level. This can significantly increase the added value of the investigation because the main questions in court concern the actions of persons not the origin of materials. The third theme "**Bringing chemical analysis to the scene**" addresses the concept of chemical identification outside the traditional laboratory. From an efficiency and safety perspective, the robust chemical analysis of drugs of abuse and explosives with rapid screening technology can be very beneficial to the criminal justice system.

Innovation theme "More from less"

Forensic trace material such as fibers, glass fragments, paint chips, and gunshot residues can be of vital importance to accurately reconstruct crime related events. In the absence of DNA or other biometric evidence, such traces can link a suspect to a crime scene. Forensic biological traces can be absent in cases where perpetrators have taken precautions by wearing face masks and gloves or in cases where the perpetrator was not in direct contact with victims or objects on the crime scene. If we consider a hit-and-run accident where the perpetrator drove off in his car after hitting the victim, paint traces can link the car to the incident and subsequently registration documents and forensic biological traces can link the suspect to the car (of course this does not necessarily prove that the suspect was driving the car when the accident occurred). In addition, trace materials provide information on the activities which resulted in their creation and/or transfer and this will be addressed in more detail in the next theme "From Source to Activity." The current theme "More from Less" addresses the challenge to chemically characterize trace material after detection and collection. We will consider fiber evidence which so far has not been discussed in detail, whereas fiber transfer constitutes important evidence especially in violent crimes. Given the intensity and the direct contact between victim and perpetrator, the mutual transfer of fibers from clothing items cannot be prevented nor be undone. If the perpetrator wears gloves to prevent leaving DNA evidence and fingermarks on the crime scene, fibers of the fabric of the gloves will inevitably be shed. But assuming that fibers of the gloves are indeed retrieved from the crime scene and that a suspect is arrested while in possession of a pair of gloves, we now need to consider what characteristic features we are going to extract and compare. Although the forensic microtrace expert has an endless supply of reference fibers (which can be obtained from the confiscated glove), there is usually very little crime scene material to work with. This can be limited to a few or even a single fiber. With very little material to begin with, it becomes very challenging to measure trace impurities for profiling purposes ("assessing the trace in a trace"). For this reason, the regular approach in forensic fiber investigation is to apply non-invasive methods aimed at classification. With an IR or Raman microscope, fibers can be visually inspected, the fiber color can be objectively assessed, and polymer matrices can be identified using spectroscopic signatures. However, fibers and associated garments are mass produced and hence establishing that the fiber found on the crime scene is of the same type and color as the glove material is somewhat incriminating, but the associated evidential value will be low. This situation is similar to the chemical identification of TNT as the main charge in the IED as discussed in **Chapter 6—Chemical Profiling, Databases, and Evidential Value**. Because TNT was found at the home of the suspect this provides some support for the prosecutor hypothesis but there is no specific information to differentiate and compare TNT samples. As for TNT, the next step would now be to develop a profiling method for fiber evidence. However, such a method should be able to robustly extract the features from a single fiber!

What fiber characteristics would allow an expert to differentiate mass produced fibers of the same type and color?

What analytical methods would be able to establish these characteristics robustly from a single fiber?

How would you set-up a research project to develop a fiber profiling method?

(This could also be discussed with fellow students as a team assignment)

The forensic microtrace experts of the Netherlands Forensic Institute recently introduced a novel method for the detailed chemical analysis of dyes used in fibers. A broad range of dyes and complex dyes mixtures is used in the fabric industry, and the dye profile could be used to differentiate fibers of the same matrix and comparable color. Furthermore, as a result of sunlight exposure and frequent washing, dyes in fabric might degrade over time creating a highly characteristic dye degradation profile that could be linked to a specific clothing item in comparison with items of the same manufacturer and production line/batch. The chemical analysis of the fiber dye composition is conducted with an LC-DAD-MS system providing full UV-vis and mass spectra of individual compounds separated on the HPLC column. A database can be constructed to identify dye-related compounds on the basis of LC retention time and UV-vis and mass spectrum. The high-resolution MS system (LTQ Orbitrap) enables the prediction of the elemental composition which aids the chemical identification of the dye constituents (for impurities and degradation products reference standards might not be available). To be able to perform this detailed chemical characterization, the forensic scientists of the NFI developed tailor made methods to extract basic and acidic dyes and direct, dispersive and reactive dyes from a single synthetic or cotton fiber. To release reactive dyes from cotton and viscose the extraction procedure involves the use of a cellulase enzyme to break the chemical bond between the dye and the fiber. Minimized extraction volumes are used to prevent excessive dilution in order to maintain sufficient sensitivity. In this manner, a substantial amount of chemical information is retrieved from a single fiber in a total run time of 78 min. Unfortunately, the method is invasive as the original trace material is lost during the extraction process. Therefore, the dye analysis can only be initiated after all traditional forensic investigations have been completed. An example of how this innovative approach assisted in an actual case investigation is provided below. This case revolved around a home robbery where the victims were forced to open their safe and were subsequently tied with a red rope that the perpetrators found in the house and belonged to the victims. Shortly thereafter the police chased a car with several individuals who were thought to be involved in the crime. During the chase, several items were thrown out of the car including a fake firearm. The police were able to arrest one of the passengers but the others ultimately escaped. During the arrest, the police officers noticed a small bunch of red fibers on a glove of the suspect. This material was secured and sent to the forensic institute together with the reference red rope which was used to tie the victims. The obvious question to the forensic experts was whether the tuft of red fibers found on the glove originated from the red cord. In the absence of DNA and fingermark evidence, the fiber evidence has become of crucial importance to investigate whether the suspect was at the home of the victims and was involved in the robbery. In addition to the findings of the regular fiber investigation (polymer matrix, physical appearance, and color assessment), the new LC-DAD-MS revealed that the dye composition in the red fiber material and the reference cord was highly comparable. To link an evidential value to such a "dye match," an extensive data collection of red dyes applied in similar synthetic fiber materials needs to be created. However, it is clear that with this advanced microtrace dye analysis, the forensic experts have provided incriminating evidence that in light of the circumstances will be difficult to refute by the defense council. With the findings of the traditional investigation, the evidential value will be much lower because of the widespread presence of synthetic fibers of the same polymer matrix and color.

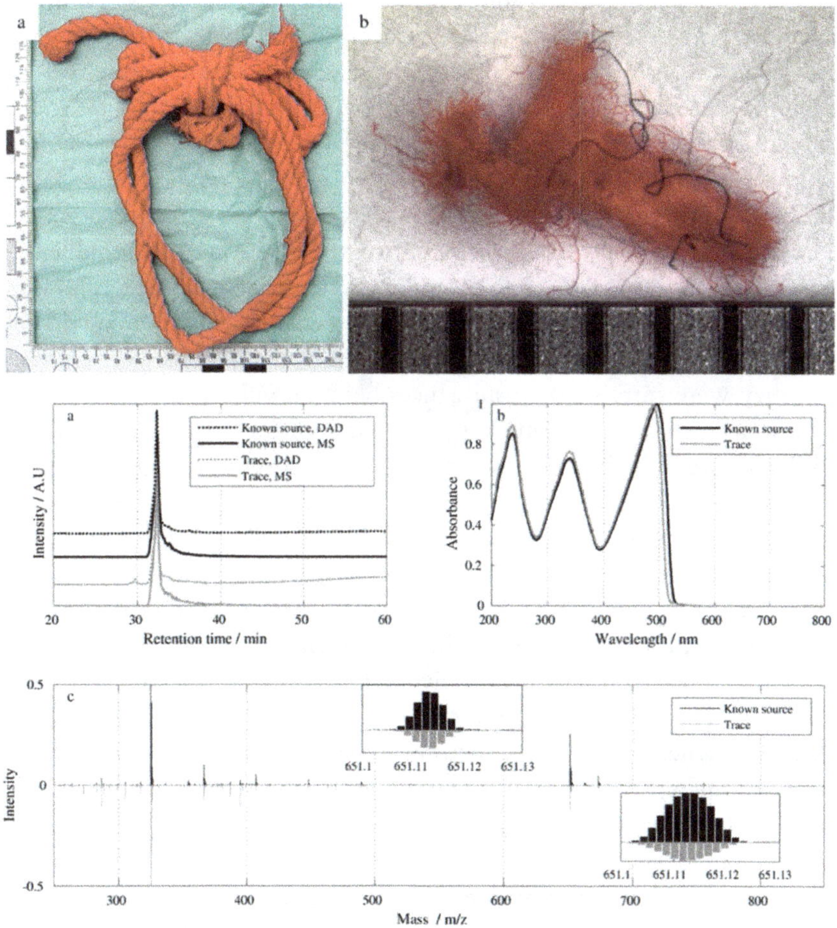

Top: Photographs of the red cord used to tie the victims (A) and the small bunch of red fibers found on the black glove of the arrested suspect (B). Bottom: Condensed representation of the LC-DAD-HRMS analysis of the dye extracted from both evidence materials showing a high degree of similarity of the chromatogram (a), and the UV-vis (b) and MS spectrum (c) of the main peak (at 32.3 min). Inserts show the high-resolution details of the main dye-related ion at 32.3 (Congo Red) and 33.8 min (unknown). With the Orbitrap the m/z value of this ion was found to be 651.1127 Da *(Reprinted with permission from Schotman et al. Forensic Science International, 278 (2017) 338–350, https://doi.org/10.1016/j.forsciint.2017.07.026)*

Innovation theme "From source to activity"

Crimes are human activities (animals can typically not be convicted) that are defined as such by law. Forensic evidence supports hypotheses and enables judicial and law enforcement professionals to reconstruct events as part of the investigation and ultimately to take (hopefully correct) decisions with respect to guilt or innocence. As we have seen throughout this book, forensic chemistry provides information on the nature, composition, and origin of evidence materials that have been secured from crime scenes or have been confiscated from suspects. Such information, although highly relevant, typically is provided at **source level**. This means that the source findings of the chemical analysis need to be coupled with activities of individuals through inference. Sometimes such a translation is rather straightforward, for instance because of the way the law is constructed. We have seen that virtually any activity with listed illicit drugs is forbidden by law including possession. Therefore, the chemical identification of a drugs of abuse (a source investigation) will almost certainly lead to a conviction. However, care has to be taken in situations in which persons were unaware that they were carrying such substances because without their knowledge they were used as drug traffickers. The same direct connection holds for DUI cases were the law clearly states maximum levels for drugs of abuse in human whole blood when driving a car. The activity is typically not contested because the suspect was arrested by the police when he or she was sitting behind the wheel of a car and basically was "caught in the act". Some care should be taken for rare cases where a driver was drugged unknowingly and was unaware of his/her intoxication. Another exceptional circumstance is a swift change of driver and passenger prior to the arrest or the actual driver fleeing a scene before the police approaches the vehicle. However, when considering microtrace evidence, the situation becomes more complex. Chemical analysis of microtraces aims to investigate whether a suspect was involved in certain activities during an incident or was present at the scene. Matching features produced with chemical profiling methods can support the assumption that microtraces are related to a source at the crime scene, especially when the features are characteristic. However, this typically does not provide any information how the microtraces were created and transferred from the source to the material that was in the possession of the suspect (*e.g.*, the presence of glass fragments from a broken window on the crime scene on the clothing of the suspect). If the defense council is not contesting the presence of the suspect at the crime scene but has an alternative explanation for the microtraces that is not crime-related, the forensic findings are not able to support either the hypothesis of the public prosecutor or the one from the defense. Developing methods that are able to provide information on **activity level** (see **Chapter 10—Reporting in the Criminal Justice System** and the discussion on **the hierarchy of propositions**) requires new scientific insights, instrumental innovations, extensive activity-based experimentation, and criminalistic interpretation. We will explore the options for the interpretation of activity-based chemical evidence for the "**shooter** versus **bystander**" dilemma in **GSR** analysis. Starting an R&D program to develop new options for interpreting GSR evidence at activity level could be a very valuable investment as crimes involving firearms and gunshot victims are considered to be high priority cases. Additionally, suspects often state that they were present when the shooting occurred but that they were not the shooter. True or not, this is an effective defense strategy because a GSR expert will never claim that the detection of GSR particles on a sampling stub proofs that the suspect fired a gun.

Why will SEM-EDX analysis of stubs used for GSR sampling most likely not provide information on activity level?

What analytical methods would be able to differentiate a shooter from a bystander on the basis of the presence and distribution of GSR?

How would you set-up a research project to develop a GSR method to provide information on the activities of a suspect during a shooting incident?

(This could also be discussed with fellow students as a team assignment)

The stubbing process using an adhesive tape is aimed at collecting and "concentrating" potential GSR particles on a defined substrate that subsequently can be scanned with SEM-EDX. As described in **Chapter 2—Analytical Chemistry in the Forensic Laboratory**, the tape surface is scanned automatically and any particles containing the elements that are characteristic for the primer composition (Pb, Ba, and Sb) in ammunition are highlighted. In this way, the GSR expert can quickly assess whether the hands or clothing of the suspect contained trace material that relates to a shooting incident. However, in this process, all information on the location of the particles is lost. Possibly the number of GSR particles on the stub could provide some indications on whether the suspect fired a gun or was merely a bystander. However, there are so many unknown factors that are governing the number of particles on the stub (sampling efficiency, sampling locations, loss of particles in subsequent stubbing, time between the shooting incident and the sampling, and activities during this interval) that a forensic expert would be very reluctant to link the number of particles to an activity. The actual position of gunshot residues and associated cloud patterns could be more informative and could result in an accurate reconstruction of the events in combination with other forensic evidence (*e.g.*, firearms investigation, DNA evidence, camera footage), tactical information, and witness statements. An example of this is the estimation of the shooting distance through the application of colorimetric agents to visualize GSR patterns. This requires a series of reference experiments to be conducted on a shooting range under controlled conditions. These conditions mimic the actual case conditions (e.g., firearm used, ammunition used, clothes worn by the victim) as much as possible while varying the shooting distance. By visual comparison of the GSR clouds, the expert is then able to estimate at what range the victim was shot. This typically works best for small ranges (less than 1 m) because at larger distance between the shooter and the victim, GSR traces on the victim will only originate from the bullet wipe as the GSR cloud that is expelled from the firearm during the discharge will not reach the victim. The visualization of GSR patterns is sometimes also applied to the suspect but experts typically do not prefer this line of investigation because of the destructive nature of the application of chemical agents and the incompatibility with the SEM-EDX screening. Additionally, the subsequent interpretation remains cumbersome for reasons listed above.

The novel instrumental development of large area XRF (X-ray fluorescence spectroscopy) imaging (**MA-XRF** or macroscopic-XRF) fueled by art science was recently identified as a promising, innovative tool for the non-invasive detection, visualization, and elemental characterization of trace evidence on physical evidence items. The chemical contrast is provided by element-specific signals and this allows the imaging of human biological stains and a range of microtraces including GSR. In a recent collaboration between Dutch art and forensic scientists, it was shown that MA-XRF is sensitive enough to detect and differentiate human whole blood and semen on the basis of the trace elements iron (500 ppm in whole blood) and zinc (150 ppm in semen). In addition, the method is very suited to image 2D GSR patterns on large clothing items such as T-shirts as is illustrated below.

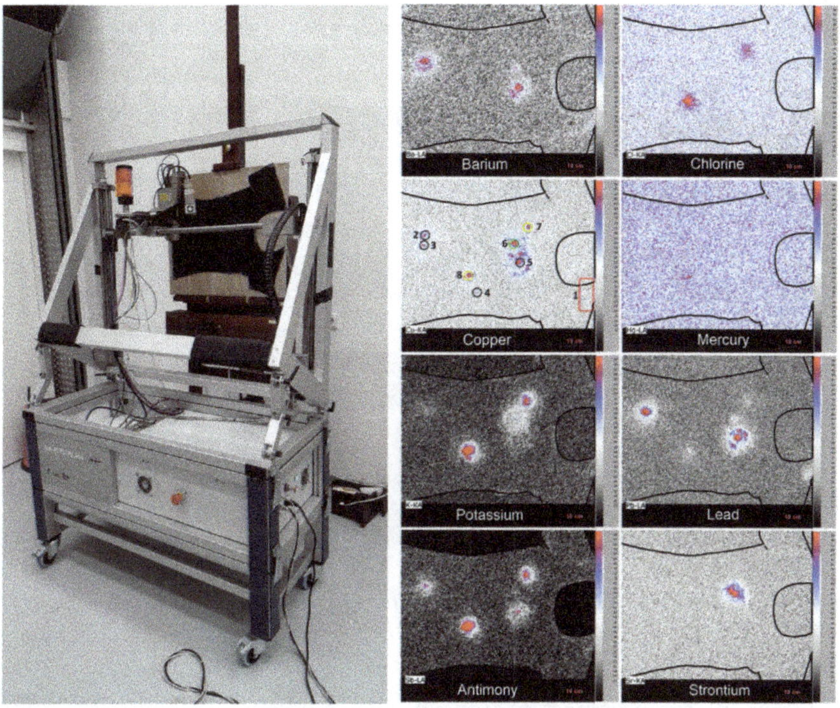

Photo from a black T-shirt being scanned with MA-XRF at the Rijksmuseum (Amsterdam, the Netherlands) and some resulting elemental images showing the GSR patterns after bullet impact of test firings conducted at the Netherlands Forensic Institute. The different elemental patterns are indicative of the various types of ammunition used while copper highlights the bullet impact ***(From Langstraat et al. Scientific Reports, 7 (2017) 15056, https://doi.org/10.1038/s41598-017-15468-5)***

An advantage of MA-XRF is its non-invasive nature. Although high energy X-rays are required to generate the elemental photon emission lines, this normally does not affect the integrity of the object and the traces. Even the genetic material present in human biological stains was found to remain mostly intact as DNA profiling was not negatively affected even after excessive exposure to radiation from the MA-XRF source. This means that elemental contrast can be generated without physically or chemically altering the evidence material leaving all other investigative options intact including SEM-EDX analysis after stubbing the areas of interest. In addition to the elemental composition of the GSR pattern also the shape and the intensity of the pattern can provide information at activity level. There is for instance a clear relation between cloud symmetry and the angle of the shot as is shown below. Additionally, in line with chemically visualized GSR patterns, the size of the cloud is indicative of the shooting distance.

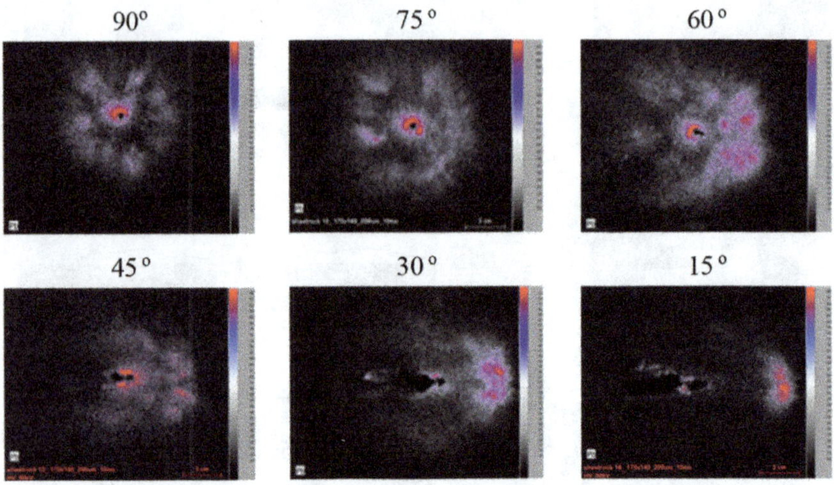

Lead XRF image of close range (5 cm) shots fired on a black T-shirt under different angles with the firearm tilted to the right *(From Langstraat et al. Scientific Reports, 7 (2017) 15056, https://doi.org/10.1038/s41598-017-15468-5)*

Another interesting aspect of using X-rays is offered by their penetrating nature, a characteristic that is used extensively in hospitals to diagnose patients using X-ray and CT scans and that has been successfully exploited for art historical and authentication studies. As canvasses were often reused to save costs, paintings were frequently covered by newer work. MA-XRF can reveal the presence of hidden paintings by selecting the elements exclusively present in the covered paint layers. This can reveal very valuable information with respect to the artist and the techniques used. From a forensic perspective, this means that elemental patterns that are hidden can be revealed using this technique despite of covering materials that usually make forensic traces of interest undetectable to the human eye and inaccessible with other spectroscopic techniques. When considering a shooting incident, GSR patterns on the clothing of the victim might become covered with blood. This severely hampers the sampling and visualization of the residues. However, when the trace pattern itself is not modified by the contact with blood, the elemental selectivity in combination with the penetration of the X-rays can reveal a GSR pattern as is illustrated below.

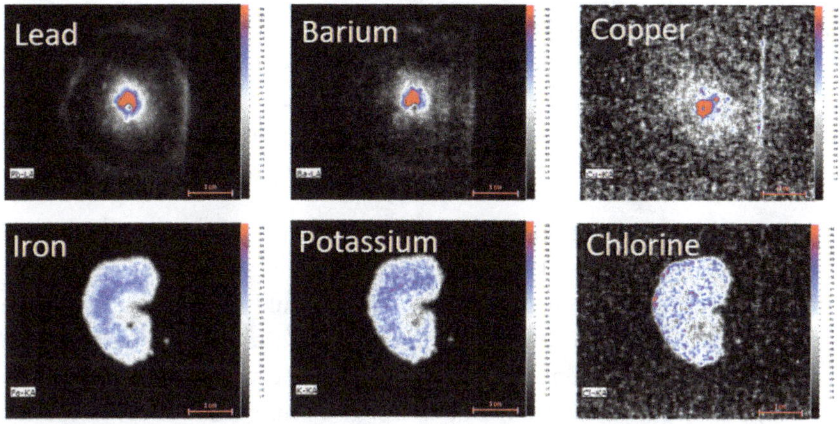

Elemental selection to visualize a GSR pattern (top row) that is covered by human whole blood (bottom row) ***(From Langstraat et al. Scientific Reports, 7 (2017) 15056, https://doi.org/10.1038/s41598-017-15468-5)***

Despite the forensic potential shown, the practical use of MA-XRF also comes with several challenges. This includes the complex, sizable, and costly instrumentation that requires trained and experienced users. However, a more remarkable unfavorable aspect is the long acquisition time required to scan a substantial area at the spatial resolution that is needed to acquire detailed images of trace patterns. Scan times exceeding 15 h are not uncommon and despite the fact that scanning can be performed automatically and unattended during out-of-office hours, this can delay the progress of the investigation. Especially in situations where technical issues occur and scans need to be repeated, more invasive but crucial investigations have to be postponed. Finally, the examples shown mostly focus on the victim and less on the shooter while the main scientific and technological challenge is to develop a forensic method that can address the question whether the suspect fired a gun or was a bystander when somebody else discharged a firearm. Irrespective of the potential of MA-XRF, the interpretation of suspect GSR patterns remains much more complex and requires extensive experimentation and accurate reconstruction of the incident. Unfortunately, no technique no matter how powerful can resolve the issue associated with the unknown factors that have affected the GSR pattern after the shooting. This especially limits the possibilities when there is a substantial time interval between the shooting incident and the arrest of the suspect, an interval during which several unknown activities of the suspect could have affected the presence and distribution of gunshot residues. However, under favorable case conditions, this new elemental imaging technique can provide promising options to interpret GSR findings at activity level, enabling an accurate reconstruction of a shooting incident. This could also involve MA-XRF scanning of clothing items or intact hand stubs of apprehended suspects.

When considering novel developments in forensic chemistry to generate forensic information at activity level, an interesting "class" of techniques aims to estimate "**the age of a trace**." This should not be mistaken with the age of the source such as techniques to reveal the age of the donor of a biological stain or a fingermark. In this instance, the aim is to estimate when the forensic trace was created, i.e., to assign a "time stamp" to it. We are basically trying to provide forensic information to address questions like "when was this fingermark placed?" or "when was this blood spatter pattern created?" The most frequent and well-known time-related question when human remains are involved pertains to the **time of death—ToD**—(or more accurately **time since death** from which the ToD can be derived as the time of the discovery of the body and subsequent autopsy is known). This is typically addressed by the forensic pathologist, forensic physician, or forensic anthropologist on the basis of **human taphonomy** (soft tissue decay and phenomena like rigor mortis and **liver mortis** also known as lividity), **body cooling,** or **forensic entomology** (study of insects and insect egg development stages on or near human remains). However, novel biochemical and bioanalysis methods have been proposed based on for instance increasing potassium levels in the vitreous humor (the gel liquid inside the human eyeball) or involving advanced proteomic approaches to identify degradation markers in biological tissues. The reason for asking a temporal question in a criminal investigation is either to accurately reconstruct the events in order to create new leads or, in case a suspect has been identified, to link this person not only to the scene of the crime but also to place him there at the time the crime was committed. Why this can be crucial is illustrated in a fictive but realistic case example. For those interested, an excellent review and analysis of temporal aspects in forensic investigations by Weyermann and Ribaux has been listed in the Further Reading section of this chapter.

A case example

After a call reporting what seems to be a struggle in an apartment, the police officers attending the scene find the lifeless body of a young woman that apparently has been stabbed to death. This is later confirmed by the forensic pathologist that reports multiple knife wounds. However, the knife used in the attack is not found at the crime scene. The crime scene officers meticulously investigate the many blood traces of which most originate from the victim. On the basis of the pattern, they, however, also find a few blood drops that could originate from the perpetrator. It is not uncommon that in a knife attack the perpetrator is also wounding him or herself. DNA profiling of these traces indeed reveal the profile of an unknown male and not that of the female victim. On the basis of tactical results, the boyfriend of the victim has become the prime suspect and he is arrested and interrogated by the police. As indicated in the law, the suspect has to provide a buccal swab reference sample and his DNA profile matches with the profile obtained from the blood traces on the scene. The couple was not living together but the boyfriend frequently stayed at the apartment of the victim. The suspect denies having assaulted his girlfriend despite the fact that intense arguments in recent times have been reported by several witnesses. He claims that he was in his house alone at the time of the attack but there are no witnesses to confirm this. When confronted with the incriminating evidence, he states that the blood drops originate from his last visit a week ago when he suddenly had a nose bleed. The police are contacting the forensic institute to consult with the experts if additional investigation is possible to check the statement of the suspect. However, no standard methods exist to answer this question.

Can you think of an innovative approach to provide forensic information that can support or refute the suspect's story? What hypotheses need to be considered? What trace features could be explored and what analytical chemical techniques would we need to develop to assess these features?

The district attorney leading the criminal investigation does not believe the story of the suspect and finds it a very unlikely coincidence that the boyfriend spilled a few blood drops a week prior to the violent crime. She also finds it strange that these blood drops were not cleaned in the meantime. However, the defense hypothesis does provide a reasonable alternative explanation for the presence of the blood drops and hence the fact that the DNA profiles match does not provide any information to distinguish the following hypothesis pair:

H_p: "*The suspect killed his girlfriend with a knife wounding himself in the process and leaving blood drops at the scene*"

H_d: "*Somebody else killed his girlfriend without any involvement of the suspect, the blood drops of the suspect were shed a week prior to the crime as the result of a sudden nose bleed*"

If we want to provide scientific, objective information that is able to address these hypotheses, a novel forensic method is required that is able to estimate the age of a human blood stain. Such a method should be able to assess characteristic features of blood that change over time. These time markers could, for instance, be related to the degradation of the bio-organic blood constituents when exposed to air. Over time, for instance, proteins present in blood might get oxidized and form new compounds that are not present in human whole blood in living individuals. However, such reactions in a single blood drop will probably be very complex and also governed by several ambient conditions such as the amount of blood, air flow, temperature (sun)light exposure, humidity, and the presence of bacteria and other microorganisms. The effect of these parameters must be carefully considered and modeled in reference experiments to create accurate blood aging models. In addition, the conditions at the crime scene need to be established to be able to provide the model with the correct parameter values to be able to estimate the "time since deposition" of the blood stains found at the crime scene.

Due to cases as described above, the forensic scientific community has been searching for novel methods that enable the estimation of the age of biological traces and fingermarks (Note: the age estimation of a fingermark is even more complex because unlike blood stains the chemical composition of a fingermark can show significant variation depending on the donor, substrate, and conditions). A robust, elegant and noninvasive approach is based on a well-known phenomenon that can be visually assessed, namely, the fact that a blood stain outside the body gradually changes color from red to brown. At the Amsterdam University Medical Center in the Netherlands, the Biomedical Engineering, and Physics department has considered this phenomenon to develop a hyperspectral imaging (HSI) method to detect blood traces on a crime scene but also to estimate the age of blood stains. With the right light source, the HSI camera can record reflectance spectra of blood stains in the UV-vis–NIR wavelength range. By unraveling the biodegradation of hemoglobin, the protein abundantly present in red blood cells to transport oxygen in the human body, accurate aging models of blood stains were developed. Hemoglobin contains iron to bind oxygen and the color change from red to brown is actually caused by the oxidation of iron to iron oxide in a way that bears resemblance to rust formation on an iron frame of a bike. Outside the human body oxyhemoglobin (containing Fe^{2+} iron) is converted to methemoglobin (containing Fe^{3+} iron) which in turn is irreversibly converted to hemichrome (methemoglobin degradation product). This composition of these three compounds changes over time and affects the shape of the reflection spectrum. If the conditions are known or can be estimated, these models can be used to estimate the age of a blood stain on the basis of a reflection spectrum as measured with HSI. The framework for blood stain age estimation using HSI, as published in Forensic Science International in 2012, is illustrated below. For over a decade, the models, camera technology, and software have been further refined and the hyperspectral camera is used several times a year on Dutch crime scenes at the request of the forensic experts of the police.

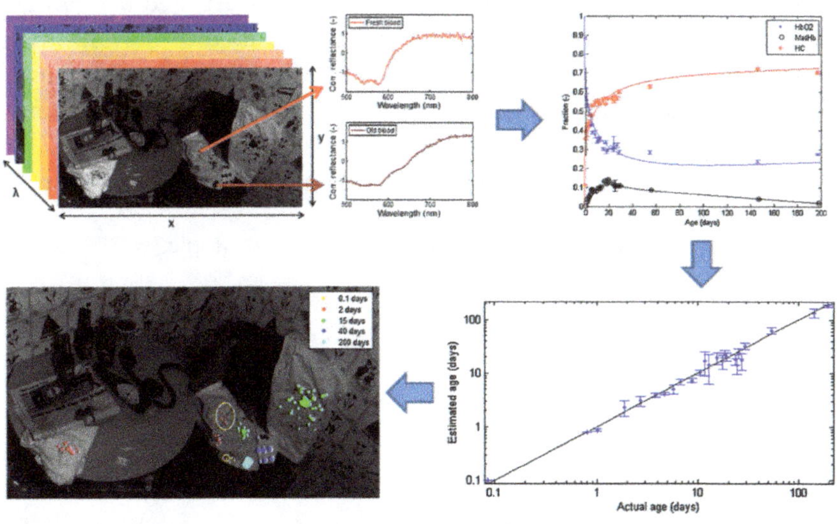

Illustration of the use of an HSI camera to non-invasively detect and estimate the age of blood stains on a crime scene. Distinct human whole blood UV-vis reflectance spectra gradually change over time as oxyhemoglobin is converted to methemoglobin which is irreversibly degraded to hemichrome. By accurately modeling this process through model experiments, the age of unknown blood traces can be estimated *(Reprinted with permission from Edelman et al. Forensic Science International, 223 (2012) 72–77, https://doi.org/10.1016/j.forsciint.2012.08.003)*

In the discussion on the age estimation of forensic traces, we briefly addressed that estimating the time of the deposition of a fingermark is extremely complex because the initial chemical composition of a fingermark is unknown, which further complicates the identification and modeling of suitable aging markers. Still a substantial body of scientific literature exists on this topic as several teams of forensic scientists attempt to tackle this "ultimate challenge." Despite its chemical complexity, a fingermark is actually are a very interesting trace from an activity perspective. The reason for this is obvious, as the mark was created while the donor was undertaking a certain activity. Ideally, a fingermark can thus be linked to the donor through the very characteristic (some claim unique, but this is scientifically incorrect) ridge pattern, while the location and chemical composition of the mark might provide an indication of the activities that lead to creation of the mark. The location can provide very obvious clues, for instance, when a fingermark is found on a firearm, a knife, or another object that can be directly linked to the crime. However, the difference between holding and actually using an object to commit a crime always needs to be appreciated. By combining forensic evidence, this can be investigated further, e.g., when a bullet in a victim can be linked to a firearm, the firearm contains fingermarks from the suspect, and GSR particles matching the ammunition were found on the hands of the suspect. However, even in this case, it could be that the suspect fired the gun not hitting the victim and then handed the gun over to another individual who subsequently shot the victim and fled the scene. This is just a simple example to illustrate the complexity when interpreting forensic evidence at activity level. In addition to the position, chemical clues in fingermarks might also prove very valuable when investigating the activities of a suspect.

This will be explored in the following fictive case example.

A case example

A very substantial police team is in search of the individuals responsible for a series of bomb attacks with what seems to be a terrorist motive. After a number of successful attacks injuring several innocent people, a recent attack failed. After the bomb ordnance squad has carefully defused an IED found at the crime scene, the forensic experts can investigate the construction including a timer, a battery, a blast cap, and a main charge. The main charge consists of TNT (tri-nitro-toluene) an explosive that typically is obtained from military sources. At the same time, there is also a tactical breakthrough in the investigation after the police receives an anonymous tip. This leads to a remote location where a garage was most likely used to prepare the IEDs. At this new crime scene, residues are found of TNT and some construction parts similar to those found on the unexploded IED. Several fingermarks are detected in the garage and one of these fingermarks match to a fingerprint of the officially registered owner of the garage. Upon police questioning, this suspect claims that he was unaware of the activities in the garage and states that the perpetrators must have broken into his garage. He also indicates that he visited the location for the last time several months ago and that on that occasion he most likely left the fingermark that has been secured by the police. The team discusses the situation with the forensic experts. Unfortunately, they see no option to estimate the age of the fingermark also because it has been dusted and tape-lifted.

Can you think of an innovative approach to provide forensic information that can support or refute the suspect's story? What hypotheses need to be considered? What trace features could be explored and what analytical chemical techniques would we need to develop to assess these features?

Also in this case the suspect provides a somewhat unlikely but plausible explanation for the presence of his trace material at the scene, which could be correct especially in the absence of other evidence with respect to extremist beliefs and links to the attacks. Again, the source level connection on the basis of a ridge pattern match equally supports both hypotheses as the defense proposition acknowledges that the suspect is the donor of the mark but provides an alternative explanation how and when it got there:

H_p: *"The suspect was involved in the terrorist attacks, the fingermark of the suspect was set during the construction of the IEDs in his garage"*

H_d: *"The suspect has no involvement in the terrorist attacks, unknown individuals prepared the IEDs in his garage without his knowledge, his fingermark predates the criminal activities"*

With no possibility to estimate the age of the fingermark, forensic experts need to explore other options. If the fingermark indeed predates the IED manufacturing, it should in principle not contain any chemical traces that relate to TNT and its handling. However, when the suspect was actively involved, then the fingermark could contain chemical residues of TNT (the absence of such clues does not necessarily mean that the suspect has no involvement, he could for instance have been wearing gloves during the handling of the chemicals, which actually is a wise thing to do as TNT is quite toxic). Scanning fingermarks for the presence of compound residues that can be attributed to the crime (explosives such as in this case but also drugs of abuse in cases involving illicit substance preparation although drug residues in fingermarks could originate either from use or handling) could reveal very important information on donor activities. But this also involves a substantial analytical chemical challenge, what technique would be able to detect a small chemical trace in a single fingermark? Dissolving the trace followed by LC-MS or GC-MS analysis is invasive and would destroy the trace and might not be sensitive enough as the extraction procedure further dilutes any trace amount present. A promising class of techniques that in recent times has been explored for forensic use is **imaging mass spectrometry** of which **MALDI-MS** is the most successful. MALDI stands for **Matrix Assisted Laser Desorption Ionization** and was discovered and developed in the period 1985–1990 as a breakthrough technique to ionize high molecular mass biomolecules, allowing for the first time the mass spectroscopic analysis of intact proteins with masses in the kDa range. The soft ionization prevents fragmentation and is realized through the use of a matrix compound that is sprayed onto the substrate surface to be analyzed. After solvent evaporation, the matrix compound forms crystals on the surface that absorb the radiation of a UV or IR laser. At the point of laser impact, the MALDI matrix molecules are excited, ionized, and transferred to the gas phase. In this process, analyte molecules are locally ionized mainly by charge transfer from the matrix species although the ionization mechanisms in MALDI are still not fully understood. Frequently used matrix compounds in MALDI-MS are sinapinic acid (SA), α-cyano-4-hydroxycinnamic acid (α-CHCA) and 2,5-dihydroxy benzoic acid (DHBA).

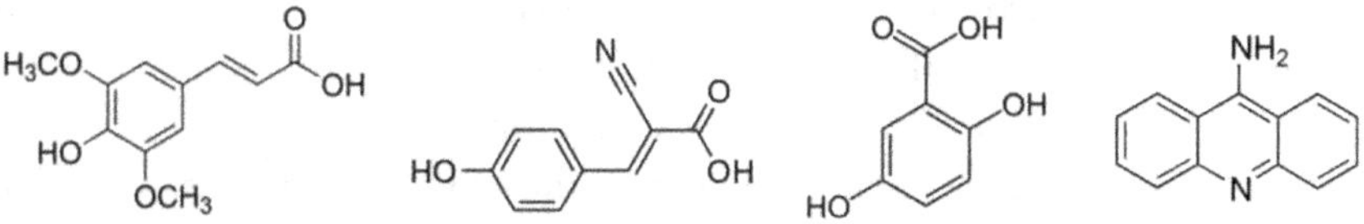

MALDI matrices that facilitate soft ionization of analytes of interest, from left to right sinapinic acid (SA), α-cyano-4-hydroxycinnamic acid (α-CHCA), 2,5-dihydroxy benzoic acid (DHBA), and 9-aminoacridine. The latter matrix can be used for the MALDI-TOF imaging of aromatic nitro explosives in negative mode

MALDI is found to be a very efficient soft ionization technique both for macromolecules and low molecular mass compounds and because ionization is induced locally with the laser, high resolution imaging is feasible with image resolution ultimately determined by the spot size of the laser. Typically, MALDI-MS instruments are based on time-of-flight (TOF) mass spectrometers because of their broad mass range, satisfactory mass resolution, and rapid scanning. When covering the entire area of a fingermark, the rapid acquisition of a mass spectrum at a given laser position is important to prevent excessive scan times. In this way, the x-y stage can swiftly move the entire fingermark in front of the laser and an image can be obtained in a reasonable time frame. This generates a data cube in which an entire mass spectrum is stored for each X−Y pixel in the image. This can be used to create chemical contrast by selecting certain ions in the TOF mass spectrum to enhance the ridge detail of the fingermark. In addition, the same ions can indicate compounds of endogenous and exogenous nature in the fingermark that can provide valuable information of the donor and his/her activities during the creation of the mark.

A recent study from forensic scientists of Albany State University (USA), showed the use of DART-MS to detect three aromatic nitro-organic explosives (TNT, picric acid, and tetryl) in a fingermark and the use of MALDI-TOF MS to subsequently image the distribution of these energetic materials in the mark. The TOF MS was operated in negative mode as is usually required for the analysis of energetic materials that typically do not form positively charged ions. The MALDI matrix compound 9-aminoacridine (structure illustrated above) provided satisfactory ionization efficiency and thus sensitivity for TNT, picric acid, and tetryl. MS/MS of the main ions was applied for compound confirmation. The authors did not indicate whether the method would be suitable for organic esters like ETN and PETN and nitro amines that do not contain an aromatic moiety in the chemical structure. Furthermore, the experiments were conducted under well controlled, favorable laboratory conditions as volunteers placed fingermarks on glass slides after contacting residues of standard solutions of the compounds of interest. Therefore, in order to use this promising methodology in the case at hand, additional method development is required to make MALDI TOF-MS applicable to latent marks that have been visualized (chemically, e.g., via cyanoacrylate fuming, or by dusting) and/or tape lifted by crime scene officers. Direct application is currently only possible for clearly visible fingermarks placed on flat glass surfaces of objects that can be transported to the laboratory. However, forensic scientists of Sheffield Hallam University have recently shown that MALDI MS imaging can be successfully integrated in common police procedures to visualize and recover fingermarks from crime scenes (https://doi.org/10.1039/c7an00218a). Possibly this approach could also be used for the detection and imaging of energetic materials such as TNT.

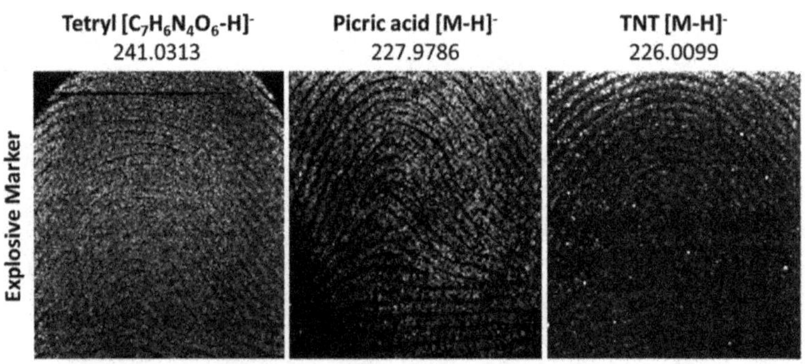

MALDI TOF MS images of volunteer fingermarks with aromatic nitro explosives placed on glass slides, negative ionization mode with 9-aminoacridine as MALDI matrix *(Reprinted with permission from Longo and Musha, Forensic Chemistry, 20 (2020) 100269, https://doi.org/10.1016/j.forc.2020.100269)*

Innovation theme "Bringing chemical analysis to the scene"

As was discussed in **Section 11.1**, forensic case work is performed by forensic experts to generate objective information that can assist in criminal investigations and court proceedings. The time required to generate that information can strongly affect its added value especially during the investigation phase when rapid availability of forensic findings can be very beneficial. It can provide new leads for or direct the investigation, which saves police capacity and, more importantly, improve the odds of solving the crime and finding the perpetrators. The so-called **golden hour** illustrates the importance of speed in a criminal investigation, rapid processing of the crime scene and the timely reporting of forensic findings. As time progresses and leads diffuse, the chance that the team will successfully conclude the criminal investigation diminishes. In high profile cases involving serial perpetrators, DNA profile or fingermark matches can prevent subsequent crimes providing that such results become available in time. Biometric evidence can also exonerate suspects and when such important information becomes immediately available, the incarceration of innocent individuals can be minimized thereby limiting the emotional distress. When considering high-volume case work such as the chemical identification of drugs of abuse or DUI cases in forensic toxicology, swift analysis in the laboratory improves efficiency and prevents backlogs and delays for the forensic institute, the police organization, and the entire criminal justice system. In general, we can state that whenever questions emerge in a criminal investigation, ideally such questions could be answered immediately by the forensic experts. However, it is generally accepted that these answers take time because evidence materials have to be transported to the forensic institute, requests have to be processed and ongoing investigations have to be finished. The shared opinion is that forensic chemical analyses need to be conducted by experts in controlled laboratory environments with dedicated, high-end analytical instrumentation to ensure the quality that is required for evidence that is admissible in a court of law (see also **Chapter 9—Quality and Chain of Custody**). However, the benefits would be substantial when through scientific insights and technological developments the chemical analysis of drugs of abuse and explosives could be conducted by law enforcement professionals directly on the crime scene or at the police station. Under the motto "***Send us your data, not your samples***", the Netherlands Forensic Institute and the Dutch Police in 2015 introduced **NFiDENT**, a revolutionary **process innovation** in the criminal justice system. With the NFiDENT approach, illustrated below, police stations were equipped with a GC-MS system allowing trained police officers to screen confiscated case samples for the presence of the most frequently occurring listed illicit drugs of abuse (cocaine, heroin, amphetamine, metamphetamine and MDMA, accounting for over 80% of the illicit drug casework in the Netherlands). This required a substantial investment in equipment and infrastructure as several GC-MS systems were installed in dedicated laboratories at police stations throughout the Netherlands. It also required a substantial investment in police capacity as police forces operating an NFiDENT system needed to have colleagues trained on the instrument, needed to ensure that their knowledge and skills were kept up to date and finally needed to free a substantial amount of their time to conduct the chemical drug screening on a daily basis. On first sight, decentralizing the chemical identification of illicit substances does not seem to be a sensible way forward. Instead of one highly specialized forensic laboratory, with NFiDENT several smaller laboratories with trained police staff need to be maintained. However, the overall picture becomes much more favorable when considering that the extensive logistic system that is required to transport the samples from the various regions to the Netherlands Forensic Institute is no longer necessary. The main driver is not so much based on cost-saving but rather to dramatically reduce the lead time of the investigation. By doing the measurements locally, the results become available on a very short time scale, which increases the efficiency of high-volume illicit drug screening on a national level and eliminates backlog (this by the way also provides cost savings). However, this efficiency gain only materializes when the findings are admissible as forensic evidence in a court of law. As discussed in **Chapters 9** and **10**, this requires fulfilling the Daubert criteria, a strict quality regime (preferably meeting the ISO17025 requirements) and a report signed by a registered forensic illicit drug expert.

Results generated with NFiDENT are court approved in the Netherlands. How is this achieved (Tip: think of the motto "*Send us your data not your samples*")?

The use of GC-MS in combination with colorimetric testing is standard practice in the identification of illicit substances. The fact that this type of high-end chemical analysis is conducted in a police laboratory is also not unique, in many countries the main forensic laboratory is part of the police organization. The innovative aspect of NFiDENT is the connection of all the police GC-MS instruments to a central server where the measurement data is stored. The registered forensic experts of the NFI have access to this data and after a thorough quality check they inspect the GC-MS runs for the presence of listed substances on the basis of observed peak shapes, retention times, and EI mass spectra, and by conducting an EI-MS library search. After completing the data analysis, the registered forensic expert signs the report that is sent electronically to all parties involved and is subsequently archived. This report is directly admissible as evidence in court allowing the police officers to quickly process the case, which is especially beneficial for relatively minor offenses involving the possession of small amounts of listed substances. Prior to the introduction of NFiDENT, the average lead time of a request for a qualitative forensic drug screening was 22 days and with the new approach the forensic report can be made available in 24 h after the analysis. In addition to this staggering reduction in lead time, the central collection of drug analysis data allows for trend analysis which can assist policy makers and forensic intelligence on the basis of chemical impurity analysis that provide insight in illicit drug manufacturing and trade.

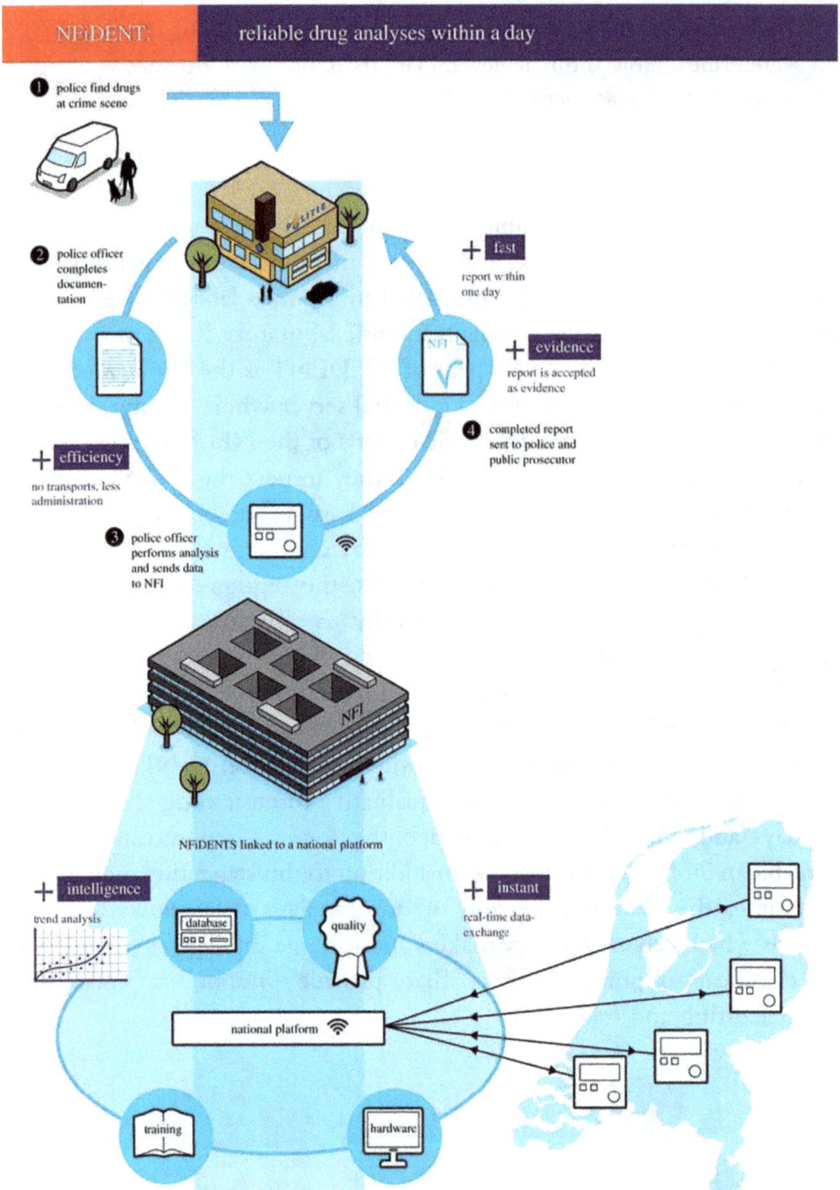

Illustration of the NFiDENT process. The police analyze a case sample on their own GC-MS system after which the data is stored on a central server. The forensic illicit drug expert inspects the data and documents the outcomes in a digital report that is electronically signed. The police retrieve the electronic report from the server to process the case. The presented evidence is admissible in the Dutch courts *(Reprinted with permission from Kloosterman et al. Philosophical Transactions of the Royal Society of London B, 370 (2015) 2014064, https://doi.org/10.1098/rstb.2014.0264)*

Although NFiDENT has provided impressive results in the Dutch criminal justice system, it does require the use of dedicated, high-end laboratory equipment, a dedicated infrastructure, and some form of local sample logistics. GC-MS instruments typically cannot be used at the scene, which indicates that confiscated, suspected drug samples still need to be brought to the police station for further analysis. Portable GC-MS instruments exist but typically not with the same functionality and specifications as the laboratory grade versions. Especially, operating a mass spectrometer at high vacuum is very challenging outside the laboratory environment and limits instrument portability given the substantial power consumption and size and weight of the vacuum pumps. However, there is undoubtedly also a wish for law enforcement to be able to perform drug and explosives screenings on the spot, e.g., at check points, during arrests and at the crime scene. This is illustrated with the extensive use of indicative testing such as colorimetric assays, breathalyzer tests and the use of ion mobility mass spectrometers at airports. However, typically a lack of selectivity can produce false positive results and therefore indicative test results in the field need to be confirmed with a subsequent analysis in the forensic laboratory. So, the ultimate challenge of the innovation theme *"Bringing Chemical Analysis to the Scene"* is to create a portable version of NFiDENT, i.e., a system that allows for a rapid, easy-to-use chemical analysis in the field by law enforcement professionals while meeting all the requirements for providing results that are admissible as forensic evidence in court. At the time that this chapter was written (September 2021), such a system was not in operation anywhere in the world.

Do you think that it is possible to develop a system that yields admissible evidence in court with a rapid, mobile analysis of drugs of abuse in the field?

What analytical methods would you select for such a system and why?

What would be the framework of the entire system?

How would you set-up a research project to develop such a system?

(This could also be discussed with fellow students as a team assignment)

In a 2020 publication the University of Lausanne (Switzerland) in collaboration with the University of Liege (Belgium) introduced a rapid (5 s!), portable qualitative and quantitative analysis of cocaine, heroin, and cannabis using a handheld NIR spectrometer in combination with a cloud-based server that receives the data from the NIR measurement and returns the results to an app on the mobile phone of the police officer conducting the analysis in the field. The need for a quantitative assessment is based on the Swiss illicit drug legislation which makes a distinction between personal use and trafficking or dealing on the basis of the total amount of pure active ingredient that was confiscated from the suspect. Server software receives the data via a secured connection, archives case-related information including the geolocation of the measurement (provided by the mobile phone of the user), performs the required data preprocessing steps (mostly SNV as such or followed by taking a first or second derivative of the signal), conducts the data analysis by applying the chemometric models that were developed on a substantial training and validation test set and finally electronically sends the results of the analysis to the officer in the field. The app on the phone displays these results in a user-friendly manner and summarizes the key information, i.e., whether the sample contains cocaine/heroin and if so, at what level or whether the sample is a CBD (<1% THC, legal) or THC type (>1% THC, illegal) cannabis. Interestingly, like the NFiDENT system, the authors also foresee the use of a more elaborate desktop application that allows for database and model management but also facilitates the use of the full collection of NIR measurements in the field for intelligence and policy purposes. The entire process, illustrated below for a cocaine sample, provided a very robust framework reporting a sensitivity (number of true positive results divided by the total number of positive samples in the validation set as determined with GC-MS) higher than 99% in combination with a 100% selectivity (number of true negative results divided by the number of negative samples in the validation set as established with GC-MS). These results are very promising because the validation samples consisted of real street samples confiscated by the Swiss police in the period 2016–2020 and representing a wide range of adulterants and active compound levels. However, the authors stress the need to continuously monitor the performance of the models through statistical quality criteria.

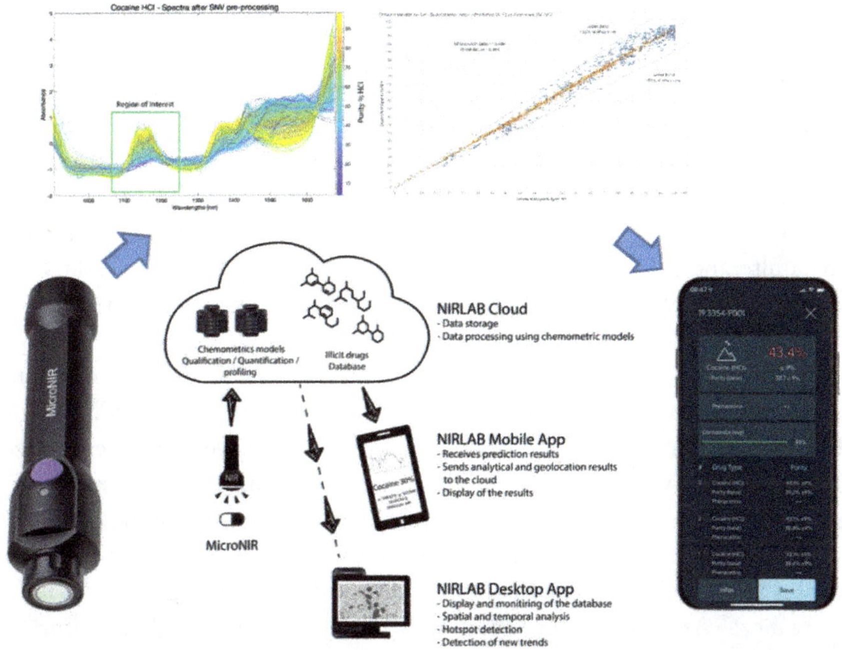

Illustration of the process developed by researchers from the University of Lausanne and the University of Liege for a robust qualitative and quantitative analysis of cocaine, heroin, and cannabis with rapid, portable, and non-invasive NIR spectroscopy *(Reprinted with permission from Coppey et al. Forensic Science International, 317 (2020) 110498, https://doi.org/10.1016/j.forsciint.2020.110498)*

What are the challenges associated with the use of portable NIR spectroscopy on a large scale by law enforcement using the presented framework?

What instrumental method would be ideally suited to tackle these challenges?

Why then is this instrumental method not used yet by law enforcement?

The main challenge associated with any portable spectroscopic technique relates to the fact that the measured spectrum reflects the entire composition of the sample. This was also discussed for lab-based spectroscopy in **Chapter 4—Qualitative Analysis and the Selectivity Dilemma**. In addition to the active ingredient of interest, the NIR reflection spectrum will also contain features of the additives. Depending on their nature and level in the sample, adulterants can thus strongly affect the measurement. As the composition of a new street sample is unknown, extensive calibration and validation datasets are required for the chemometric models to be able to accurately account for all possible variations in the spectral features to accurately and precisely establish if and how much of the illicit substance is present in the sample. This also makes the models very sensitive to changes in adulterant compositions over time. If for instance a new additive is introduced in street formulations and this additive is not included in the training data sets, this could lead to erroneous results for new case samples containing this additive. After this novel additive is discovered and identified in the forensic laboratory, new samples containing this additive will have to be added to the NIR database, models will have to be recalibrated and additional validation will have to be undertaken. If street sample formulations change frequently over time, keeping the databases and models up to date could be a laborious and challenging process. Another challenge relates to the wide range of compounds that have been registered as illicit substances. As discussed in **Chapter 4**, the list of hard and soft drugs is extensive and is continuously expanding due to the introduction of NPS that over time are replaced by successors when they are legally banned. Ideally, methods are able to detect all listed compounds or at least cover a majority that includes all compounds that occur frequently in case samples. Having a very effective portable method that only works for a limited number of drugs of abuse is not practical from the perspective of the law enforcement officer. The idea of portable analysis quickly loses its attraction when a whole range of techniques have to be applied on the scene or when laboratory analysis is still required to complete the screening. Furthermore, not every listed substance and associated street formulations might be suitable for NIR analysis, and the ones that are suitable will require their own reference data set and tailor made chemometric models. Furthermore, every substance will require its own reference data set and tailor made chemometric models. As the method includes more compounds, maintaining the data and the models will become more laborious as well.

An instrumental method that in principle is ideally suited to cover a wide range of analytes of interest while not being affected by the product formulation is **mass spectrometry**! The reason for this is that mass selectivity can be used for both compound selection and mitigation of adulterant interference. Based on triple quad mass spectrometry, MRM schemes can be designed that quickly "cycle" through an extensive list of compounds on the basis of their precursor ions while the product ions produced in the collision cell can be used to confirm the identity and perform a quantitative analysis (involving suitable calibration schemes) if necessary. As long as the frequently occurring adulterants do not generate ions of the same m/z value as the precursor ion and do not cause excessive ion suppression, the MS results will not be affected by the actual composition of the case sample making the method very resistant to formulation changes in the illicit drug market. It is to be expected that a wide range of illicit substances can be included in one MRM screening method and that a single rapid method can thus be developed covering all relevant listed substances. As many drugs of abuse contain a primary, secondary, or tertiary amine functionality (an important exception being THC), most analytes of interest can easily be protonated (*i.e.*, forming a $[M\text{-}H]^+$ precursor ion) in an ambient ionization process such as electrospray ionization. The approach is then very similar to the LC-MS method that was discussed for the analysis of DUI whole blood samples in **Chapter 5—Quantitative Analysis and the Legal Limit Dilemma**. While it is currently still inconceivable that a complete LC-MS setup could be used in the field, let alone be reduced to a portable format, interesting progress has been made in the development of mobile and even portable stand-alone mass spectrometers. Most fieldable MS setups are ion trap instruments because an ion trap can be miniaturized and can be operated at a relatively high pressure compared to other mass spectrometers. Another advantage of the ion trap is that it also offers MS/MS (and even MS^n) capability. Recently, scientists from the University of Purdue and Illinois (USA) introduced a versatile platform to interface various ambient ionization techniques directly to a commercially available fieldable ion trap MS. The low-cost constructions allows for a rapid change between **paper spray ionization** (**PSI**) and related techniques (all based on electrospray) and **atmospheric pressure chemical ionization** (**APCI**). Where electrospray is suited for the analysis of most illicit drugs, APCI is more frequently used for the MS analysis of explosives as nitro organic compounds are more difficult to ionize and require an operation in negative mode. As illustrated below, the use of **ambient ionization ion trap mass spectrometry** ("LC-MS without the LC") for rapid fieldable analysis of drugs of abuse is very promising. However, the main challenge for robust application of fieldable mass spectrometry is of a technological nature. All mass spectrometers need to be operated at low pressures thus requiring pumps and considerable electrical power supplies. At the same time, mass spectrometers are precision instruments that exhibit high analyte sensitivity thus requiring a carefully controlled environment. However, in the coming decades, considerable technological advances are expected that will open up the possibilities for rapid, portable, MS-based chemical identification of drugs and explosives!

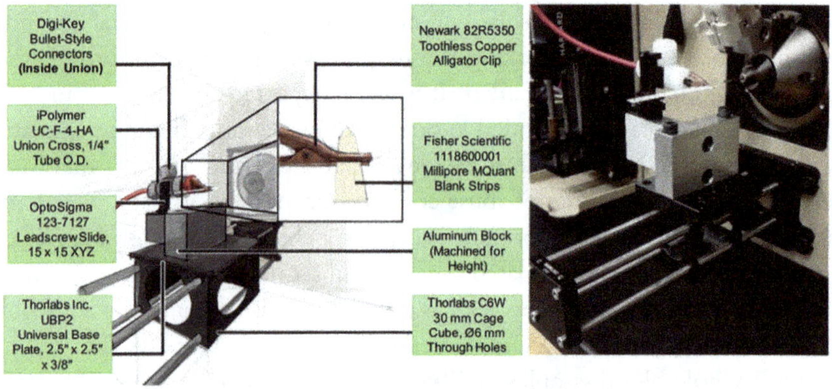

Construction for applying PSI to a fieldable ion trap MS *(Reprinted from Fedick et al. Instruments, 2 (2018) 5, https://doi.org/10.3390/instruments2020005)*

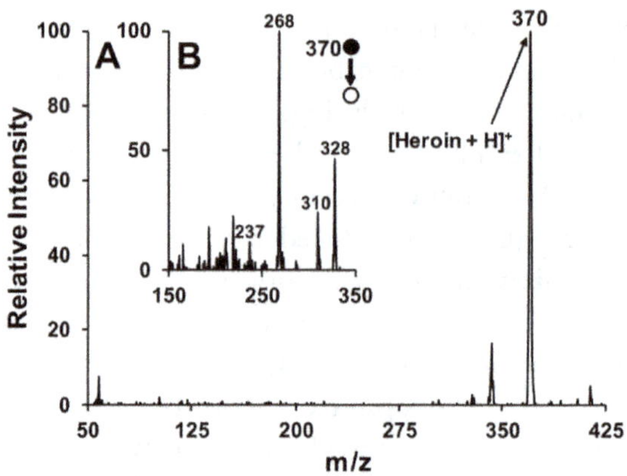

PSI—ion trap mass spectrometry analysis of heroin showing the protonated precursor ion at m/z 370 (A) and the associated MS/MS fragmentation (B) that can be used for chemical identification *(Reprinted from Fedick et al. Instruments, 2 (2018) 5, https://doi.org/10.3390/instruments2020005)*

11.5 The end of a journey

Those readers who have studied the book in detail and started with **Chapter 1**, working their way through the various sections, have now reached the finish line. Congratulations with this remarkable achievement! It makes sense that this journey, that started with the very fundamentals of the use of analytical chemistry in the forensic laboratory, ends with a chapter on how to innovate and develop valuable methods that can make the difference in criminal investigations. It is the ambition of every forensic expert and scientist to use the latest scientific insights and technological possibilities to contribute to a just and safe society. This has always been a perpetual cycle where science and technology "pushes" (new methods and instruments) and the criminal justice system "pulls" (with new requests and questions). It is exactly how the famous forensic pioneers from the early 20th century, like Co van Ledden Hulsebosch in the Netherlands, envisioned how science could give a voice to the silent witnesses so important nowadays in the reconstruction of crimes.

Co van Ledden Hulsebosch (second from the right), Dutch forensic science pioneer, instructing police officers from Indonesia in "the art of forensic investigation" (translated from the Dutch text on the chalk board on the photo) during a course that took place in Amsterdam in the period September–December 1921 more than 100 years prior to the release of this book.

Further reading

van Asten, A.C., 2014. On the added value of forensic science and grand innovation challenges for the forensic community. Sci. Justice 54, 170–179. https://doi.org/10.1016/j.scijus.2013.09.003.

Kloosterman, A., Mapes, A., Geradts, Z., van Eijk, E., Koper, C., van den Berg, J., Verheij, S., van der Steen, M., van Asten, A.C., 2015. The interface between forensic science and technology: how technology could cause a paradigm shift in the role of forensic institutes in the criminal justice system. Phil. Trans. R. Soc. B 370, 2014064. https://doi.org/10.1098/rstb.2014.0264.

Weyermann, C., Ribaux, O., 2011. Situating forensic traces in time. Sci. Justice 52, 68–75.

Exercises

The chapters in this academic course book do not contain individual exercises. The reason for this is that questions are integrated in the text and form a crucial and integral part of the learning process. By allowing sufficient time to process the questions and form independent thoughts on the matter, the student will gain a deep understanding of the presented material. However, the course as taught at the University of Amsterdam of course includes a final exam that the students must pass. In this section, a number of interesting questions have been collected from exams and re-exams from recent years. Associated answers are also provided but on separate pages to allow readers to first formulate their own answers. This Exercise section will allow readers to test their knowledge and insight and might spark ideas for tutors that use the book for their own course when they prepare exam questions. The questions are ordered according to the chapters in the book. The author also frequently creates exam questions on the basis of recently published articles. The students are informed beforehand that a certain article will be the topic of an exam questions and they can read and study the article during the course, they are allowed to ask questions on the paper and they are allowed to discuss it together and help each other. Of course, they do not know the exam question beforehand, but the question will link the article to the content discussed in one of the chapters of the book. This approach makes the examination relevant and dynamic and is recommended to tutors using the CAFE book for their academic forensic chemistry curriculum.

Exam questions without answers

Question 1

(a) US forensic pioneer Dr Paul Kirk stated "*criminalistics is the science of individualization*" He distinguished three consecutive steps as part of the forensic investigation: detection/identification—classification—individualization. Explain/clarify these steps for an alleged shoe mark found at a home were a violent robbery was committed and for which a suspect was arrested and his shoes confiscated and sent to a forensic laboratory.

(b) Now describe these steps for the following forensic evidence types and illustrate by giving an example for each step:
- Explosives
- Paint
- Glass

(c) For each of these forensic evidence types choose an analytical chemistry technique that you would select for the individualization step. Explain why this technique is capable of individualization (i.e. forensic comparison of the evidence against reference material related to a suspect or to the crime)
- Explosives
- Paint
- Glass

(d) Is in your opinion individualization also possible in Forensic Toxicology? Explain why/why not

(e) Is in your opinion individualization also possible for Gun Shot Residues? Explain why/why not

(f) In a very important article Michael Saks and Jonathan Koehler introduce the term "individualization fallacy". Explain what error a forensic or legal expert is making when committing this fallacy. Does this mean that forensic analytical chemistry methods for conducting individualization studies should not be used in criminal investigations? Explain your answer.

(g) A company is receiving threat letters. These A4 letters contain printed text and explicit graphical pen drawings. After a tactical police investigation, a former employee that was recently sacked is arrested. As the police prepare for a search of the home of the suspect, you as a questioned documents expert are consulted. Mention three evidence items (not considering any digital evidence) you would advise the police to retrieve from the house and explain why. Then for each evidence item, select a suitable analytical chemistry technique you would use for a forensic comparison with the threat letters.

Question 2

During a routine check with drug detection dogs in the Rotterdam harbor, the police finds a suspected container. Using an X-ray scanner, police experts discover a hidden area containing a substantial number of sealed packages. An indicative colorimetric test on the content (a white powder) of one of the packages gives a positive result for cocaine. In total, 255 packages are retrieved of which 100 have a white taping and 155 have a brown taping. The entire seizure is brought to the forensic laboratory for further investigation and you as a forensic expert in illicit drug analysis are requested to conduct the chemical analysis and report on the forensic findings.

1. Describe the difference between the two forms of cocaine: the salt form and so-called crack or freebase form. How would a criminal produce crack from the salt form?
2. The most widely used colorimetric test for cocaine is the so-called Scott's test developed in the 1970s. Why is this test considered to be of an indicative nature only and listed as a Category C technique by SWGDRUG (Scientific Working Group for the analysis of seized DRUGs)?
3. The list of Category C techniques also includes UV-vis spectroscopy. On the basis of selectivity, which technique would you prefer for cocaine, the Scott's test, or UV-vis? Or do you feel that both techniques are equally selective? Explain your answer.
4. Street samples of cocaine are typically not very pure but contain several adulterants. How would this affect the Scott's test and the UV-vis analysis? Which of these two techniques is most robust when handling street samples? Or do you feel that both techniques are equally robust? Explain your answer.
5. Using the hypergeometric distribution table provided in **Chapter 3** what would be your sampling plan for this case (95% confidence, k=0.9).
6. The public prosecutor in charge of the investigation has been using a practical protocol for drug smuggling cases based on the following formula: $n = 20 + 0.1*(N-20)$. What would be the implications for your sampling plan with respect to the number of analyses (n) given the total sample size (N)? Irrespective of the amount of work involved, there is a scientific reason to stick to the statistical approach. What would be your explanation to the public prosecutor?
7. The chemical analysis results provide a positive identification of cocaine in all selected samples with the brown taped packing. How should you now report these findings given the fact that not analyzing the entire seizure creates uncertainty?

8. The chemical analysis also shows that the material consists of cocaine with a 95% purity (uncertainty associated with the purity can be neglected). After carefully removing the packaging, the contents are weighed of 30 units with a brown taping. The average weight of the brown units is 510 gr with a standard deviation of 63 gr. Using the student t distribution table ($\alpha = 0.01$) calculate the confidence interval for the total amount of cocaine in the units with the brown tape.

Question 3

The police are investigating a suspicious fire that destroyed a home in a big city center. The residents discovered the fire in time and could escape with only minor injuries. After the fire has been extinguished and the scene has been cleared, police specialists start their investigation. They take several fire debris samples using special glass jars and label the samples with special evidence stickers to mark them as forensic evidence. The samples are brought to the forensic institute for further investigation. The request is to analyze these samples for the presence of residues of ignitable liquids.

1. What is the preferred analytical chemistry method to analyze ignitable liquids and ignitable liquid residues in fire debris? Explain why.
2. Sample preparation is crucial when analyzing ignitable liquid residues in fire debris. Chose the preferred option by forensic laboratories of the pairs listed below and explain your choice:
   - **i.** Solvent extraction vs head space sampling
   - **ii.** Dynamic vs static head space sampling
   - **iii.** Direct headspace analysis vs head space trapping
   - **iv.** Head space trapping vs SPME (Solid Phase Micro Extraction)
3. The interpretation of analytical fire debris data is complex and very difficult to automate. Typically, the data are interpreted by a forensic expert who arrives at a personal conclusion if and what ignitable liquid residue is present in the sample.
   - **i.** Explain the difference between a TIC (Total Ion Chromatogram) and EIC (Extracted Ion Chromatogram)
   - **ii.** Why is EIC used by the forensic expert to analyze fire debris data?
   - **iii.** Why are ion fragments m/z 43, 57, 71,... indicative of n-alkanes? Draw the associated fragment structures
   - **iv.** Why is ion fragment m/z 91 indicative of aromatic compounds? Draw the associated fragment structure

4. Chemical analysis and forensic interpretation reveals that several fire debris samples from the fire site contain traces of gasoline.
   i. Does the presence of gasoline in the samples prove that the fire was deliberate and thus that arson was committed? Motivate your answer.
   ii. Does the presence of gasoline in the samples support the hypothesis that the fire was deliberate and thus that arson was committed? Motivate your answer.
   iii. If instead of gasoline, traces of lamp oil were found in the fire debris samples, how would this affect the weight of the evidence when considering arson versus an accident/technical failure? Motivate your answer.

Question 4

In many countries XTC (MDMA, 3,4-methylenedioxymethamphetamine) is a popular but illicit party drug that is orally administered as a pill. Depicted below is a typical Electron Impact Mass Spectrum of MDMA (C11H15NO2, Mw = 193 g/mol) as measured on a GC-MS system. The structures of the characteristic m/z=58 and m/z=135 fragment ions are depicted in the figure for reference.

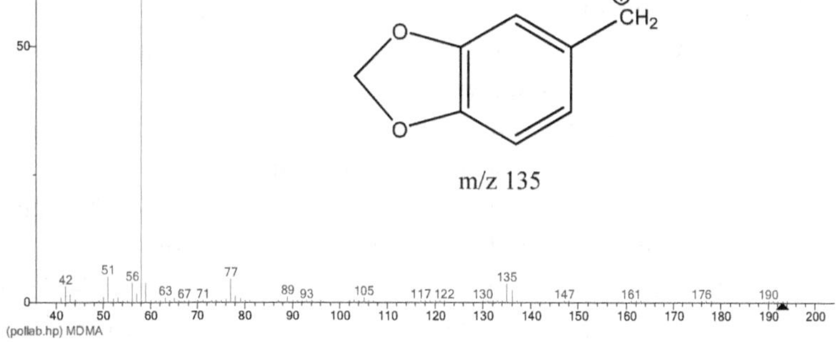

Now sketch the electron impact mass spectra of the following three MDMA-related compounds (the sketch can be limited to the nominal masses of the two characteristic mass fragments similar to MDMA). For these two fragments, draw the molecular structures similar to the MDMA example (nominal masses: H=1, C=12, N=14, O=16):

1. MDA, 3,4-methylenedioxyamphetamine, $C_{10}H_{13}NO_2$, Mw = 179 g/mol

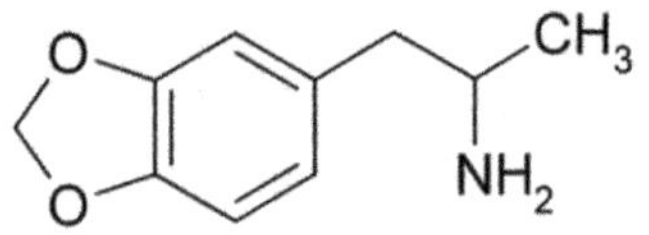

2. MDEA, 3,4-methylenedioxy-N-ethylamphetamine, $C_{12}H_{17}NO_2$, Mw = 207 g/mol

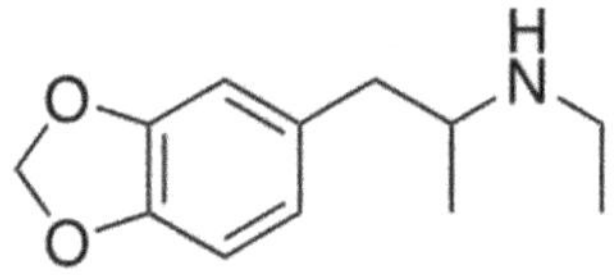

3. 2,3-MDMA, 2,3-methylenedioxymethamphetamine, $C_{11}H_{15}NO_2$, Mw = 193 g/mol

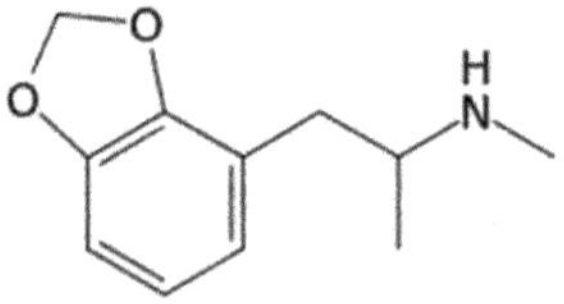

4. Whereas MDMA is a listed and thus illegal drug in many countries, MDA, MDEA, and 2,3-MDMA are typically not in legal systems with lists of drugs of abuse. If we would mistakenly misidentify these compounds as MDMA than that could lead to a wrongful conviction on basis of the chemical analysis!

i. Which of the three compounds poses the greatest risk of a misidentification on the basis of the mass spectrum?

ii. What could still prevent a misidentification in a GC-MS analysis when two compounds have near identical mass spectra?

iii. If such a misidentification cannot be prevented with GC-MS what analytical strategy could be employed to ensure correct chemical identification?

Question 5

During a traffic screening operation by the police, a driver tests positive for several drugs of abuse (indicative saliva test). The driver is taken into custody and complies with a blood test. Blood samples are taken and sent to the forensic institute for further analysis.

1. Which forensic expertise area deals with the analysis of drugs of abuse, medicines and metabolites in human matrices such as whole blood?
2. Which analytical technique is predominantly used for the analysis of trace levels of drugs of abuse, medicines, and metabolites in whole blood?
3. Explain the principle of triple Quad Mass Spectrometry
4. Would you prefer high-resolution mass spectrometry without the option of fragmentation or triple quad mass spectrometry with nominal mass resolution for the trace analysis of drugs in whole blood? Explain your answer.
5. For an accurate quantitative trace analysis of drugs of abuse and associated metabolites in human whole blood, the use of deuterated standards is essential. Explain in your own words the function of such standards
6. Deuterated standards are very expensive. As a cost-saving measure, a laboratory analyst proposes to use one deuterated amphetamine standard to quantitatively analyze all amphetamine variants (amphetamine, methamphetamine, MDMA, etc.). Explain in your own words what risk is associated with implementing this measure.
7. During method development experts try to create a robust method and maximize its precision and accuracy. Explain or illustrate the difference between precision and accuracy.
8. Which type of error is associated with a method with poor precision and which type of error is associated with a method with poor accuracy?
9. In the case of the driver with the positive saliva test, the forensic institute reports the following findings for the blood analysis:
   - MDMA: <c> = 40 μg/L, CI (α = 0.01) = +/- 20%
     (Legal limit single use: 50 mg/L, Legal limit combined use: 25 mg/L)
   - Cocaine: <c> = 12 μg/L, CI (α = 0.01) = +/- 25%
     (Legal limit single use: 50 μg/L, Legal limit combined use: 10 μg/L)

   (<c> = average concentration measured, CI = relative confidence interval for the analysis). Should the judge in this case convict the suspect? Provide an argumentation for the verdict.

Question 6

In a country, a series of explosive attacks have been conducted on a chain of supermarkets. The attacks have occurred in various cities. The explosions all occurred at night causing no serious injuries and causalities but resulting in huge infrastructural damage. No suspects have been apprehended at this stage. Witnesses and video cameras have recorded several individuals carrying and placing plastic bags and then driving off in a car. You are heading the forensic team that is assisting a special tactical team of the police that is investigating and trying to solve this sudden outbreak of targeted attacks.

1. As head of the forensic team, you can pick a team of five experts that will be working on the case non-stop to assist the tactical team. Which five expertise areas will you select to represent your team? Explain your choices.
2. Forensic investigations into these crimes will have priority over all other investigations to ensure that forensic results become available instantly. Why is it so important that forensic results are provided instantly?
3. So far all attacks were successful, that is, lead to an explosion causing extensive damage to the property. This severely complicates the forensic investigation. Your team is asked to assist with the crime scene investigation. What type of traces would you secure and why?
4. What is the risk of securing traces on a post-explosion crime scene that could potentially be attributed to the IED?
5. One of the key questions of the tactical team is whether all attacks are connected and are conducted by a single group. This seems to make sense given the specific targets. Could the forensic investigations also be used to investigate the potential link between the incidents? Explain your answer.
6. All the tactical findings and forensic evidence indicate that the attacks are connected and conducted by a single group of individuals. The main focus of the police is now to find these individuals. Excluding potential DNA and fingermark evidence, how could the forensic explosives experts assist in this effort?
7. We now assume that the Police in the near future will obtain a viable lead that results in the apprehension of several individuals. With a search warrant, they search the property of the suspects. What kind of evidence would you tell the police to look for and secure? Explain your answer.

Question 7

In the dunes, the human remains of an unknown victim are discovered in a shallow grave. The victim has been put in the grave some time ago as all soft tissue has decayed; however, the remains include hair, teeth, nails, and bones. There is no missing person report that matches with the discovery and hence there are no tactical leads to direct the investigation. The police discuss the options for investigation with the forensic experts. The primary challenge is to identify the victim.

1. What is first method of choice to identify the victim (not involving forensic analytical chemistry?)
2. Assuming that the applied method is successful, what is required to arrive at an identification of the victim?

Unfortunately, this approach turns out to be unsuccessful, the victim is still unknown and there are no tactical leads. The police explore the options for further investigation through Isotope Ratio Mass Spectrometry (IRMS) with the experts of the forensic institute.

3. Explain in your own words the principles of IRMS.
4. How can IRMS be used in the forensic investigation of manmade materials such as explosives?
5. What is meant with “you are what you eat?” in human provenancing in forensic science?
6. What is an “isoscape” and how are isoscapes used in human provenancing in forensic science?
7. Present a plan to the police to use IRMS for reconstructing the origin and recent whereabouts of the unknown victim. Which samples have to be collected from the victim and sent to the lab for isotope analysis? Due to capacity restrictions and other priorities only three samples can be analyzed. Which samples would you select?
8. How could the police use the isotope-based information generated in this investigation to solve the case and identify the victim?

In addition to recent whereabouts and origin, the police would also like to know the age of the victim. They discuss this request with the experts of the forensic institute.

9. What isotope-based technique could be used to estimate the age of the victim?
10. For a correct estimation of the victim age two different samples will have to be analyzed. What samples would you select and explain (or illustrate) why.

Question 8

An unknown perpetrator has recently sent powder threat letters to several institutions. Letters have been delivered to a hospital, a law firm and a publisher and no clear motive seems to exist. The police have indicated that the letters contain a toxic powder and a card with text. In light of the investigation no further details have been given. A picture of such a letter has been shared with the media to allow institutions to recognize the letters before opening them. All letters have a single fake sender address. The evidence consists of an envelope containing a card with handwritten text and roughly 10 grams of a fine white powder. The following questions put you in the role of the forensic experts at the forensic institute that is currently investigating the powder letters to support the Police in their investigation to find the perpetrator and stop the threat.

1. Before starting the forensic investigation, we have to make sure that it is safe to handle the evidence in the forensic laboratory. Therefore, the first task is to remove the powder, identify the material, and assess its toxicity. How do we ensure that we can determine the identity of the material in a safe manner and how do we prevent analyst exposure and dangerous contamination in the laboratory?
2. What would your analytical strategy be and what techniques would you use to establish the identity of the material (roughly 10 gr of a fine white powder)? Explain your choices.
3. Investigating threat letters is an interdisciplinary effort, several experts collaborate to study the evidence and determine a plan of approach. Doing the investigations in the right order is critical to prevent contamination and optimize the results of the forensic investigations. Place the following investigations in the right order and explain your choice:
   - DNA and fingermarks (on the envelope and the card)
   - Chemical identification of the powder
   - Handwriting investigation (of the text on the card)
   - Ink analysis (of the handwritten text on the card)
   - Paper analysis (of the card and the envelope)
4. When looking for DNA traces and fingermarks, forensic experts focus on the sticky (inner) side of the stamp and the card with written text and not so much on the envelope. What is the reason for this?

Unfortunately, no DNA and fingermarks are found on the threat letters. Apparently, the perpetrator is careful not to leave such traces. Please exclude DNA and fingermark investigations in the remaining questions.

5. The Police initially have no tactical leads and plan a meeting with the forensic experts of the NFI. They want to know if the forensic evidence can support their hypothesis that the threat letters are all related and originate from a single individual/group. What forensic investigations would you do to answer this question? Explain your choice.
6. Both tactical and forensic investigations indicate that all threat letters are related and are prepared and sent from a single source. However, the police do not have any tactical leads and no suspects in custody. The police ask the forensic experts of the forensic institute if the results of the forensic investigations could provide any information to direct the tactical investigation. Do we now resort to classification or individualization methods? Explain your answer and give an example of a forensic finding that could provide a useful lead for the police
7. After an intensive tactical investigation, the Police finally obtain a meaningful lead. As they plan the arrest of the suspect and a subsequent search of his house (for which they have obtained a search warrant from the investigative judge) they discuss the action with the forensic experts of the NFI. The Police want to know what evidence to secure when searching the house. What do you tell the police to look for and explain why.

Question 9

This question is about the high degree of integrity, quality and objectivity that is required for the work of the forensic expert in the criminal justice system.

1. In many countries the forensic laboratory is part of the police force (e.g. in Sweden, Finland, Germany, Norway, France). In the Netherlands, the NFI is a separate agency in the criminal justice system which reports directly to the Ministry of Justice and Security despite the fact that requests are received from the Police and the Public Prosecution Office. What is the reason for this?
2. In the United Kingdom, the forensic market has been privatized, that is, there are no national forensic services within the government but instead forensic investigations are conducted by commercial laboratories that compete for assignments from the police and the public prosecution office. Can you mention two benefits and two drawbacks of an open forensic market?
3. Now for any of the following situations indicate what you as an expert would do and report.

Case 1

After the discovery of what might seem an IED (Improvised Explosive Device) the Bomb Ordnance Squad arrives at the scene and investigate the device. They conclude that is a functional device and decide to eliminate the threat by means of a controlled explosion off-site. This explosion is triggered with a special explosive from the squad supplies. After the neutralization, post-explosion swabs are taken by the police experts and sent to the forensic institute for further investigation.

Case 2

The police are conducting a high-profile investigation into an organized crime association. On the basis of a novel chemical profiling method the forensic institute has been able to link a large shipment of drugs to a production location. The public prosecutor in charge of the investigation asks the expert to report the conclusions but to leave out all details of the methodology because of the pending investigation.

Case 3

In a criminal investigation of a potential extortion case, the forensic institute is asked to perform a chemical analysis of a white substance that has been found in a threat letter received by a victim. After examining the letter in more detail the expert to his surprise discovers that he knows the victim. They are members of the board of a local tennis club.

4. Sampling and sample preparation need to be carefully controlled to ensure reliable results in analytical chemistry. This is especially important in forensic science where crime scenes and evidence items can show a high degree of variability. For each requirement listed below indicate what kind of error can be made when this requirement is not met and give an example in a forensic context.

Sampling and sample preparation for chemical analysis needs to be

1. Representative
2. Non-invasive
3. Hygienic/sterile

Exam questions with answers

Question 1

1. US forensic pioneer Dr Paul Kirk stated "criminalistics is the science of individualization." He distinguished three consecutive steps as part of the forensic investigation: detection/identification—classification—individualization. Explain/clarify these steps for an alleged shoe mark found at a home were a violent robbery was committed and for which a suspect was arrested and his shoes confiscated and sent to a forensic laboratory.

Detection/Identification:
The mark found at the home was made by a shoe

Classification:
The shoe is a size nine Nike Air Max of type x

Individualization:
The mark was made by the left shoe of the pair of Nike Air Max taken from the suspect

2. Now describe these steps for the following forensic evidence types and illustrate by giving an example for each step:

Explosives

Detection/Identification:
Establishing the chemical identity of an explosive compound
The yellow material retrieved from the crime scene consists of pure TNT

Classification:
Determining the type of explosive material based on e.g. typical composition
The material found at the crime scene is Semtex type 1A

Individualization:
Relating a home-made explosive material found at the crime scene to a batch produced at the home of the suspect.
The TATP from the improvised explosive device originates from the batch of TATP found at the home of the suspect

Paint

Detection/Identification:
Establishing that a microtrace on a substrate corresponds to dried paint. The trace retrieved from the bumper of the car is a red paint chip.

Classification:
Determining the type of paint based on its composition and color. The red paint chip corresponds to Akzo Nobel Brilliant Red no 42 – clear coat.

Individualization:
Relating a paint trace to a specific object carrying that paint. The red paint chip is originating from the bike of the victim.

Glass

Detection/Identification:
Establishing that a particulate material is made of glass The particle found on the sweater of the suspect is a glass fragment.

Classification:
Determining the type of glass based on its composition. The particle found on the sweater of the suspect is originating from the window of a Mercedes car.

Individualization:
Relating a glass particle to a specific glass object. The particle found on the sweater of the suspect is originating from the window of the Mercedes car of the victim.

3. For each of these forensic evidence types choose an analytical chemistry technique that you would select for the individualization step. Explain why this technique is capable of individualization (i.e., forensic comparison of the evidence against reference material related to a suspect or to the crime).

Several options are possible, examples are given below.

Explosives

IRMS – through isotope analysis chemically identical materials can be distinguished typically based on origin of production and raw materials used.

Paint

Py-GC-MS — Small variations in compositional and drying and decomposition processes can be mapped with detailed chemical analysis after pyrolysis.

Glass

LA-ICP-MS — on a batch level trace elemental impurities as measured with LA-ICP-MS can differ yielding characteristic profiles for forensic comparison.

4. Is in your opinion individualization also possible in Forensic Toxicology? Explain why/why not

No, in forensic toxicology drugs are analyzed at ppb level in complex human biological matrices, this does not allow impurity drug profiling.

5. Is in your opinion individualization also possible for Gun Shot Residues? Explain why/why not

No, in this case the problem is related to the chaotic and destructive nature of the combustion process yielding the typical GSR particles. The chemical composition is dominated by this process and provides little leads to the original gun powder batch.

6. In a very important article Michael Saks and Jonathan Koehler introduce the term "individualization fallacy." Explain what error a forensic or legal expert is making when committing this fallacy. Does this mean that forensic analytical chemistry methods for conducting individualization studies should not be used in criminal investigations? Explain your answer.

The individualization fallacy is made when on the basis of a characteristic chemical profile an absolute statement is made that the evidence must originate from the reference material/batch. Such a statement is only scientifically valid if all other potential sources have been considered and can be ruled out. This does not mean that individualization methods and results are flawed. The key is to formulate accurately and precisely: the more characteristic a method is the higher the evidential value, i.e. the stronger the evidence supports the same-source hypothesis, i.e. the more likely the evidence becomes when considering the same-source hypothesis. It however never excludes the different-source proposition.

7. A company is receiving threat letters. These A4 letters contain printed text and explicit graphical pen drawings. After a tactical police investigation, a former employee that was recently sacked is arrested. As the police prepare for a search of the home of the suspect, you as a questioned documents expert are consulted. Mention three evidence items (not considering any digital evidence) you would advise the police to retrieve from the house and explain why. Then for each evidence item, select a suitable analytical chemistry technique you would use for a forensic comparison with the threat letters

1. Paper

XRF/LA-ICP-MS/IRMS: on the basis of the elemental profile and light isotope ratios the paper from the threat letters can be compared to the paper supply at the home of the suspect. This can provide a link between the letter and the paper in the suspect's home.

2. Printer

First a test print needs to be made on the confiscated printer. Then the ink lines can be compared with e.g. LA-ICP-MS (trace elements) or with LC-MS to analyze the dye system. This can provide a link between the letter and the printer in the suspect's home.

3. Pen

First the pen needs to be used to make a test drawing.

Then the ink lines can be compared with e.g. LA-ICP-MS (trace elements) or with LC-MS to analyze the dye system

Question 2

During a routine check with drug detection dogs in the Rotterdam harbor, the Police find a suspected container. Using an X-ray scanner, police experts discover a hidden area containing a substantial number of sealed packages. An indicative colorimetric test on the content (a white powder) of one of the packages gives a positive result for cocaine. In total, 255 packages are retrieved of which 100 have a white taping and 155 have a brown taping. The entire seizure is brought to the forensic laboratory for further investigation and you as a forensic expert in illicit drug analysis are requested to conduct the chemical analysis and report on the forensic findings.

1. Describe the difference between the two forms of cocaine: the salt form and so-called crack or freebase form. How would a criminal produce crack from the salt form?

Cocaine is mostly encountered as its HCl salt form. This means that in an acidic environment the amine group is protonated, and cocaine is precipitated/crystallized as Cl salt. Freebase or so-called crack cocaine is the uncharged species where the amine group is not protonated/charged.

To produce crack cocaine from its HCl salt, cocaine salt needs to be dissolved in an aqueous environment. By solvent-solvent extraction the cocaine can be extracted to a non-miscible organic solvent layer when the pH of the aqueous solution is increased.

2. The most widely used colorimetric test for cocaine is the so-called Scott's test developed in the 1970s. Why is this test considered to be of an indicative nature only and listed as a Category C technique by SWGDRUG (Scientific Working Group for the analysis of seized DRUGs)?

Colorimetric tests are based on a specific chemical reaction in which the drug to be detected reacts with the reagents of the test to form a colored reaction product yielding a clear visual signal of its presence. Although the chemistry involved can be quite selective, chemical similar compounds or compounds containing the same functional groups can give rise to the same color formation. Hence colorimetric tests are not selective enough for robust chemical identification.

3. The list of Category C techniques also includes UV-vis spectroscopy. On the basis of selectivity which technique would you prefer for cocaine, the Scott's test or UV-vis? Or do you feel that both techniques are equally selective? Explain your answer.

UV-vis usually provides semi-characteristic spectra for compounds containing aromatic rings. Many compounds in our natural and man-made chemical environment contain aromatic rings and can give rise to similar UV-vis spectra. Because in the Scott's test specific chemistry is involved in the formation of a Co complex, the indicative test is more selective and should be preferred over UV-vis.

4. Street samples of cocaine are typically not very pure but contain several adulterants. How would this affect the Scott's test and the UV-vis analysis? Which of these two techniques is most robust when handling street samples? Or do you feel that both techniques are equally robust? Explain your answer.

Again, the colorimetric test is to be preferred. UV-vis would give a mixed spectrum as several additives will show absorption in the UV range. As UV-vis spectra are not very selective this will make it very hard to identify the active compound. If the colorimetric test is not affected too much by the intrinsic color of the mix and the chemistry involved is selective enough the presence of cocaine could still be effectively be indicated.

5. Using the hypergeometric distribution table provided in Chapter 4 what would be your sampling plan for this case (95% confidence, k=0.9)

The white and brown taped packages need to be considered as separate collections with separate sampling plans:
White taped packages: N=100 > k=0.9, 95% confidence > n=23
Brown taped packages: N=155 > k=0.9, 95% confidence > n=26
(round to next highest N in the table!)

6. The public prosecutor in charge of the investigation has been using a practical protocol for drug smuggling cases based on the following formula: n = 20 + 0.1*(N-20). What would be the implications for your sampling plan with respect to the number of analyses (n) given the total sample size (N)? Irrespective of the amount of work involved, there is a scientific reason to stick to the statistical approach. What would be your explanation to the public prosecutor.

White taped packages: N=100 > n=20+0.1*(100-20)=28 Brown taped packages: N=155 > n=20+0.1*(155-20)=33,5=34

With this approach more samples will have to be analyzed leading to more lab work. However, the more important argument here is the scientific foundation of the Hypergeometric distribution enabling statements regarding the uncertainty associated with analyzing only a part of the entire sample set.

7. The chemical analysis results provide a positive identification of cocaine in all selected samples with the brown taped packing. How should you now report these findings given the fact that not analyzing the entire seizure creates uncertainty?

On the basis of the findings after the analysis of 26 samples it can be stated with 95% confidence that at least 90% of the brown taped packages will contain cocaine.

8. The chemical analysis also shows that the material consists of cocaine with a 95% purity (uncertainty associated with the purity can be neglected). After carefully removing the packaging, the contents are weighed of 30 units with a brown taping. The average weight of the brown units is 510 gr with a standard deviation of 63 gr. Using the student t distribution table ($\alpha = 0.01$) calculate the confidence interval for the total amount of cocaine in the units with the brown tape.

n = 30, μ = 510 gr, s = 63 gr, t = 2.756 (df ***= 30–1 = 29)***
Average weight per unit: 478.3–541.7 gr
N = 155
Total weight: 74.1–84.0 kg Purity = 95%
Total amount of cocaine: 70.4–79.8 kg

Question 3

The police are investigating a suspicious fire that destroyed a home in a big city center. The residents discovered the fire in time and could escape with only minor injuries. After the fire has been extinguished and the scene has been cleared, police specialists start their investigation. They take several fire debris samples using special glass jars and label the samples with special evidence stickers to mark them as forensic evidence. The samples are brought to the forensic institute for further investigation. The request is to analyze these samples for the presence of residues of ignitable liquids.

1. What is the preferred analytical chemistry method to analyze ignitable liquids and ignitable liquid residues in fire debris? Explain why.

Gas Chromatography with Mass Spectrometric Detection (GC-MS). Ignitable liquids typically are oil based, refined products containing complex mixtures of volatile and semi-volatile hydrocarbons of limited polarity. GC-MS is a very suitable technique for such compounds and this is the main analytical technique used in the oil industry.

2. Sample preparation is crucial when analyzing ignitable liquid residues in fire debris. Chose the preferred option by forensic laboratories of the pairs listed below and explain your choice:

i. Solvent extraction vs head space sampling

Head space sampling: fire debris constitutes a complex and unknown matrix, solvent extraction could lead to fouling of the GC-MS injector and column.

ii. Dynamic vs static head space sampling

Static head space sampling: although dynamic head space is more sensitive, static head space allows a better control of the amount analyzed and multiple analyses of the same fire debris sample.

iii. Direct headspace analysis vs head space trapping

Head space trapping: direct head space is less sensitive because only a limited head space volume can be injected directly onto a GC-MS system. Especially for samples containing trace amount of ignitable liquid residue this would result in a false negative.

iv. Head space trapping vs SPME (Solid Phase Microextraction)

Head space trapping: although both techniques are used successfully in forensic laboratories, most experts prefer the use of head space trapping because it less sensitive for sampling conditions and because SPME sampling is compound (volatility, polarity) selective.

3. The interpretation of analytical fire debris data is complex and very difficult to automate. Typically, the data are interpreted by a forensic expert who arrives at a personal conclusion if and what ignitable liquid residue is present in the sample.

i. Explain the difference between a TIC (Total Ion Chromatogram) and EIC (Extracted Ion Chromatogram)

A TIC chromatogram shows the overall ion current from the mass spectrometer as function of time thereby serving as a universal detector like a FID (Flame Ionization Detector) showing all organic compounds eluting from the column. In EIC certain masses are selected from the TIC to provide a filter to only show compounds from the same class that share typical ion fragments

ii. Why is EIC used by the forensic expert to analyze fire debris data?

Due to the fire debris, head space compositions can be complex containing many compounds both from the matrix and the ignitable liquid residue (if present). By using EIC the expert can 'zoom in' on certain compound classes to reveal patterns that are typical for an ignitable liquid.

iii. Why are ion fragments m/z 43, 57, 71,... indicative of n-alkanes? Draw the associated fragment structures

N-alkanes typically show severe fragmentation upon EI (electron impact) ionization giving no to a very low molecular ion signal. The smaller ion fragments are very characteristic though with a mass difference of 14 corresponding to a CH2 unit:

CH3–CH2–CH2+ (m/z 43)
CH3–CH2–CH2–CH2+ (m/z 57)
CH3–CH2–CH2–CH2–CH2+ (m/z 71)

iv. Why is ion fragment m/z 91 indicative of aromatic compounds? Draw the associated fragment structure (3 points)

Unsubstituted aromatic compounds can be recognized through the characteristic tropylium ion (C7H7+):

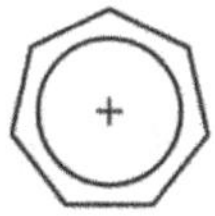

(d) Chemical analysis and forensic interpretation reveals that several fire debris samples from the fire site contain traces of gasoline.

i. Does the presence of gasoline in the samples prove that the fire was deliberate and thus that arson was committed? Motivate your answer.

No, the presence of ignitable liquid in a fire debris sample can never provide absolute proof of arson. If gasoline happened to be present at the location where the fire started than traces of gasoline can also be found for an accidental fire.

ii. Does the presence of gasoline in the samples support the hypothesis that the fire was deliberate and thus that arson was committed? Motivate your answer.

Yes, if considering a house it is not very likely that gasoline is present for reasons other than starting a fire (please note that this depends on the location of the fire, the situation is very different for for instance a garage).

iii. If instead of gasoline, traces of lamp oil were found in the fire debris samples, how would this affect the weight of the evidence when considering arson versus an accident/technical failure? Motivate your answer.

As lamp oil is found more frequently in houses than gasoline, the weight of the evidence when considering arson vs accidental fire would be significantly reduced. Please note that the evidence is still incriminating (i.e. supports the arson hypothesis) as not all house inventories were accidental fires start contain lamp oil.

Question 4

In many countries, XTC (MDMA, 3,4-methylenedioxymethamphetamine) is a popular but illicit party drug that is orally administered as a pill. Depicted below is a typical Electron Impact Mass Spectrum of drug MDMA (C11H15NO2, Mw = 193 g/mol) as measured on a GC-MS system. The structures of the characteristic m/z=58 and m/z=135 fragment ions are depicted in the figure for reference.

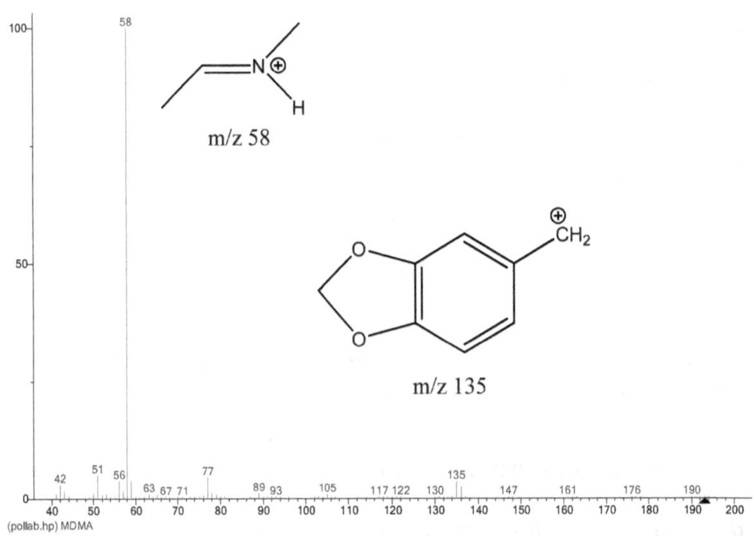

Now sketch the electron impact mass spectra of the following three MDMA-related compounds (the sketch can be limited to the nominal masses of the two characteristic mass fragments similar to MDMA). For these two fragments, draw the molecular structures similar to the MDMA example (nominal masses: H=1, C=12, N=14, O=16):

(a) MDA, 3,4-methylenedioxyamphetamine, C10H13NO2, Mw = 179 g/mol

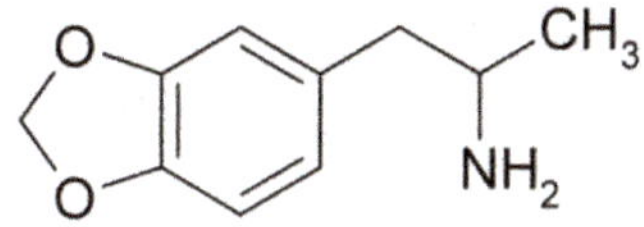

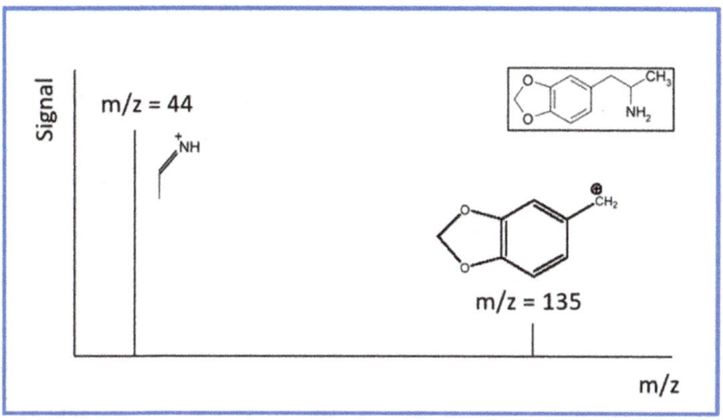

(b) MDEA, 3,4-methylenedioxy-N-ethylamphetamine, C12H17NO2, Mw = 207 g/mol

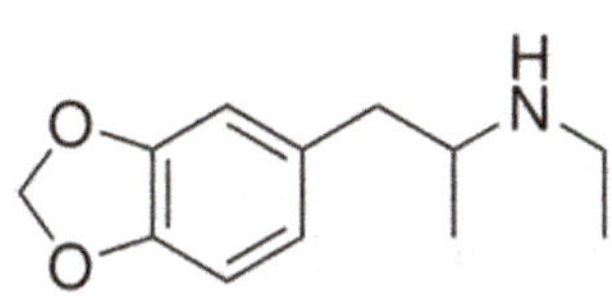

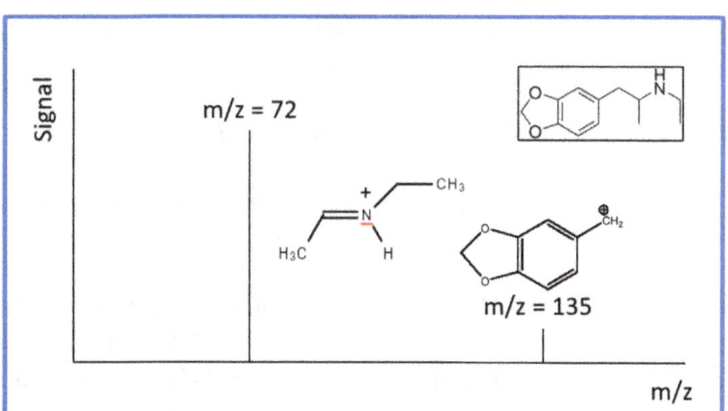

(c) 2,3-MDMA, 2,3-methylenedioxymethamphetamine, C11H15NO2, Mw = 193 g/mol

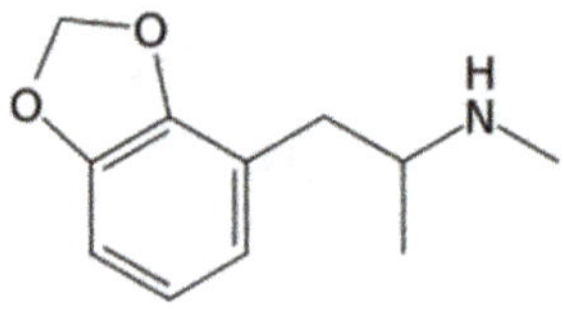

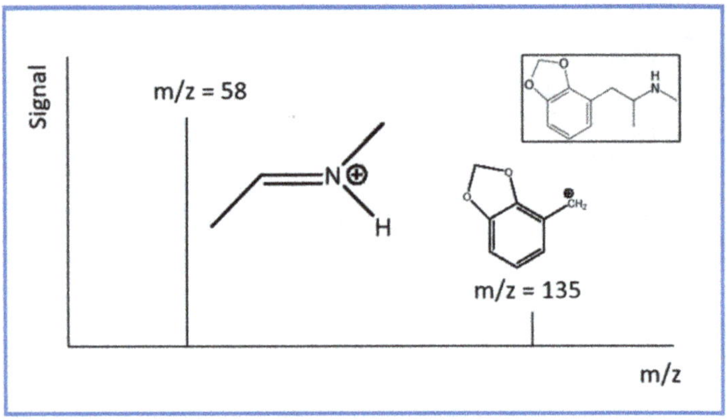

(d) Whereas MDMA is a listed and thus illegal drug in many countries, MDA, MDEA, and 2,3-MDMA are typically not in legal systems with lists of drugs of abuse. If we would mistakenly misidentify, these compounds as MDMA than that could lead to a wrongful conviction on basis of the chemical analysis!

i. Which of the three compounds poses the greatest risk of a misidentification on the basis of the mass spectrum?

2,3-MDMA, 2,3-methylenedioxymethamphetamine

ii. What could still prevent a misidentification in a GC-MS analysis when two compounds have near identical mass spectra?

2,3-MDMA and 2,4-MDMA might show sufficient difference in volatility and interaction with the stationary phase of the GC column to be fully separated with GC-MS. The retention time of 2,3-MDMA would then be significantly different from the 2,4-MDMA standard or library value and would trigger an alarm with the experts.

iii. If such a misidentification cannot be prevented with GC-MS what analytical strategy could be employed to ensure correct chemical identification?

As these types of isomers test positive in a colorimetric test and share the same retention time and MS spectrum, additional analytical techniques would have to be employed to differentiate them. The use of Raman or IR spectroscopy could reveal different spectroscopic properties or the structure could be elucidated with NMR. However, because drug samples can consist of multiple compounds (cutting agents) the best option is to use GC in combination with a spectroscopic detector, e.g. GC-IR or GC-VUV

Question 5

During a traffic screening operation by the police, a driver tests positive for several drugs of abuse (indicative saliva test). The driver is taken into custody and complies with a blood test. Blood samples are taken and sent to the forensic institute for further analysis.

(a) Which forensic expertise area deals with the analysis of drugs of abuse, medicines and metabolites in human matrices such as whole blood?

Forensic toxicology

(b) Which analytical technique is predominantly used for the analysis of trace levels of drugs of abuse, medicines, and metabolites in whole blood?

Liquid chromatography with mass spectrometric detection (LC-MS)

(c) Explain the principle of triple Quad mass spectrometry

In the triple Quad mass spectrometer, the first quadrupole is used to select the target (or parent) ion. The ions at the selected m/z value then enter the second quadrupole operated at higher pressure to induce fragmentation (CID or collision induced dissociation). These fragments are then analyzed in the third quadrupole. One fragment ion is used for quantitative analysis, at least two fragment ions are used for identification/confirmation of the target analyte. By rapid cycling multiple target analytes can be monitored at any given time.

(d) Would you prefer high resolution mass spectrometry without the option of fragmentation or triple quad mass spectrometry with nominal mass resolution for the trace analysis of drugs in whole blood? Explain your answer.

Typically in forensic practice, triple quad mass spectrometry is preferred for several options: The instrument is cheaper, triple Q MS provides more accurate quantitative analysis and (most importantly) there is additional selectivity through the fragmentation process allowing the differentiation of isomers. Isomers always share the same mass irrespective of the resolution of the MS!

(e) For an accurate quantitative trace analysis of drugs of abuse and associated metabolites in human whole blood the use of deuterated standards is essential. Explain in your own words the function of such standards

The deuterated standard behaves chemically almost identical to the target analyte. Hence with the use of deuterated standards the loss of analyte in the sample pretreatment and ionization efficiency fluctuations in the MS can be accurately compensated.

(f) Deuterated standards are very expensive. As a cost-saving measure, a laboratory analyst proposes to use one deuterated amphetamine standard to quantitatively analyze all amphetamine variants (amphetamine, methamphetamine, MDMA, etc.). Explain in your own words what risk is associated with implementing this measure.

By using a chemically different species as an internal standard there is an enhanced risk of significant systematic error in the analysis. Concentrations will be too high or too low depending on whether the internal standard shows additional or less loss of compound or sensitivity compared to the target analyte. Before taking this cost saving measure additional validation experiments are required.

(g) During method development experts try to create a robust method and maximize its precision and accuracy. Explain or illustrate the difference between precision and accuracy.

A precise method: the measured average value has a low standard deviation
An accurate method: the measured average value is close to the true value

(h) Which type of error is associated with a method with poor precision and which type of error is associated with a method with poor accuracy?

Poor precision: high random error
Poor accuracy: high systematic error

(i) In the case of the driver with the positive saliva test, the forensic institute reports the following findings for the blood analysis:

- MDMA: <c> = 40 μg/L, CI ($\alpha = 0.01$) = +/- 20%

(Legal limit single use: 50 mg/L, Legal limit combined use: 25 mg/L)

- Cocaine: <c> = 12 μg/L, CI ($\alpha = 0.01$) = +/- 25%

(Legal limit single use: 50 μg/L, Legal limit combined use: 10 μg/L)

(<c> = average concentration measured, CI = relative confidence interval for the analysis). Should the judge in this case convict the suspect? Provide an argumentation for the verdict.

The judge should convict the suspect if a false positive threshold of 0.5% is applied ($\alpha = 0.01$) in relation to the combined use limits (which applies in this case given the fact that two drugs of abuse are found in the blood of the suspect). For cocaine the lower limit of the CI corresponds to 12 − (0.25*12) = 9 μg/L which is below the combined use limit of 10 μg/L. However, for MDMA the lower limit of the CI corresponds to 40 − (0.2x40) = 32 μg/L which is above the combined use limit of 25 μg/L.

Question 6

In a country a series of explosive attacks have been conducted on a chain of supermarkets. The attacks have occurred in various cities. The explosions all occurred at night causing no serious injuries and casualities but causing huge infrastructural damage. No suspects have been apprehended at this stage. Witnesses and video cameras have recorded several individuals carrying and placing plastic bags and then driving off in a car. You are heading the forensic team that is assisting a special tactical team of the police that is investigating and trying to solve this sudden outbreak of targeted attacks.

(a) head of the forensic team, you can pick a team of five experts that will be working on the case nonstop to assist the tactical team. Which 5 expertise areas will you select to represent your team? Explain your choices.

Forensic explosives experts (to investigate and reconstruct the IEDs involved), Forensic DNA experts (to retrieve perpetrator DNA profiles), Forensic Fingermark experts (to retrieve perpetrator fingermarks), Forensic Facial Recognition experts (to compare and identify suspects in video footage), Digital Forensics experts (to secure video data, check mobile phone connections etc) and/or maybe Forensic Mictrotrace experts (glass, fibers).

(b) Forensic investigations into these crimes will have priority over all other investigations to ensure that forensic results become available instantly. Why is it so important that forensic results are provided instantly?

The series of attacks is ongoing and the tactical team has no leads nor any suspects. Every new attack leads to substantial damage and risk of casualties. So the tactical team can use any source of forensic information to create leads and stop the attacks before execution.

(c) So far all attacks were successful, that is, lead to an explosion causing extensive damage to the property. This severely complicates the forensic investigation. Your team is asked to assist with the crime scene investigation. What type of traces would you secure and why?

First of all, we need to identify the type of explosive involved, we do this by taking swabs of areas that were close to the center of the explosion. We are also looking for parts and remnants of what could have been part of the IED (Improvised Explosive Device) as this could give us a clue of the construction of the IED and materials used. Finally, we need to secure all the video footage from security cameras and any other interesting source of digital data (e.g. wifi network, mobile phone connections).

(d) What is the risk of securing traces on a postexplosion crime scene that could potentially be attributed to the IED?

An explosion is a very destructive process causing severe damage to the property but also to the many products in the shops. Because we are not sure how the IED was constructed and which parts were used in the bomb construction, there is a serious risk that we secure traces that we think are connected to the IED but in reality originate from the shop and have nothing to do with the IED.

(e) One of the key questions of the tactical team is whether all attacks are connected and are conducted by a single group. This seems to make sense given the specific targets. Could the forensic investigations also be used to investigate the potential link between the incidents? Explain your answer

Yes, forensic evidence can be very useful in this respect assuming that the perpetrators have a similar mode of operandus for each attack using similar IEDs and materials. First, we could look into the trace explosive investigation, are the same energetic materials used in each attack? If so this already provides support that the attacks are connected. In addition, we can look into remnants of IED parts, are the same type of traces found on various crime scenes? Finally, we focus on the security video footage and other digital data. Are the same individuals involved, is the same car used?

(f) All the tactical findings and forensic evidence indicate that the attacks are connected and conducted by a single group of individuals. The main focus of the police is now to find these individuals. Excluding potential DNA and fingermark evidence, how could the forensic explosives experts assist in this effort?

We need to attempt to classify some of the evidence found that is possibly related to the IED. Can we reconstruct from the remnants what kind of parts and items have been used in the IED? If we can deduce a brand and model of a certain part than maybe this turns out to be quite exclusive and rare allowing police investigators to check potential stores and sources of this material.

(g) We now assume that the Police in the near future will obtain a viable lead that results in the apprehension of several individuals. With a search warrant, they search the property of the suspects. What kind of evidence would you tell the police to look for and secure? Explain your answer.

First look for energetic materials and possibly complete IEDs or materials that could be part of the IEDs. This can be compared to the post-explosion findings. Always check all digital evidence and devices which could reveal preparations for the attacks (driving routes, addresses of the shops, text messages etc). Finally secure the clothing and any other items (including a car) that could have been used in the attacks. These items might contain traces of the explosion and of material from the shops (glass particles, paint etc).

Question 7

In the dunes, the human remains of an unknown victim are discovered in a shallow grave. The victim has been put in the grave some time ago as all soft tissue has decayed; however, the remains include hair, teeth, nails, and bones. There is no missing person report that matches with the discovery and hence there are no tactical leads to direct the investigation. The police discuss the options for investigation with the forensic experts. The primary challenge is to identify the victim.

1. What is first method of choice to identify the victim (not involving forensic analytical chemistry?)

Forensic DNA profiling

2. Assuming that the applied method is successful, what is required to arrive at an identification of the victim?

A matching reference DNA profile in a suitable database

Unfortunately, this approach turns out to be unsuccessful, the victim is still unknown and there are no tactical leads. The police explore the options for further investigation through Isotope Ratio Mass Spectrometry (IRMS) with the experts of the forensic institute.

3. Explain in your own words the principles of IRMS

With IRMS small variations in the isotope ratio of light (H, C, N and O) and heavy (e.g. Pb and Sr) elements can be accurately measured. For light isotopes all organic matter of the sample is converted via catalytic reactors into gasses (CO, CO_2, N_2, H_2), these gases are separated on a packed GC column and then in the mass spectrometer the isotopic intensities are measured. Small variations (expressed as ‰) in isotopic ratios are observed due to chemical and physical processes and these variations can also be used for forensic investigations.

4. How can IRMS be used in the forensic investigation of manmade materials such as explosives?

For pure compounds or for materials with highly similar impurity profiles, IRMS can allow forensic comparison on the basis of isotope ratios. Taking very pure crystalline TATP (tri-acetone-tri-peroxide) peroxide explosive as an example: when TATP found at a crime scene shows significant differences in ^{13}C, ^{2}H and ^{18}O values compared to material found at the home of a suspect, this suspect material can be ruled out as its origin. For matching δ values there is moderate support for the common origin hypothesis. By studying shifts in isotopic ratio values during the synthesis of TATP the forensic expert can also investigate potential links between the precursor acetone and TATP addressing the question whether a given source of acetone was used to produce a given batch of TATP.

5. What is meant with "you are what you eat?" in human provenancing in forensic science?

This famous statement refers to the fact that isotopic signatures in human material such as hair, teeth, bone and nail are determined by the nutrients consumed as these form the essential building blocks for the human body. These signatures thus reflect eating habits and geographic residence.

6. What is an "isoscape" and how are isoscapes used in human provenancing in forensic science?

An isoscape is a world or regional map showing the distribution of isotopic ratios of a given element in a given matrix of form. Like air pressure in a meteorological map (isobar) an isoscape shows the boundaries/areas of equal isotope ratio. These maps can be used to pinpoint the geographic origin of an unknown individual on the basis of measured isotope ratios in body material. Depending on the growth rate and development of this material, the origin applies to a certain age of the individual.

7. Present a plan to the police to use IRMS for reconstructing the origin and recent whereabouts of the unknown victim. Which samples have to be collected from the victim and sent to the lab for isotope analysis? Due to capacity restrictions and other priorities, only three samples can be analyzed. Which samples would you select?

To get most information the samples should be selected to cover a large time span in the life time of the unknown victim, e.g. hair (last year), bone (last 20-25 years) and teeth (at age 7-15). These samples should be analyzed for light isotopes and compared against Dutch, European and world isoscape maps to investigate where the unknown victim originated from and how long the victim had resided in the Netherlands prior to death.

8. How could the police use the isotope-based information generated in this investigation to solve the case and identify the victim?

IRMS human provenancing data can provide useful leads for further tactical investigation. If for instance an area of origin is found outside the Netherlands, the police can contact local authorities abroad to check for missing persons in that area.

In addition to recent whereabouts and origin, the police would also like to know the age of the victim. They discuss this request with the experts of the forensic institute.

9. What isotope-based technique could be used to estimate the age of the victim?

C14 radioactive isotope analysis can be used to estimate the age of an individual by exploiting the C14 bomb spike, the increase in C14 levels in earth's atmosphere due to nuclear bomb testing in between 1950-1960.

10. For a correct estimation of the victim age, two different samples will have to be analyzed. What samples would you select and explain (or illustrate) why.

Because of the shape of the bomb spike, a given ^{14}C δ value can correspond to two periods in time. By analyzing human material of recent (hair, nail) and past (teeth, bone) production and by determining which material has the highest value, the correct year can be estimated.

Question 8

An unknown perpetrator has recently sent powder threat letters to several institutions. Letters have been delivered to a hospital, a law firm and a publisher and no clear motive seems to exist. The police have indicated that the letters contain a toxic powder and a card with text. In light of the investigation, no further details have been given. A picture of such a letter has been shared with the media to allow institutions to recognize the letters before opening them. All letters have a single fake sender address. The evidence consists of an envelope containing a card with handwritten text and roughly 10 grams of a fine white powder. The following questions put you in the role of the forensic experts at the forensic institute that is currently investigating the powder letters to support the Police in their investigation to find the perpetrator and stop the threat.

1. Before starting the forensic investigation, we have to make sure that it is safe to handle the evidence in the forensic laboratory. Therefore, the first task is to remove the powder, identify the material, and assess its toxicity. How do we ensure that we can determine the identity of the material in a safe manner and how do we prevent analyst exposure and dangerous contamination in the laboratory?

Removal of the powder and first screening needs to be done in full protection at the scene. Mobile spectroscopy (Raman, IR) and the use of sensors (radioactivity!) can be applied on the scene for a first assessment. The powder than needs to be removed from the letter and placed in a secure container. Possibly the letter and envelope need to be decontaminated (without affecting the various traces). The container needs to be stored at a safe location. From this location small samples can be taken and solutions of low concentration can be prepared for analysis in the forensic laboratory. It is important the minimize the amount of agent in the laboratory.

2. What would your analytical strategy be and what techniques would you use to establish the identity of the material (roughly 10 gr of a fine white powder)? Explain your choices.

If you need to apply a broad screening while having sufficient intact material, you can apply a strategy similar to organic and inorganic explosives analysis. Starting with techniques with require minimal sample handling and are non-invasive you can select XRD, XRF (elemental composition), Raman and IR spectroscopy. When the composition is inorganic a technique like IC-MS can reveal the ions present. When the composition is organic you can resort to LC-MS, GC-MS or when available NMR.

3. Investigating threat letters is an interdisciplinary effort, several experts collaborate to study the evidence and determine a plan of approach. Doing the investigations in the right order is critical to prevent contamination and optimize the results of the forensic investigations. Place the following investigations in the right order and explain your choice:
   - DNA and fingermarks (on the envelope and the card)
   - Chemical identification of the powder
   - Handwriting investigation (of the text on the card)
   - Ink analysis (of the handwritten text on the card)
   - Paper analysis (of the card and the envelope)

1. Chemical identification of the powder (to assess the threat involved and determine a safe route for the forensic investigation)
2. DNA and fingermarks (very important evidence for which we need to prevent contamination)
3. Handwriting investigation (can be very characteristic)
4. Ink analysis & paper analysis (not very characteristic and invasive)

4. When looking for DNA traces and fingermarks forensic experts focus on the sticky (inner) side of the stamp and the card with written text and not so much on the envelope. What is the reason for this?

Traces on the outside of the envelope are not necessarily crime or perpetrator related. In the logistics process several innocent individuals such as the postman may have handled the letter and leave their fingermarks and biological material. All traces on the inside of the envelope and the stamp cannot have been touched by anybody else but the perpetrator (assuming that the letter has not been not opened and the content handled).

Unfortunately, no DNA and fingermarks are found on the threat letters. Apparently, the perpetrator is careful not to leave such traces. Please exclude DNA and fingermark investigations in the remaining questions.

5. The Police initially have no tactical leads and plan a meeting with the forensic experts of the NFI. They want to know if the forensic evidence can support their hypothesis that the threat letters are all related and originate from a single individual/group. What forensic investigations would you do to answer this question? Explain your choice.

When establishing a link between cases we typically resort to individualization methods. We can for instance attempt a chemical impurity profile or elemental impurity profile of the fine white powder. In addition, we can look and compare the handwriting that is a characteristic feature of every individual. Other options are elemental trace analysis of the paper and the ink of the handwritten text.

6. Both tactical and forensic investigations indicate that all threat letters are related and are prepared and sent from a single source. However, the police do not have any tactical leads and no suspects in custody. The police ask the forensic experts of the forensic institute if the results of the forensic investigations could provide any information to direct the tactical investigation. Do we now resort to classification or individualization methods? Explain your answer and give an example of a forensic finding that could provide a useful lead for the police

Without reference/comparison material, individualization methods typically are not very useful to support the police investigation. However, classification methods provide information on product type and brand. Such information could be useful in combination with tactical information or in case the product type and brand are relatively rare. If for instance the paper investigation reveals features of a very expensive and rare type of paper only sold by a limited number of exclusive shops.

7. After an intensive tactical investigation, the Police finally obtain a meaningful lead. As they plan the arrest of the suspect and a subsequent search of his house (for which they have obtained a search warrant from the investigative judge), they discuss the action with the forensic experts of the NFI. The Police want to know what evidence to secure when searching the house. What do you tell the police to look for and explain why.

We now want to resort to individualization methods to provide strong evidence that links the evidence to reference materials found in the suspect's home (or that clearly shows that such a link does not exist). Of course, digital evidence can also be crucial but this will be more difficult to link to the letter because of the handwritten card. We ask the police to confiscate material that can be compared against the threat letters, i.e. a fine white powder, envelopes, stamps, paper, pens, examples of handwriting of the suspect.

Question 9

This question is about the high degree of integrity, quality, and objectivity that is required for the work of the forensic expert in the criminal justice system.

1. In many countries, the forensic laboratory is part of the police force (e.g. in Sweden, Finland, Germany, Norway, France). In the Netherlands, the NFI is a separate agency in the criminal justice system which reports directly to the Ministry of Justice and Security despite the fact that requests are received from the Police and the Public Prosecution Office. What is the reason for this?

The reason that the NFI is positioned independently is to ensure that the forensic reports and expert statements can be as neutral and objective as possible. In the Dutch legal system it is of critical importance that forensic findings are used appropriately irrespective whether it is incriminating or exculpatory for the suspect(s)

2. In the United Kingdom, the forensic market has been privatized, that is, there are no national forensic services within the government but instead forensic investigations are conducted by commercial laboratories that compete for assignments from the police and the public prosecution office. Can you mention two benefits and two drawbacks of an open forensic market?

Benefits:

1. ***Value for money! In a commercial market, providers need to offer competitive pricing***
2. ***An open market allows high-end laboratories to provide forensic services even if this is not their main focus, in this way special techniques can be applied***
3. ***Clear agreements can be made with respect to number of products and delivery times of forensic services***

Drawbacks:

1. ***Reduced quality and thoroughness as commercial labs compete for the market and try to offer products as cheaply and effectively as possible***
2. ***Fragmented market where several laboratories provide forensic services can lead to additional complexity***
   - ***It is difficult to maintain the chain of custody***
   - ***It is difficult for experts to maintain a neutral/independent position***

3. Now for any of the following situations indicate what you as an expert would do and report.

Case 1

After the discovery of what might seem an IED (Improvised Explosive Device), the Bomb Ordnance Squad arrives at the scene and investigates the device. They conclude that is a functional device and decide to eliminate the threat by means of a controlled explosion off-site. This explosion is triggered with a special explosive from the squad supplies. After the neutralization, postexplosion swabs are taken by the police experts and sent to the forensic institute for further investigation.

The course of the crime scene investigation and the actions of the Bomb Ordnance Squad need to be carefully considered by the expert and need to be used for the interpretation in the report. An external explosive not associated with the original IED has been used. This can lead to wrong conclusions if post explosive residues of this external explosive are detected and assumed to originate from the IED.

Case 2

The police are conducting a high-profile investigation into an organized crime association. On the basis of a novel chemical profiling method, the forensic institute has been able to link a large shipment of drugs to a production location. The public prosecutor in charge of the investigation asks the expert to report the conclusions but to leave out all details of the methodology because of the pending investigation.

The forensic expert should (in a friendly manner) refuse to report according to these instructions. Forensic reports need to be sufficiently detailed to allow defense lawyers and judges to use this information in a court case.

Case 3

In a criminal investigation of a potential extortion case, the forensic institute is asked to perform a chemical analysis of a white substance that has been found in a threat letter received by a victim. After examining the letter in more detail, the expert to his surprise discovers that he knows the victim. They are members of the board of a local tennis club.

The forensic expert should stop the investigation immediately and should report this accidental connection to management. It is decided to assign another expert to the case and the situation is reported to police and legal experts in the case.

4. Sampling and sample preparation need to be carefully controlled to ensure reliable results in analytical chemistry. This is especially important in forensic science where crime scenes and evidence items can show a high degree of variability. For each requirement listed below indicate what kind of error can be made when this requirement is not met and give an example in a forensic context.

Sampling and sample preparation for chemical analysis needs to be.......

1. Representative

It the sample is not representative systematic errors can occur in a quantitative analysis (or false positive/negative results in a qualitative analysis)

Example:
A drug sample is more heterogeneous than it seems, when the expert takes a single, small sample accidently this predominantly contains an adulterant. Consequently, the quantitative drug analysis leads to a systematically too low active ingredient level and this results in a severe underestimation of the total amount of drugs in an illegal shipment.

2. Non-invasive

If the sampling or sampling processes are invasive the sample composition is altered leading to new compounds being introduced or concentrations of existing compounds being altered. This can lead to false negative and false positive results

Example:

Sample preparation of a forensic toxicological blood sample leads to the accidental formation of a metabolite from the drug present in the blood. This leads to a wrong interpretation by the forensic toxicologists with respect to the cause of death.

3. Hygienic/sterile

If the conditions are not hygienic/sterile contaminations can occur introducing new compounds in a forensic sample due to the investigation

Example:

After visiting the shooting range a GSR experts accidently carries gunshot residues particles and transfers this to a stub in a case that needs to be investigated with SEM-EDX. The stub originally did not contain GSR.

Copyright and image licenses

The cartoon style visualizations have been created with freeware photos and cartoons downloaded from Pixabay (https://pixabay.com/nl/) and have been modified with the Clip2Comic app (https://apps.apple.com/us/app/clip2comic-caricature-maker/id876328355). In addition, photo material from the Netherlands Forensic Institute has been used with permission of the institute. This material has been modified such that NFI employees are not shown or are not recognizable to ensure their privacy. Freeware chemical structures and photos from famous scientists have been retrieved from Wikipedia (https://www.wikipedia.org/). Material copied from publications either originate from open access publications ('*Reprinted from...*') or are reproduced with the permission of the publisher ('*Reprinted with permission from...*'). For replicated material from scientific publications, corresponding references are provided. A few illustrations retrieved from Wikipedia require a reference and the associated details are listed below. Many figures and illustrations have been created by the author and have been specifically designed for this book. If you want to use this material for your own research and education, permission from Elsevier has to be requested. In the unlikely and unintentional event that this book contains material that requires a license or has not received adequate attribution, please contact the author so he can resolve the issue.

Chapter 4

Molecular structure of PCP:
Wikipedia
Acdx—Own work, based on en:Image:PCP.svg
CC BY-SA 3.0
File:Phencyclidine structure.svg
Created: 17 October 2009

Molecular structure of Flakka:
Wikipedia
Arrowsmaster—Own work
CC BY-SA 3.0
File:Alpha-Pyrrolidinopentiophenone.svg
Created: 28 May 2014

Molecular structure of Acetyl Fentanyl:
Wikipedia
Aethyta—Own work

File:Acetylfentanyl.svg
Created: 25 January 2016

Chapter 6

Photo of melting TNT:
Wikipedia
Daniel Grohmann—Own work

File: Tání TNT při 81 °C.JPG
Created: 12 December 2013

Chunks of explosive-grade TNT:
Wikipedia
Daniel Grohmann—Own work

File: Trinitrotoluen.JPG
Created: 11 December 2013

Chapter 7

Detonation of 16 tons of explosives at the Nevada Test Site—Big Explosives Experimental Facility (BEEF). WATUSI is one of the largest activities conducted at BEEF by Los Alamos National Laboratory. The 36,600 pound TNT-equivalent, research and development activity was designed to validate low-end detection capabilities associated with ground motion and the acoustic footprint of an explosives detonation.

File:NTS-BEEF-WATUSI.jpg
Uploaded: 13 September 2012

Abbreviations

AAS	Atomic Absorption Spectroscopy
ACN	Acetonitrile
ACS	Activated Carbon Strip
AES	Atomic Emission Spectroscopy
AFIS	Automated Fingerprint Identification Systems
AN	Ammonium Nitrate
ANFO	Ammonium Nitrate Fuel Oil
APCI	Atmospheric Pressure Chemical Ionization
AS	Auto Scaling
ASTM	American Society for Testing and Materials
ATR	Attenuated Total Reflection
AUC	Area Under the (ROC) Curve
BPA	Blood Pattern Analysis
CAFE	Chemical Analysis for Forensic Evidence
CAI	Case Assessment and Interpretation
CBRN	Chemical, Biological and Radiological, and Nuclear Agent
CE	Capillary Electrophoresis
CHCA	α-cyano-4-hydroxycinnamic acid
CIAAW	Commission on Isotopic Abundances and Atomic Weights
CID	Collision Induced Dissociation
CITES	Convention on International Trade in Endangered Species of Wild Fauna and Flora
CSI	Crime Scene Investigation
CSO	Crime Scene Officer(s)
DART	Direct Application in Real-Time
DCM	Dicholoromethane
DESI	Desorption Electrospray Ionization
DFT	Density Functional Theory
DHBA	2,5-dihydroxy benzoic acid
DHSS	Direct Head Space Sampling
DHSST	Direct Head Space Sampling with Trapping
DUI	Driving Under the Influence
DVB	Divinylbenzene
EDTA	Ethylene-Diamine-Tetra-Acetic acid
EDX	Energy-dispersive X-ray spectroscopy
EI	Electron Impact Ionization/Electron Ionization
EMCDDA	European Monitoring Centre for Drugs and Drug Addiction
ENFSI	European Network of Forensic Science Institutes
FA	Fluor-Amphetamine
FDA	Fire Debris Analysis
FIFO	First In-First Out
FSS	Forensic Science Service
FT-ICR-MS	Fourier-Transform Ion Cyclotron Resonance Mass Spectrometry
FTIR	Fourier Transform Infrared Spectroscopy

GC-AES	Gas Chromatography with Atomic Emission Detection
GC-FID	Gas Chromatography with Flame Ionization Detection
GC-IR	Gas Chromatography with Infrared Detection
GC-MS	Gas Chromatography with Mass Spectrometric Detection
GC	Gas Chromatography
GCxGC	Comprehensive Gas Chromatography
GIS	Geographic Information System
GSR	Gunshot Residues
HME/HMEs	Home-Made Explosive(s)
HS	Head Space
HSI	Hyper Spectral Imaging
IC	Ion Chromatography
IED	Improvised Explosive Device
ILAC	International Laboratory Accreditation Cooperation
IMS	Ion Mobility Spectrometry
IR	Infrared Spectroscopy
IRMS	Isotope Ratio Mass Spectrometry
ISO	International Organization for Standardization
IUPAC	International Union of Pure and Applied Chemistry
LA-ICP-MS	Laser Ablation Inductively Coupled Plasma Mass Spectrometry
LAESI	Laser Ablation Electrospray Ionization
LC-MS	Liquid Chromatography with Mass Spectrometric Detection
LC-PDA-MS	Liquid Chromatography with Photo Diode Array and Mass Spectrometric Detection
LC-UV	Liquid Chromatography with Ultra-Violet Detection
LC	Liquid Chromatography
LCL	Lower Control Limit
LCxLC	Comprehensive Liquid Chromatography
LD50	Lethal Dose 50%
LDA	Linear Discriminant Analysis
LoD	Limit of Detection
LoQ	Limit of Quantitation (or Limit of Quantification)
LR	Likelihood Ratio (quantitative measure of the evidential strength)
m/z	mass to charge ratio
MA-XRF	macroscopic-XRF (XRF imaging)
MA	Methamphetamine
MALDI	Matrix Assisted Laser Desorption Ionization
MC	Mean Centering
MDMA	3,4-Methylenedioxymethamphetamine (XTC, ecstasy)
MLR	Multiple Linear Regression
MRM	Multiple Reaction Monitoring
MS	Mass Spectrometry
NAA	Neutron Activation Analysis
NC	Nitro-cellulose
NFI	Netherlands Forensic Institute
NG	Nitro-glycerine
NIDA	National Institute on Drug Abuse

NIPALS	Nonlinear Iterative PArtial Least Squares
NIR	Near Infrared Spectroscopy
NIST	National Institutes of Standards and Technology
NMR	Nuclear Magnetic Resonance Spectroscopy
OSAC	The Organization of Scientific Area Committees for Forensic Science
PCA	Principal Component Analysis
PCR	Principal Component Regression
PDMS	Polydimethylsiloxane
PEG	Polyethylene glycol
PETN	Pentaerythritol tetranitrate
PLS	Partial Least Squares (regression)
PPE	Personal Protective Equipment
PPPO	poly(2,6-diphenyl-p-phenylene oxide (trade name: Tenax)
PSI	Paper Spray Ionization
Py-GC-MS	Pyrolysis Gas Chromatography with Mass Spectrometric Detection
QDA	Quadratic Discriminant Analysis
Raman	Raman Spectroscopy
RI	Refractive Index
rmp	random match probability
ROC	Receiver Operating Characteristics
RP	Reversed Phase
RPG	Rocket Propelled Grenade
RSD	Relative Standard Deviation
SA	sinapinic acid
SAM	Spectral Angle Mapping
SBSE	Stir Bar Sorptive Extraction
SEM	Scanning Electron Microscopy
SFC	Supercritical Fluid Chromatography
SIE	Selected Ion Extraction
SIMCA	Soft Independent Modelling of Class Analogies
SIMS	Secondary Ion Mass Spectrometry
SLA	Service Level Agreement
SNV	Standard Normal Variate
SPE	Solid Phase Extraction
SPME	Solid Phase Micro Extraction
SRM	Selective Reaction Monitoring
SVD	Singular Value Decomposition
SWG/SWGs	Scientific Working Group(s)
TATP	Tri-Acetone-Tri-Peroxide
TD	Thermal Desorption
TDS	Thermal Desorption System
TIC	Total Ion Current
TLC	Thin Layer Chromatography
TNT	2,4,6 Tri-Nitro-Toluene
ToD	Time of Death
TOF-SIMS	Time of Flight-Secondary Ion Mass Spectrometry
TTP	Trace Transfer and Persistence

UCL	Upper Control Limit
UN	United Nations or Urea Nitrate
UNODC	United Nations Office on Drugs and Crime
USA	United States of America
UV-vis	Ultra-Violet – visible Spectroscopy (fixed wavelength and PAD – photo array detection)
UvA	University of Amsterdam
VCDT	Vienna Canyon Diablo Troilite
VMD	Vacuum Metal Deposition
VPDB	Vienna Pee Dee Belemnite
VSMOW	Vienna Standard Mean Ocean Water
VUV	Vacuum Ultra-Violet
WHO	World Health Organization
XRD	X-ray Diffraction Spectroscopy
XRF	X-ray Fluorescence Spectroscopy

Index

Note: 'Page numbers followed by '*f*' indicate figures those followed by '*t*' indicate tables and 'b' indicate boxes.'

G

H

www.ingramcontent.com/pod-product-compliance
Lightning Source LLC
Chambersburg PA
CBHW071721250226
40257CB00024B/558
9780128207154